In this series original contributions of mathematical research in all fields are contained, such as

— research monographs
— collections of papers to a single topic
— reports on congresses of exceptional interest for mathematical research.

This series is aimed at promoting quick information and communication between mathematicians of the various special branches.

In diese Reihe werden Originalbeiträge zu allen Gebieten der mathematischen Forschung aufgenommen wie

— Forschungsmonographien
— Sammlungen von Arbeiten zu einem speziellen Thema
— Berichte von Tagungen, die für die mathematische Forschung besonders aktuell sind.

Die Reihe soll die schnelle Information und gute Kommunikation zwischen den Mathematikern der verschiedenen Fachgebiete fördern.

Manuscripts
in English and German comprising at least 100 pages and not more than 500 pages can be admitted to this series. With respect to a quick publication the manuscripts are reproduced photomechanically. Authors who are interested in this series please turn directly to the 'Akademie-Verlag'. Here you will get more detailed information about the form of the manuscripts and the modalities of publication.

Manuskripte
in englischer und deutscher Sprache, die mindestens 100 Seiten und nicht mehr als 500 Seiten umfassen, können in diese Reihe aufgenommen werden. Im Interesse einer schnellen Publikation werden die Manuskripte auf fotomechanischem Weg reproduziert. Autoren, die an der Veroffentlichung entsprechender Arbeiten in dieser Reihe interessiert sind, wenden sich bitte direkt an den Akademie-Verlag. Sie erhalten dort genauere Informationen über die Gestaltung der Manuskripte und die Modalitäten der Veröffentlichung.

Parametric Optimization and Related Topics

Mathematical Research · Mathematische Forschung

Wissenschaftliche Beiträge
herausgegeben von der
Akademie der Wissenschaften der DDR
Karl-Weierstraß-Institut für Mathematik

Band 35

Parametric Optimization and Related Topics

Parametric Optimization and Related Topics

edited by

Jürgen Guddat
Hubertus Th. Jongen
Bernd Kummer
František Nožička

Akademie-Verlag Berlin 1987

Herausgeber:

Prof. Dr. Jürgen Guddat
Humboldt-Universität Berlin
Prof. Dr. Hubertus Th. Jongen
TH Twente Enschede
Doz. Dr. Bernd Kummer
Humboldt-Universität Berlin
Prof. Dr. František Nožička
Charles University Prague, Czechoslovakia, und Humboldt-Universität Berlin

---

Die Titel dieser Schriftenreihe werden vom Originalmanuskript der Autoren reproduziert.

ISSN 0138-3019
ISBN 3-05-500337-3

Erschienen im Akademie-Verlag Berlin, DDR-1086 Berlin, Leipziger Straße 3-4

Lizenznummer: 202 · 100/408/87
Printed in the German Democratic Republic
Gesamtherstellung: VEB Kongreß- und Werbedruck, 9273 Oberlungwitz
Lektor: Dr. Reinhard Höppner
LSV 1085
Bestellnummer: 763 723 6 (2182/35)
05200

PREFACE

The papers contained in this volume are based on talks given at the international conference "Parametric Optimization and Related Topics" held during October 7 - 12, 1985 in Plaue (Thuringia, GDR), which was organized by the Humboldt University Berlin.

Parametric optimization is a branch of mathematical programming which has grown fairly fast during the last decade and there is still a stormy development. What are the reasons for the broad interest and continuous progress in this field? In our opinion the following aspects are most essential: Parameter dependent optimization problems arise in a quite natural way in the model-building of various practical problems; parametric optimization is considered to be an interface between subdisciplines of mathematical programming, like stochastic optimization, multi-objective optimization, duality theory, solution methods, multi-level optimization, ill-posed problems, semi-infinite optimization; there are very close relations to other mathematical fields like approximation theory.

This book presents full and survey papers on the basic theory of parametric optimization from different points of view and on the applications and interrelations mentioned above.

All articles were refereed. The editors wish to express their gratitude to the various referees. Our thanks are also due to R. Schultz for supporting the editors. Further, we would like to thank B. Bank and D. Klatte for the translation of the papers by E.G. Belousov and I.I. Eremin, respectively. We are thankful to Ch. Dobers, Ch. Jacob, J. Kerger, Ch. Reimann, S. Schmidt and M.Willenberg for their assistance in the final preparation of this volume. Our appreciation must also be directed to our publishing house, Akademie-Verlag, who has continued to be extremely encouraging and helpful, our special thanks are due to R. Höppner and G. Reiher.

Berlin, August 1986

The Editors.

Contents

REGULARITY THEOREMS IN SENSITIVITY THEORY WITH NONSMOOTH DATA

A. Auslender*

ABSTRACT : This is a survey concerning local and global regularity theorems in sensitivity theory with nonsmooth data. For a system of inequalities and equalities bounds and the best bound are given for the quotient of the distance of an infeasible point and the norm of the residual vector.

I INTRODUCTION In recent years, a considerable amount of effort has been expended in analysis of the behaviour of the optimal set of a mathematical program under perturbations on the objective function and the constraints with smooth data. Among particularly useful tools are regularity theorems (see, in particular Robinson [14], [15], Ioffe [7]). The aim of this paper is to give a short survey about some results concerning regularity theorems with nonsmooth data. Section II is devoted to local regular theorems and Section III to global regular results. For the following, $\mathbb{R}^n$ is the usual-vector space of n-tuples with the usual inner product denoted by $(.,.)$.

II LOCAL REGULARITY THEOREM Let W be an open set in $\mathbb{R}^q$ and let f_i, g_j $i = 1,\dots m$, $j = 1,\dots p$ be real-valued locally Lipschitzian functions defined on $\mathbb{R}^n \times W$. Let

$$C(w) = \{x \in \mathbb{R}^n : f_i(x,w) \leqslant 0 \quad \forall i = 1,\dots m,\ g_j(x,w) = 0 \quad \forall j = 1,\dots p\}.$$

The following definition of regularity is an extension of that given by Robinson [14] in the differentiable case and by Ioffe [7] in the non-perturbed case.

Definition 1.1 The point $\bar{x} \in C(\bar{w})$ is called regular at $\bar{w}$ if there exists an open neighbourhood V of $\bar{x}$, an open neighbourhood W^* of $\bar{w}$ ($W^* \subset W$) and a constant c such that for each $x \in V$, each $w \in W^*$ we have

$$d(x,C(w)) \leqslant c \max(f_i^+(x,w),\ |g_j(x,w)| \ \big| \ i = 1,\dots m,\ j = 1,\dots p) \tag{1}$$ ■

In this formula $d(x,C(w))$ denotes the distance of x to $C(w)$ and $f_i^+ = \max(f_i,0)$. For giving sufficient conditions in order to obtain regular points, one must generalize the Mangasarian-Fromovitz condition to the nondifferentiable case. For this let us recall what is $\partial_x^* f_j(x,w)$ the partial generalized gradient of f_j at (x,w) introduced by Hiriart-Urruty [5] :

$$\partial_x^* f_j(x,w) = \text{co}\{ \lim_{(x_i,w_i) \to (x,w)} \nabla_x f_j(x_i,w_i) \mid (x_i,w_i) \in E_{f_j} \} \tag{2}$$

In this formula E_{f_j} is the set in $\mathbb{R}^n \times W$ of points where f_j is differentiable, $\nabla_x f_j(x_i,w_i)$ denotes the gradient of the function $f_j(.,w_i)$ at x_i and co(A) the convex hull of A. Now we give a condition that generalizes the Mangasarian-Fromovitz condition to the nondifferentiable case and which is closely related to the condition given by Hiriart-Urruty [5].

* Département de Mathématiques Appliquées, Université de Clermont II, B.P. 45, 63170 Aubière, FRANCE.

Definition 1.2 Let $\bar{x} \in C(\bar{w})$ and let $I(\bar{x},\bar{w}) = \{i : f_i(\bar{x},w) = 0\}$. The point $\bar{x}$ is said to satisfy the extended Mangasarian-Fromovitz condition for the set $C(\bar{w})$ if

1) For each set $\{c_i, d_j,\ i \in I(\bar{x},w),\ j = 1,2,\dots p\}$ with $c_i \in \partial_x^* f_i(\bar{x},\bar{w})$, $d_j \in \partial_x^* g_j(\bar{x},\bar{w})$ there exists a vector h such that

$$(c_i,h) < 0 \quad \forall i \in I(\bar{x},\bar{w}),\ (d_j,h) = 0 \quad \forall j = 1,2,\dots p.$$

2) Each set $\{d_j\ j = 1,2,\dots p\}$ with $d_j \in \partial_x^* g_j(\bar{x},\bar{w})$ is composed of linearly independent vectors.

Theorem 1. [Th. 1.1 [1]] Suppose that $\bar{x} \in C(\bar{w})$ satisfies the extended Mangasarian-Fromovitz condition for the set $C(\bar{w})$ and that $C(w)$ is nonempty for each $w \in W_o$ where W_o is an open neighbourhood of $\bar{w}$ in W. Then $\bar{x}$ is regular at $\bar{w}$. ■

One can ask if it is really necessary to assume in Theorem 1 that $C(w) \neq \Phi$ for all w in a neighbourhood W_o. In fact this is not necessary and the proof was given by Klatte in a private communication.

Theorem 2. [Klatte [9]] Suppose that $\bar{x} \in C(\bar{w})$ satisfies the extended Mangasarian-Fromovitz condition for the set $C(\bar{w})$ then there exists an open neighbourhood w_o of W such that $C(w)$ is nonempty for each $w \in W_o$.

Remark: In the differentiable case Theorem 1 and 2 were given by Robinson [14] and in the convex case by Robinson in [15].

When w is a fixed parameter a sufficient condition to obtain regularity was given in [7] by Ioffe. This condition was then expressed to obtain the Ljusternik theorem and a new proof of Hoffman's inequality in linear programming.

III GLOBAL REGULARITY THEOREMS For certain number of applications it would be interesting to obtain inequality (1) valid not only for x belonging to a neighbourhood V of $\bar{x}$ but for all x in $\mathbb{R}^n$. Such a point would be called a global regular point. We shall investigate now in this section this problem.

3.1 Preliminaries Let f_i $i = 1,2,\dots,m$ be now closed proper convex functions defined on $\mathbb{R}^n$ and let $a_i \in \mathbb{R}^n$, $b_i \in \mathbb{R}$ for $i = m+1 \dots p$. Let

$$C = \{x: f_i(x) \leq 0\ i = 1 \dots m;\ (a_j,x)-b_j = 0\ \ j = m+1 \dots p\}.$$

Suppose that C is nonempty. Let $||.||_\beta$ be any norm on $\mathbb{R}^n$ and $||\ ||_{\beta*}$ its dual norm, that is

$$||x||_{\beta*} = \max\ ((x,y)|\ ||y||_\beta = 1).$$

In this section we consider also a monotonic norm γ that is

$$|x_i| \leq |y_i|\ \ i = 1 \dots m \Rightarrow ||x||_\gamma \leq ||y||_\gamma.$$

In particular, the p-norms $||\ ||_p$ with $1 \leq p \leq +\infty$ are monotonic norms.

Let $d_\beta(x,C)$ be the β-distance of x to C :

$$d_\beta(x,C) = \min(||x-y||_\beta\ |\ y \in C)$$

and define $\theta(x)$ as the vector in $\mathbb{R}^p$ with components :

$$\{f_i^+(x)\ \ i = 1,2,\dots,m,\ (a_j,x)-b_j\ \ j = m+1 \dots p\}$$

with the notation $a^+ = \max(a,0)$. We want to find a constant $k_{\beta,\gamma}$ such that

$$d_\beta(x,C) \leq k_{\beta,\gamma}||\theta(x)||_\gamma \quad \forall x \in \mathbb{R}^n \tag{3}$$

Such a constant $k_{\beta,\gamma}$ is well known for linear equalities

$$A\bar{x} = b \tag{4}$$

where A is a given (n,n) real nonsingular matrix and b a nonzero vector in $\mathbb{R}^n$. Indeed we have obviously

$$||x-\bar{x}||_\beta \leqslant ||A^{-1}||_\beta \; ||Ax-b||_\beta \tag{5}$$

where $||C||_\beta = \max(||Cu||_\beta | \; ||u||_\beta = 1)$ so that

$$k_{\beta,\beta} = ||A^{-1}||_\beta . \tag{6}$$

In this case $||A^{-1}||_\beta$ is the best bound since there exists $x \neq \bar{x}$ such that inequality (5) becomes an equality.

The problem of finding such a bound $k_{\beta,\gamma}$ for inequality (3) was also investigated in 1952 by Hoffman [6] for linear inequalities and in [11] a new explicit expression was derived for that bound for a system of linear inequalities and equalities. In [13] Robinson extended this bound to a system of convex inequalities which defines a bounded feasible region with a nonempty interior. In [10] Mangasarian has extended this bound to a system of differentiable convex inequalities but he relaxed the boundness assumption on the feasible region. For this he introduced a new concept which seems to be important. It is an asymptotic constraint qualification defined as follows : "Let g be a differentiable convex function from $\mathbb{R}^n$ into $\mathbb{R}^m$, let S^o and S be defined by

$$\Phi \neq S^o = \{x: g_j(x) < 0 \quad j \in J_1, \; g_j(x) \leqslant 0 \; j \in J_2\} \subset S = \{x: g(x) \leqslant 0\}$$

where $J_1 \cup J_2$ is a partition of $\{1,2,\ldots m\}$ such that g_j is nonlinear for $j \in J_1$ and linear for $j \in J_2$, then the asymptotic constraint qualification on S is :

"For each nonempty $I \subset \{1,2,\ldots m\}$ and each sequence of points $\{x^j\} \subset S$ such that $g_i(x^j) = 0 \;\; i \in I$ and $\nabla g_i(x^j) \;\; i \in I$ are linearly independent, each accumulation point $(\bar{\nabla} g_i \;\; i \in I_o, \; \bar{\nabla} g_i \;\; i \in I_1, \; \bar{\nabla} g_i \;\; i \in I_2)$ of the sequence $\{\nabla g_i(x^j)/||\nabla g_i(x^j)|| \;\; i \in I_o, \; \nabla g_i(x^j) \;\; i \in I_1, \; \nabla g_i(x^j) \;\; i \in I_2\}$ satisfies for some $z \in \mathbb{R}^n$

$$(\bar{\nabla} g_i, z) \; > 0 \quad \forall i \in I_o \cup I_1, \; (\bar{\nabla} g_i, z) \geqslant 0 \quad \forall i \subset I_2$$

where $I_o \cup I_1 \cup I_2$ is a partition of I such that the sequence $\{\nabla g_i(x^j)\}$ is unbounded for $i \in I_o$ and bounded for $i \in I_1$, $I_1 \subset J_1$, $I_2 \subset J_2$."

Under these assumptions Mangasarian proved that there exists a real constant $\mu^M_{\beta,\gamma}$ such that

$$d_\beta(x,S) \leqslant \mu^M_{\beta,\gamma} \quad ||g(x)^+||_\gamma \quad \forall x \in \mathbb{R}^n \tag{7}$$

and an explicit formula for $\mu^M_{\beta,\gamma}$ is given.

This result was generalized by Auslender and Crouzeix in [2] when the feasible set C is defined by functions f_i which are not necessarily differentiable functions defined on the whole space $\mathbb{R}^n$ but only proper closed functions such that

$$\text{int}(\text{dom } f_i) \supset C \qquad \forall i = 1,2, \ldots m.$$

Recall that $\text{int}(\text{dom } f_i)$ denotes the interior of the effective domain of f_i. Inequality (7) was used by Mangasarian to obtain a condition number for S. Consider again the linear system given by (4). Inequality (5) leads to the concept of condition number $\text{cond}(A) = ||A^{-1}||_\beta \; ||A||_\beta$ which provides useful bounds for the relative error $\frac{||x-\bar{x}||_\beta}{||\bar{x}||_\beta}$ in terms of $\frac{||Ax-b||_\beta}{||b||_\beta}$.

Actually we obtain from (5)

$$\frac{||x-\bar{x}||_\beta}{||\bar{x}||_\beta} \leq ||A^{-1}||_\beta \; ||A||_\beta \; \frac{||Ax-b||_\beta}{||b||_\beta} \tag{8}$$

It is well known that the condition number is a fundamental concept in the numerical analysis of linear systems. To obtain such a number, the best bound must be found. Such a problem can be generalized for inequality (3) and in [2] the "best" bound is given. We shall now present the results given in [2] , their extensions to perturbed sets C(u) depending on a parameter, and some applications.

3.2 <u>The global regularity theorem</u> : <u>the unperturbed case</u> As for the differentiable case, some assumptions are needed for obtaining a finite bound $k_{\beta,\gamma}$. It is supposed that Slater's condition is satisfied, that is :

a) there exist $\tilde{x}$, $\varepsilon_i > 0$ such that :

$$f_i(\tilde{x}) < -\varepsilon_i \quad i = 1,2,\ldots,m, \quad (a_i,\tilde{x}) = b_i \quad i = m+1 \ldots p$$

b) the vectors a_i are linearly independent.

When C is noncompact, it is assumed the following condition called the asymptotic constraint qualification condition (A.Q.C.).

"Let $\{p_n\}$ be a sequence such that

$$p_n \in C \quad \forall n, \quad \lim_{n\to\infty} ||p_n|| = +\infty$$

Let $I = \{i: \exists n_o: f_i(p_n) = 0 \quad \forall n \geq n_o\}$ and $c_i^n \in \partial f_i(p_n)$

($\partial f_i(p_n)$ denotes the subdifferential of f_i at p_n). Let I_o be the set of indices $i \in I$ such that the sequence $\{c_i^n\}$ is bounded, $I_o^c = I\backslash I_o$.

Let $\bar{c} = (\bar{c}_i^n, i \in I)$ be a limit point of the sequence

$$\{c_i^n \; i \in I_o, \; \frac{c_i^n}{||c_i^n||} \quad i \in I_o^c\}.$$

Let $\alpha = (\alpha_{m+1}, \ldots \alpha_p)$ with $\alpha_i = \pm 1$ and let $S_\alpha = co(\bar{c}_i \; i \in I, \; \alpha_{m+1}a_{m+1}, \ldots \alpha_p a_p)$ (co denotes the convex hull). Then the asymptotic constraint qualification is :

$$0 \notin S_\alpha \text{ for all } S_\alpha \text{ when } \{p_n\}, \{c_i^n\} \; i \in I, \text{ vary".}$$

<u>Remark</u>: A.Q.C. is an extension of the condition given by Mangasarian for the differentiable case; in this case it coincides with Mangasarian's condition when the linear inequalities are equalities.

Let now $N_C(y)$ be the usual normal cone of C at y (definition page 15 [16]) and set for $y \in bd(C)$ (the boundary of C)

$$M(y) = \{d: -d \in \partial||y^*||_{\beta^*}, \; -y^* \in N_C(y), \; ||y^*||_{\beta^*} = 1\}.$$

<u>Theorem 3</u>. (Th. 2 [2]) Suppose that Slater's condition is satisfied and suppose either C is compact or A.Q.C. is satisfied. Let $(f_i^+)'(y;d)$ the directional derivative of f_i at direction d:

$$(f_i^+)'(y;d) = \lim_{\lambda\to 0^+} \frac{f_i^+(y+\lambda d)-f_i^+(y)}{\lambda} \quad \text{and set}$$

$$k^*_{\beta,\gamma} = \inf_{y\in bd(C)} \; \inf_{d\in M(y)} ||\{(f_i^+)'(y;d) \; i\in I(y), (a_i,d) \; i = m+1\ldots p\}||_\gamma, \; k_{\beta,\gamma} = \frac{1}{k^*_{\beta,\gamma}}$$

then $k^*_{\beta,\gamma} > 0$ and $k_{\beta,\gamma} = \sup_{x \notin C} \dfrac{d_\beta(x,C)}{||\theta(x)||_\gamma}$

The proof is based on the use of the normal cone at each boundary point. The A.Q.C. assumption is introduced for the analysis of the behaviour of asymptotic boundary points.

Remark. The bound $k^M_{\beta,\gamma}$ found by Mangasarian in [10] was not the best. Taking the example given by Mangasarian in [10] it is easy to verify that

$$\sqrt{2} = k_{\infty,2} < k^M_{\infty,2} = 2\sqrt{2}$$

3.3 A global regularity theorem : the perturbed case Theorem 3 can be extended to perturbed sets. For simplicity consider only the case with inequalities and vertical perturbations. Then $C(u)$ the perturbed convex set at $\bar{u} = 0$ is defined by :

$$C(u) = \{x: f_i(x) \leq u_i \quad i = 1,2, \ldots m\}.$$

Suppose that Slater's condition is satisfied and let W be a compact neighbourhood of 0 such that

1) $\mathrm{int}(\mathrm{dom} f_i) \supset C(u) \quad \forall u \in W$ (9)

2) $f_i(x) \pm u_i \leq -\frac{\varepsilon}{2} : \quad \forall u \in W$ (10)

Then $C(u)$ is nonempty for each $u \in W$; Slater's condition is satisfied for $C(u)$ and it is well known that $C(.)$ is continuous on W. To obtain a perturbed version of Theorem 1 A.Q.C. is changed slightly and becomes A.Q.C.P. :

"Let $\{p_n\}$, $\{u^n\}$ be sequences such that

$$p_n \in C(u^n),\ u^n \in W,\ \lim_{n\to\infty} ||p_n|| = +\infty.$$

Let $I = \{i: \exists n_o: f_i(p_n) = u^n_i \ \forall n \geq n_o\}$ and $c^n_i \in \partial f_i(p_n)$.
Let I_o be the set of indices $i \in I$ such that the sequence $\{c^n_i\}$ is bounded, set $I^c_o = I \setminus I_o$. Let $\bar{c} = (\bar{c}_i,\ i \in I)$ be a limit point of the sequence $\{c^n_i,\ i \in I_o,\ \dfrac{c^n_i}{||c^n_i||}\ \ i \in I^c_o\}$ and $S = \mathrm{co}(\bar{c}_i,\ i \in I)$ then A.Q.C.P. is

"$0 \notin S$ for all S when $\{p_n\}$, $\{c^n_i\}$ $\{u^n_i\}$ vary" ".

Let us now denote for $y \in \mathrm{bd}(C(u))$, $I(y,u) = \{i: f_i(y) = u_i\}$

$$M(y,u) = \{d: \exists y^* \in -N_{C(u)}(y),\ ||y^*||_{\beta^*} = 1,\ -d \in \partial||y^*||_{\beta^*}\}$$

$$k^*_{\beta,\gamma}(u) = \inf_{y \in \mathrm{bd}(C(u))} \ \inf_{d \in M(y,u)} ||\{(f^+_i)'(y;d),\ i \in I(y,u)\}||_\gamma$$

$$k^*_{\beta,\gamma}(W) = \inf(k^*_{\beta,\gamma}(u) \mid u \in W);\ k_{\beta,\gamma}(W) = \frac{1}{k^*_{\beta,\gamma}(W)}$$

$$\theta(x,u) = \{(f_i(x)-u_i)^+ \ i = 1,2,\ldots,m\}.$$

Theorem 4. (Th. 4 [2]) Suppose that Slater's condition is satisfied and that W is a compact neighbourhood of 0 such that (9) and (10) are verified. Suppose that either $C(0)$ is compact or A.Q.C.P. is satisfied then $k^*_{\beta,\gamma}(W) > 0$ and

$$\sup_{x \notin C(u) \quad u \in W} \frac{d_\beta(x,C(u))}{||\theta(x,u)||_\gamma} = k_{\beta,\gamma}(W).$$

The proof is analogous to that of Theorem 3, except for some minor changes.
In [2] also two applications of Theorem 4 are given about the Lipschitzian behaviour of C(.) and the ε-optimal set in sensitivity theory.

Corollary 5 (Corollary 5 [2]) Let $\Delta_\infty(A,B)$ be the Hausdorff distance between the sets A and B

$$\Delta_\infty(A,B) = \max \{ \sup_{a \in A} \inf_{b \in B} ||a-b||_\infty, \sup_{b \in B} \inf_{a \in A} ||a-b||_\infty \}$$

then under the assumptions of Theorem 4 we have :

$$\Delta_\infty(C(u),C(u')) \leq k_{\infty,\infty}(W) \; ||u-u'||_\infty \quad \forall u, u' \in W.$$

Remark: The Lipschitzian behaviour of C(.) on a compact neighbourhood of the origin was proved by Psenichny in [12] when C(0) is compact.
As another consequence it was obtained in [2] a corollary proved by other authors with other methods (Eremin and Astafiev [4], Klatte and Kummer [8], Blatt [3]):

Corollary 6 (Corollary 6 [2]) Let f_0 be a real-valued function defined on $\mathbb{R}^n$, $\varepsilon > 0$ and set

$$g(u) = \inf(f_0(x)|x \in C(u)), \; M_\varepsilon(u) = \{x \in C(u) : f_0(x) \leq g(u) + \varepsilon\}.$$

Suppose that $M_0(0)$ is compact and nonempty and that Slater's condition is satisfied for C(0), then $M_\varepsilon(.)$ is Lipschitzian in a neighbourhood of 0.

REFERENCES

[1] A. Auslender, Stability in Mathematical Programming with nondifferentiable data, Siam J. Control and Optimization, Vol.22, N°2, March 1984.

[2] A. Auslender and J.-P. Crouzeix, Global Regularity Theorems. Technical Report, Université de Clermont II.

[3] H.P. Blatt, Lipschitz-Stabilitität von Optimierungs und Approximationsaufgaben, Numerical Methods of Approximation Theory, ISNM Birkhaüser Verlag Basel (1980) pp. 9-28.

[4] I. I. Eremin and N.N. Astafiev, Introduction to the theory of linear and convex programming, Nauka, Moscow (in Russian) (1976).

[5] J.-B. Hiriart-Urruty, Refinements of necessary optimality conditions in nondifferentiable programming I, Appl. Math. Opt. 5 (1979) pp. 63-82.

[6] A.J. Hoffman, On approximate solutions of systems of linear inequalities, J. Rest. Nat. Bur. Standards, 49 (1952) pp. 263-265.

[7] A.D. Ioffe, Regular points of Lipschitz functions, Trans.Amer.Math.Soc. 251 (1979) pp. 61-69.

[8] D. Klatte and B. Kummer, Stability properties of infima and optimal solutions of parametric optimization problems, Proceedings of the IIASA Workshop on nondifferentiable optimization, Sopron, Hungary (1984).

[9] D. Klatte, On a regularity theorem. Private communication.

[10] O.L. Mangasarian, A condition number for differentiable convex inequalities, MOR, Vol.10, N°2 (May 1985).

[11] O.L. Mangasarian, A condition number of linear inequalities and linear programs. Methods of operations research. Proc. 6th Symposium. Oper. Res. Augsburg, 7-9 September 1981. Bamberg and O. Opitz ed Verlagsgruppe Athenaum/Hain/Scriptor/Hanstein/Konigstein (1981) pp. 3-15.

[12] Psenichny, Convex analysis and extremal problems, Nauka, Moscow (in Russian) (1982).

[13] S.M. Robinson, An application for error bounds for convex programming in linear space, SIAM J. Control Opt., 13 (1975) pp. 271-273.

[14] S.M. Robinson, Stability theory for systems of inequalities Part II : Differentiable nonlinear systems, SIAM J. Numer. Anal. 13 (1976) pp. 497-512.

[15] S.M. Robinson, Regularity and stability for convex multivalued functions, MOR, Vol.1, N°2 (1976) pp. 130-143.

[16] R.T. Rockafellar, Convex analysis, Princeton University Press (1970).

NONLINEAR PARAMETRIC INTEGER PROGRAMMING

by

BERND BANK and REINHARD MANDEL*

Abstract

Qualitative stability results for nonlinear parametric integer optimization problems without boundedness requirements will be presented. The theory developed is mostly concerned with problems of the following general class of (mixed-) integer optimization problems

$$P(p,b) \inf \{ f(x) + p^T x \mid x \in M(b), x_1, \ldots, x_s \in Z \}, \quad (p,b) \in \mathcal{E}$$

with

(1) $M(b) = (C(b) + V) \subset R^n$

and $0 \leq s \leq n$, where the sets $C(b)$ are given by an upper semicontinuous compact-valued multifunction C and V is a polyhedral cone. We consider quadratic and convex polynomial functions, respectively, to be the parameter independent function $f(x)$ of the problem $P(p,b)$.

In particular, we study constraint sets of the form

(2) $M(b) = \{ x \in R^n \mid f_i(x) \leq b_i, \ i = 1, \ldots, m \}$,

where the functions $f_i(x)$ are given by quasiconvex polynomials. Moreover, we derive conditions under which these sets belong to the general class (1).

* Humboldt-University
Sektion Mathematik
PSF 1297
Berlin (GDR)
1086

Introduction

Stability theory and the parametrization of optimization problems are of great importance both in theory and applications. During the last decade a remarkable success has been made on this topic for nonlinear programming. But, if the parametric nonlinear program contains integer requirements on the variables (or on some of them), then up to now, only a few results can be listed.

One reason for this situation is that one has to handle the difficulties of nonlinear programming as well as those which arrise by the fact that only little is known about (mixed-) integer solutions of non-linear equations and inequalities and their structure. Even for systems of linear inequalities in (mixed-) integer variables first comprehensive considerations were published only in the 70-th (FIOROT 1972, BOWMAN/BURDET 1974, CHARNES/GRANOT 1975, BOROSH/TREYBIG 1976).

However, for parametric (mixed-) integer linear problems fairly deep qualitative and quantitative results are known (MEYER 1975, BANK 1977, BLAIR/JEROSLOW 1977,1979).

Similarly as in nonlinear programming this results could be applied in the theory of integer programming and to the design of algorithms (e.g. penaltyfunctions and duality; JEROSLOW 1978, BLAIR/JEROSLOW 1981, TIND/WOLSEY 1981, BANK/MANDEL in BANK/BELOUSOW et al 1986).

Remarkable results on parametric (mixed-) integer nonlinear programming (without compactness requirements on the constraint sets) are derived by HANSEL 1980 und MANDEL 1985.

In this paper we shall present a qualitative stability theory for the following parametric nonlinear (mixed-) integer optimization problem:

$$P(p,b) \quad \inf \left\{ f(x) + p^T x \mid x \in M(b),\ x_1,\dots,x_s \in Z \right\},$$

where $M(b)$ is a subset of R^n, $s \in [0,n]$, integer, Z the set of all integer numbers and $p \in R^n$ and $b \in R^m$ are parameter vectors. Two kinds of functions $f: R^n \longrightarrow R$ will be considered: quadratic functions and convex polynomials. The constraint sets $M(b)$ in question are given as the images of multifunctions $M : \Lambda \longrightarrow 2^{R^n}$ ($\Lambda \subset R^m$) which contain, as a particular case, the one defined by linear constraints with perturbed right-hand sides.

We shall make use of the following notations and definitions (see e.g. BANK et.al.):

- R^n denotes the n-dimensional euclidean space;
- $A \subset R^n$, $\varepsilon > 0$, then $\mathcal{U}_\varepsilon A$ denotes the ε-neighbourhood of A;
- A multifunction $\Gamma : \Lambda \longrightarrow 2^{R^n}$ is said to be at a point $b^o \in \Lambda$ ($\subset R^m$) upper semicontinuous (according to HAUSDORFF), briefly u.s.c., if

  $$\forall \varepsilon > 0 \quad \exists \delta = \delta(\varepsilon) > 0 : \Gamma(b) \subset U_\varepsilon \Gamma(b^o) \; \forall b \in U_\delta(b^o),$$

  and lower semicontinuous (according to BERGE), briefly l.s.c., if

  $$\left.\begin{array}{l} \forall \Omega \subset R^n, \ \Omega \text{ open} \\ \Gamma(b^o) \cap \Omega \neq \emptyset \end{array}\right\} \Longrightarrow \begin{array}{l} \exists \delta = \delta(\Omega) > 0 : \\ M(b) \cap \Omega \neq \emptyset \; \forall b \in U_\delta(b^o) \end{array} .$$

  We say that Γ is continuous at b^o if it is both l.s.c. and u.s.c. at b^o.
- A function $f : R^n \longrightarrow R$ is said to be quasiconvex if

  $$x,y \in R^n, \ \alpha \in (0,1) \Longrightarrow f(\alpha x + (1-\alpha)y) \leq \max\{f(x), f(y)\}.$$

Now we can specify the multifunction M :

(1) $M(b) \underset{Df}{=} C(b) + V$

where (i) $C(b) \neq \emptyset$ and compact,

(ii) $C : \Lambda \longrightarrow 2^{R^n}$ is u.s.c.,

(iii) $V \subset R^n$ is a convex polyhedral cone, with $0 \in V$,

and

(2) $M(b) \underset{Df}{=} \{ x \in R^n \mid f_i(x) \leq b_i, \ i = 1,\ldots,m \}$

where $f_i : R^n \longrightarrow R$, $i = 1,\ldots,m$ are quasiconvex polynomials and

$$\Lambda \underset{Df}{=} \{ b \in R^m \mid M(b) \neq \emptyset \}.$$

By a result of KUMMER 1977 we know that the multifunction M defined by

$$M(b) = \{ x \in R^n \mid Ax \leq b \}, \; \Lambda = \text{dom } M$$

belongs to the class (1) of multifunctions (and, obviously, it belongs to (2), too). In the following we show that also a multifunction M given by (2) under the additional assumption (CPC) (see below) belongs to the class (1).

The following three aspects of a stability theory for the problem P(p,b) are treated here:

- Existence and stability of feasible points;
- Continuity properties of the value function of P(p,b), of its optimal sets and its level-sets;
- Existence of optimal points.

Obviously, the case of parametric (mixed-) integer linear programming is a special case of the general problems considered in this paper. A first deeper study of the here included (mixed-) integer problem with a quadratic objective function and linear constraints is due to HANSEL 1980 (see also BANK/HANSEL 1984) and can be considered as one of our bases.

For shortness we have to restrict ourselves to some essential proofs. The proofs omitted here the reader can find in the monograph BANK/MANDEL 1986/87, to appear, which contains a comprehensive representation of the theory of parametric (mixed-) integer programming and some of its applications.

Some of the proofs can also be found in the monograph BANK/BELOUSOV et.al. 1986.

The paper is devided into two sections, the first one is devoted to the constraint sets, and the second one contains stability and existence results.

1. The constraint sets.

Let the parametric constraint sets of P(p,b) be defined by

(3) $G(b) = \{ x \in M(b) \mid x_1,\dots,x_s \in Z \}, \quad b \in \Lambda,$

where we consider the cases that M(b) is given in the form (1) and (2). If we denote by conv G(b) the convex hull of (3), then there are two multifunctions related to G(b), namely, the constraint set-mapping $G : B \longrightarrow 2^{R^n}$ and the convex hull-mapping $G_{conv} : B \longrightarrow 2^{R^n}$, where the domain of both mappings is given by

$$B \underset{Df}{=} \{ b \in R^m \quad G(b) \neq \emptyset \} .$$

(a) M(b) is of the form (1).

Stability properties of the constraint sets G(b) from (3) where the involved multifunction M is given by (1), i.e. $M(b) = C(b) + V$, $b \in B$ depend on assumptions to be posed on the convex polyhedral cone V. We say that a convex polyhedral cone V satisfies

(MIG) if V has a generating system

$$\{v^1,\dots,v^r\} \subset \{v \in R^n \mid v_1,\dots,v_s \in Z\}$$

and,

(IG) if V has a generating system

$$\{v^1,\dots,v^r\} \subset Z^n .$$

The following definition of the uniform distribution of (mixed-) integer points in subsets of the space R^n can be used if more general constraint sets than considered here are to be investigated. The points of $A_s \underset{Df}{=} \{ x \in A \mid x_1,\dots,x_s \in Z \}$ are said to be uniformly distributed in $A \subset R^n$ if $A_s \neq \emptyset$ and $\delta \underset{Df}{=} \sup_{x \in A_s} d(x,A) < \infty$.

Concerning the distribution of (mixed-) integer points the following Theorem 1., which may be considered as a generalization of a well-known approximation theorem of KRONECKER, holds true:

Theorem 1.

Let $C \subset R^n$ be a bounded set and V a convex cone in R^n, then the existence of a (mixed-) integer relatively interior point in A = C + V implies that the points of A_s are uniformly distributed in A if the (mixed-) integer points of the affine hull aff A are uniformly distributed in aff A.

One easily sees that the (mixed-) integer points in A_s are uniformly distributed if V is a polyhedral cone satisfying (MIG) and A_s is a non-empty set.

Theorem 2. will show that one obtains upper semicontinuity of the multifunctions G as well as G_{conv} if the assumption (MIG) is required for the cone V included in (1). Under these circumstances the common domain Λ of both multifunctions is a closed set.

Theorem 2.

Let M(b) be of the form (1) and V satisfy the assumption (MIG), then:

(i) There exist two compact-valued u.s.c. multifunctions

$K : B \longrightarrow 2^{R^n}$ and $K_{conv} : B \longrightarrow 2^{R^n}$

such that, for all $b \in B$,

$G(b) = K(b) + W$ and $G_{conv}(b) = K_{conv}(b) + V$,

where either $W = \{ v \in V \mid v_1, \dots, v_s \in Z \}$ or

$$W = \{ u \in R^n \mid u = \sum_{i=1}^{r} \beta_i v^i,\ \beta_i \in Z,\ \beta_i \geq 0,\ i = 1, \dots, r \}$$

if $v^1, \dots, v^r$ are (mixed-) integer generators of V;

(ii) The multifunctions G and G_{conv} are u.s.c. on B;

(iii) The domain B is closed if the domain Λ of M is closed.

(iv) The points of $G(b) \neq \emptyset$ are uniformly distributed in M(b).

Proof.

The multifunctions K and K_{conv} given by

$$(4) \quad K(b) \underset{Df}{=} \{ x \in R^n \mid x = y + \sum_{i=1}^{r} \alpha_i v^i,\ 0 \leq \alpha_i \leq 1,\ i = 1, \dots, r; \\ y \in C(b),\ x_1, \dots, x_s \in Z \}$$

and

$$(5) \quad K_{conv}(b) \underset{Df}{=} \text{conv } K(b),$$

respectively, are defined on B. Using now the u.s.c. property of the compact-valued multifunction C and the fact that the set $Z^s \times R^{n-s}$ is closed one obtains the upper semicontinuity of K, which the same property of K_{conv} is an immediate consequence of. The desired representation $G(b) = K(b) + W$ is in both cases easy to show. For an arbitrary $b \in B$ one has

$$\begin{aligned} \operatorname{conv} G(b) &= \operatorname{conv} (K(b) + W) = \\ &= \operatorname{conv} K(b) + \operatorname{conv} W = \\ &= \operatorname{conv} K(b) + V, \end{aligned}$$

which completes the proof of (i). The remaining statements (ii), (iii) and (iv) are simple sonsequences of (i). #

In order to guarantee the continuity of the constraint set-mapping G given by (3) and (1) it does not suffice to impose only the stronger hypothesis that the included multifunction C is continuous as it is shown by the simple example

$$G(b) = \{ x \in Z \mid 0 \leq x \leq b \}.$$

We have

$$G(b_t) = \{0\} \quad \text{and} \quad G(b_o) = \{0,1\}$$

if

$$b_t \underset{Df}{=} 1 - \frac{1}{t} \longrightarrow b_o = 1.$$

However, for the case of constraint sets G(b) given by (3) and (2) we shall show that if the assumptions (CPC) (see below) and (MIG) are imposed it is possible to identify subsets of B on which the multifunction G is continuous.

b) M(b) is of the form (2).

First, we list some elementary properties of the quasiconvex polynomials $f_i(x)$, $i = 1,\dots,m$, which appear in the sets

$$M(b) = \{ x \in R^n \mid f_i(x) \leq b_i,\ i = 1,\dots,m \}.$$

If a quasiconvex polynomial f(x) is constant (or strictly decreasing or strictly increasing) on a line $S_o = \{ x \in R^n \mid x = x^o + tu,\ t \in R \}$ (i.e., f has the respective property as a polynomial of $t \in R$), then f also shows this property on each line $S = \{ x \in R^n \mid x = x^1 + tu,\ t \in R \}$ which is parallel to S_o. If the function f is even a convex

one, then the above properties imply that f is a linear function of $t \in R$, where the coefficient of t does not depend on $x^1 \in R^n$.
On the basis of these elementary properties of quasiconvex polynomials and using known results from nonlinear parametric programming (see e.g. BANK et.al. 1982, Theorems 3.1.6., 3.1.5. and 3.1.3.) one obtains that the multifunction $M : \Lambda \rightarrow 2^{R^n}$ given by (2) is l.s.c. on Λ.
This fact and further properties of M are listed in the following Theorem 3.

<u>Theorem 3.</u>

Let $M : \Lambda \rightarrow 2^{R^n}$ be the multifunction given by (2). Then

(i) M is l.s.c. on ;

(ii) Λ is a closed set;

(iii) all sets M(b), $b \in \Lambda$, have the same recession cone V, which is a convex polyhedral cone satisfying (IG) if all coefficients of the polynomials f_i are rational numbers;

(iv) M is continuous on Λ if V is a linear subspace.

It is important to remark that the multifunction M given by (2) is not upper semicontinuous in general (for a counterexample see SHIRONIN 1980, BANK/BELOUSOW et.al. 1986). However, Theorem 5. includes a more general case than treated in (iv) of Theorem 3. for which the upper semicontinuity of M is ensured.
We will clarify under which conditions a multifunction M given by (2) belongs to the class of such ones defined by (1). The following statement on the continuity of the value function of quasiconvex polynomial optimization problems with only right-hand side perturbations of the constraints gives a tool for such a purpose.

Theorem 4.

If $f(x)$ is a quasiconvex polynomial function and $M(b)$ is given by (2), then the value function $\varrho : \Lambda \rightarrow R$ defined by

$$\varrho(b) \underset{Df}{=} \inf \{ f(x) \mid x \in M(b) \}, \; b \in \Lambda ,$$

is either continuous on Λ or $-\infty$ for all $b \in \Lambda$.

For the common recession cone $V = O^+M(b)$, $b \in \Lambda$, we may always underly a representation of the form

$$(6) \qquad V = \{ x \in R^n \mid d^{j^T} x \leq 0, \; j = 1,\ldots,r \} .$$

If we do so, then the condition

$$(7) \qquad \sup \{ d^{j^T} x \mid x \in M(b) \} < \infty, \; j = 1,\ldots,r$$

is necessary for the existence of a representation of $M(b)$, $b \in \Lambda$ in the form (1).

The next theorem also shows that the validity of (7) for a single $b^0 \in \Lambda$ ensures the existence of such a representation.

Theorem 5.

If $M(b)$ is of the form (2) and (7) holds for some $b^o \in \Lambda$, then there exists a continuous multifunction $C : \Lambda \to 2^{R^n}$ such that

(8) $\quad M(b) = C(b) + V, \quad b \in \Lambda$

holds and, the sets $C(b)$ are compact and convex.

Proof.

Let H be the subspace of maximal dimension contained in the recession cone V. Then the closed convex sets $M(b)$, $b \in \Lambda$ can be written in the form

$$M(b) = (M(b) \cap H^{\perp}) + H$$

where $H^{\perp}$ denotes the orthogonal complement of the subspace H.
If $V \cap H^{\perp} = \{0\}$, then the non-empty sets $M(b) \cap H^{\perp}$ are compact and convex. By Theorem 2. (iv), the multifunction C having the latter sets as images is the desired one in this case.
Therefore we have to consider the case $V \cap H^{\perp} \neq \{0\}$. Our aim will be to construct a continuous function $t : \Lambda \to R$ such that the multifunction $C : \Lambda \to 2^{R^n}$ defined by

(9) $\quad C(b) \underset{Df}{=} \{x \in M(b) \mid a^T x \geq t(b)\} \cap H^{\perp}$

with $a \underset{Df}{=} \sum_{j=1}^{r} d^j$ possesses the property (8) (here the vectors d^j are taken from the underlying representation (6) of the recession cone V). The choice of the vector a guaranties the compactness of the sets $C(b)$ (their convexity is obvious) and, the continuity of t and Theorem 3. (iv) imply that C is a continuous mapping. In order to construct the function t we define

$$\varphi_j(b) \underset{Df}{=} \sup \{ d^{j^T} x \mid x \in M(b) \}, \quad j = 1, \ldots, r,$$

$$M^*(b) \underset{Df}{=} \{x \in R^n \mid d^{j^T} x \leq \varphi_j(b), \ j = 1, \ldots, r\}, \quad b \in \Lambda.$$

Theorem 4. and the fact that the condition (7) is satisfied for b^o imply that the value functions φ_j, $j = 1,\dots,r$ are continuous on Λ. For the polyhedra $M^*(b)$, $b \in \Lambda$ we have

$$M(b) \subset M^*(b) \quad \text{and} \quad O^+M^*(b) = V$$

and, by a result of KUMMER 1977 we know that there is a continuous function $t^* : \Lambda \to R$ such that the mapping $C^* : \Lambda \to 2^{R^n}$ given by

$$C^*(b) \underset{Df}{=} \left\{ x \in R^n \mid a^T x \geq t^*(b) \right\} \cap H^{\perp}$$

is continuous (and, obviously compact-valued) which permites the representation

$$M^*(b) = C^*(b) + V, \qquad b \in \Lambda$$

(i.e. the claimed proposition of the theorem is true for the case of linear constraints). KUMMER constructs such a function t^* (using the duality theory of linear programming) in the form

$$(10) \qquad t^*(b) \underset{Df}{=} \min_{1 \leq l \leq q} z(b)^T (w^l - v^l)$$

where

$$z(b) \underset{Df}{=} \left(\varphi_1^*(b), \dots, \varphi_r^*(b) \right)^T$$

with $\varphi_j^*(b) = \min \left\{ d^{j^T} x \mid x \in M^*(b) \right\}$

and where $\begin{pmatrix} v^l \\ w^l \end{pmatrix}$, $l = 1,\dots,q$ denote all the vertices of the polyhedron

$$W \underset{Df}{=} \left\{ \begin{pmatrix} v \\ w \end{pmatrix} \in R^{2r} \mid (D^T, -D^T) \begin{pmatrix} v \\ w \end{pmatrix} = -a, \ \begin{pmatrix} v \\ w \end{pmatrix} \geq 0 \right\}$$

where $D^T = (d^1,\dots,d^r)$.

Using the above notations we proceed with the proof of the general case.

Let $\{u^1,\dots,u^c\}$ be a system of generators of the polyhedral cone $V \cap H^{\perp}$ and we define for the quasiconvex polynomials f_j, $j = 1,\dots,m$

$$I(f_j) \underset{Df}{=} \left\{ i \in \left\{ 1,\dots,c \right\} \mid -u^i \notin V \right\}$$

$$I \underset{Df}{=} \left\{ j \in \left\{ 1,\dots,m \right\} \mid I(f_j) \neq \emptyset \right\}$$

$$t_{ij}(b) \underset{Df}{=} \min\left\{ \alpha \mid \alpha \geq 0,\ f_j(k+\alpha u^i) \leq b_j \ \forall k \in C^*(b) \right\}$$

for $j \in I$ and $i \in I(f_j)$.

The index set I is non-empty because of $V \cap H^{\perp} \neq \{0\}$. We will use the functions $t_{ij}(b)$ in order to construct the function t(b) in question and first we show that each function $t_{ij}(b)$ is continuous on Λ. Therefore, we assume $t_{ij}(b)$ not to be upper semicontinuous. Then we have

$$\exists \left\{ b^l \right\} \subset \Lambda \ \exists \varepsilon > 0 :$$

$$b^l \longrightarrow b^o \in \Lambda \text{ and } t_{ij}(b^l) > t_{ij}(b^o) + \varepsilon \ , \ l = 1,2,\dots \ .$$

Hence, there are points $k^l \in C^*(b^o)$ with

$$f_j(k^l + (t_{ij}(b^o) + \frac{\varepsilon}{2}) \ u^i) > b_j^l \ , \quad l = 1,2,\dots \ .$$

Underlying without loss of generality that $k^l \longrightarrow k^* \in C^*(b^o)$ it follows that

$$f_j(k^* + (t_{ij}(b^o) + \frac{\varepsilon}{2}) \ u^i) \geq b_j^o \ .$$

Since the definitions of $I(f_j)$ and $i \in I(f_j)$ cause $f_j(x)$ to decrease on $\left\{ x \in R^n \mid x = k^* + \mu u^i,\ \mu \geq 0 \right\}$, we have $f_j(k^* + t_{ij}(b^o) \ u^i) > b_j^o$, which yields a contradiction to the definition of $t_{ij}(b^o)$. Therefore $t_{ij}(b)$ must be upper semicontinuous.

In order to show that $t_{ij}(b)$ is lower semicontinuous on Λ we assume that

$$\exists \left\{ b^l \right\} \subset \Lambda \ \exists \varepsilon > 0 :$$

$$b^l \longrightarrow b^o \in \Lambda \text{ and } t_{ij}(b^l) < t_{ij}(b^o) - \varepsilon \ , \ l = 1,2,\dots \ .$$

Assume without loss of generality that

$$t_{ij}(b^l) \longrightarrow \bar{t} = t_{ij}(b^o) - \varepsilon \ .$$

Let $k \in C^*(b^o)$ be arbitrarily fixed. Since C^* is l.s.c., there is a sequence $\{k^l\}$ with $k^l \in C^*(b^l)$ and $k^l \longrightarrow k$.
From $f_j(k^l + t_{ij}(b^l)u^i) = b_j^l$ one obtains $f_j(k + \bar{t}\, u^i) \leq b_j^o$.
Therefore, it holds

$$f_j(k + \bar{t}\, u^i) \leq b_j^o \quad \forall\, k \in C^*(b^o) \ .$$

This contradicts the definition of $t_{ij}(b^o)$ since $\bar{t} \leq t_{ij}(b^o) - \varepsilon$.
Using the continuous function

$$t_1(b) \underset{Df}{=} \max \left\{ t_{ij}(b) \mid j \in I,\ i \in I(f_j) \right\}$$

we can define the desired continuous function t by

$$(11) \qquad t(b) \underset{Df}{=} t^*(b) + t_1(b) \sum_{i=1}^{c} a^T u^i, \quad b \in \Lambda \ .$$

It remains to show that the sets C(b) defined by (9) have the property (8).
Obviously, it suffices to show that

$$C(b) + V \supset M(b) \ , \quad b \in \Lambda \ ,$$

holds. Let $b \in \Lambda$ and $x \in M(b)$ be arbitrarily fixed, then the point x may be represented in the form

$$x = k + \sum_{i=1}^{c} \alpha_i u^i + h \quad \text{with} \quad k \in C^*(b),\ h \in H \text{ and } \alpha_1, \dots, \alpha_c \geq 0.$$

We set

$$\alpha^* = t_1(b) \ ,$$

$$k^* = k + \sum_{i=1}^{c} \min(\alpha_i, \alpha^*) u^i$$

$$u^* = \sum_{i=1}^{c} \max(0, \alpha_i - \alpha^*) u^i$$

and obtain the representation

$$x = k^* + u^* + h \quad \text{with} \quad u^* + h \in V,\ k^* \in H^{\perp} \ .$$

Hence it suffices to show

(12) $a^T k^* \geq t(b)$ and $f_j(k^*) \leq b_j$, $j = 1,\ldots,m$.

We can estimate as follows

$$a^T k^* = a^T k + \sum_{i=1}^{c} \min(\alpha_i, \alpha^*)\, a^T u^i \geq t^*(b) + \sum_{i=1}^{c} t_i(b) a^T u^i = t(b)$$

since $k \in C^*(b)$, $t_i(b) \geq \min(\alpha_i, \alpha^*) \geq 0$ and $a^T u^i \leq 0$.

In order to show the second part of (12) we distinguish between two cases. Let j be arbitrarily fixed.

Case 1.: $\alpha_i > \alpha^* \Longrightarrow i \notin I(f_j)$.

Here one has

$$f_j(k^*) = f_j(k + \sum_{i \in I(f_j)} \alpha_i u^i + \sum_{i \notin I(f_j)} \min(\alpha_i, \alpha^*) u^i).$$

Since f_j is constant on each line having a direction u^i, $i \notin I(f_j)$, we can conclude

$$f_j(k^*) = f_j(k + \sum_{i=1}^{c} \alpha_i u^i) = f_j(x) \leq b_j.$$

Case 2.: $\exists\, i^* \in I(f_j) : \alpha_{i*} > \alpha^*$.

With $\alpha^* \geq t_{i*j}(b)$ one obtains

$$f_j(k^*) = f_j(k + \sum_{i=1}^{c} \min(\alpha_i, \alpha^*) u^i) \leq$$

$$\leq f_j(k + \min(\alpha_{i*}, \alpha^*) u^{i^*}) =$$

$$= f_j(k + \alpha^* u^{i^*}) \leq b_j \quad . \qquad \#$$

Remark 1.

Theorem 5. shows that a multifunction M given by (2) possesses a representation (8) with a continuous multifunction C if and only if the following condition is satisfied:

(CPC) There exists a $b^o \in \Lambda$ and a compact set $C_o \subset R^n$ such that

$$M(b^o) = C_o + 0^+M(b^o) .$$

Obviously, if M from (2) satisfies (CPC) then the multifunction M belongs to the class of multifunctions defined by (1).

The following example shows that the representation (8) of M(b) for all $b \in \Lambda$ essentially depends on the property of the constraints $f_j(x)$ to be quasiconvex polynomials. The multifunction M defined by

$$M(b) = \{ x \in R^2 \mid e^{x_1} \leq x_2, x_1 \leq b_1 , - x_2 \leq b_2 \}$$

is continuous on $\Lambda = R^2$. For all b with $b_2 < 0$ there exists a representation of M(b) as a sum of a compact set and the recession cone

$$V = \{ u \in R^2 \mid u_1 \leq 0, u_2 \geq 0 \}$$

but, for $b_2 \geq 0$ such a representation is impossible.

By Remark 1. we know from Theorem 2. that the multifunction G given by (3) and (2) is u.s.c. on its domain, if one imposes the assumptions (CPC) on the apping M and (MIG) on the recession cone of M(b).
But, in order to be able to guarantee the continuity of this mapping G one has to restrict it to certain subsets $B(b^o)$ of the domain B.
Let $b^o \in B$ be a fixed point. We define

$$(13) \qquad B(b^o) \underset{Df}{=} \{b \in B \mid P_{x_1,\dots,x_s} G(b) = P_{x_1,\dots,x_s} G(b^o)\}$$

where $P_{x_1,\dots,x_s}$ denotes the projection into the subspace of R^n corresponding to the integer variables $x_1,\dots,x_s$.

One immediately shows that the following hold, for all fixed $b^o \in B$:

(i) $B(b^o) \neq \emptyset$;

(ii) $b^1 \in B(b^o) \Longrightarrow B(b^1) = B(b^o)$;

(iii) $b^1 \in B \setminus B(b^o) \Rightarrow B(b^1) \cap B(b^o) = \emptyset$

Theorem 6.

Let G be given by (3) with M from (2). If M satisfies (CPC) and if the recession cone $V = 0^+M(b)$ fulfils (MIG) then

(i) G is u.s.c. on B;

(ii) G is continuous on $B(b^o)$ for an arbitrarily fixed $b^o \in B$.

Proof. (i) is clear by the remarks preceeding this theorem.

(ii) Because of (i) we have only to show the l.s.c. property of G on $B(b^o)$. Let be

$$b^* \in B(b^o), \quad \{b^t\} \subset B(b^o) \quad \text{and} \quad b^t \longrightarrow b^* .$$

Obviously, if suffices to verify that for each arbitrarily fixed $x^* \in G(b^*)$ there exists a sequence $\{x^t\} \subset R^n$ with $x^t \in G(b^t)$ for almost all t such that $x^t \longrightarrow x^*$. Hence, let $x^* \in G(b^*)$ be fixed. We consider the multifunction $G(x_1^*,\dots,x_s^*)$ defined on $B(b^o)$ by

$$G(x_1^*,\dots,x_s^* ; b) \underset{Df}{=} \{x \in M(b) \mid x_j = x_j^* , j = 1,\dots,s\} .$$

We have $G(x_1^*,\dots,x_s^* ; b^t) \neq \emptyset \ \forall t$ because of the definition of $B(b^o)$ and, it holds that $x^* \in G(x_1^*,\dots x_s^* ; b^*)$. Since the functions

$$f_i^*(x_{s+1},\dots,x_n) \underset{Df}{=} f_i(x_1^*,\dots,x_s^* , x_{s+1},\dots,x_n) , i = 1,\dots,m$$

are quasiconvex polynomials in the variables $x_{s+1},\dots,x_n$ one concludes from Theorem 3.(i) that $G(x_1^*,\dots,x_s^*)$ is l.s.c. on $B(b^o)$. Hence there is a sequence $\{x^t\} \subset R^n$ such that

$$x^t \in G(x_1^*,\dots,x_s^* ; b^t) \ \forall t \quad \text{and} \quad x^t \longrightarrow x^* .$$

Because of $G(x_1^*,\dots,x_s^* ; b^t) \subset G(b^t)$ by $\{x^t\}$ the desired sequence is obtained. #

The following example shows that the validity of the proposition (ii) of Theorem 6. essentially depends on the underlying description of the involved multifunction M.

For the compact-valued continuous multifunction M given by

$$M(b) = \{ x \in R^2 \mid x_2 \le 1,\ 2x_1 - x_2 \le 1,\ -2x_1 - x_2 \le 1,\ (1 - b)\, x_1 + b x_2 = -b \}$$

and the constraint set mapping G by

$$G(b) = \{ x \in M(b) \mid x_1 \in Z \}$$

one has

$$B \equiv \Lambda = (-\infty, \infty).$$

Further, $B(0) = (-\infty, \infty) \setminus \{-1, 1/3\}$. Since $G(0) = \{ x \in R^2 \mid x_1 = 0,\ -1 \le x_2 \le 1 \}$ and $G(b) = \{ (0,-1) \}$ for all $b \notin \{-1, 1/3\}$ the multifunction G is not l.s.c. at b = 0.

Only for the linear case

$$M(b) = \{ x \in R^n \mid A x \le b \},\ b \in \Lambda$$

with a rational matrix A a deeper characterization of the sets $B(b^o)$ (13) is known (BANK 1977). For example, this sets are star-shaped and in general neither open nor closed. A possible form is shown in the figure standing beside.

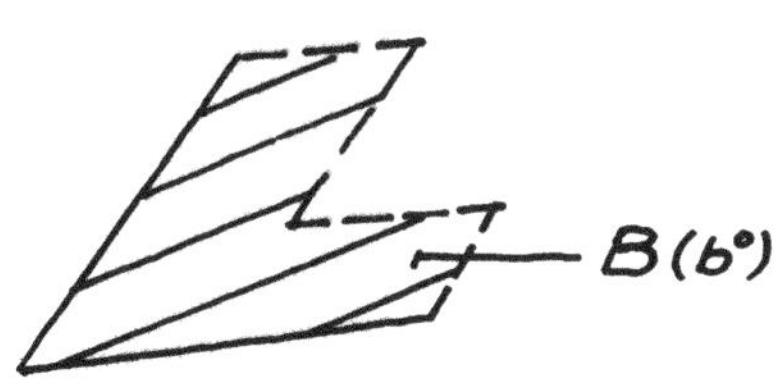

The example illustrates two extremal cases:

$$M(b) = \{ x \in R^2 \mid 2x_1 + x_2 \le b_1,\ x_1 + 2x_2 \le b_2 \}.$$

Let $b^o = (2,2)^T$. If $G(b) = \{ x \in M(b) \mid x_1 \in Z \}$ then one obtains $B(b^o) = R^2$ and, if both the variables are required to be integer then one has

$$B(b^o) = \{ b \in R^2 \mid 2 \leq b_1 < 3,\ 2 = b_2 < 3 \} .$$

The next theorem, is particularly devoted to the question whether a set G(b) (3) with M(b) of the form (2) is non-empty. Thereby, only the constraint polynomials bounded from below on M(b) are important. We define:

$$I^* \underset{Df}{=} \{ i \in \{1,\dots,m\} \mid \inf \{ f_i(x) \mid x \in M(b) \} > -\infty \} \ (b \in \Lambda),$$

$$M^*(b) \underset{Df}{=} \{ x \in R^n \mid f_i(x) \leq b_i,\ i \in I^* \} .$$

Because of Theorem 4. the index set I^* does not depend on the choice of $b \in \Lambda$.

The mapping $M^* : \Lambda \rightarrow 2^{R^n}$ defined by $M^*(b)$ is called the stable mapping to M.

Taking into account the monotonicity- and constancy-properties of quasiconvex polynomials (see above) one can see that the recession cone V^* of the stable sets $M^*(b)$ is a linear subspace and, by Theorem 3.(iv), the stable mapping M^* is continuous on Λ .

Theorem 7.

(i) If the (mixed-) integer points of the stable sets $M^*(b)$ are uniformly distributed, then the set G(b) of feasible points is non-empty.

(ii) If the recession cone V^* of the stable sets satisfies the assumption (MIG), then the domain $B = \{ b \in R^m \mid G(b) \neq \emptyset \}$ is a closed set.

The assumption of the statement (ii) ensures that the (mixed-)integer points of $M^*(b)$ are uniformly distributed if there exists a (mixed-) integer point (Theorem 2.iv).

For illustration we consider

$$M(b) = \{ x \in R^2 \mid \sqrt{2}\, x_1 - x_2 \leq b_1,\ -\sqrt{2}x_1 + x_2 \leq b_2,\ -x_1 \leq b_3 \} .$$

Then one gets

$$M^*(b) = \{ x \in R^2 \mid \sqrt{2}x_1 - x_2 \leq b_1,\ -\sqrt{2}x_1 + x_2 \leq b_2 \}$$

and

$$V^* = \{ u \in R^2 \mid \sqrt{2}u_1 - u_2 = 0 \} .$$

Obviously, the subspace V^* does not satisfy (IG).

For the feasible set

$$G(b) = \{ x \in M(b) \mid x_1, x_2 \in Z \}$$

it follows:

- If $b^0 = (0,0,-1)^T$, then $G(b^0) = \emptyset$ but, $0 \in M^*(b^0)$.
- If $b^t = (\frac{1}{t}, \frac{1}{t} - 1)^T$, then $G(b^t) \neq \emptyset$

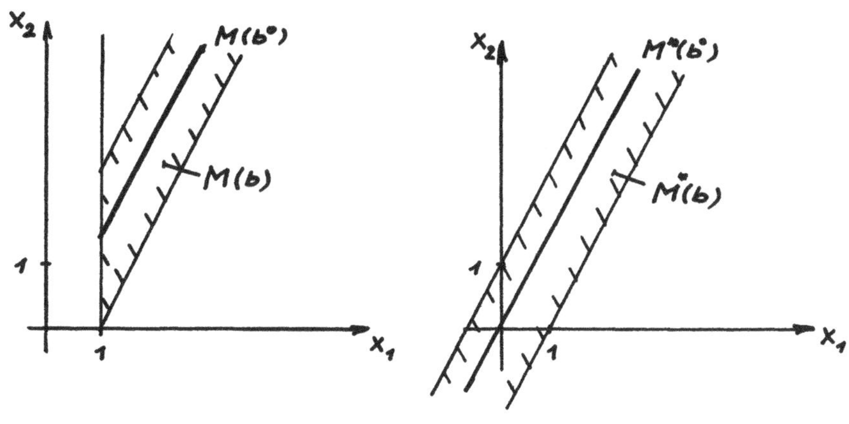

2. Stability and Existence.

In this section we deal with stability properties of the parametric problem

$P(p,b)$ $\inf \{ f(x) + p^T x \mid x \in G(b) \}$, $(p,b) \in R^n \times R^m$,

where $G(b)$ is given by (3) and the involved $M(b)$ is of the general form (1).

We distinguish between the two cases

A. $f(x)$ is a quadratic form on R^n

B. $f(x)$ is a convex polynomial on R^n .

The valuefunction $\varphi : \mathcal{E} \longrightarrow R$ of $P(p,b)$ is defined by

$$(14) \qquad \varphi(p,b) \underset{Df}{=} \inf \{ f(x) + p^T x \mid x \in G(b) \}$$

$$\mathcal{E} \underset{Df}{=} \{ (p,b) \in R^n \times B \mid \varphi(p,b) > -\infty \}.$$

Further, the multifunction $\psi : \mathcal{E} \longrightarrow 2^{R^n}$ given by the optimal sets

$$\psi(p,b) \underset{Df}{=} \{ x \in G(b) \mid f(x) + p^T x = \varphi(p,b) \}$$

is called the optimal set-function.

The aim now is to characterize the continuity behaviour of the value-function φ, the optimal set-function ψ and the ε -optimal set-function $\tilde{\psi}$ given by

$$(15) \qquad \tilde{\psi}(p,b,\varepsilon) \underset{Df}{=} \{ x \in G(b) \mid f(x) + p^T x \le \varphi(p,b) + \varepsilon \}$$

where $\varepsilon \ge 0$. The level set-mapping $N : \mathcal{E}_N \longrightarrow 2^{R^n}$ given by

$$(16) \qquad N(p,b,a) \underset{Df}{=} \{ x \in G(b) \mid f(x) + p^T x \le a \}$$

and

$$\mathcal{E}_N \underset{Df}{=} \{ (p,b,a) \in \mathcal{E} \times R \mid N(p,b,a) \ne \emptyset \}$$

plays an important role for stability considerations.

The following simple example of an integer linear parametric problem shows that neither the value function is upper semicontinuous nor the optimal set-function is u.s.c. if the constraint set-mapping is not continuous.

$P(b) \quad \inf \{ -x \mid 0 \leq x \leq b, \ x \in Z \}.$

For $\{ b_t \}$ with $b_t \underset{Df}{=} 1 - \frac{1}{t} \longrightarrow b_o = 1$ one has

$$\varphi(b_t) = 0 \quad \text{and} \quad \varphi(b_o) = -1$$

$$\psi(b_t) = \{0\} \quad \text{and} \quad \psi(b_o) = \{1\} .$$

In order to guarantee the above properties for φ and ψ one has to restrict the constraint set-mapping to sets $B(b^o)$ (13) (see Theorem 6.)

A. $f(x) = x^T D x$, D a symmetric matrix.

First we cite a result do to HANSEL 1980 for the set

(17) $\quad \tilde{V} \underset{Df}{=} \{ x \in V \mid x^T D x = 0 \}.$

Lemma 1.

Let $x^T D x \geq 0$ for all $x \in V$. Then it holds:

(i) The set $\tilde{V}$ (17) is a union of a finite number of polyhedral cones.

(ii) There exists a representation of $\tilde{V}$ as under (i) announced such that all needed polyhedral cones fulfil the (IG) condition if V does so and if D is a rational matrix.

A proof of Lemma 1. can be found in BANK/HANSEL 1984.

In the following we will say that

"V satisfies (MIG) (or (IG))"

if $x^T D x \geq 0$ for all $x \in V$ and if each of the polyhedral cones whose union yields $\tilde{V}$ fulfils the (MIG) - (or (IG) -) condition.

If the constraint set G(b) is a fixed set, then one obtains the following result.

Theorem 8.

Let $b \in B$ be fixed and let V satisfy (MIG). Then

(i) $E(b) \underset{Df}{=} \{ p \in R^n \mid \varphi(p,b) > -\infty \}$ is a convex polyhedron;

(ii) the value function $\varphi(\cdot, b)$ is continuous on E(b).

The existence of optimal points and the upper semicontinuity of the optimal set for the particular case of fixed constraint sets will not be investigated here; these results can easily be derived from the following more general considerations.

The following lemma is a powerful tool for treating the general quadratic problem P(p,b) with constraint sets G(b) given by (3) and M(b) of the form (1). It should be noted that this lemma has been established first in HANSEL 1980 for the case of linear (mixed-)integer constraints. Because of Theorem 6. we can use a similar proof by taking the representation of $G(b^t)$ into account. The same is true for the Theorems 9. and 10. .

Lemma 2.

Let $\mathcal{E} \neq \emptyset$ and let $\tilde{V}$ satisfy (MIG) and further let $(p^t, b^t, a_t) \in \mathcal{E}_N$, $t = 1,2,\dots$ be a sequence with $\lim(p^t, b^t, a_t) = (p^o, b^o, a_o)$ where $b^o \in B$. Then, for every sequence $\{x^t\} \subset R^n$ having the properties $x^t \in N(p^t, b^t, a_t)$ and $\| x^t \| \longrightarrow \infty$ there is a vector $v \in V_s$, $v \neq \emptyset$ satisfying both $v^T D v = 0$ and there is a subsequence $\{y^r\}$ of the sequence $\{y^t\}$ defined by $y^t \underset{Df}{=} x^t - v$ such that the following three statements hold

(i) $y^r \in N(p^r, b^r, a_r)$;

(ii) $\| y^r \| < \| x^r \|$;

(iii) For every natural number k there is an index r(k) with

$$x^r - k\, v \in N(p^r, b^r, a_r) \quad \forall\, r > r(k) .$$

Finally, if $\tilde{V}$ satisfies (IG), then v can be considered to be an integer vector.

<u>Proof.</u> We can suppose that the convex polyhedral cone V is described in the form

$$V = \{ v \in R^n \;/\; d^{i^T} v \leq 0,\; i = 1,\dots,q \}$$

and that the elements of $\{x^t\}$ may be represented by

(18) $$x^t = z^t + \alpha_t\, u^t ,$$

where $u^t \in V$, $\| u^t \| = 1$, $z^t \in C(b^t)$, $\alpha_t \to \infty$.

Without loss of generality we can assume that $u^t \to u^o$, where $u^o \in V$ and $\|u^o\| = 1$. Because of the upper semicontinuity of C and since $C(b^o) \neq \emptyset$ is a compact set, the sequence $\{z^t\}$ must be bounded. The inequality

$$x^{t^T} D\, x^t + p^{t^T} x^t = z^{t^T} D\, z^t + p^{t^T} z^t + \alpha_t (2D\, z^t + p^t)^T u^t + $$
$$+ \alpha_t^2\, u^{t^T} D\, u^t \leq a_t$$

yields $u^{o^T} D\, u^o = 0$ if $u^{t^T} D\, u^t \geq 0$ is taken into account. Using the index set $I \underset{Df}{=} \{ i \in \{ 1,\dots,q \} \,/\, d^{i^T} u^o = 0 \}$

we obtain

$$u^o \in \tilde{V} ,\quad \| u^o \| = 1 ,\quad d^{i^T} u^o = 0 \quad \forall\, i \in I .$$

Since $\tilde{V}$ satisfies (MIG) and because of $d^{i^T} v \leq 0$ for all $i \in I$ and all $v \in V$ there must be a vector $v \in R^n$ whose components $v_1,\dots,v_s$ are integers such that

$$v \in \tilde{V} ,\quad \| v \| \geq 1 ,\quad u^{o^T} v \geq \tfrac{1}{2} ,\quad d^{i^T} v = 0 \quad \forall\, i \in I.$$

if $\tilde{V}$ satisfies (IG), then v can be considered as an all integer vector.

Setting now $y^t \underset{Df}{=} x^t - v$ one has

$$\|y^t\|^2 = \|x^t\|^2 + \|v\|^2 - 2x^{t^T}v = \|x^t\|^2 + \|v\|^2 - 2z^{t^T}v - 2\alpha_t u^{t^T}v .$$

and, taking into account that $\alpha_t \longrightarrow \infty$ $v^Tu^o \geq \frac{1}{2}$ and $u^t \longrightarrow u^o$ hold, we obtain without loss of generality that $\|y^t\| < \|x^t\|$, which proves proposition (ii).

For a fixed t we now consider the points $x = x^t - \beta v$, $\beta \geq 0$. Substituing (18) one has $x = z^t + \alpha_t u^t - \beta v$, $\beta \geq 0$, and by $z^t \in C(b^t)$ it follows that

$$z^t + \alpha_t u^t - \beta v \in C(b^t) + V$$

if it is ensured that $\alpha_t u^t - \beta v$ belongs to V . In order to discuss the latter question we distinguish between two cases.

(1) If $d^{i^T} v = 0$, then we have

$$d^{i^T}(\alpha_t u^t - \beta v) = \alpha_t d^{i^T} u^t \leq 0 \quad \forall \beta \geq 0 .$$

2) If $d^{i^T} v < 0$, then the choice of v implies that $d^{i^T} u^o < 0$ holds. By $u^t \longrightarrow u^o$ we have without loss of generality

$$d^{i^T} u^t \leq \frac{1}{2} d^{i^T} u^o < 0 \quad \text{and, by}$$
$$d^{i^T}(\alpha_t u^t - \beta v) = \alpha_t d^{i^T} u^t - \beta d^{i^T} v \leq \frac{1}{2} \alpha_t d^{i^T} u^o - \beta d^{i^T} v$$

we finally obtain

$$d^{i^T}(\alpha_t u^t - \beta v) \leq 0 \quad \forall \beta \in \left[0, \frac{1}{2} \alpha_t \frac{d^{i^T} u^o}{d^{i^T} v}\right].$$

For every natural number k it holds that the components $x_j^t - k v_j$, $j = 1,\ldots,s$, are integer numbers. Now, by $\alpha_t \longrightarrow \infty$ the discussion of the two cases above implies that without loss of generality $x^t - v \in G(b^t)$ and for every natural k there is an index r(k) such that

$$x^t - k\, v \in G(b^t) \quad \forall\, t > r(k).$$

The proof of (i) and (iii) will be completed by estimating as follows

$$(x^t - k\, v)^T D(x^t - k\, v) + p^{t^T}(x^t - k\, v) =$$

$$= x^{t^T} D\, x^t + p^{t^T} x^t - k(2D\, x^t + p^t)^T v \leq$$

$$\leq x^{t^T} D\, x^t + p^{t^T} x^t \leq a_t \quad \forall\, k \geq 0 ,$$

where we took into account that $(p^t, b^t) \in \mathcal{E}$ implies the inequality

$$(2D\, x^t + p^t)^T v \geq 0 .$$

#

Theorem 9.

If $\tilde{V}$ satisfies (MIG) then:

(i) P(p,b) has an optimal solution if and only if $(p,b) \in \mathcal{E}$

(ii) The value function is lower semicontinuous on $\mathcal{E}$.

Proof.

Both statements can be shown by applying Lemma 2.; we show only (i). Obviously, one only has to verify that $(p,b) \in \mathcal{E}$ implies the existence of an optimal point of P(p,b). Let $(p,b) \in \mathcal{E}$ be a fixed parameter pair. Setting

$$a_o = \varphi(p,b) \quad \text{and} \quad a_t \underset{\text{df}}{=} a_o + \frac{1}{t}$$

we can choose points

$$x^t \in N(p,b,a_t) \; ; \quad t = 1,2,\ldots$$

of minimal norm. By Lemma 2. the sequence $\{x^t\}$ must have an accumulation point belonging to $N(p,b,a_o)$. #

We remark that the value function is also continuous under the assumption that the cone V satisfies MIG, but the existence of an optimal point cannot be ensured in this case as the following example illustrates.

$$\inf \left\{ (x_1 - x_2 \sqrt{2})^2 \mid x_i \geq 1 , \quad x_i \in Z , \quad i = 1,2 \right\} = 0$$

but there is no optimal point.

Under stronger assumptions one obtains, by application of Lemma 2., the following result on the level set-mapping N.

<u>Theorem 10.</u>

Let the matrix D be rational and let $\tilde{V}$ satisfy (IG). When $(p^t, b^t, a_t) \in \mathcal{E}_N$, $t = 1,2,\ldots$ is a sequence with $(p^t, b^t, a_t) \longrightarrow (p^o, b^o, a_o)$, where $(p^o, b^o) \in \mathcal{E}$, then

$\forall \ \varepsilon > 0 \quad \exists \ t(\varepsilon) : N(p^t, b^t, a_t) \subset U_\varepsilon \ N(p^o, b^o, a_o) \ \forall \ t > t(\varepsilon)$
holds for the level sets.

Hence, it is clear that the level set-mapping N is u.s.c. on $\mathcal{E}_N$ and $N(p^o, b^o, a_o) \neq \emptyset$ if D is rational and $\tilde{V}$ satisfies (IG).

Theorem 11. obtained as a simple consequence of Theorem 10. expresses that the ε-optimal set-function $\tilde{\psi}$ given by (15) has the property to be u.s.c. on a certain subset $\mathcal{E}^* \times R_+$ if besides the hypotheses of Theorem 10. one additionally requires that the value function φ is continuous on the subset $\mathcal{E}^* \subset \mathcal{E}$.

Theorem 11.

Let D be a rational matrix and let $\tilde{V}$ satisfy (IG). If, moreover, $\mathcal{E}^*$ is a subset of $\mathcal{E}$ such that φ is continuous on $\mathcal{E}^*$, then the ε-optimal set-function $\tilde{\psi}$ is u.s.c. on $\mathcal{E}^* \times R_+$.

On account of Theorem 9.(ii) the required continuity of the value function φ in Theorem 11. means in fact that φ has to be upper semi-continuous on $\mathcal{E}^*$. But, this is ensured if the constraint set-mapping G is continuous. By Theorem 6.(ii) we know that G from (3) is continuous on a subset $B(b^0)$ (13) of B = dom G if the involved multifunction M is given by (2) which fulfils (CPC) and its recession cone V satisfies (MIG). Thus we define

(19) $$\mathcal{E}^* =_{Df} \{(p,b) \in \mathcal{E} \mid b \in B(b^0)\}$$

for an arbitrarily fixed $b^0 \in B$.

Theorem 12.

Let G(b) be given by (3) with an included multifunction M of the form (2) for which (CPC) holds. Further, let $b^0 \in B$ be arbitrarily fixed. Then

(i) the value function φ is continuous on $\mathcal{E}^*$ given by (19) if the recession cone V of M(b) satisfies (MIG);

(ii) the ε-optimal set-function $\tilde{\psi}$ is u.s.c. on $\mathcal{E}^* \times R_+$ if the matrix D is rational and the recession cone V of M(b) satisfies (IG).

B. f(x) is a convex polynomial without a linear term.

This subsection is devoted to parametric (mixed-)integer problems P(p,b) where the objective function is a convex polynomial with a variable linear part.

The polyhedral cone

$$(20) \qquad \tilde{V} \underset{Df}{=} V \wedge 0^+ \{ x \in R^n \mid f(x) \leq 0 \}$$

plays a similar role as the set (17) (denoted by the same symbol) in the previous subsection A.

From the introductory remarks of the subsection 1.(b) one concludes that

$0^+ \{ x \in R^n \mid f(x) \leq 0 \}$ is a linear subspace and it holds that

$f(x+kv) = f(x) \quad \forall\, x \in R^n \; \forall\, k \in R \quad \forall\, v \in 0^+ \{ x \mid f(x) \leq 0 \}$.

By Theorem 4. this recession cone satisfies (IG) if all the coefficients of f(x) are rational numbers.

The results listed in the following theorem can be derived by proceeding, in a way, as in the subsection A., in particular one can set up a lemma analogously to Lemma 2.

Theorem 13.

Let the polyhedral cone $\tilde{V}$ (20) satisfy (MIG). Then we have:

(i) The set

$$E(b) \underset{Df}{=} \{ p \in R^n \,/\, \varphi(p;b) > -\infty \}, \ b \in B$$

is the fixed polyhedral cone

$$\{ p \in R^n \,/\, p^T v \geq 0, \ v \in \tilde{V} \} .$$

(ii) The value function $\varphi(\cdot, b)$, $b \in B$ is continuous on $E(b)$.

(iii) φ is l.s.c. on $\mathcal{E}$.

(iv) $P(p,b)$ has an optimal solution if and only if $(p,b) \in \mathcal{E}$.

(v) The level set-mapping N is u.s.c. on $\mathcal{E}_N$.

In order to guarantee the continuity of the value function φ (and also the u.s.c.-property of the ε-optimal set-function $\tilde{\psi}$) one has to consider constraint sets (3) where the involved multifunction M is of the form (2) satisfying the same assumptions as imposed in Theorem 12.

Theorem 14.

Let G(b) be given by (3) with an included multifunction M of the form (2) for which (CPC) holds and the recession cone V satisfies (MIG). If $b^0 \in B$ is an arbitrarily fixed parameter point then

(i) the value function φ is continuous on the set $\mathcal{E}^*$ given by (19) and,

(ii) the ε-optimal set-function $\tilde{\psi}$ is u.s.c. on $\mathcal{E}^* \times R_+$.

REFERENCES

Allende, S.(1980). Globale Stabilitätsuntersuchungen rein- und gemischtzahliger parametrischer Optimierungsprobleme. Diss.(A), Humboldt-Universität Berlin.

Bank, B (1977). Qualitative Stabilitätsuntersuchungen rein- und gemischtganzzahliger linearer parametrischer Optimierungsprobleme. Diss.(B), Humboldt-Universität Berlin.

Bank, B., E.G.Belousow, R.Mandel, J.N.Ceremnych, V.M.Shironin (1986). Matematyceskije programmirowanje: voprozy rasreshimosti i ustoycivosti. Iszd. MGU, Moskwa (russian).

Bank, B., R.Hansel (1984). Stability of mixed-integer quadratic programming problems. Mathematical Programming Study 21, 1-17 North-Holland.

Bank, B., R.Mandel (1986/87). Nonlinear integer programming. Akademie-Verlag Berlin, to appear.

Bank, B., R.Mandel, K.Tammer (1979). Parametrische Optimierung und Aufteilungsverfahren. In: Lommatzsch, K.(ed.). Anwendungen der linearen parametrischen Optimierung. Akademie-Verlag, Berlin 1982.

Bank, B., J.Guddat, D.Klatte, B.Kummer and K.Tammer (1982). Non-linear Parametric Optimization. Akademie-Verlag, Berlin.

Belousov, E.G.(1977). Vvedenie v vypuklyj analiz i celocislennoe programmirovanie. Izdat.Moskovskogo universiteta, Moskva.

Blair, C.E., R.G.Jeroslow (1977). The value function of a mixed integer program: I. Discrete Mathematics, 25, 121-138.

Blair, C.E., R.G.Jeroslow (1979). The value function of a mixed integer program: II. Discrete Mathematics, 25, 7-19.

Blair, C.E., R.G. Jeroslow (1981). An exact penalty method for mixed-integer programs. Mathematics of operations research, 6, No.1, 14-18.

Blum, E., W.Oettli (1972). Discret proof of the existence theorem for quadratic programming. Operations Research, 20, 165-167.

Borosh, I., L.B.Treybig (1976). Bounds on positive integral solutions of linear Diophantine equations. Proc.Amer.Math.Soc. 55, 299-304.

Bowman, V.J., C.-A.Burdet (1974). On the general solution to systems of mixed-integer linear equations. SIAM J.Appl.Math.26, No.1.

Charnes, A., F.Granot (1975). Existence and representation of diophantine and mixed diophantine solutions to linear equations and inequalities. Discrete Mathematics, 11, 233-248.

Eaves, C.R. (1971). On quadratic programming. Management Science, Vol.17, 698-711.

Fiorot, J.Ch. (1972). Generation of all integer points for given sets of linear inequalities. Math.Programming, 3, 276-295.

Geoffrion, A.M., R.Nauss (1977). Parametric and postoptimality analysis in integer linear programming. Management Science, Vol.23, No.5, 453-466.

Hansel, R. (1980). Stabilität gemischt-ganzzahliger Optimierungsprobleme. Diss.(A), Humboldt-Universität, Berlin.

Jeroslow, R.G. (1978). Cutting Plane Theory: Algebraic Methods. Discrete Mathematics, 23, 121-150.

Khachiyan, L.G. (1982). Vypuklost i algoritmiseskaja slosmost reshenij sadac polinomialnowo programmirowanija. Usw.AN SSSR, Techniceskaja Kibernetika, No.6. (russian).

Klatte, D.(1984). Beiträge zur Stabilitätsanalyse nichtlinearer Optimierungsprobleme. Diss.(B), Humboldt-Universität, Berlin.

Kummer,B. (1978). Stability and weak duality in convex programming without regularity. Preprint 1978; inzwischen erschienen in: Wiss. Z. der Humboldt-Univ. Berlin, Math.-Nat.R., XXX, 5, 381-386.

Kummer, B. (1977). Globale Stabilität quadratischer Optimierungsprobleme. Wiss.Z. der Humboldt-Univ.Berlin, Math.-Nat.R. XXVI, 5, 565-569.

Mandel, R. (1983). Zur Konstruktivität bei (gemischt-) ganzzahligen Optimierungsaufgaben. Math.Operationsforschung u. Statistik, Ser.Optimization, 12, Heft 1, 33-39.

Mandel, R. (1985). Beiträge zur Theorie ganzzahliger Optimierungsaufgaben. Dissertation (B), Humboldt-Universität zu Berlin.

Meyer, R.R. (1976). Mixed-integer minimization models for piecewise linear functions of a single variable. Discrete Math., 16, 163-171.

Meyer, R.R. (1975). Integer and mixed integer programming models: General properties. J.Optim. Theory Appl., 16, 191-206.

Radke, M. (1975). Sensitivity analysis in discrete optimization. Working paper 240, Western Management Science Institute, Univ. of California, Los Angeles.

Rockafellar, R.T. (1970). Convex analysis. Princeton University Press, Princeton.

Schironin, V.M. (1980). O raspolozenii celocislennych tocek v neogranicennych vypuklych mozestvach euklidovo prostranstava. Diss.(Kand. fiz-mat nauk), Ökon.Fakultät, Moskauer Staatl.Un.

Tind, J., L.A.Wolsey (1981). An elementary survey of general duality theory in mathematical programming. Math.Prog.21, 241-261.

STABILITY OF POLYNOMIAL OPTIMIZATION PROBLEMS

E. G. Belousov*)

Abstract.

The paper deals with the Hausdorff-continuity of the optimal set and the optimal value for optimization problems with a cubic polynomial objective function and linear constraints when perturbations of the linear part in the objective function and in the right hand sides of the constraints are allowed.

Introduction.

The following optimization problem shall be considered in the paper

$$P(p,\lambda): \max \{f(x) + (Bx+b)p \;/\; Ax \leq \lambda\}$$

where $p \in E_q$, $\lambda \in E_m$ are parameters, $f(x)$ is a fixed polynomial, B and A are fixed matrices of the sizes (q,n) and (m,n), respectively and, $b \in E_q$ is a fixed vector.

In the paper the Hausdorff continuity of the optimal set and the maximum function of $P(p,\lambda)$ is treated if the parameters p and λ vary. We use the notations:

$f_p(x) = f(x) + (Bx+b)p$; $A(\lambda) = \{x \in E_n \;/\; A\,x \leq \lambda\}$;

$\mathcal{E} = \{(p,\lambda) \in E_q \times E_m \;/\; \sup_{x \in A(\lambda)} f_p(x) < +\infty\}$,

the set on which P(p,b) is bounded;

$P(\lambda) = \{p \in E_q \;/\; (p,\lambda) \in \mathcal{E}\}$; $\Lambda = \{\lambda \in E_m \;/\; A(\lambda) \neq \emptyset\}$;

$\tilde{\Lambda} = \{\lambda \in E_m \;/\; A(\lambda) \neq \emptyset,\ P(\lambda) \neq \emptyset\}$

$\mathring{\Lambda} = \{\lambda \in \tilde{\Lambda} / \exists \hat{\lambda} \in \tilde{\Lambda} \;:\; \lambda < \tilde{\lambda}\}$, the "interior" of the set $\tilde{\Lambda}$.

In [1], [2] it was shown that the problem $P(p,\lambda)$ is solvable for all $(p,\lambda) \in \mathcal{E}$ if f(x) is a polynomial of degree ≤ 3. Therefore, the mappings

- $\varphi(p,\lambda) = \max_{x \in A(\lambda)} f_p(x)$, the maximum function of $P(p,\lambda)$,
- $\psi(p,\lambda) = \{x \in A(\lambda) \;/\; f_p(x) = \varphi(p,\lambda)\}$, the optimal set of $P(p,\lambda)$

are well-defined on $\mathcal{E}$.

For the case λ = const we know from [1] that φ is continuous and Hausdorff-upper semicontinuous (briefly: H-u.s.c.) on $P(\lambda)$ if f(x) is a cubic polynomial.

*) Moscow State University, Faculty of Economics, Leninskiegori 119899 Moscow

The aim of this paper is to derive a sufficient stability condition for a cubic problem $P(p,\lambda)$. We shall show that the function φ is continuous at $(p,\lambda)\in\mathcal{E}$ if $\lambda\in\mathring{\Lambda}$ and, that ψ is H-u.s.c. at such a point (see Theorem 1. and the corresponding corollary). Theorems 2. and 3. are dedicated to the Lipschitzian continuity of the maximum function φ on a bounded set $S\subset\mathcal{E}$ if the polynomial $f(x)$ in $P(p,\lambda)$ is cubic or quadratic or concave or convex.

Analogous questions are treated in many papers. For example, quadratic objective functions were explored in [3] - [6] and concave polynomials of degree n in [7].

Main Part.

The central propositions are derived by application of the Lemmata 0., 1. and 2. The following notations let be introduced:

$f_k(x)$ polynomials on E_n

x_k, $k=1,2,\dots$ a sequence of points

x_o a fixed point with $x_o \neq x_k$

$$t_k = |x_k - x_o|, \quad r_k = \frac{x_k - x_o}{|x_k - x_o|} \quad ; \Longrightarrow x_k = x_o + r_k t_k$$

$$\Delta_k = f_k(x_k) - f_k(x_o).$$

Further, we make use of the abbreviations: $u_k > v \longrightarrow u_o$ and $u_k < v \longrightarrow u_o$ if $\underline{\lim}\, u_k \geq u_o$ and $\overline{\lim}\, u_k \leq u_o$, respectively, for any sequence of numbers u_k and a number u_o.

Lemma 0. Let $f_k(x)$ be a polynomial of degree $k \leq 3$ and $x_k \in E_n$, $k = 1,2,\dots$. Then

(1) if the conditions

$$\frac{\Delta_k}{t_k^2} > v \longrightarrow 0, \quad \frac{\nabla f_k(x_k) r_k}{t_k} < v \longrightarrow 0, \quad \nabla(\nabla f_k(x_o) r_k) r_k < \gamma < 0$$

are satisfied then there is a number $\rho > 0$ and a natural number N exists such that

$$f_k(x_k + \tau r_k) > f_k(x_k) + (\rho t_k)\tau + \rho\tau^2 + \frac{\rho}{t_k}\tau^3$$

for $k > N$ and $\tau > 0$

(2) Let the conditions of (1) be satisfied for $t_k > \mu > 0$. Then

$$f_k(x_k + \tau_k r_k) - f_k(x_k) > v \longrightarrow 0$$

where $\tau_k = T_k/t_k$, $0 \leq t_k \leq T_o$.

(The constants γ, ρ, μ, T_o are arbitrarily fixed positive numbers). For a proof the reader is referred to [1] .

Lemma 1. Let

$$A(\lambda) = \{x \in E_n \,/\, Ax \leq \lambda\} \cap L$$
$$f_p(x) = f(x) + (Bx+b)p$$

be where L is a fixed affine subspace and f(x) a fixed polynomial of a degree ≤ 3. Further, let be λ_k a sequence satisfying

$$\lambda_k \leq \bar{\lambda} \leq \hat{\lambda} \quad \text{where } P(\hat{\lambda}) \neq \emptyset.$$

Finally, let $p_k \in P(\lambda_k)$ and $x_k \in A(\lambda_k)$ be sequences such that

$$\inf_k \frac{f_{p_k}(x_k)}{|x_k|} > -\infty \quad |x_k| \longrightarrow +\infty, \quad \frac{x_k}{|x_k|} \longrightarrow r.$$

Then

$$\nabla(\nabla(\nabla f(x)r)r)r = 0, \quad \nabla(\nabla f(x)r)r = 0 \; \forall x \in L, \tag{1}$$

i.e. the polynomial $f_p(x)$ is linear on every line having the direction r and which is contained in L.

Proof. Since $\inf \frac{f_{p_k}(x_k)}{|x_k|} > -\infty$ it must be that $\nabla(\nabla(\nabla f(x)r)r)r \geq 0$.

On the other hand $\mathcal{E} \neq \emptyset$ implies $\nabla(\nabla(\nabla f(x)r)r)r \leq 0$ such that the first equation of (1) is clear. From $P(\hat{\lambda}) \neq \emptyset$ follows

$$\nabla(\nabla f(x)r)r \leq 0 \quad \forall x \in A(\hat{\lambda}). \tag{2}$$

Assume, that there is an $x_o \in L$ such that $\nabla(\nabla f(x_o)r)r < 0$. Let $p_o \in P(\hat{\lambda})$. Using the notations

$$r_k = \frac{x_k - x_o}{|x_k - x_o|}, \quad t_k = |x_k - x_o|$$

$$\Delta_k = f_{p_o}(x_k) - f_{p_o}(x_o) = f_{p_k}(x_k) + (p_o - p_k)B_k x_k + (p_o - p_k)b - f_{p_o}(x_o),$$

$$\nabla f_{p_o}(x_o)r_k = \nabla f(x_o)r_k + p_o B r_k$$

we obtain

$$\frac{\Delta_k}{t_k^2} > v \longrightarrow 0, \quad \frac{\nabla f_{p_o}(x_o)r_k}{t_k} \longrightarrow 0,$$

$$\nabla(\nabla f_{p_o}(x_o)r_k)r_k = \nabla(\nabla f(x_o)r_k)r_k \longmapsto \nabla(\nabla f(x_o)r)r < 0.$$

By Lemma 0. there exist a $\rho > 0$ and a natural number N such that

$$f_{p_o}(x_k + \tau r_k) > f_{p_o}(x_k) + \rho t_k \tau, \quad k > N, \quad \tau > 0.$$

Because of $p_o \in P(\hat{\lambda})$ the half-lines $\{x \,/\, x = x_k + \tau r_k, \tau \geq 0\}$ intersect the boundary of the set $A(\hat{\lambda})$. Consequently, there is a face of $A(\hat{\lambda})$ containing an infinite number of points $y_k = x_k + \tau_k r_k$ where

$$\tau_k = \min_{i\in J(k)} \frac{\hat{\lambda}^i - a_i x_k}{a_i r_k}, \quad J(k) = \{i \in \{1,\dots,m\} / a_i r_k > 0\} = \emptyset.$$ [1)]

Since $a_i r_k \leq 0$ for all $i \in \{1,\dots,m\}$, we have $a_i r_k \longrightarrow 0$ if $i \in J(k)$ and, further, $t_k \longrightarrow \infty$ by $a_i x_k \leq \bar{\lambda}_i < \hat{\lambda}^i$. Therefore, it follows

$$f_{p_o}(y_k) > f_{p_o}(x_k) + \varrho t_k \tau_k =$$
$$= f_{p_k}(x_k) + (p_o - p_k)Bx_k + (p_o - p_k)b + \varrho t_k \tau_k \longrightarrow + \infty$$

which contradicts to $p_o \in P(\hat{\lambda})$.

The function $\nabla(\nabla f(x)r)r$, linear with respect to x, is non-negative on L and vanishes (by (2)) for all $x \in A(\hat{\lambda}) \subset L$, implying (1) which completes the proof.

Corollary.

If $P(\lambda) \neq E_q$ for some $\lambda \in \mathring{\Lambda}$ then, defining

$$R(L) = \{r \in L - L \;/\; |r| = 1,\ Ar \leq 0,\ \nabla(\nabla f(x)r)r = 0,\ x \in L\},$$

it holds $R(L) \neq \emptyset$.

Remark.

If $P(p, \lambda)$ is a quadratic problem then Lemma 1. holds under weaker assumptions. Let be $A(\lambda) = \{x \in E_n \;/\; Ax \leq \lambda\}$ and $f_p(x) = f(x) + p^T x$ with a polynomial f of degree ≤ 2. Let λ_k be a sequence and $p_k \in P(\lambda_k)$ abbounded sequence and $x_k \in A(\lambda_k)$ satisfying

$$\inf_k \frac{f_{p_k}(x_k)}{|x_k|} > -\infty, \quad |x_k| \longrightarrow +\infty, \quad \frac{x_k}{|x_k|} \longrightarrow r.$$

Then $\nabla(\nabla f(x)r)r = 0$, i.e. $f(x)$ is linear on each line having the direction r.

Indeed, if $\nabla(\nabla f(x)r)r < 0$ then $\frac{f_{p_k}(x_k)}{|x_k|} \longrightarrow -\infty$ which contradicts to the assumptions. On the other hand, if $\nabla(\nabla f(x)r)r > 0$ would be true then $P(\lambda) = \emptyset$, for every λ, would follow.

Lemma 2. Let

$$A(\lambda) = \{x \in E_n \;/\; Ax \leq \lambda\},$$
$$f_p(x) = f(x) + (Bx+b)p$$

where $f(x)$ is a fixed polynomial of degree ≤ 3. Finally, let the sequences λ_k and $p_k \in P(\lambda_k)$ be bounded such that

$$\lambda_k \leq \bar{\lambda} < \hat{\lambda} \quad \text{where } P(\hat{\lambda}) \neq \emptyset.$$

Then

$$\sup_k \varphi(p_k, \lambda_k) < +\infty.$$

---

1) By a_i the i-th row of the matrix A is denoted and by b^i the i-th component of the vector b; these notations will also be used below.

Proof. The proof proceeds by induction on the dimension of the polyhedron $A(\hat{\lambda})$. If $\dim A(\hat{\lambda}) = 0$ then the proposition is obvious. Let $\dim A(\hat{\lambda}) > 0$. Assume that there is a sequence $x_k \in A(\lambda_k)$ with

$$f_{p_k}(x_k) \longrightarrow \infty, \quad |x_k| \longrightarrow \infty, \quad \frac{x_k}{|x_k|} \longrightarrow r.$$

By Lemma 1., $\nabla(\nabla f(x)r)r = 0$ for all $x \in E_n$. Furthermore, from $p_k \in P(\lambda_k)$ one concludes that

$$\nabla f_{p_k}(x)r \leq 0, \quad x \in A(\lambda_k).$$

Two cases are possible: 1) $Ar = 0$ and 2) $Ar \neq 0$.

1) If $Ar = 0$ then $A(\lambda)$ contain all lines with the direction r. Then $\nabla f_{p_k}(x)r = 0$ for all $x \in A(\lambda_k)$ and, consequently $f_{p_k}(x) = \text{const}$ on each line with direction r. Therefore, in the intersections $A(\lambda_k) \cap \{x \in E_n \,/\, rx = 0\}$ one finds an infinite number of points y_k of the form

$$y_k = x_k + \tau_k r, \quad f_{p_k}(y_k) = f_{p_k}(x_k) \longrightarrow +\infty.$$

But, this is a contradiction to the assumption of induction when it is applied to the sets $A(\lambda) \cap \{x \in E_n \,/\, rx = 0\}$ with $\dim(A(\lambda) \cap \{x \in E_n \,/\, rx = 0\}) < \dim A(\hat{\lambda})$.

2) $Ar \neq 0$, i.e. $\check{J} = \{i \in \{1,\dots,m\} / a_i r < 0\} \neq \emptyset$.

We consider an auxilary polyhedron

$$\check{A}(\check{\lambda}) = \{x \in E_n \,/\, a_i x \leq \check{\lambda}^i, \; i \in \check{J}\}$$

where the numbers $\check{\lambda}^i$, $i \in \check{J}$ are fixed such that $\check{\lambda}^i < \lambda_k^i$ for all k. For all $i \in \check{J}$ we have

$$0 > a_i r = \lim \frac{a_i x_k}{|x_k|} \quad \text{implying } a_i x_k \longrightarrow -\infty.$$

Consequently, $x_k \in \check{A}(\check{\lambda})$ for all k greater as a certain k_o. The half-lines $\{x \,/\, x = x_k - \tau r, \tau \geq 0\}$ intersect the boundary of $\check{A}(\check{\lambda})$. Therefore, there is a face Γ of the polyhedron $\check{A}(\check{\lambda})$ containing an infinite number of points y_k of the form

$$y_k = x_k - \tau_k r, \quad \tau_k \geq 0, \quad f_{p_k}(y_k) \geq f_{p_k}(x_k) \longrightarrow +\infty.$$

But this is a contradiction to the assumption of induction applied to $A(\lambda) \cap \operatorname{aff} \Gamma$ since

$$\dim(A(\hat{\lambda}) \cap \operatorname{aff} \Gamma) < \dim A(\hat{\lambda}).$$

Thus, Lemma 2. is shown.

Theorem 1. Let

$$A(\lambda) = \{x \in E_n \,/\, Ax \leq \lambda\}$$

$$f_p^j(x) = f^j(x) + (B_j x + b_j)p, \quad j \in J \text{ (arbitrary index set)}$$

where $f^j(x)$ are polynomials of degree ≤ 3. Further let $p_k \in P^j(\lambda_k)$ and $x_k \in A(\lambda_k)$ be such that

$$p_k \longrightarrow p_o, \lambda_k \longrightarrow \lambda_o < \hat{\lambda} \text{ where } P^j(\hat{\lambda}) \neq \emptyset, \; f^j_{p_k}(x_k) > v \longrightarrow s^j, \; j \in J.$$

Then a sequence $x_k' \in A(\lambda_o)$ exists such that

$$|x_k - x_k'| \longrightarrow 0, \; f^j_{p_o}(x_k') \geq s^j, \; j \in J.$$

(Remark: By Lemma 2, $s^j < \infty$, $j \in J$; moreover, $s^j > -\infty$, $j \in J$ may be supposed.)

Proof. The proof can be carried out in an analogous way as in case of Lemma 2. One proceeds by induction on the dimension of $A(\hat{\lambda})$. Since the mapping $\lambda \longrightarrow A(\lambda)$ is H-l.s.c. one finds for every $x \in A(\lambda_o)$ a sequence of points $z_k \in A(\lambda_k)$ with $|x - z_k| \longrightarrow 0$. From this latter fact follows

$$\nabla f^j_{p_o}(x)r = \lim \nabla f^j_{p_k}(z_k)r \quad \forall x \in A(\lambda_o) \tag{3}$$

if one takes Lemma 1. into account.

As in the proof of Lemma 2. two cases, $Ar = 0$ and $Ar \neq 0$, has to be considered. In the first case one concludes the existence of infinitely many points y_k such that

$$y_k = x_k + \tau_k r, \; f^j_{p_k}(y_k) = f^j_{p_k}(x_k) > v \longrightarrow s^j$$

and the inductions assumption implies that there are points $y_k' \in \{x \in E_n \,/\, rx = 0\} \cap A(\lambda_k)$ fulfilling $|y_k - y_k'| \longrightarrow 0$ and $f^j_{p_o}(y_k') \geq s^j$. Then the sequence $x_k' = y_k' - \tau_k r$ possesses the desired properties.
In the case $Ar \neq 0$ one finds (see the corresponding part in the proof of Lemma 2.) in a face Γ of $\check{A}(\check{\lambda})$ an infinite number of points y_k with

$$y_k = x_k - \tau_k r, \; \tau_k \geq 0, \; f^j_{p_k}(x_k) \leq f^j_{p_k}(y_k) > v \longrightarrow s^j$$

where $\tau_k = \min_{i \in \check{J}} \dfrac{a_i x_k - \lambda^i}{a_i r} \longrightarrow +\infty$ (since $a_i x_k \longrightarrow +\infty$ for all $i \in \check{J}$).

By Lemma 2. $\nabla f^j_{p_k}(y_k)r = \nabla f^j_{p_k}(x_k)r \longrightarrow 0$ for all $j \in J$. Employing the induction's assumption one finds a sequence $y_k' \in \Gamma \cap A(\lambda_o)$ such that

$$|y_k - y_k'| \longrightarrow 0, \; f^j_{p_o}(y_k') \geq s^j, \; \nabla f^j_{p_o}(y_k')r \geq 0, \; j \in J$$

and, taking (3) into account, it follows that the sequence $x_k' = y_k' + \tau_k r$ has the desired properties which completes the proof.

Corollary. When the sequences λ_k, $p_k \in P(\lambda_k)$ and $x_k \in A(\lambda_k)$ (for the problem $P(p, \lambda)$) have the following properties

$$p_k \longrightarrow p_o, \; \lambda_k \longrightarrow \lambda_o < \hat{\lambda} \text{ where } P(\hat{\lambda}) \neq \emptyset;$$

$$f_{p_k}(x_k) - \varphi(p_k, \lambda_k) \longrightarrow 0,$$

then $\varphi(p_k, \lambda_k) \longrightarrow \varphi(p_o, \lambda_o)$ and one can find a sequence $x'_k \in \Psi(p_o, \lambda_o)$ with $|x_k - x'_k| \longrightarrow 0$ (i.e. φ is continuous at (p_o, λ_o) and Ψ H-u.s.c. at (p_o, λ_o)).

<u>Proof.</u> Let $\varphi(p_k, \lambda_k) \longrightarrow S$. By Theorem 1. for every sequence $x_k \in A(\lambda_k)$ satisfying $f_{p_k}(x_k) \longrightarrow S$ there is a sequence $x'_k \in A(\lambda_o)$ for which $|x_k - x'_k| \longrightarrow 0$ and $f_{p_o}(x'_k) \geqq S$ hold. On the other hand, we have $S \geqq \varphi(p_o, \lambda_o)$, since $\varphi(p, \lambda)$ is lower semicontinuous on $\mathcal{E}$ (which is obvious). Therefore,

$$f_{p_o}(x'_k) = S = \varphi(p_o, \lambda_o) \text{ and } x'_k \in \Psi(P_o, \lambda_o).$$

The following two examples show that the assumption "$\lambda_k \longrightarrow \lambda_o < \hat{\lambda}$ where $P(\hat{\lambda}) \neq \emptyset$" is essential for the validity of Theorem 1.

<u>Example 1.</u>

Let

$$A(\lambda) = \{x \in E_2 \,/\, x_2 \leqq \lambda\}, \quad f_p(x) = x_1^2 x_2 + p_1 x_1 + p_2 x_2.$$

One finds that

$$\mathcal{E} = \{(p, \lambda) \,/\, p_2 \geqq 0, \lambda < 0\} \cup \{(p, \lambda) \,/\, p_1 = 0,\ p_2 \geqq 0,\ \lambda = 0\}.$$

For the sequences

$$\lambda_k < 0,\ \lambda_k \longrightarrow 0,\ p_k = (p_{1k}, 0) \longrightarrow 0 \text{ (i.e. } (p_k, \lambda_k) \longrightarrow (0,0))$$

one obtains

$$\varphi(0,0) = 0;\ \Psi(0,0) = \{x \in E_2 \,/\, x_1 = 0\} \cup \{x \in E_2 \,/\, x_2 = 0\}$$

$$\varphi(p_k, \lambda_k) = -\frac{p_{1k}^2}{4\lambda_k} \quad ; \quad \Psi(p_k, \lambda_k) = \left(-\frac{p_{1k}}{2\lambda_k}, \lambda_k\right),$$

showing that $\varphi(p, \lambda)$ is not continuous at $(0,0)$.

<u>Example 2.</u>

Let $A(\lambda)$ be as in Example 1. and $f_p(x) = x_1^2 x_2 - (x_2+1)^2 + p_1 x_1 + p_2 x_2$. One confirms that

$$\mathcal{E} = \{(p, \lambda) \,/\, \lambda < 0\} \cup \{(p, \lambda) \,/\, p_1 = 0,\ \lambda = 0\}.$$

We consider the sequence $(p_k, \lambda_k) \in \mathcal{E}$ with $\lambda_k = -\alpha_k$, $p_k = (2\beta_k, 0)$, where $\alpha_k \longrightarrow 0$, $\alpha_k > 0$, $\beta_k \longrightarrow 0$, i.e. $(p_k, \lambda_k) \longrightarrow (0,0)$. Obviously, $\varphi(0,0) = 0$ and $\Psi(0,0) = (0,-1)$. For all $\alpha \neq 0$ one has

$$f_{p_k}(x) = (x_2+\alpha)x_1^2 - \alpha\left(x_1 - \frac{\beta_k}{\alpha}\right)^2 - (x_2+1)^2 + \frac{\beta_k^2}{\alpha}$$

and, for k sufficiently large the inequality $f_{p_k}(x_k) \leqq \frac{\beta_k^2}{\alpha}$ follows, from which one gets $\varphi(p_k, \lambda_k) \geqq \frac{\beta_k^2}{\alpha_k} - 1$ for the same k.

If α_k and β_k satisfy $\frac{\beta_k^2}{\alpha_k} < 2$ then $\varphi(p_k, \lambda_k) > 1$. Obviously, then

$$f_{p_k}(x) \le 2\beta_k x_1 \quad \text{for all } x \in A(\lambda_k)$$

which implies that

$$|x_1| \quad \frac{1}{2|\beta_k|} \text{ if } x \in \Psi(p_k, \lambda_k).$$

But, this latter fact indicates that neither Ψ is continuous at (0,0) nor Ψ is H-u.s.c. at this point.

The following two theorems are devoted to assensitivity characterization of the problem $P(p,\lambda)$ in terms of Lipschitzian continuity of the maximum function $\varphi(p,\lambda)$ on a set $S \subset E_n \times E_m$.

As usual we call a function $\varphi(p,\lambda)$ Lipschitzian on a set $S \subset E_n \times E_m$ if there is a number $\alpha = \alpha(S)$ such that for every arbitrary points (p,λ), $(p', \lambda') \in S$ holds:

$$|\varphi(p,\lambda) - \varphi(p', \lambda')| \le \alpha \quad |(p,\lambda)-(p', \lambda')|.$$

For a fixed (m,n)-matrix A the following Property L is well-known from [4] for the mapping $\lambda \longrightarrow \{x \in E_n \;/\; Ax \le \lambda\} = A(\lambda)\ \lambda \in \Lambda$.

Property L. There exists a number $\alpha_0 = \alpha_0(A)$ such that

$$d(A(\lambda), A(\lambda')) \le \alpha_0 |\lambda - \lambda'| \quad \forall \lambda, \lambda' \in \Lambda,$$

where

$$d(A(\lambda), A(\lambda')) = \max\{\sup_{x \in A(\lambda)} \inf_{y \in A(\lambda)} |x-y|;\ \sup_{x \in A(\lambda')} \inf_{y \in A(\lambda)} |x-y|\},$$

i.e. the Hausdorff-distance between $A(\lambda)$ and $A(\lambda')$.

Theorem 2. Let

$$A(\lambda) = \{x \in E_n \;/\; Ax \le \lambda\} \text{ and,}$$
$$f_p(x) = f(x) + p^T x$$

where $f(x)$ is a polynomial of degree ≤ 3. Further, let $S \subset \mathcal{E}$ be a bounded set such that there exists a vector $\lambda_0 = \lambda_0(S) \in \mathring{\Lambda}$ with $\lambda \le \lambda_0$ for each point $(p,\lambda) \in S$. Then the maximum function $\varphi(p,\lambda)$ of $P(p,\lambda)$ is Lipschitzian on S.

Proof. First we show (by contradiction) that there exists a bounded set $Q = Q(S)$ satisfying

$$Q \cap \Psi(p,\lambda) \ne \emptyset \quad \forall (p,\lambda) \in S. \tag{4}$$

Assume the contrary. Then one can find a sequence $x_k \in \Psi(p_k, \lambda_k)$, $(p_k, \lambda_k) \in S$ such that

$$|x_k| = \inf_{x \in \Psi(p_k, \lambda_k)} |x|, \quad |x_k| \longrightarrow \infty, \quad \frac{x_k}{|x_k|} \longrightarrow r, \quad Ar \le 0.$$

The contradiction shall be constructed in a few steps.

1. First we show, that $f_p(x)$ is linear on each line having the direction r. We have $\inf_k \varphi(p_k, \lambda_k) > -\infty$. Indeed, if the opposite would be true, then there would be a subsequence $\{k'\} \subset \{k\}$ satisfying

$$\lim \varphi(p_{k'}, \lambda_{k'}) = -\infty, \; p_{k'} \longrightarrow p_o, \; \lambda_{k'} \longrightarrow \lambda_o \tag{5}$$

and, for each point $x_o \in A(\lambda_o)$ the Property L. causes the existence of a sequence $y_{k'} \in A(\lambda_{k'})$ with $|y_{k'} - x_o| \longrightarrow 0$ which implies that

$$\varphi(p_{k'}, \lambda_{k'}) \geq f_{p_{k'}}(y_{k'}) \longrightarrow f_{p_o}(x_o)$$

contradicting to (5). Therefore,

$$\inf_k \varphi(p_k, \lambda_k) = \inf_k f_{p_k}(x_k) > -\infty$$

and, taking Lemma 1. into account it follows that $f_p(x)$ is linear on each line having the direction r.

2. Next, we show that actually $f_{p_k}(x) = \text{const}$ on every line $x = x_k + tr$, $-\infty < t < \infty$. Obviously, $f_{p_k}(x)$ cannot strictly increase on such lines because this would contradict to $x_k \in \Psi(p_k, \lambda_k)$ and $Ar \leq 0$.
The polynomials $f_{p_k}(x)$ cannot strictly decrease on such lines, since this immediately would yield a contradiction to $x_k \in \Psi(p_k, \lambda_k)$ if $Ar = 0$. If $Ar \neq 0$ would be then there would be rows a_i of A for which $a_i r < 0$ implying $a_i x_k \longrightarrow -\infty$. Therefore, for an arbitrarily fixed t_o would exist a k_o such that

$$a_i(x_k - t_o r) \leq \lambda_k^i \quad i=1,\dots,m, \; k \geq k_o.$$

i.e.

$$x_k - t_o r \in A(\lambda_k), \; k \geq k_o.$$

For such $k \geq k_o$ we would obtain that

$$f_{p_k}(x_k - t_o r) > f_{p_k}(x_k)$$

which would contradict to $x_k \in \Psi(p_k, \lambda_k)$. Hence, we have obtained that $f_{p_k}(x) \equiv \text{const}$ on each line of the form $x = x_k + t\,r$, $-\infty < t < \infty$ and $x_k - t_o r \in \Psi(p_k, \lambda_k)$ for k sufficiently large. But this contradicts to the definition of x_k because, for t sufficiently small $(0 < t < t_o)$, the inequality

$$|x_k - tr|^2 = x_k x_k - 2t x_k r + t^2 rr < |x_k|^2$$

holds.
Consequently, (4) is proven.
Now, let $(p, \lambda), (p', \lambda') \in \dot{S}$ be. By Property L. there is a number $\alpha_o = \alpha_o(A)$ such that for each $x \in \Psi(p, \lambda)$ a point $x' \in A(\lambda')$ can be found for which

$$|x - x'| \leq \alpha_o |\lambda - \lambda'|$$

holds. Further, for a number θ $(0<\theta<1)$ we have

$$
\begin{aligned}
\varphi(p',\lambda') \geq fp(x) = \; & fp(x) + \nabla f(x+\theta(x'-x))(x'-x) - \\
& - p'x' + px = \\
& = \varphi(p,\lambda) + \nabla f(x+\theta(x'-x))(x'-x) + \\
& + p'(x'-x) + x(p'-p).
\end{aligned} \tag{6}
$$

Let the following numbers $\beta, \gamma, \delta, \nu, \alpha$ be defined by

$$
\beta = \sup_{x\in Q} |x| \;, \quad \gamma = \sup_{(p,\lambda)\in S} |p| \;, \quad \delta = \sup_{\substack{(p,\lambda)\in S \\ (p',\lambda')\in S}} |\lambda-\lambda'|
$$

$$
\nu = \sup_{x\in Q+\alpha\delta B} |\nabla f(x)| \;, \text{ where } B = \{x\in E_n \,/\, |x| \leq 1\} \;,
$$

$$
\alpha = \sqrt{2}\ \max\{\alpha_0(\nu+\gamma), \beta\} \;.
$$

Then we obtain from (6) that

$$
\begin{aligned}
& \varphi(p,\lambda) - \varphi(p',\lambda') \leq \nabla f(x+\theta(x'-x))(x-x') + \\
& + p'(x-x') + x(p-p') \leq \nu|x-x'| + \gamma|x-x'| + \beta\,|p-p'| \\
& \leq (\nu+\gamma)\alpha_0|\lambda-\lambda'| + \beta|p-p'| \leq \\
& \leq \alpha|(p,\lambda) - (p',\lambda')| \,.
\end{aligned}
$$

The last inequality is valid for each pair (p,λ), $(p',\lambda')\in S$ which proves Theorem 2.

Theorem 3. Let

$A(\lambda) = \{x\in E_n \,/\, Ax \leq \lambda\}$ and $fp(x) = f(x) + p^T x$ be where $f(x)$ can be polynomial of degree ≤ 2 or a fixed concave or convex polynomial of an arbitrary degree. Then the maximum function $\varphi(p,\lambda)$ of the problem $P(p,\lambda)$ is Lipschitzian on each bounded set $S\subset\mathcal{E}$ ($\mathcal{E}$, the set of boundedness of $P(p,\lambda)$)

Proof. Theorem 3. can be shown in a similar way as it was applied for Theorem 2. One has only to take into account the Remark related to Lemma 1 and the following Lemma 3.

Lemma 3. If $f(x)$ is a concave polynomial and $x_k\in E_n$ a sequence satisfying

$$
\inf_k \frac{f(x_k)}{|x_k|} > -\infty, \quad |x_k| \to \infty, \quad \frac{x_k}{|x_k|} \to r.
$$

Then the polynomial $f(x)$ is linear on each line having the direction r.

Proof. Since the inequality $r\dfrac{x_k}{|x_k|} > \alpha > 0$ is valid for sufficiently large k there is a number $\beta > 0$ such that

$$
\frac{f(x_k) + \beta r x_k}{|x_k|} = \frac{f(x_k)}{|x_k|} + \beta r\frac{x_k}{|x_k|} \geq 0
$$

implying $f(x_k) + \beta r x_k \geq 0$.

Consequently, for sufficiently large k,

$$x_k \in B = \{x \in E_n \ / \ f(x) + \beta r x \geq 0\} .$$

It is well-known (see e.g. [7]) that the closed convex set B contains all half-lines of the form $x = x_k + tr$, $t \geq 0$. The polynomial f is linearly increasing on such half-lines from which one completes the proof.

Remark. One can show that Theorem 3. is not true for general convex polynomial problems, even not when p or λ are fixed.

References

[1] Andronow, W.G.; Belousov, E.G.: Ob ustoičivosti polinomialnych optimisazionnych sadač s lineinymi ograničeniami. Isv. AN USSR, Technyčeskaja Kibernetika, 1985, No. 3. (russian)

[2] Andronow, W.G.; Belousov, E.G. and Shironin, W.M.: O rasrešimosti sadači polinomialnogo programmirovania, Isv. AN USSR, Technyčeskaja Kibernetika, 1982, No. 4 (russian)

[3] Kummer, B.: Globale Stabilität quadratischer Optimierungsprobleme. Wissenschaftliche zeitschrift der Humboldt-Univ. Berlin, Math.-Nat. R. XXVI, 1977,5

[4] Klatte, D.: On the Lipschitz behaviour of optimal solutions in parametric problems of quadratic optimization and linear complementarity Working Paper, 83-121, Austria, 1983.

[5] Bank, B.; Hansel, R.: Stability of mixed-integer quadratic programming problems. - Mathematical Programming Study, 1984, 21.

[6] Bank B., Guddat J., Klatte, D., Kummer B. and Tammer,K.: Non-Linear Parametric Optimization., Berlin, Akademie-Verlag, 1982.

[7] Belousov, E.G.: Vvedenie v vypuklyi analisis i zeločislennoje programmirovanie. Moscow, Isd.-vo MGU, 1977.

ON THE EXISTENCE OF LIPSCHITZ - CONTINUOUS AND DIFFERENTIABLE SELECTIONS FOR MULTIFUNCTIONS

Gerlinde Dommisch [1)]

Abstract: The results of this paper concern the existence of Lipschitz-continuous or C^∞- selections and differentiable ε - approximations for multifunctions which assign a nonempty subset of the Euclidean space R^n to each element of a metric space T. Various types of sufficient conditions for the existence of such selections are given. Applications of the results in the fields of parametric optimization, differential inclusions and generalized equations are added.

1. INTRODUCTION

The present paper is concerned with the existence of various types of selections for multivalued functions.
Let (T,d_T) be a metric space, R^n the n - dimensional Euclidean space.
We regard multifunctions

$$\Gamma : T \longrightarrow 2^{R^n}$$

which assign a nonempty subset $\Gamma(t)$ of R^n to each element $t \in T$.
A function $f: T \longrightarrow R^n$ having the property $f(t) \in \Gamma(t)$ for each $t \in T$ is called a selection for the multifunction Γ.
The results presented in this paper give sufficient conditions for the existence of Lipschitz - continuous or C^∞ - selections and differentiable ε - approximations.
In paragraph 2 multifunctions Γ with convex image sets $\Gamma(t)$ are regarded. If Γ is Lipschitz - continuous with compact convex image sets then a Lipschitz - continuous selection exists. If T is a subset of a normed space we can find a Lipschitz - continuous selection for Γ which goes through a given point of graph Γ.
A C^∞ - selection for Γ can be found if T is a subset of R^m, $\Gamma(t)$ is convex for each $t \in T$ and $\Gamma^-(x) := \{t \in T \mid x \in \Gamma(t)\}$ is open in T for each $x \in R^n$.
In the proofs we use the relations between compact convex subsets of the space R^n and their linear support functionals and the principle of partition of unity.
In paragraph 3 the convexity assumption for the image sets of Γ is dropped. Now only in the case that T is onedimensional we can ensure the existence of a Lipschitz - continuous selection for a Lipschitz - continuous multifunction. There is a counterexample due to D.KLATTE / B.KUMMER which shows that a closed Lipschitz - continuous multifunction of the unit ball in R^2 into itself may have no fixed point and (consequently) no continuous selection.

1) Humboldt - Universität zu Berlin, Sektion Mathematik
DDR-1086 Berlin PSF 1297

Another example (by E.MICHAEL) shows that even if T is onedimensional a continuous multifunction with compact, but nonconvex image sets need not have a continuous selection. However, in this case we can construct a continuously differentiable ε - approximation for Γ which goes through a given point of graph Γ.
The last paragraph deals with some applications of the stated results in the fields of parametric optimization , differential inclusions and generalized equations.

NOTATIONS AND DEFINITIONS

Let R^n denote the n - dimensional Euclidean space with the inner product $c'x = \sum_{i=1}^{n} c_i x_i$ and the norm $\|x\| = \sqrt{x'x}$.
Let B be the closed unit ball in R^n, 2^{R^n} the set of all nonempty subsets of R^n.
cl A stands for the closure of a subset A of a metric space, $U_\varepsilon(\,.\,)$ denotes the open ε - neighbourhood of a point or a subset of a metric space.
For each multifunction $\Gamma: T \longrightarrow 2^{R^n}$
the mapping $\Gamma^-: R^n \longrightarrow 2^T$
is defined by $\Gamma^-(x) = \{t \in T \mid x \in \Gamma(t)\}$,
and graph Γ denotes the set $\{(t,x) \in T \times R^n \mid x \in \Gamma(t)\}$.
A multifunction $\Gamma : T \longrightarrow 2^{R^n}$ is said to be continuous at a point $t_o \in T$ if for each $\varepsilon > 0$ there exists a $\delta > 0$ such that

$$d_H(\Gamma(t), \Gamma(t_o)) < \varepsilon \qquad \text{for each } t \text{ from } U_\delta(t_o),$$

where d_H denotes the Hausdorff - distance.
Γ is called Lipschitz - continuous at t_o if a number L exists such that

$$d_H(\Gamma(t), \Gamma(t_o)) \leq L \cdot d_T(t, t_o) \qquad \text{for all } t \in T \tag{1}$$

L is called Lipschitz - module. If property (1) is fulfilled for all $t_o \in T$ with the same number L Γ is called Lipschitz - continuous.

MULTIFUNCTIONS WITH CONVEX IMAGE SETS

Let Γ be a multifunction which maps a metric space (T, d_T) into the set of all nonempty, compact and convex subsets of R^n.
Γ can be characterized by a functional

$$p: R^n \times T \longrightarrow R \quad , \text{ where } \quad p(c,t) := \max_{x \in \Gamma(t)} c'x$$

p is sublinear (and hence subdifferentiable) in c, and

$$\Gamma(t) = \partial_c p(0,t) = \{x \in R^n \mid c'x \le p(c,t) \quad \forall c \in R^n\}$$

holds for each $t \in T$. (∂_c denotes the subgradient with respect to c.) Γ is Lipschitz - continuous with module L if and only if for each $c \in R^n$ the functional $p(c,.) : T \longrightarrow R$ is also Lipschitz - continuous with module $\|c\| \cdot L$.

Proposition 1: Let T be a metric space and $\Gamma : T \longrightarrow 2^{R^n}$ a Lipschitz-continuous multifunction (Lipschitz - module L) with nonempty, compact and convex image sets.
Then Γ has a selection which is Lipschitz - continuous with module $n \cdot L$.

Proof: (based on an idea of B.KUMMER)
For every fixed $t \in T$ we approximate the subdifferentiable function $p(.,t)$ by a differentiable function $p_\varepsilon(.,t)$ such that for small values of ε the gradient $\nabla_c p_\varepsilon(0,t)$ belongs to $\partial_c p(0,t) = \Gamma(t)$.

Because $f(.) := \nabla_c p_\varepsilon(0,.)$ is Lipschitz - continuous it is the selection we wanted.
In order to construct $p_\varepsilon(c,t)$ we need a function $h: R \longrightarrow R$ with the following properties:

a) $h(r) \ge 0 \quad \forall r \ge 0, \quad h(r) = 0 \quad \forall r \ge 1$

b) h is continuously differentiable

c) $\int_0^1 h(r) \cdot v_n'(r)\, dr = 1$ ($v_n(r)$ denotes the volume of the n - dimensional Euclidean ball of radius r.)

d) $h'(r) \le 0 \quad \forall r \ge 0$

e) $\nabla_z h(\|z\|)$ exists continuously for each $z \in R^n$.

These assumptions are fulfilled e.g. by the function

$$h(r) = \begin{cases} \alpha \cdot [1 + \cos(\pi \cdot r^2)] & \text{for } r \in [-1,1] \\ 0 & \text{for } |r| \ge 1 , \end{cases}$$

where $\alpha = \left[\int_0^1 [1 + \cos(\pi \cdot r^2)] \cdot v_n'(r)\, dr \right]^{-1}$.

Now we define the approximations p_ε:

$$p_\varepsilon(c,t) := \int_{R^n} p(c-y,t) \cdot h_\varepsilon(\|y\|)\, dy \tag{2}$$

where $h_\varepsilon(r) := \frac{1}{\varepsilon^n} h(\frac{r}{\varepsilon})$.

Since $h_\varepsilon(\|y\|)$ vanishes for $\|y\| > \varepsilon$, only the ball $\varepsilon \cdot B$ remains as the field of integration in (2).
From property c) of h we derive

$$\int_{R^n} h_\varepsilon(\|z\|)\, dz = \int_{\varepsilon \cdot B} h_\varepsilon(\|z\|)\, dz = \int_0^1 h(r) \cdot v_n'(r)\, dr = 1 \tag{3}$$

i.e. formula (2) represents a weighted mean of the values of $p(\tilde{c},t)$

for $\tau \in c + \varepsilon \cdot B$.

The transformation $c-y = -z$ provides the second representation of p_ε:

$$p_\varepsilon(c,t) = \int_{R^n} p(-z,t)\cdot h_\varepsilon(\|c+z\|)\, dz$$

Property e) of h implies

$$\nabla_c p_\varepsilon(c,t) = \int_{R^n} p(-z,t)\cdot \nabla_c h_\varepsilon(\|c+z\|)\, dz$$

and, in particular,

$$\nabla_c p_\varepsilon(0,t) = \int_{R^n} p(-z,t)\cdot \nabla_z h_\varepsilon(\|z\|)\, dz$$

Because of the special form of h_ε the expression $\nabla_c p_\varepsilon(0,t)$ does not depend on ε . So we can define

$$f(t) = \nabla_c p_1(0,t) = \nabla_c p_\varepsilon(0,t) \qquad \forall \varepsilon \in [0,1]$$

It is easy to show that $p_\varepsilon(.,t)$ is convex.

Now we have to estimate $\| p_\varepsilon(.,t) - p(.,t)\|$. First we remark that $p(.,t)$ is uniformly continuous on $2\cdot B$. Therefore, for each $t \in T$ and each $\varepsilon \in [0,1]$ there exists a positive number $\delta_t(\varepsilon)$ with the properties

$$\Delta_t(c,y) := | p(c,t) - p(c-y,t)| \leq \delta_t(\varepsilon) \quad \forall c \in B,\ \forall y \in \varepsilon \cdot B$$

$$\text{and} \quad \delta_t(\varepsilon) \longrightarrow 0 \ \text{ if } \ \varepsilon \longrightarrow 0 .$$

For each $c \in B$ we obtain by formula (3)

$$\begin{aligned} | p_\varepsilon(c,t) - p(c,t)| &= \Big| \int_{R^n} (p(c-y,t) - p(c,t))\cdot h_\varepsilon(\|y\|)\, dy \Big| \\ &\leq \int_{R^n} | p(c-y,t) - p(c,t)| \cdot h_\varepsilon(\|y\|)\, dy \\ &\leq \int_{Y_1} \delta_t(\varepsilon)\cdot h_\varepsilon(\|y\|)\, dy + \int_{Y_2} \Delta_t(c,y)\cdot h_\varepsilon(\|y\|)\, dy, \end{aligned}$$

where $Y_1 := \{ y \in R^n \mid \Delta_t(c,y) \leq \delta_t(\varepsilon)\}$ and $Y_2 := R^n \setminus Y_1$.

The last integral vanishes since $\Delta_t(c,y) > \delta_t(\varepsilon)$ implies $\|y\| > \varepsilon$ and $h_\varepsilon(\|y\|) = 0$.

Thus, for each $t \in T$,

$$\| p_\varepsilon(.,t) - p(.,t)\|_B \leq \delta_t(\varepsilon) \tag{4}$$

Now let $N_B(.)$ denote the map which assigns to every $c \in B$ its normal cone on the unit ball B, and let graph $\partial_c p(.,t)$ be the graph of the subdifferentials of $p(.,t)$ restricted to B.

For p_ε we have graph $\partial_c p_\varepsilon(.,t)$ = graph $\nabla_c p_\varepsilon(.,t)$.

By a result from R.SCHULTZ about the stability of the subdifferentials of convex functions from formula (4) we can conclude for each $t \in T$

$$d_H(\mathrm{graph}(\nabla_c p_\varepsilon(\cdot,t) + N_B(\cdot)), \mathrm{graph}(\partial_c p(\cdot,t) + N_B(\cdot))) \leq \sqrt{2\delta_t(\varepsilon)} \qquad (5)$$

(norm in R^{n+n}: $\|(y,y^*)\|_o = \max\{\|y\|, \|y^*\|\}$)

We consider the point $(0, \nabla_c p_\varepsilon(0,t)) \in \mathrm{graph}(\nabla_c p_\varepsilon(\cdot,t) + N_B(\cdot))$.

According to (5) there exists a point (y,y^*) which belongs to $\mathrm{graph}(\partial_c p(\cdot,t) + N_B(\cdot))$ with

$$\|(0, \nabla_c p_\varepsilon(0,t)) - (y,y^*)\|_o \leq \sqrt{2\cdot\delta_t(\varepsilon)} \qquad (6)$$

If we choose ε sufficiently small then $\|y\| \leq \sqrt{2\cdot\delta_t(\varepsilon)} < 1$ follows and we obtain $N_B(y) = \{0\}$ and $y^* \in \partial_c p(y,t)$.

Since $p(\cdot,t)$ is sublinear,

$$\partial_c p(c,t) \subseteq \partial_c p(0,t) \quad \text{holds for each } c \in B.$$

Now formula (6) shows that for each $t \in T$

$$d(\nabla_c p_\varepsilon(0,t), \partial_c p(0,t)) \leq \sqrt{2\cdot\delta_t(\varepsilon)}$$

Because $f(t) = \nabla_c p_\varepsilon(0,t) \quad \forall \varepsilon \in (0,1]$ we can conclude

$$d(f(t), \partial_c p(0,t)) = 0$$

and, since $\partial_c p(0,t) = \Gamma(t)$ is compact, $f(t) \in \Gamma(t)$ for each $t \in T$ follows. Hence f is a selection for Γ.

Last not least we prove the Lipschitz-continuity of f:

For $t,s \in T$ we have

$$\begin{aligned}\|f(t) - f(s)\| &= \|\nabla_c p_1(0,t) - \nabla_c p_1(0,s)\| \\ &\leq \int_{R^n} |p(-z,t) - p(-z,s)| \cdot \|\nabla_z h(\|z\|)\| \, dz \\ &\leq L \cdot d_T(t,s) \cdot \int_{R^n} \|z\| \cdot \|\nabla_z h(\|z\|)\| \, dz\end{aligned}$$

To calculate the last integral we apply the properties of h:

$$\begin{aligned}\int_{R^n} \|z\| \, \|\nabla_z h(\|z\|)\| \, dz &= \int_B \|z\| \cdot \|\nabla_z h(\|z\|)\| \, dz \\ &= \int_0^1 r \cdot |h'(r)| \cdot v_n'(r) \, dr \\ &= -\int_0^1 r \cdot h'(r) \cdot v_n'(r) \, dr \\ &= -\big[r\, v_n'(r) \cdot h(r)\big]_0^1 + \int_0^1 h(r) \cdot (v_n'(r) + r \cdot v_n''(r)) \, dr\end{aligned}$$

Since $r \cdot v_n''(r) = (n-1) \cdot v_n'(r)$ for $n \geq 1$, we have

$$\int_{R^n} \|z\|\cdot\|\nabla_z h(\|z\|)\| \; dz = \int_0^1 h(r)\cdot n\cdot v_n'(r) \; dr = n$$

Hence, f is really Lipschitz-continuous with module $n\cdot L$, and our proof is complete.

Another proof for the existence of a Lipschitz-continuous selection for Lipschitz-continuous multifunctions was given by J.P. AUBIN / A. CELLINA under the additional assumption that the image sets $\Gamma(t)$ are included in a bounded subset of R^n, but the Lipschitz-module of the selection they constructed is greater than $n\cdot L$.

The technique of the proof of proposition 1 can also be used for proving the existence of differentiable selections (under the supposition that for each $c \in B$ $p(c,.)$ is differentiable) and L^q-selections (if $p(c,.) \in L^q(T)$).

Another interesting question is whether a Lipschitz-continuous selection exists which goes through a given point of the graph of Γ. The answer "yes" is a consequence of the next result:

Lemma 2: We suppose that T is a subset of a normed space and $\Gamma: T \longrightarrow 2^{R^n}$ is a Lipschitz-continuous multifunction (Lipschitz-module L) with closed convex image sets. Let (t_o,x_o) be a point of graph Γ and k a real number greater than $\sqrt{\frac{3}{2}}$.

Then the multifunction $G_k: T \longrightarrow 2^{R^n}$ with

$$G_k(t) = \{x \in \Gamma(t) \mid \|x - x_o\| \leq k\cdot L\cdot\|t - t_o\|\}$$

is Lipschitz-continuous with module $L(k)\cdot L$, where

$$L(k) := \frac{2k^2+2k}{2k^2-3}\cdot\sqrt{k^2-1} + 1$$

Proof: Without loss of generality we take $t_o = 0$, $x_o = 0$ and $L = 1$.
Let u(t) be that element of $\Gamma(t)$ which minimizes $\|x\|$ over $\Gamma(t)$.
Because Γ is Lipschitz-continuous we have

$$\|u(t)\| \leq \|t\| < k\cdot\|t\| \quad \text{and hence} \quad u(t) \in G_k(t).$$

Let x be a point of $G_k(t) \subseteq \Gamma(t)$. We have to show

$$d(x,G_k(s)) \leq L(k)\cdot\|t-s\| \quad \text{for each} \quad s \in T.$$

The Lipschitz-continuity of Γ implies the existence of a point $y \in \Gamma(s)$ with the property $\|y-x\| \leq \|t-s\|$. If $\|y\| \leq k\cdot\|s\|$, i.e. $y \in G_k(s)$, then the assertion follows. We have to investigate the case $\|y\| > k\cdot\|s\|$.

We regard the plane spanned by the points 0, y and u(s). The intersection point v of the line through y and u(s) with the sphere of radius $k\cdot\|s\|$ belongs to $G_k(s)$.

Hence $d(x,G_k(s)) \leq \|x-y\| + \|y-v\| \leq \|t-s\| + \|y-v\|$

By simple geometrical means (we have to regard several right-angled triangles) we can show that

$$\|y-v\| \leq \frac{2k(k+1)\cdot\sqrt{k^2-1}}{2k^2-3}\cdot\|t-s\|$$

From that the assertion follows.

It should be remarked that the Lipschitz-module of G_k does not depend on the position of the point (t_o,x_o).
From Lemma 2 we can conclude the following result:

Proposition 3: Let T be a subset of a normed space and $\Gamma: T \longrightarrow 2^{R^n}$ a Lipschitz-continuous multifunction (Lipschitz-module L) with convex closed image sets.
Then for each point $(t_o,x_o) \in \operatorname{graph}\Gamma$ there is a Lipschitz-continuous selection f for Γ with Lipschitz-module $(\frac{12}{5}\sqrt{3}+1)\cdot L$ and the property $f(t_o) = x_o$.

Proof: We construct $G(t) = \{x \in \Gamma(t) \mid \|x-x_o\| \leq 2\cdot\|t-t_o\|\}$

G is Lipschitz-continuous with module $L(2)\cdot L = (\frac{12}{5}\sqrt{3}+1)\cdot L$.

Due to proposition 1 G has a Lipschitz-continuous selection f (Lipschitz- module $L(2)\cdot L\cdot n$).
Because of $G(t_o) = x_o$ we have $f(t_o) = x_o$.

Proposition 3 can be generalized for a finite number of points (t_i,x_i) of graph Γ, but its clear that the Lipschitz-module of the selection increases with the number of points.

The next results state the existence of Lipschitz-continuous (or C^∞-) selections under other types of assumptions. The proofs use the principle of partition of unity.

Proposition 4: Let T be a compact metric space and $\Gamma: T \longrightarrow 2^{R^n}$ a multifunction with nonempty convex image sets. Further we suppose that for each $x \in R^n$ $\Gamma^-(x)$ is open.
Then Γ has a Lipschitz-continuous selection.

Proof: The sets $\Gamma^-(x)$, $x \in R^n$ form an open covering of T. Since T is compact there is a finite set $\{x_1,\dots,x_N\} \subset R^n$ such that

$$T = \bigcup_{k=1}^{N} \Gamma^-(x_k)$$

The sets $A_k := T \setminus \Gamma^-(x_k)$ are closed and $\bigcap_{k=1}^{N} A_k = \emptyset$ holds.

$d_k(t) := d(t,A_k)$ is Lipschitz-continuous with module 1.

Now we define $d(t) = \sum_{k=1}^{N} d_k(t)$, which is Lipschitz - continuous with module N.

Since $\bigcap_{k=1}^{N} A_k = \emptyset$ we have $\alpha := \inf_{t\in T} d(t) > 0$.

Now we set $f(t) := \sum_{k=1}^{N} \frac{d_k(t)}{d(t)} \cdot x_k$

Since T is compact there exists a constant M such that

$$\left| \frac{d_k(t)}{d(t)} - \frac{d_k(s)}{d(s)} \right| \leq \frac{1}{\alpha^2} \cdot M \cdot \|t-s\|$$

Hence $|f(t) - f(s)| \leq \sum_{k=1}^{N} \left| \frac{d_k(t)}{d(t)} - \frac{d_k(s)}{d(s)} \right| \cdot \|x_k\|$

$$\leq \frac{N \cdot M}{\alpha^2} \left(\max_{k=1..N} \|x_k\| \right) \cdot \|t-s\| ,$$

i.e. F is Lipschitz - continuous.

$d_k(t) > 0$ for some k means $t \notin A_k$, hence $t \in \Gamma^-(x_k)$ and $x_k \in \Gamma(t)$, i.e. f(t) is a convex combination of a finite number of points of $\Gamma(t)$ and consequently, since $\Gamma(t)$ is convex, f(t) belongs to $\Gamma(t)$.

A sufficient condition for the fact that for all $x \in R^n$ $\Gamma^-(x)$ is open is the lower semicontinuity of Γ in connection with the nonempty interior of the image sets of Γ.

If we replace in proposition 4 the condition that T is a compact metric space by the supposition that T is a closed subset of R^m we can prove the existence of a C^∞- selection for the multifunction Γ:

Proposition 5: We assume that T is a closed subset of R^m, $\Gamma: T \longrightarrow 2^{R^n}$ is a multifunction with convex image sets and that for each $x \in R^n$ $\Gamma^-(x)$ is relatively open in T. Then a C^∞- selection for Γ exists.

Proof: The sets $\tilde{\Gamma}^-(x) := \Gamma^-(x) \cup (R^m \setminus T)$, $x \in R^n$ form an open covering of the R^m. Due to R.NARASIMHAN there exists a C^∞- partition of unity subordinate to this covering, i.e. there is a collection of C^∞- functions $\{F_x\}_{x\in R^n}$, where $F_x: R^m \longrightarrow R$ with the properties

(1) $F_x(t) \geq 0 \quad \forall t \in R^m, \forall x \in R^n$

(2) $\text{Supp } F_x := \text{cl}\{\bar{x} \in R^m \mid F_x(\bar{x}) \neq 0\} \subseteq \tilde{\Gamma}^-(x) \quad \forall x \in R^n$

(3) The family $\{\text{Supp } F_x\}_{x\in R^n}$ is locally finite.

(4) $\sum_{x\in R^n} F_x(t) = 1 \quad \forall t \in R^m$

We define $f(t) := \sum_{x\in R^n} F_x(t) \cdot x$.

Obviously f is a C^∞- function. For each $t \in T$ from $F_x(t) > 0$ we can conclude $t \in \tilde{\Gamma}^-(x) = \Gamma^-(x) \cup (R^m \setminus T)$, hence $t \in \Gamma^-(x)$ and $x \in \Gamma(t)$.
Because of the properties of F_x f(t) is a convex combination of a finite number of points of $\Gamma(t)$. Hence $f(t) \in \Gamma(t)$ follows, and the proof is complete.

3. MULTIFUNCTIONS WITHOUT CONVEXITY ASSUMPTIONS

In contrast to chapter 2 now the convexity assumption for the image sets of Γ is dropped.
The first example (cf. D.KLATTE / B.KUMMER) shows that in general for the existence of a continuous selection the convexity assumption is necessary.

Example 1: We regard the unit ball B in R^2 and a multifunction

$\Gamma: B \longrightarrow 2^B$ with

$$\Gamma(x) := \begin{cases} \text{bd } B & \text{for } \|x\| \leq r \\ \text{bd } B \setminus \{y \in R^2 \mid \|y-u(x)\| < \|x\| - r\} & \text{for } \|x\| > r \end{cases}$$

where $u(x) := \frac{x}{\|x\|}$ for $x \neq 0$ and $r \in (0,1)$ fixed.

It is possible to show that Γ is Lipschitz - continuous with module $1 + \frac{1}{r}$, but Γ has no fixed point and hence a continuous selection for Γ does not exist.

In the case that T is onedimensional we can prove the following

Proposition 6: We suppose that $T = [0,1]$, $\Gamma: T \longrightarrow 2^{R^n}$ is a multi-function with closed image sets.
If Γ is continuous then for every $\varepsilon > 0$ and every (t_o, x_o) from graph Γ a function $f \in C^\infty(T)$ exists with $f(t_o) = x_o$ and $d(f(t), \Gamma(t)) \leq \varepsilon$.
If Γ is Lipschitz - continuous then for every point (t_o, x_o) from graph Γ a Lipschitz - continuous selection f with $f(t_o) = x_o$ exists.

Proof: Without loss of generality we take $t_o = 0$.
We choose an integer N such that

$$d_H(\Gamma(t), \Gamma(s)) \leq \frac{\varepsilon}{3} \quad \text{for all} \quad t,s \in [0,1] \quad \text{with } |t-s| \leq \frac{1}{N}$$

and regard $t_k = \frac{k}{N}$ for $k = 1,\ldots,N$.

Now we can find points x_k, $k = 1,\ldots,N$, with the property

$$\|x_k - x_{k-1}\| \leq \frac{\varepsilon}{3}, \quad k = 1,\dots,N.$$

By connecting this points we define a piecewise linear continuous function f_N:

$$f_N(t) = (1-\lambda)\cdot x_k + \lambda\cdot x_{k-1} \quad \text{for } t = t_k - \frac{\lambda}{N},\ \lambda\in[0,1],\ k=1,\dots,N$$

The distance between $\Gamma(t)$ and $f_N(t)$ can be estimated in the following way (for $t = t_k - \frac{\lambda}{N}$, $\lambda\in[0,1]$):

$$d(f_N(t),\Gamma(t)) \leq d(x_k,\Gamma(t)) + \|x_k - f_N(t)\|$$

$$\leq d_H(\Gamma(t),\Gamma(t_k)) + \lambda\cdot\|x_k - x_{k-1}\|$$

$$\leq \frac{\varepsilon}{3} + \lambda\cdot\frac{\varepsilon}{3} = (1+\lambda)\cdot\frac{\varepsilon}{3}$$

Using x_{k-1} instead of x_k we get the estimation

$$d(f_N(t),\Gamma(t)) \leq (2-\lambda)\cdot\frac{\varepsilon}{3}$$

Consequently

$$d(f_N(t),\Gamma(t)) \leq \frac{\varepsilon}{3}\cdot\min\{1+\lambda,\ 2-\lambda\} \leq \frac{\varepsilon}{2} \quad \text{for all } t\in[0,1]$$

The continuous function f_N we can approximate by a C^∞-function f such that $f(0) = x_0$ and $\|f(t) - f_N(t)\| \leq \frac{\varepsilon}{2}$ for each $t\in T$.

Hence $d(f(t),\Gamma(t)) \leq \varepsilon$, and the first part of proposition 6 is proved.

The proof of the second part one can find in a paper of H.HERMES.

The next example (by E.MICHAEL) shows that even if T is onedimensional a continuous multifunction with compact image sets need not have a continuous selection.

Example 2: Let $T = [0,1]$ and

$$Z = \left\{(x,y)\in R^2 \;\middle|\; \begin{array}{ll} y = \sin\frac{1}{x} & \text{for } x\in(0,1] \\ y\in[-1,1] & \text{for } x = 0 \end{array}\right\}$$

The multifunction $\Gamma: T \longrightarrow 2^{R^2}$ is defined by

$$\Gamma(t) = \{(x,y)\in Z \mid x\in[\tfrac{t}{2}, t]\}$$

For each $t\in T$ the image set $\Gamma(t)$ is nonempty and compact, but for $t > 0$ nonconvex. Γ is continuous, but since $Z = \bigcup_{t\in T}\Gamma(t)$ is not connected Γ has no continuous selection. We only can find a differentiable ε-approximation for Γ.

The following table gives a survey of the results presented in this paper.

$\Gamma: T \longrightarrow 2^{R^n}$, $\Gamma(t) \neq \emptyset$ $\forall t \in T$, $(t_o, x_o) \in$ graph Γ

Suppositions				Results
T	Γ	$\Gamma(t)$	$\Gamma^-(t)$	
metric	Lipsch. L	compact convex		Proposition 1: Lipschitz - continuous selection (Lipschitz - module $n \cdot L$)
subset of a normed space	Lipsch. L	closed convex		Proposition 3: Lipschitz - continuous selection through (t_o, x_o)
dim T = 2	Lipsch. L > 1	compact		Example 1: no continuous selection
[0,1]	Lipsch. L	closed		Proposition 6: Lipschitz - continuous selection through (t_o, x_o), module L
[0,1]	contin.	compact		Example 2: no continuous selection
[0,1]	contin.	closed		Proposition 6: ε - approximation $f \in C^\infty(T)$ through (t_o, x_o)
metric compact		convex	open	Proposition 4: Lipschitz - continuous selection
metric compact	l.s.c.	convex int $\Gamma(t) \neq \emptyset$		Conclusion of proposition 4: Lipschitz - continuous selection for int $\Gamma(.)$
closed subset of R^m		convex	open	Proposition 5: C^∞ - selection

4. APPLICATIONS

In the field of parametric optimization many results concerning the behavior of the optimal set mapping are known. Using the propositions stated in this paper one can find new results about the existence of optimal selections (i.e. selections of the optimal set mapping) or differentiable ε - optimal solution functions, for instance the following:

We regard the parametric optimization problem

$$P(t): \quad \varphi(t) := \inf \{ f(x,t) \mid g_i(x,t) \leq 0, \ i = 1,\dots,p, \ x \in R^n \}$$

where $t \in T \subseteq R^m$, T closed, and $f, g_i : R^n \times T \longrightarrow R$.

Now we define the optimal set mapping

$$\Psi(t) := \{ x \in R^n \mid g_i(x,t) \leq 0, \ i = 1,\dots,p, \ f(x,t) = \varphi(t) \} ,$$

the set of Slaterpoints of the problem P(t)

$$S(t) := \{ x \in R^n \mid g_i(x,t) < 0, \ i = 1,\dots,p \}$$

and the ε - optimal set

$$\Psi_\varepsilon(t) := \{ x \in R^n \mid g_i(x,t) < 0, \ i = 1,\dots,p, \ f(x,t) < \varphi(t) + \varepsilon \} .$$

Proposition 7: We suppose that

(1) the functions f, g_i, $i = 1,\dots,p$, are continuous,

and that for each $t \in T$

(2) $f(.,t)$, $g_i(.,t)$, $i = 1,\dots,p$, are convex,

(3) $\Psi(t)$ is nonempty and compact,

(4) $S(t)$ is nonempty.

Then for each $\varepsilon > 0$ there is a C^∞ - function $x(.): T \longrightarrow R^n$ with $x(t) \in S(t)$ and $f(x(t),t) < \varphi(t) + \varepsilon$ for each $t \in T$.

Proof: From supposition (2),(3) and (4) we can conclude that for each $t \in T$ $\Psi_\varepsilon(t)$ is nonempty and convex.

From (1),(2) and (3) the lower semicontinuity of the infimumfunction $\varphi(.)$ follows.

Because φ is lower semicontinuous and the functions f and g_i, $i = 1 \dots p$ are continuous the sets $\Psi_\varepsilon^-(x) = \{ t \in T \mid x \in \Psi_\varepsilon(t) \}$ are open in T for each $x \in R^n$.

Hence the assumptions of proposition 5 are fulfilled, i.e. there exists a C^∞ - selection for Ψ_ε.

In the field of differential inclusions we can apply the stated results to replace the differential inclusion by a differential equation and, in this way, to ensure the existence of continuously differentiable solutions:

Let $F: [0,1] \times R^n \longrightarrow 2^{R^n}$ be a Lipschitz - continuous multifunction with convex closed image sets and $(0, x_0, y_0) \in$ graph F.

We regard the differential inclusion

$$\dot{x}(t) \in F(t,x(t)) \ , \quad t \in [0,1]$$

Due to proposition 3 the multifunction F has a Lipschitz - continuous selection $f\colon [0,1] \times R^n \longrightarrow R^n$ with $f(0,x_0) = y_0$.
The differential equation

$$\dot{x}(t) = f(t,x(t)), \quad x(0) = x_0, \quad t \in [0,1]$$

has a unique solution $x(.)\colon [0,1] \longrightarrow R^n$ which is also a solution for the differential inclusion.
$x(.)$ has the properties $x(0) = x_0$, $\dot{x}(0) = y_0$ and $\dot{x}(.)$ is Lipschitz-continuous.
Using proposition 5 under the corresponding suppositions on F we can ensure the existence of a C^∞- solution of the differential inclusion.

The third application concerns generalized equations $0 \in \Gamma(t)$.

Proposition 8: We suppose that T is a closed subset of the R^m, $\Gamma\colon T \longrightarrow 2^{R^n}$ is a Lipschitz - continuous multifunction (Lipschitz - module L) with closed convex image sets. Further we suppose that $T_0 = \{t \in T \mid 0 \in \Gamma(t)\}$ is nonempty.
Then a Lipschitz - continuous selection f for Γ exists with $f(t) = 0$ for each $t \in T_0$.

Proof: We set

$$\tilde{\Gamma}(t) = \begin{cases} \{0\} & \text{for } t \in T_0 \\ \Gamma(t) \cap 2\cdot L\cdot d(t,T_0)\cdot B & \text{for } t \notin T_0 \end{cases}$$

and show that $\tilde{\Gamma}$ fulfilles the suppositions of proposition 1:

1) For each $t \in T$ $\tilde{\Gamma}(t)$ is nonempty:
If $t \in T_0$ we have $\tilde{\Gamma}(t) = \{0\}$. Now we regard $t \notin T_0$. It is easy to show that the set T_0 is closed. Hence $\varepsilon := d(t,T_0) > 0$.
There exists a point $t_0 \in T_0$ with $\|t_0 - t\| = \varepsilon$.
Since Γ is Lipschitz - continuous with module L we have

$$\Gamma(t_0) \subseteq \mathrm{cl}\,(U_{L\cdot\varepsilon}(\Gamma(t)) \ ,$$

i.e. there is a point $y \in \Gamma(t)$ with $\|y\| \leq L\cdot\varepsilon$, hence $y \in \tilde{\Gamma}(t)$.
2) Obviously for each $t \in T$ $\tilde{\Gamma}(t)$ is convex and compact.
3) $\tilde{\Gamma}\colon T \longrightarrow 2^{R^n}$ is Lipschitz - continuous with module $L(2)\cdot L$:
We fix $t,s \in T$.
If $t \in T_0$ then $\tilde{\Gamma}(t) = \{0\}$ and hence

$$d_H(\tilde{\Gamma}(t),\tilde{\Gamma}(s)) \leq 2\cdot L\cdot d(s,T_0) \leq 2\cdot L\cdot\|t-s\| \leq L(2)\cdot L\cdot\|t-s\|$$

If $t,s \notin T_0$ we choose $t_0 \in T_0$ with $d(t,T_0) = \|t-t_0\|$ and construct a multifunction $G\colon T \longrightarrow 2^{R^n}$ with

$$G(r) = \{x \in \Gamma(r) \mid \|x\| \leq 2\cdot L\cdot\|t-t_0\|\}$$

Due to Lemma 2 G is Lipschitz-continuous with module $L(2)\cdot L$, hence

$$d_H(G(t),G(s)) \leq L(2)\cdot L\cdot\|t-s\|$$

Further we have $G(t) = \tilde{\Gamma}(t)$ and $\tilde{\Gamma}(s) \subseteq G(s)$, hence

$$\tilde{\Gamma}(s) \subseteq U_\delta(\tilde{\Gamma}(t)) \quad \text{for each} \quad \delta > L(2)\cdot L\cdot\|t-s\|$$

Symmetrically we can show

$$\tilde{\Gamma}(t) \subseteq U_\delta(\tilde{\Gamma}(s)) \quad \text{for each} \quad \delta > L(2)\cdot L\cdot\|t-s\|$$

Consequently we can conclude

$$d_H(\tilde{\Gamma}(t),\tilde{\Gamma}(s)) \leq L(2)\cdot L\cdot\|t-s\|$$

From 1), 2) and 3) by proposition 1 the existence of a Lipschitz-continuous selection f for $\tilde{\Gamma}$ follows (Lipschitz-module $L(2)\cdot L\cdot n$), i.e. $f(t) \in \Gamma(t)$ for each $t \in T$ and $f(t) = 0$ for each $t \in T_0$. Hence T_0 is the solution set of the equation $0 = f(t)$.

Proposition 8 shows that the solution set of a generalized equation with a Lipschitz-continuous multifunction has the same behavior as the solution set of a "normal" equation with a Lipschitz-continuous function.

REFERENCES

Aubin,J.P.,Cellina,A.: Differential Inclusions. Springer-Verlag, Berlin 1984.

Bank,B., Guddat,J., Klatte,D., Kummer,B., Tammer,K.: Non-Linear Parametric Optimization. Akademie-Verlag, Berlin 1982.

Dommisch,G.: Über die Existenz Lipschitz-stetiger und differenzierbarer Auswahlfunktionen für mehrwertige Abbildungen. Diss. (A), Sektion Mathematik der Humboldt-Universität zu Berlin, 1986.

Hermes,H.: On continuous and measurable selections and the existence of solutions of generalized differential equations. Proc. Amer. Math. Soc. 29(1971)3, 535 - 542.

Klatte,D.,Kummer,B.: On the (Lipschitz-) continuity of solutions of parametric optimization problems. In: Proceedings der 16. Jahrestagung "Mathematische Optimierung" Sellin/Rügen 1984, Seminarbericht Nr. 64 der Sektion Mathematik der Humboldt-Univ. zu Berlin, 1984.

Kummer,B.: Generalized Equations: Solvability and Regularity. Math. Progr. Study 21(1981), 199 - 212.

Michael,E.: Continuous selections I - III, Ann. of Math. 63-65(1956 - 1957), 361-381, 563-580, 375-390.

Narasimhan,R.: Analysis on Real and Complex Manifolds. North-Holland Publ. Co., Amsterdam 1968.

Schultz,R.: Abschätzungen für Kuhn-Tucker-Punkte gestörter konvexer Optimierungsaufgaben. Diss. (A), Sektion Mathematik der Humboldt-Univ. zu Berlin, 1984.

Yosida,K.: Functional Analysis. Springer-Verlag, Berlin, Heidelberg, New York 1971.

ON SOME CONNECTIONS BETWEEN PARAMETRIC AND STOCHASTIC PROGRAMMING

Jitka Dupačová [1)]

ABSTRACT

To perform stability studies for decision models of stochastic programming, different methods have been developed in which parametric programming and nonlinear programming stability methods complemented by statistical approaches play an essential role. We shall concentrate to the stability of optimal solutions.

Two approaches will be discussed in detail: solving stochastic programs with estimated parameters and those whose distribution is perturbed by contamination.

1 Introduction

Quite a large class of stochastic programming decision problems can be transformed into the following mathematical programming problem

$$\begin{aligned} &\text{maximize} \quad g_0(x;F) \\ &\text{subject to} \quad g_i(x;F) \geq 0,\ 1 \leq i \leq m,\ g_i(x;F) = 0,\ m+1 \leq i \leq m+p,\ x \in X, \end{aligned} \tag{1}$$

where $X \subset R^n$ is a given nonempty set. The functions g_i, $0 \leq i \leq m+p$, do not depend on random parameters directly but by means of their distribution F only and their form is given by the chosen model, such as the stochastic program with recourse or with probabilistic constraints.

The optimal solution $x(F)$ and the optimal value of the objective function in (1) depend on the chosen type of model and its parameters as well as on the distribution F which is usually assumed to be completely known and independent on the chosen decision x. In real-life situations, however, the distribution F is hardly known completely and, in addition, in course of numerical solution of problem (1) F is often approximated by simpler distributions. The obtained numerical results should be thus at least complemented by additional information about sensitivity of the obtained solution of (1) to eventual changes in the distribution. In the robust case, a small change in the distribution should cause a small change of the optimal solution.

We shall give an introduction to the methods which are relatively easily applicable and closely connected with deterministic parametric programming results and we shall concentrate to the stability of optimal solutions.

An *ad hoc* idea could be to study stability of the optimal solution of (1) with respect to the underlying distribution F directly. However, the space of probability measures endowed with a metric corresponding to the weak topology is not a linear one, so that many results of parametric programming are not applicable directly.

The stability of the optimal solution $x(F)$ with respect to the parameters of the distribution F can be studied to a certain extent through the methods of parametric programming

[1)]Department of Statistics, Charles University, Sokolovská 83, 18600 Prague 8 and
IIASA, A-2361 Laxenburg, Austria

(see e.g. [1], [11]). Also by introducing additional parameters to stochastic programming models one can in certain extent cope with the incomplete information about the distribution F and/or its parameters. Garstka [12] studied the stochastic linear program with recourse with random right-hand sides whose mean value μ and variance matrix $\Sigma = BB^T$ were perturbed in a special way by means of two scalar parameters. Properties of the perturbed program

$$\begin{array}{ll} \text{minimize} & c^T x + E\{q^T y\} \\ \text{subject to} & x \in X,\ Ax + My = k_1\mu + k_2 B\xi,\ y \geq 0 \end{array}$$

with ξ random, $E\xi = 0$ and var $\xi = I$ provide among others theoretical means for obtaining bounds on the difference between the optimal value of the objective function for the true parameter values (i.e., for $k_1 = k_2 = 1$) and for an alternative choice of $k_1 \neq 1$ or $k_2 \neq 1$. Similarly, Seppälä [21] introduced one scalar parameter in his stochastic multi-facility problem in order to eliminate the estimation of the correlation coefficients. The parameter weighs the totally correlated case and the uncorrelated one. Nadeau and Theodorescu [16] used a scalar parameter to weigh the objective function $g_0(x;F)$ with that corresponding to the worst possible choice of distribution. The theory of parametric programming has been successfully used in many ways to get stability results for the optimal value of (1), see e.g. the recent working papers [2], [15], [20]. Results on stability of optimal solutins, however, can be achieved under substantially more restrictive assumptions only. In addition, for the subsequent statistical techniques differentiability is needed.

In this paper, two approaches will be presented:

(i) the statistical sensitivity analysis with respect to estimated parameters of the distribution F; and

(ii) the contamination method which provides a sufficiently flexible and relatively easily applicable tool to the sensitivity analysis as well as to postoptimality studies for the considered type of stochastic programming problems and which covers the above mentioned examples.

The previously obtained results [5], [6], [8] will be unified and generalized.

In both of the mentioned cases, the dependence of program (1) on the distribution F will be expressed via its dependence on a real parameter vector, say y, which reflects different assumptions on the (incomplete) knowledge of the distribution F and/or its parameters. For the resulting parametric program

$$\begin{array}{ll} \text{maximize} & g_0(x;y) \\ \text{subject to} & g_i(x;y) \geq 0,\ 1 \leq i \leq m, \\ & g_i(x;y) = 0,\ m+1 \leq i \leq m+p,\ x \in X, \end{array} \qquad \mathrm{M(y)}$$

knowledge of classical results on stability will be assumed. More explicitly, we shall use repeatedly the first order differentiability of the (local) maximizer $x(y)$ of $M(y)$ in a neighborhood of a parameter vector y_0 for which the (local) maximizer $x(y_0)$ of $M(y_0)$ satisfies the linear independence condition, strict complementarity conditions and the second order sufficient condition (see e.g. [9]). For $M(y)$, we shall denote by

$$L(x,u,v;y) = g_0(x;y) + \sum_{i=1}^{m} u_i g_i(x;y) + \sum_{i=1}^{p} v_i g_{m+i}(x;y)$$

the Lagrange function, by $w(y)=[x(y), u(y), v(y)]$ a Kuhn-Tucker point of $M(y)$ and by $I(y) \subset \{1,\dots,m\}$ the corresponding set of indices of the active inequality constraints. Under the mentioned conditions, we have for $y \in O_\varepsilon(y_0)$

$$\frac{\partial w_I(y)}{\partial y} = -D^{-1}(y)B(y), \tag{2}$$

where $w_I(y)=[x(y), u_i(y), i \in I(y), v(y)]$ is the reduced Kuhn-Tucker point and

$$D(y) = \begin{bmatrix} \nabla^2_{xx}L(w;y) & \nabla_x g_I(x;y) \\ [\nabla_x g_I(x;y)]^T & 0 \end{bmatrix}_{[w(y);y]}$$

$$B(y) = [\nabla^2_{yx}L(w;y), \nabla_y g_I(x;y)]^T_{[w(y);y]}$$

with $g_I(x;y)=\{g_i(x;y), i \in I(y), g_i(x;y), m+1 \le i \le m+p\}$. The remaining components $\frac{\partial w(y)}{\partial y}$ (i.e., $\frac{\partial u_i(y)}{\partial y}$ for $i \notin I(y)$) equal to zero. The optimal value of the objective function in $M(y)$ is twice continuously differentiable for $y \in O_\varepsilon(y_0)$; we have namely that

$$\nabla_y g_0(x(y); y) = \nabla_y L(w; y)\Big|_{[w(y);y]} . \tag{3}$$

The assumptions listed above are crucial for differentiability of the Kuhn–Tucker points whereas for differentiability of the (local) maximum value $g_0(x(y); y)$ in $M(y)$, the assumptions can be weakened essentially (see e.g. [10], [19] and the discussion in Section 2).

2 Estimated parameters

Assume now that the considered distribution F is known to belong to a parametric family of distributions, say $F \in \{F_y, y \in Y\}$. Using $g_i(x;y), 0 \le i \le m+p$ instead of $g_i(x;F_y), 0 \le i \le m+p$, respectively, we can rewrite program (1) in the form of $M(y)$. Our aim is to solve program (1) for the true distribution corresponding to the true parameter vector, say $\eta \in Y$, however, our decision can only be based on the knowledge of an estimate, say $y^N \in Y$, of η. As a result, the substitute program $M(y^N)$ is solved instead of $M(\eta)$. The values of the derivatives $\frac{\partial w(y^N)}{\partial y}$ give an insight into sensitivity of the solution respective to small parameter changes (see e.g. Armacost and Fiacco [1] for the stochastic cattle-feed problem). Having in mind the stochastic nature of the parameter values, the results of parametric programming have been complemented by statistical approaches (see Dupačová [4] – [8]).

Theorem 1. *Let y^N be an asymptotically normally distributed estimate of the true parameter vector η that is based on the sample size N*

$$\sqrt{N}(y^N - \eta) \sim N(0,\Sigma)$$

with a known variance matrix Σ. Let for $M(\eta)$ the linear independence condition, strict complementary conditions and the second order sufficient condition be fulfilled. Then the optimal solution $x(y^N)$ of $M(y^N)$ is asymptotically normal

$$\sqrt{N}(x(y^N) - x(\eta)) \sim N(0,V)$$

with the variance matrix

$$V = \frac{\partial x(\eta)}{\partial y} \Sigma \left[\frac{\partial x(\eta)}{\partial y} \right]^T$$

where $\frac{\partial x(\eta)}{\partial y}$ *is the corresponding* (n,q) *submatrix of (2).*

As the asymptotic normality of an estimate is a rather mild assumption, it is the differentiability of the optimal solution $x(\eta)$ which is crucial for the validity of Theorem 1. In this respect, the strict complementarity conditions play an important role. When replacing this assumption by the strong second order sufficient condition [17] one can still get differentiability of the corresponding (local) optimal values of the objective function in $M(\eta)$, see e.g. [14] and [19]. As to the differentiability of the local maximizers, the situation is quite different. It is possible to get directional derivatives of the Kuhn-Tucker points in any direction under the mentioned strong second order sufficient condition (Jittorntrum [14]) and we shall use this result later in connection with the contamination method (see Section 3). The most general result on differentiability of the local maximizers is due to Robinson [18]: Under the strong second order sufficient condition, nondegenerated local maximizers in $M(\eta)$ are Bouligand differentiable. It means, inter alia, that for $y \in O_\varepsilon(\eta)$, the trajectory $x(y)$ can be approximated by a conic function. The approximation function is affine (i.e., $x(\eta)$ is differentiable) if and only if the strict complementarity conditions are fulfilled.

This discussion implies the following conclusions regarding eventual *generalizations of Theorem 1*: Without strict complementarity conditions but under the strong second order sufficient condition we can easily get asymptotic normality of the local optimal values $g_0(x(y^N); y^N)$ of $M(y^N)$ but we cannot expect a similar result for the (local) maximizer $x(y^N)$ unless strict complementarity conditions are fulfilled.

The best result one can get without assuming strict complementarity conditions is the assertion that the asymptotic distribution of $\sqrt{N}(x(y^N) - x(\eta))$ is the same as that of $\sqrt{N}\,\Delta x(\eta; x(y^N) - x(\eta))$ where $\Delta x(\eta; z)$ denotes the Bouligand derivative of the local maximizer at η applied to z.

3 Contaminated distributions

Throughout this section, the functions $g_i, 0 \le i \le m+p$ in (1) will be assumed to depend *linearly* on the distribution F. This assumption is evidently satisfied for the stochastic programs with recourse as well as for those with probablistic constraints and in all cases where g_i, $0 \le i \le m+p$, are expectations of suitable functions related to the starting formulation of the stochastic programming problem. Furthermore, we shall assume that $X = R^n$; it means only that the original deterministic constraints and the induced ones have been incorporated into the explicit constraints of (1) with $g_i(x;F)$ independent on F, of course.

The local behavior of the optimal solution $x(F)$ of the program (1) with respect to small changes of the distribution F can be studied via t-contamination of the distribution F by a suitably chosen distribution G, i.e., instead of F, distributions of the form

$$F_t = (1-t)F + tG, \; 0 \le t \le 1 \tag{4}$$

will be considered. In (4), F_t is called *distribution F t-contaminated by distribution G*.

Due to our assumption, the original stability problem thus reduces to that linearly perturbed by a scalar parameter $t \in \langle 0,1 \rangle$:

$$\begin{array}{lll} \text{minimize} & (1-t)\, g_0(x;F) + t g_0(x;G) & \\ \text{subject to} & (1-t)\, g_i(x;F) + t g_i(x;G) \geq 0, & 1 \leq i \leq m\,, \\ & (1-t)\, g_i(x;F) + t g_i(x;G) = 0, & m+1 \leq i \leq m+p\,. \end{array} \tag{5}$$

In principle, it is possible to get the trajectory of the optimal solutions $x(F_t)$, $0 \leq t \leq 1$; for an appropriate method see e.g. [13].

We shall aim to obtain the Gâteaux differential $dx(F;G-F)$ of the optimal solution of (1) in the direction of $G-F$. To get the explicit results, one has to check the differentiability and regularity assumptions mentioned in Section 1 and to compute matrices $B(O)$, $D(O)$ corresponding to the contamination parameter $t=0$.

Theorem 2. *For the program*

$$\begin{array}{lll} \text{minimize} & g_0(x;F) & \\ \text{subject to} & g_i(x;F) \geq 0, & 1 \leq i \leq m\,, \\ & g_i(x;F) = 0, & m+1 \leq i \leq m+p \end{array} \tag{6}$$

assume:

(i) $g_i(\bullet;F): R^n \to R^1$ *are twice continuously differentiable,* $0 \leq i \leq m+p$

(ii) *the Kuhn-Tucker conditions of the first and second order, the linear independence condition and the strict complementarity conditions are fulfilled for* $w(F) = [x(F), u(F), v(F)] \in R^n \times R^m_+ \times R^p$,

(iii) *there is a neighborhood* $O(x(F)) \subset R^n$ *on which* $g_i(\bullet;G)$, $0 \leq i \leq m+p$ *are twice continuously differentiable.*

Then

(a) *There is a neighborhood* $O(w(F)) \subset R^n \times R^m_+ \times R^p$, *a real number* $t_0 > 0$ *and a continuous function* $w: \langle 0, t_0) \to O(w(F))$, $w(0) = w(F)$ *such that for any* $t \in \langle 0,t_0)$, $w(t) = [x(t), u(t), v(t)]$ *is the Kuhn-Tucker point of (5) for which the second order sufficient condition, the linear independence condition and the strict complementarity conditions are fulfilled.*

(b) *The Gâteaux differential* $dw(F;G-F)$ *of the Kuhn-Tucker point* $w(F)$ *of (6) in the direction of* $G-F$ *is given by*

$$dw_I(F;G-F) = -\begin{bmatrix} \nabla^2_{xx} L(w(F);F) & \nabla_x g_I(x(F);F) \\ [\nabla_x g_I(x(F);F)]^T & 0 \end{bmatrix}^{-1} \bullet \begin{bmatrix} \nabla_x L(w(F);G) \\ g_I(x(F);G) \end{bmatrix}. \tag{7}$$

The remaining components of $dw(F;G-F)$, *which correspond to the non-active constraints in (6), equal to 0.*

Proof is a straightforward application of the well known stability results. We took the liberty of adopting the notation to our case; namely

$$\nabla_x L(w(F);G) = \nabla_x g_0(x(F);G) + \sum_{i=1}^{m} u_i(F) g_i(x(F);G) + \sum_{i=m+1}^{p} v_i(F)\, g_i(x(F);G)\,,$$

$w_I(F) = [x(F); u_i(F),\ i \in I(F), v(F)]$ with $I(F) \subset \{1,\ldots,m\}$ containing the indices of the active inequality constraints $g_i(x(F);F) = 0$, $g_I(x(F);G)$, and $g_I(x(F);F)$ are vectors consisting of components $g_i(x(F);G)$ and $g_i(x(F);F)$ for $i \in I(F)$, $m+1 \le i \le m+p$, respectively, $\nabla_x g_I(x(F);F)$ is a matrix consisting of columns $\nabla_x g_i(x(F);F)$ for $i \in I(F)$ and $m+1 \le i \le m+p$.

Due to the fact that (5) is linearly perturbed, we have

$$\nabla_{xt}^2 L(w;t) = \nabla_x L(w;G) - \nabla_t L(w;F) ,$$

$$\nabla_t g_I(x;t) = g_I(x;G) - g_I(x;F) ,$$

so that

$$\nabla_{xt}^2 L(w(F);t) = \nabla_x L(w(F);G)$$

$$\nabla_t g_I(x(F);t) = g_I(x(F);G) .$$

Theorem 3. *Let assumptions of Theorem 2 hold true.*

(a) *Let the matrix*

$$L = \nabla_{xx}^2 L(w(F);F)$$

be nonsingular. Then the Gâteaux differential of the isolated local maximizer $x(F)$ of (6) in the direction of $G-F$ is given by

$$dx(F;G-F) = -C^{-1}\nabla_x L(w(F);G)$$

where

$$C^{-1} = [I - L^{-1}P(P^T L^{-1}P)^{-1}P^T]L^{-1}$$

$$P = \nabla_x g_I(x(F);F)$$

and I is the n-dimensional unit matrix.

(b) *Let the matrix $P = \nabla_x g_I(x(F);F)$ be of rank n. Then the Gâteaux differential of the isolated local maximizer $x(F)$ of (6) in the direction of $G-F$ is given by*

$$dx(F;G-F) = -(P^T)^{-1} g_I(x(F);G) .$$

Proof follows from (7) by well known formulas for inversion of the matrix

$$D(0) = \begin{pmatrix} L & P \\ P^T & 0 \end{pmatrix}$$

which is nonsingular and which contains the nonsingular square submatrix L in the case of (a) or the nonsingular square submatrix P in case (b). See also [9] for a similar approach.

The assumptions of strict complementarity again play an essential role in the proof of Theorem 2. They guarantee that the interval $(0,t_0)$ on which $w(t)$ is the Kuhn-Tucker point of (5) is nonempty. Alternatively, the strict complementarity conditions can be replaced by the strong second order sufficient condition. Thanks to the fact, that we have a scalar parameter only and that we are in fact interested in the right-hand derivatives of the optimal solution with respect to the parameter at the given point $t = 0$, the result of Jittorntrum [14] applied to our problem gives the desired assertion on the Gâteaux differential.

Denote $I^+(F) = \{ i \in I(F) : u_i(F) > 0 \}$, $I^0(F) = \{ i \in I(F) : u_i(F) = 0 \}$. (Under strict complementarity conditions, $I^0(F) = \phi$.)

Theorem 4. *Let in assumptions of Theorem 2, strict complementarity conditions be replaced by the strong second order sufficient condition. Then:*

(a) *There is a neighborhood* $O(w(F)) \subset R^n \times R^m_+ \times R^p$, *a real number* $t_0 > 0$ *and a continuous function* $w: \langle 0, t_0) \to O(w(F))$, $w(0) = w(F)$ *such that for any* $t \in \langle 0, t_0)$, $w(t) = [x(t), u(t), v(t)]$ *is the Kuhn-Tucker point of (5) for which the strong second order sufficient condition and the linear independence condition are fulfilled.*

(b) *There is a set R of indices such that*

$$I^+(F) \subset R \subset I^+(F) \cup I^0(F) = I(F),$$

for which the nonzero components of the Gâteaux differential $dw(F; G-F)$ *are given by*

$$dw_R(F;G-F) = -\begin{bmatrix} \nabla^2_{xx} L(w(F);F) & \nabla_x g_R(x(F);F) \\ [\nabla_x g_R(x(F);F)]^T & 0 \end{bmatrix}^{-1} \cdot \begin{bmatrix} \nabla_x L(w(F);G) \\ g_R(x(F);G) \end{bmatrix}.$$

For the sake of completeness let us mention the corresponding results on the differentiability of the optimal value $g_0(x(t); t)$:

Theorem 5. *Under assumptions of Theorem 2 or Theorem 4, the optimal value of the objective function in (5) is continuously differentiable on a neighbourhood of* $t = 0$ *and*

$$\frac{d}{dt} g_0(x(t); t)\Big|_{t=0} = L(w(F); G) - L(w(F); F) .$$

Proof in the first case, the proof follows from the classical result (3) and in the second case, it follows from the application of the relevant result by Jittorntrum [14] to program (5).

Remark. For fixed constraints, i.e., for $g_i(x; F)$ independent on F for $1 \le i \le m + p$ we get obviously

$$\frac{d}{dt} g_0(x(t); t\Big|_{t=0} = g_0(x(F); G) - g_0(x(F); F) .$$

To study the effect of contamination of F by distributions belonging to a specified *set* G of distributions one can concentrate to those distributions which give maximum/minimum local change of the optimal value of the objective function, i.e., to G^* or G^{**} such that

$$\sup_{G \in \mathcal{G}} g_0(x(F); G) = g_0(x(F); G^*) \quad \text{and} \quad \inf_{G \in \mathcal{G}} g_0(x(F); G) = g_0(x(F); G^{**})$$

what links with the approach of [16]. The existence of the extremal distributions G^*, G^{**} has been proved for various sets G and functions g_0, see e.g. [3], [7].

For specific decision models of stochastic programming, we can get correspondingly the specific form of the assumptions as well as the explicit formulas for the Gâteaux differentials. The assumptions can be subdivided into three categories:

(A) The basic model assumptions, including the absolute continuity of the distribution F.

(B) The general assumptions such as the existence of the Kuhn-Tucker point for which strict complementarity conditions are fulfilled.

(C) The assumptions of differentiability, the linear independence condition and the second order sufficient condition, which can be fitted to the considered model.

For examples see [5] – [8].

References

[1] Armacost, R.L., and A.V. Fiacco: Computational experience in sensitivity analysis for nonlinear programming. *Mathematical Programming* **6** (1974) 301–326.

[2] Cipra, T.: Confidence regions for linear programs with random coefficients. WP-86-20 IIASA – Laxenburg 1986.

[3] Dupačová, J.: Minimaxová úloha stochastického lineárního programování a momentový problém. Ekonomicko – matematický obzor 13 (1977) 279–307. [Extended abstract in ZAMM 58 (1978) T466–T467.]

[4] Dupačová, J.: Stability in stochastic programming with recourse. *Acta Univ. Carol., Math. et Phys.* **24** (1983) 23–34.

[5] Dupačová, J.: Stability in stochastic programming with recourse – estimated parameters *Mathematical Programming* **28** (1984) 23–34.

[6] Dupačová, J.: Stability in stochastic programming with recourse -- contaminated distributions. *Mathematical Programming Studies* – to appear.

[7] Dupačová, J.: Stochastic programming with incomplete information. WP-85-08, IIASA Laxenburg 1986.

[8] Dupačová, J.: Stability in stochastic programming – probabilistic constraints. In: Arkin, V.I., A. Shiriaev, R. Wets (eds). Proceedings of International Conference on Stochastic Optimization, Kiev. Lecture Notes in Control and Information Sciences, Springer-Verlag, Berlin 1986.

[9] Fiacco, A.V.: Introduction to sensitivity and stability analysis in nonlinear programming. Academic Press, New York 1983.

[10] Fiacco, A. and J. Kyparisis: Sensitivity analysis in nonlinear programming under second order assumptions. In: Baghi, A. and H. Th. Jongen (eds): Systems and Optimization. Lecture Notes in Control and Information Sciences 66, Springer Verlag, Berlin 1985, 74–97.

[11] Friedrich, B. and K. Tammer: Untersuchungen zur Stabilität einiger Ersatzprobleme der stochastichen Optimierung in Bezug auf Änderungen des Zugrunde gelegten Zufallsvektors. *Wiss. Z. der HU Berlin, Math.-Naturwiss. Reihe* 1981, No 5, 373–375.

[12] Garstka, S.J.: Distribution functions in stochastic programs with recourse: a parametric analysis. *Mathematical Programming* **6** (1974) 339–351.

[13] Gfrerer, H., J. Guddat and Hj. Wacker: A globally convergent algorithm based on imbedding and parametric optimization. *Computing* **30** (1983) 255–252.

[14] Jittorntrum, K.: Solution point differentiability without strict complementarity in nonlinear programming. *Mathematical Programming Study* **21** (1984) 127–138.

[15] Kall, P.: On approximations and stability in stochastic programming. Research Report, Universität Zürich, February 1986.

[16] Nadeau, R. and R. Theodorescu: Restricted Bayes strategies for programs with simple recourse. *Operations Research* **28** (1980) 777–784.

[17] Robinson, S.M.: Strongly regular generalized equations. *Mathematics of Operations Research* **5** (1980) 43–62.

[18] Robinson, S.M.: Local structure of feasible sets in nonlinear programming, Part III: Stability and sensitivity. Comp. Sci. Technical Report #552, University of Wisconsin - Madison 1984.

[19] Rockafellar, R.T.: Directional differentiability of the optimal value function in a nonlinear programming problem. *Mathematical Programming Study* **21** (1984) 213–226.

[20] Römisch, W. and A. Wakolbinger: Obtaining convergence rates for approximations in stochastic programming. Bericht Nr. 308, Inst. für Mathematik, J. Kepler Universität Linz, February 1986.

[21] Seppälä, Y.: On a stochastic multi-facility location problem. *AIIE Trans.* **7** (1975) 56–62.

METHODS OF PARAMETRIZATION IN THE ANALYSIS OF IMPROPER MATHEMATICAL PROGRAMMING PROBLEMS

I. I. Eremin[1)]

Introduction.

The role of models of mathematical programming (MP) for the realization of economic planning problems is well-known. In these problems the constraints which are given, for example, as a system of inequalities and equations, and which reflect restrictions of resources, technological restrictions, environmental restrictions, demands from directive rules etc., may be inconsistent (contradictory). Such models may arise for the following reasons: deficiency of resources, increasing of plan demands, missing of reserve capacities of production, inaccuracy of economical information, consideration of contradictory directives, including standards with respect to negative influences of production on the environment and so on. The occurrence of contradictory models is quite usual in practical situations. In the simplest models it makes no special difficulties to overcome these contradictions by means of certain corrections of the model. Such corrections can be made, for example, by weakening the constraints, by omitting parts of the constraints, by correcting data etc.. The study of more complicated models reflecting much more difficult situations leads to the necessity of considering the phenomenon of improperly posed (contradictory) models, and therefore it is necessary to develop a corresponding theory (first of all the duality theory) as well as numerical methods and their computer implementation.

Consider the MP problem

$$\sup \{f_0(x) \mid f_j(x) \leq 0 \quad (j=1,\dots,m),\ x \geq 0\}, \tag{0.1}$$

and let $F(x,u) = f_0(x) - \sum_{j=1}^{m} u_j f_j(x)$ be its Lagrange function.

Problem (0.1) is equivalent to

1) Academy of Science USSR, Inst. of Math.Mech.
Sverdlovsk, Kovalevskaja str.16

$$\sup_{x \geq 0} \inf_{u \geq 0} F(x,u) =: \bar{v} . \tag{0.2}$$

The dual problem of problem (0.2) is defined by

$$\inf_{u \geq 0} \sup_{x \geq 0} F(x,u) =: \underline{v} . \tag{0.3}$$

If both (0.1) and (0.3) are solvable (i. e., $\underline{v}$ is finite and there is some $[\bar{x},\bar{u}] \geq 0$: $F(\bar{x},\bar{u}) = \bar{v}$), and if $\underline{v} = \bar{v}$, then the problem (0.1) is said to be <u>proper</u>; otherwise it is said to be <u>improper</u>. In particular, if $\bar{v} = -\infty$ or if the system of constraints of (0.1) is inconsistent, then (0.1) is improper.
If (0.1) is a linear programming (LP) problem,

$$\sup \{ (c,x) \mid Ax - b \leq 0,\ x \geq 0 \} , \tag{0.4}$$

then, in this case, the equivalent problem (0.3) becomes

$$\inf \{ (b,u) \mid A^T u - c \geq 0,\ u \geq 0 \} . \tag{0.5}$$

Let now the feasible sets in (0.4) and (0.5) be denoted by M and M^*, respectively. The solvability of problem (0.4) (and of problem (0.5)) is equivalent to the condition that $M \neq \emptyset$ and $M^* \neq \emptyset$. The cases (1) $M = \emptyset$, $M^* \neq \emptyset$, (2) $M \neq \emptyset$, $M^* = \emptyset$, (3) $M = \emptyset$, $M^* = \emptyset$ correspond to problem (0.4) being improper. In dependence on the occurrence of the cases (1), (2) or (3), problem (0.4) is called <u>improper of 1st, 2nd or 3rd type</u>, respectively.

To overcome the improperness of some MP problem, one could imbed this problem into one or the other class of parametric programs. For instance, problem (0.1) can be carried over to the class of problems

$$\sup \{ f_o(x) - (q,x) \mid f_j(x) \leq p_j \quad (j=1,\dots,m),\ x \geq 0 \} , \tag{0.6}$$

where $p = [p_1,\dots,p_m]^T$ and $q = [q_1,\dots,q_n]^T$ are parameters. Given some set K_o of parameters $\Delta = [p,q]$, we define the set of <u>effective parameters</u> by

$$K = \{ \Delta \in K_o \mid (0.6) \text{ is proper} \} .$$

The choice of a concrete point $\Delta = [p,q]$ is, however, subjected to one or the other requirement, like the requirement of minimizing some <u>quality function of correction</u> of problem (0.1). Let such a quality functional be denoted by $\varphi(\Delta)$. Then the following problem arises

$$\inf \{ \varphi(\Delta) \mid \Delta \in K \} . \tag{0.7}$$

In view of the program (0.4) the corresponding parametric program (0.6) is of the form

$$L_\Delta \sim \sup \{ (c-q,x) \mid Ax \leq b+p,\ x \geq 0 \}, \tag{0.8}$$

an analogous imbedding of problem (0.5) leads to the problem

$$L_\Delta^* \sim \inf \{ (b+p,u) \mid A^T u \geq c-q,\ u \geq 0 \}. \tag{0.9}$$

The problems (0.8) and (0.9) are dual to each other in the usual sense. Setting $K_o = \{ \Delta = [p,q] \geq 0 \}$, we have

$$K = \{ \Delta \in K_o \mid M(p) \neq \emptyset,\ M^*(q) \neq \emptyset \},$$

where $M(p) = \{ x \geq 0 \mid Ax \leq b+p \}$ and $M^*(q) = \{ u \geq 0 \mid A^T u \geq c-q,\ u \geq 0 \}$. This representation of K trivially follows from classical duality results on solvable LP problems.

Above we have discussed the approach of correcting improper MP problems by parametrization procedures. Developing concrete algorithms in the framework of this approach requires a specific theory for improperly posed problems. As it is known, the duality theory is a global part of the theory of mathematical programming. Concerning improper problems the classical duality theory does not provide enough information and contents, hence one needs, in this case, a new approach to develop a duality theory. First we shall briefly present the problem of duality.

1. Duality [1].

In mathematical programming, the duality theory plays an essential role. Duality allows to get deeper into the mathematical nature of the problems, and it is a generator of efficient methods for solving these problems. Duality for improper MP problems plays the same role. The duality scheme is the following: To each pair of dual MP problems P and P^* a uniform scheme π assigns a pair of problems C and $C^{\#}$ which approximate P and P^*, and which are dual to each other (in the sense that a duality theorem in the standard formulation holds). This can be formally described by

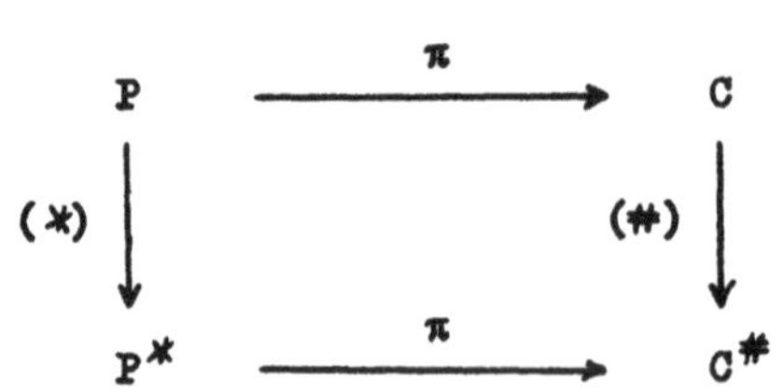

One of the possible realizations of this scheme with respect to the problems (0.4) and (0.5) has the form

$$C: \quad \sup \Big\{(c,x) - \sum_{j=1}^{m_o} R_j \|(A_j x - b^j)^+\|_j \;\Big|\; A_o x \le b^o,\ x \ge 0,\ \|x^i\|_i \le r_i \quad (i=1,\dots,n_o)\Big\},$$

$$C^{\#}: \quad \inf \Big\{(b,u) + \sum_{i=1}^{n_o} r_i \|(c^i - B_i^T u)^+\|_i^* \;\Big|\; B_o^T u \ge c^o,\ u \ge 0,\ \|u^j\|_j \le R_j \quad (j=1,\dots,m_o)\Big\}.$$

Here $R_j \ge 0$ $(j=1,\dots,m_o)$, $r_i \ge 0$ $(i=1,\dots,n_o)$ are real parameters, $\alpha^+ = \max\{0,\alpha\}$, $[\alpha_1,\dots,\alpha_k]^+ = [\alpha_1^+,\dots,\alpha_k^+]$, $\|\cdot\|_j$, $\|\cdot\|_i$ is any collection of monotonic norms (i. e., $x \ge y \ge 0 \Rightarrow \|x\| \ge \|y\|$), and $\|\cdot\|_j^*$, $\|\cdot\|_i^*$ are the associated dual norms which are monotonic, too. Further $A_o, A_1, \dots, A_{m_o}$ and $B_o, B_1, \dots, B_{n_o}$ are submatrices of A such that

$$A = \begin{bmatrix} A_o \\ A_1 \\ \vdots \\ A_{m_o} \end{bmatrix} = [B_o \;\; B_1 \;\; \dots \;\; B_{n_o}],$$

and $c^T = [c^o, c^1, \dots, c^{n_o}]$, $x^T = [x^o, x^1, \dots, x^{n_o}]$ are the decompositions of c and x, corresponding to the decomposition of A into submatrices B_i, and $b^T = [b^o, b^1, \dots, b^{m_o}]$, $u^T = [u^o, u^1, \dots, u^{m_o}]$ are the decompositions of b and u, corresponding to the decomposition of A into submatrices A_j $(i=0,1,\dots,n_o;\ j=0,1,\dots,m_o)$.

Assume that $M_o = \{x \ge 0 \mid Ax_o \le b^o\} \ne \emptyset$ and $M_o^* = \{u \ge 0 \mid B_o^T u \ge c^o\} \ne \emptyset$. Some of the submatrices B_i and A_j may vanish; for example, if $A_1, \dots, A_{m_o}$ vanish, then $A = A_o$. We denote by $f_R(x)$ and $f_r^{\#}(x)$ the objective functions of the problems C and $C^{\#}$ and by M_r and $M_R^{\#}$ the corresponding feasible sets. We use the symbols Arg C and Arg $C^{\#}$ and the symbols opt C and opt $C^{\#}$ to denote the optimal solution sets and the optimal values of C and $C^{\#}$, respectively.

<u>Theorem 1.1. [1].</u>

1) It holds that $f_R(\bar{x}) \le f_r^{\#}(\bar{u})$ for each $\bar{x} \in M_r$ and each $\bar{u} \in M_R^{\#}$, and $f_R(\bar{x}) = f_r^{\#}(\bar{u})$ implies that $\bar{x} \in \operatorname{Arg} C$ and $\bar{u} \in \operatorname{Arg} C^{\#}$.
2) Suppose C is solvable and there is some $\bar{x} \in M_o$ such that $\|\bar{x}^i\|_i < r_i$ $(i=1,\dots,n_o)$. Then $C^{\#}$ is solvable and opt C = opt $C^{\#}$.

We note that the converse variant of assertion 2) of Theorem 1.1 is also true.

We consider now several particular realizations of the problems C and $C^{\#}$:

C_0: $\max\{(c,x) - (R,(Ax-b)^+) \mid 0 \leq x \leq r\}$,

$C_0^{\#}$: $\max\{(b,u) + (r,(c-A^Tu)^+) \mid 0 \leq u \leq R\}$;

C_1: $\max\{(c,x) - R_0\|(Ax-b)^+\| \mid x \geq 0\}$,

$C_1^{\#}$: $\min\{(b,u) \mid A^Tu \geq c,\ u \geq 0,\ \|u\|^* \leq R_0\}$;

C_2: $\max\{(c,x) \mid Ax \leq b,\ x \geq 0,\ \|x\| \leq r_0\}$,

$C_2^{\#}$: $\min\{(b,u) + r_0\|(c-A^Tu)^+\|^* \mid u \geq 0\}$.

Here R_0 and r_0 are positive real numbers, and $\|\cdot\|$ and $\|\cdot\|^*$ are monotonic norms.

<u>Theorem 1.2.[1].</u>

1) The problems C_0 and $C_0^{\#}$ are always solvable, and their optimal values coincide.
2) If $\bar{u} \in M^* = \{u \geq 0 \mid A^Tu \geq c\}$ and $\|\bar{u}\|^* < R_0$, then C_1 and $C_1^{\#}$ are solvable, and their optimal values coincide.
3) If $\bar{x} \in M = \{x \geq 0 \mid Ax \leq b\}$ and $\|\bar{x}\| < r_0$, then C_2 and $C_2^{\#}$ are solvable, and their optimal values coincide.

These results on the duality for improper mathematical programming problems have been extended to the case of Banach spaces in [2].

A. A. Vatolin has proposed a more general duality scheme for improper LP problems in finite-dimensional spaces. This will now be presented. A function Ω defined on R^k is called <u>feasible</u> if the following two conditions are fulfilled:

1) $\Omega(\cdot)$ is a closed convex function whose effective domain $\operatorname{dom}\Omega$ is a subset of $R_+^k := \{z \geq 0 \mid z \in R^k\}$,
2) $\Omega(\cdot)$ is a monotonically non-decreasing function on R_+^k and $\Omega(0) = 0$.

Given a function Ω on R^k, we define a function Ω^+ by

$$\Omega^{\oplus}(z^*) = \begin{cases} \sup\limits_{(z)} \{(z^*,z) - \Omega(z)\}, & z^* \in R_+^k \\ +\infty, & z^* \notin R_+^k \end{cases}.$$

Let now R, λ_i, μ_j, ξ_i, η_j be a fixed system of positive parameters satisfying

$$\lambda_i \xi_i = \mu_j \eta_j = R^{-1} \quad (i=1,\dots,n_o;\ j=1,\dots,m_o) .$$

Consider an arbitrary feasible function Φ of ν variables, where $\nu = n + m - \nu_1 - \nu_2$, and ν_1, ν_2 are the dimensions of x^o and u^o, respectively. Then the following problems arise:

$$D : \quad \sup \{(c,x)-R\,\Phi(\lambda_1 x^1,\dots,\lambda_{n_o} x^{n_o},\mu_1(A_1 x-b^1)^+,\dots,\mu_{m_o}(A_{m_o}x-b^{m_o})^+) \mid A_o x \leq b^o,\ x \geq 0\}$$

$$D^{*}: \quad \inf \{(b,u)+R\,\Phi^+(\xi_1(c^1-B_1^T u)^+,\dots,\xi_{n_o}(c^{n_o}-B_{n_o}^T u)^+,\eta_1 u^1,\dots,\eta_{m_o}u^{m_o}) \mid B_o^T u \geq c^o,\ u \geq 0\} .$$

<u>Lemma:</u> If Ω is a feasible function, then $\Omega^{\oplus}$ is feasible, too, and $(\Omega^{\oplus})^{\oplus} = \Omega$ holds.

We note that the relation $(\Phi^{\oplus})^{\oplus} = \Phi$ guarantees that $(D^{*})^{*} = D$. Setting $\nu_3 = n - \nu_1$ and $\nu_4 = m - \nu_2$, we consider the decomposition $z = [z^1,z^2]$, $z^1 \in R^{\nu_3}$, $z^2 \in R^{\nu_4}$, of the vector $z \in R^{\nu}$ to be fixed. Given any function Φ on R^{ν}, we then define

$$\bar{\Phi}(z) = \Phi(z^1,(z^2)^+) .$$

Further let

$$\Gamma(x) = [\lambda_1 x^1,\dots,\lambda_{n_o} x^{n_o},\mu_1(A_1 x-b^1),\dots,\mu_{m_o}(A_{m_o}x-b^{m_o})].$$

The symbol ri W means the relative interior of a set W.

Now we present a basic duality theorem concerning the problems D and D^{*}.

<u>Theorem 1.3.</u> Let opt D be finite, and suppose that the following regularity condition is fulfilled:

$$\exists \bar{x},\ A_o\bar{x} \leq b^o,\ \bar{x} \geq 0,\ \Gamma(\bar{x}) \in \operatorname{ri}\operatorname{dom}\bar{\Phi} .$$

Then opt D = opt D^{*} and Arg $D^{*} \neq \emptyset$.

In the following we shall take up the questions of duality for improperly posed problems of convex programming (CP). Let (0.1) be a CP problem, i. e., the functions $-f_o(x), f_1(x), \dots, f_m(x)$ are supposed to be convex. This problem will be denoted by the symbol P. To the problem P we assign, similarly to the case of improper linear programs, the problem

$$\sup \{f_o(x) - \sum_{j=1}^{m_o} R_j \|F_j^+(x)\|_j \mid F_o(x) \leq 0,\ x \geq 0,\ \|x^i\|_i \leq r_i \quad (i=1,\dots,n_o)\} ; \tag{1.1}$$

where $[F_o(x),\dots,F_{m_o}(x)] = [f_1(x),\dots,f_m(x)]$, $R_j > 0$, $r_i > 0$ $(j=1,\dots,m_o;\ i=1,\dots,n_o)$, and the sense of the remaining symbols is the same as above. The norms $\|\cdot\|_j$, $\|\cdot\|_i$ and their dual norms are supposed to be monotonic. Set

$$M(r) = \{x \geq 0 \mid \|x^i\|_i \leq r_i \quad (i=1,\dots,n_o)\} ,$$

$$M^{\#}(R) = \{u \geq 0 \mid \|u^j\|_j \leq R_j \quad (j=1,\dots,m_o)\} ,$$

and consider the problems

$$\sup_{x\in M(r)} \inf_{u\in M^{\#}(R)} \Phi(x,u) =: \bar{\bar{\Phi}} , \tag{1.2}$$

$$\inf_{u\in M^{\#}(R)} \sup_{x\in M(r)} \Phi(x,u) =: \underline{\Phi} , \tag{1.3}$$

where $\Phi(x,u) = f_o(x) - \sum_{j=1}^{m} u_j f_j(x)$, and $u = [u_1,\dots,u_m]$ and $\{u^j \mid j=0,1,\dots,m_o\}$ are connected by $u = [u^o, u^1, \dots, u^{m_o}]$.

It can be shown that Arg (1.1) = Arg_x (1.2), cf. [1, § 23]. Therefore it is natural to say that the problem (1.3) is dual to (1.1).

<u>Theorem 1.4. [1].</u> If the problem (1.1) is solvable and satisfies some regularity condition, then

$$\text{Arg } (1.3) \neq \emptyset \quad \text{and} \quad \text{opt } (1.1) = \bar{\bar{\Phi}} = \underline{\Phi} .$$

2. Methods of approximation (correction)

As mentioned above, one possibility of correcting an improper program is that of parametrization. The idea is to imbed an initial problem P, say the problem (0.1), into a parametric class of problems $\{P(\Delta)\}_\Delta$, $\Delta \in K_o$. This means that for some $\Delta_o \in K_o$, it holds that $P(\Delta_o) = P$. Given a set K (cf. the Introduction), one can define a correction problem in the form (0.7), i. e.

$$\inf \{\varphi(\Delta) \mid \Delta \in K\},$$

where $\varphi(\Delta)$ is some function indicating the quality of approximation of the improper problem P. The parametric problem

$$\sup \{(c-q,x) \mid (A+H)x \leq b+p,\ x \geq 0\}, \tag{2.1}$$

where $K_0 = \{\Delta = [q,p,H] \mid [q,p] \geq 0\}$ may serve as an example of a class $\{P(\Delta)\}$ associated with the problem (0.4). As a quality function φ we could choose $\varphi(\Delta) = (\bar{u},p) + (\bar{v},q) + \|H\|$, where $\bar{u} > 0$, $\bar{v} > 0$.

It is easy to study the case of the parametrization (0.8) for $\varphi(\Delta) = (\bar{u},p) + (\bar{v},q)$ and $K_0 = \{\Delta = [q,p] \geq 0\}$. As mentioned above, we obviously have in this situation

$$K = \{\Delta \in K_0 \mid M(p) \neq \emptyset,\ M^*(q) \neq \emptyset\},$$

where $M(p) = \{x \geq 0 \mid Ax \leq b+p\}$, $M^*(q) = \{u \geq 0 \mid A^T u \geq c-q\}$. The approximating problem can be written in the form

$$\min \{(\bar{u},p) + (\bar{v},q) \mid Ax \leq b+p,\ x \geq 0,\ A^T u \geq c-q,\ u \geq 0,\ [p,q] \geq 0\}.$$

This problem is equivalent to the pair of LP problems

$$\min \{(\bar{u},p) \mid Ax \leq b+p,\ [x,p] \geq 0\}, \tag{2.2}$$

$$\min \{(\bar{v},q) \mid A^T u \geq c-q,\ [u,q] \geq 0\}. \tag{2.3}$$

The programs (2.2) and (2.3) are always solvable. Provided that $M^* \neq \emptyset$, the combination of the correction method and of the solution of the corrected problem can be realized, in our case, by

$$\sup \{(c,x) - R_0((Ax-b)^+,\bar{u}) \mid x \geq 0\}, \tag{2.4}$$

where R_0 has to be sufficiently large. If $\tilde{x} \in \operatorname{Arg}(2.4)$, then $\tilde{p} = (A\tilde{x},b)^+ \in \operatorname{Arg}(2.2)$. The condition $M^* \neq \emptyset$ is equivalent to the condition $\tilde{q} \in \operatorname{Arg}(2.3) \Longrightarrow \tilde{q} = 0$. If, on the other hand, the problems (0.4) and (0.5) satisfy the conditions $M \neq \emptyset$ and $M^* = \emptyset$, then one has $\tilde{p} \in \operatorname{Arg}(2.2) \Longrightarrow \tilde{p} = 0$, and the search for $\tilde{q} \in \operatorname{Arg}(2.3)$ and for the optimal solution $\tilde{u}$ may be carried out via the problem

$$\min \{(b,u) + r_0((c-A^T u)^+,\bar{v}) \mid u \geq 0\}, \tag{2.5}$$

where r_0 has to be sufficiently large. If $\tilde{u} \in \operatorname{Arg}(2.5)$, then $\tilde{q} = (c-A^T\tilde{u})^+ \in \operatorname{Arg}(2.3)$. The condition $M \neq \emptyset$ is equivalent to the condition $\tilde{p} \in \operatorname{Arg}(2.2) \Longrightarrow \tilde{p} = 0$.

In the paper of A. A. Vatolin [3] (cf. also [1, § 26], the problem (0.7) has been studied in the case of the parametrizing problem (0.1) in the form (0.6) under the assumption $\varphi(\Delta) = \|\Delta\|$.

Methods of iterative approximations of improper LP problems have been proposed by L. D. Popov, cf. [1, § 19]. The case of parametrization (2.1) under the assumption $\varphi(\Delta) = \|\Delta\|^2$, $\Delta = [q,p,H]$, has been studied in [1, § 13].

A method of iterative regularization for correcting improper LP problems has been suggested by V. D. Skarin. Now we are going to consider this method. The pair of dual problems (0.4) and (0.5) will be replaced by the problem of finding saddle points $[x^\sigma, u^\sigma]$ of the regularized Lagrange function

$$L_\sigma(x,u) = (c,x) - (u, Ax-b) - \alpha\|x\|^2 + \beta\|u\|^2$$

on the domain $R^n_+ \times R^m_+$, where $[\alpha,\beta] = \sigma > 0$. The symbols L_Δ, L^*_Δ, K will be used in the same sense as agreed upon in the Introduction. Let $\Delta = [q,p]$ and

$$\bar{\Delta} = \operatorname{Arg\,min} \{ \|\Delta\|^2 \mid \Delta \in K \} .$$

<u>Theorem 2.1.</u> For any real number $\sigma > 0$ it holds

$$\|(Ax^\sigma - b - \bar{p})^+\| \le \beta^{1/2} \chi_1(\alpha,\beta) ,$$

$$\|(c - \bar{q} - A^T u^\sigma)^+\| \le \alpha^{1/2} \chi_2(\alpha,\beta) ,$$

$$\max \{ |(c-\bar{q}, x^\sigma) - f^*|, |(b+\bar{p}, u^\sigma) - f^*| \} \le \gamma\, \chi_3(\alpha,\beta),$$

where $\chi_1(\alpha,\beta) = \alpha^{1/2}\|x^*\| + \vartheta(\alpha,\beta)$, $\chi_2(\alpha,\beta) = \beta^{1/2}\|u^*\| + \vartheta(\alpha,\beta)$, $\chi_3(\alpha,\beta) = \max \{\alpha^{1/2}\|x^*\|\, \chi_2(\alpha,\beta), \beta^{1/2}\|u^*\|\, \chi_1(\alpha,\beta)\}$, $\gamma = \max\{\|x^*\|, \|u^*\|\}$, $\vartheta(\alpha,\beta) = (\alpha\|x^*\|^2 + \beta\|u^*\|^2)^{1/2}$, and x^*, u^* are normal solutions of the problems $L_{\bar{\Delta}}$ and $L^*_{\bar{\Delta}}$, respectively, and $f^* = (c-\bar{q}, x^*)$.

The suggested approach allows to generate various iterative procedures for correcting improper LP problems. Let us consider the following algorithm, which is one of the possible realizations of the principle of iterative regularization for mathematical programming problems. Given the problems (0.4), suppose that it is improper of 1st type. Further we assume that, instead of the data A, b, c their approximations A^δ, b^δ, c^δ with some error $\max\{\|b^\delta - b\|, \|c^\delta - c\|, \|A^\delta - A\|\} \le \delta$ are known, where $\delta \ge 0$, $\|A = (a_{ij})\|^2 = \sum_{i,j=1}^{m,n} a_{ij}^2$. Let the function $L^\delta_\sigma(x,u)$ be defined in such a way that in the function $L_\sigma(x,u)$ the components A, b, c are replaced by A^δ, b^δ, c^δ. Further let $g^\delta_\sigma(x) = \min_{u \ge 0} L^\delta_\sigma(x,u)$.

Choose any $x^1 \in R^n_+$, and define

$$x^{k+1} = [x^k + \mu_k \nabla g^{\delta_k}_{\sigma_k}(x^k)]^+, \; k=1,2,\ldots; \; \sigma_k > 0, \mu_k > 0, \; \delta_k > 0. \quad (2.6)$$

Theorem 2.2. Suppose that the sequences $\{\sigma_k = [\alpha_k, \beta_k]\}$, $\{\delta_k\}$ and $\{\mu_k\}$, which are generated by the method (2.6), satisfy the conditions

$$\lim \alpha_k = \lim \frac{\delta_k}{\alpha_k \beta_k} = \lim \frac{\mu_k}{\alpha_k \beta_k^2} = 0, \quad \frac{\beta_{k+1}}{\alpha_{k+1}} \leq \frac{\beta_k}{\alpha_k}, \; \forall k,$$
$$\sum_{k=1}^{+\infty} \alpha_k \mu_k = \infty, \; \sum_{k=1}^{\infty} \frac{\mu_k^2}{\beta_k^2} < \infty, \; \sum_{k=1}^{\infty} \frac{\mu_k \delta_k}{\beta_k} < \infty, \; \sum_{k=1}^{\infty} (\alpha_k \beta_k)^{1/2} \mu_k < \infty \quad (2.7)$$

Then $\lim x^k = x^*$.

For example, sequences $\{\alpha_k\}$, $\{\beta_k\}$, $\{\delta_k\}$, $\{\mu_k\}$ which satisfy (2.7), are

$$\alpha_k = k^{-0,1}, \quad \beta_k = k^{-0,2}, \quad \delta_k = k^{-0,4}, \quad \mu_k = k^{-0,9}.$$

In the duality scheme introduced in 1., the problems C and C^* depend on the function of approximation of the initial mathematical programming problem. We find the same in the present situation. This will be illustrated for the problem (0.1) in the case that (0.1) is a CP problem being improper of 1-st type (i. e., it is supposed that $M = \{x \geq 0 \mid f_j(x) \leq 0 \;\; (j=1,\ldots,m)\} = \emptyset$, $M^* = \{u \geq 0 \mid \sup_{x \geq 0} \Phi(x,u) < +\infty\} \neq \emptyset$). Under these assumptions we consider a partial realization of the problem (1.1), namely

$$\sup \{f_o(x) - R_o \| F_1^+(x) \| \mid F_o(x) \leq 0, \; x \geq 0\}, \quad (2.8)$$

where $[f_1(x),\ldots,f_m(x)] = [F_o(x), F_1(x)]$, $R_o > 0$, and $\|\cdot\|$ is an arbitrary monotonic norm in the space of the dimension of the vector $F_1(x)$. Its dual problem in the sense of "#" has the form

$$\inf_{\substack{\|u^1\|^* \leq R_o \\ u=[u^o,u^1] \geq 0}} \; \sup_{x \geq 0} \; \Phi(x,u) =: \Phi, \quad (2.9)$$

where $\Phi(x,u) = f_o(x) - \sum_{j=1}^{m} u_j f_j(x)$.

In accordance with Theorem 1.4 the solvability of problem (2.8) and a regularity condition (i. e. $\exists \bar{x} \geq 0$: $F_o(\bar{x}) < 0$) imply that

$$\text{opt}\ (2.8) = \underline{\Phi} = \min_{\substack{\|u^1\|^* \le R_o \\ u \ge 0}} \ \min_{x \ge 0} \Phi(x,u) .$$

At this place, it is necessary to refer to the approximation character of problem (2.8) in relation to problem (0.1). This character will become clear by

Theorem 2.3. Suppose that (0.1) and (2.8) satisfy the conditions stated before. Then, for sufficiently large $R_o > 0$,

$$\text{opt}\ (2.8) = \max \{ f_o(x) \mid x \in \tilde{M} \} ,$$

where $\tilde{M} = \text{Arg} \min \{ \|F^+(x)\| \mid F_o(x) \le 0, \ x \ge 0 \}$.

3. Committee solutions of improper problems of optimization and diagnosis [1,4,5,6]

3.1. The notion of a p-committee

Many models of planning and control, of forecasting, of the classification of objects and situations can be described in the form of the following abstracts constraints:

$$x \in M_j \quad (j \in J), \quad x \in X \subset E , \qquad (3.1)$$

where E is any linear space.

In practice a system of the form (3.1) is often inconsistent. The reason for the inconsistency may be not only the inaccuracy of the information involved in the model, but also antagonisms being inherent to the situation modelled. Therefore, various kinds of generalized notions of a solution of the system (3.1) have to be used (cf. [1]), where the choice of a suitable notion depends on the subject of the model under consideration. In the present case the following concept is used: By a p-committee of the system (3.1) (with $0 \le p < 1$) we shall understand a finite subset K of X such that

$$|K \cap M_j| > p|K| \quad (\forall j \in J) ,$$

where $|K|$ denotes the number of elements of K.

3.2. Existence of p-committees.

Theorem 3.1. Consider system (3.1), and suppose $|J| < \infty$. If for any collection of s sets M_j the intersection of these s sets contains a point of X, and if $\frac{s}{|J|} > p$, then there exists a p-committee $K \subset X$ of the system (3.1).

Further results concern systems of inequalities in a real linear space X. Consider the system

$$f_j(x) < a_j \qquad (j=1,\ldots,m) \quad . \tag{3.2}$$

In the case of $p = 1/2$ a p-committee is simply called a committee. In other words, by a committee of the system (3.2) we understand a finite set K having the property that, for each inequality in (3.2), there is a subset of K containing more than $|K|/2$ elements which satisfy this inequality.

Theorem 3.2. Let the functions $f_1,\ldots,f_m$ be linear. Then the system (3.2) has a committee if and only if any two inequalities of the system are consistent.

Theorem 3.3. Let $f_j \in \{R^n \longrightarrow R^1\}$, $j=1,\ldots,m$, be not necessarily linear, but differentiable at the point q. Then there exists a committee of the system (3.2) for $X = R^n$, if $a_j \geq f_j(q)$ $(\forall j)$ and $\nabla f_j(q) \neq -\alpha \nabla f_i(q)$ $(\forall i, \forall j, \forall \alpha > 0)$, where $\nabla f_j(q)$ denotes the gradient of f_j at q.

Now we briefly discuss a problem of discriminant analysis. Let $A,B \subset X$ be finite subsets of a real linear space, and let F be any class of functions. The problem is to find a function $f \in F$ such that

$$f(x) > 0 \quad (x \in A), \qquad f(x) < 0 \quad (x \in B) .$$

A committee $K \subset F$ of this system is called a committee of functions of the class F, separating A and B.

Theorem 3.4. There exists a committee of affine functions separating A and B if and only if $A \cap B = \emptyset$.

3.3. Further results

There are known estimates of the number of members of a minimal committee and methods for constructing minimal committees for systems of convex piecewise-linear inequalities. Committees are applied to the solution of problems of discriminant analysis, to mathematical models of classification of objects, to the estimation of the evidence of symptoms, to the solution of optimization problems with inconsistent systems of constraints.

References

[1] Eremin, I. I., V. D. Mazurov and N. N. Astaf'ev: Improper problems of linear and convex programming. (In Russian). Nauka, Moscow, 1983.

[2] Eremin, I. I. and A. A. Vatolin: Duality for improper infinite-dimensional problems of linear and convex programming. (In Russian). In: Methods of approximation of improper mathematical programming problems. (In Russian). Uraln.Nauchn. Tsentr. Akad. Nauk SSSR, Sverdlovsk 1984.

[3] Vatolin, A. A.: Method of approximation of improper convex programming problems. (In Russian), In: Improper optimization problems. (In Russian). Uraln. Nauchn. Tsentr. Akad. Nauk SSSR, Sverdlovsk 1982.

[4] Mazurov, Vl. D.: On the construction of a committee of a system of convex inequalities. (In Russian). Kibernetika, 1967, No. 2, 56 - 59.

[5] Mazurov, Vl. D., A. I. Krivonogov, V. S. Kazantsev, N. O. Sachkov and N. G. Beletskii: Committees in decision-making. (In Russian). Kibernetika, 1984, No. 1, 90 - 95.

[6] Chernikov, S. N.: Contraction of finite systems of linear inequalities. (In Russian). Doklady Akad. Nauk SSSR, Ser. A, 1969, No. 1, 32 - 35.

CONTINUOUS SELECTIONS FOR SEMI-INFINITE OPTIMIZATION

Thomas Fischer*

Abstract:

The aim of this paper is to define a lower-semicontinuous submap of the optimal-set-mapping. This gives a characterization for the existence of a continuous selection. A consequence is the equivalence of two-lower-semicontinuity and the existence of a continuous selection. An application to best Chebyshev approximation theory is given.

1. Introduction

Let T be a compact Hausdorff space, N a positive integer, $p \in \mathbb{R}^N$ and $B \in C(T,\mathbb{R}^N)$. Here $C(T,\mathbb{R}^N)$ denotes the space of continuous, $\mathbb{R}^N$-valued functions on T endowed with the following norm

$$||B|| := \sup_{t \in T} ||B(t)||_2 \qquad , \; B \in C(T,\mathbb{R}^N),$$

where $||.||_2$ is the Euclidean norm on $\mathbb{R}^N$. For each element $b \in C(T) := C(T,\mathbb{R})$, we consider the following optimization problem:

$$\text{minimize} \quad \langle p,x \rangle$$

$$\text{subject to } x \in \mathbb{R}^N \text{ and}$$
$$\langle B(t),x \rangle \le b(t) \quad \text{for each } t \in T,$$

where $\langle .,. \rangle$ denotes the Euclidean product on $\mathbb{R}^N$.

Examples for this kind of semi-infinite optimization problem can be found e.g. in Brosowski and Fischer [3].

The feasible set is given by

$$Z_b := \{x \in \mathbb{R}^N | \; \langle B(t),x \rangle \le b(t) \text{ for each } t \in T\}$$

and the optimal set by

$$P_b := \{v \in Z_b | \; \langle p,v \rangle \le \langle p,x \rangle \text{ for each } x \in Z_b\}.$$

The solvability set is

*Johann-Wolfgang-Goethe-Universität, Fachbereich Mathematik, Robert-Mayer-Str. 6-10, 6000 Frankfurt 90, FRG.

$$\mathsf{L} := \mathsf{L}_{B,p} := \{b\in C(T) \mid P_b \neq \emptyset\}$$

and for a subset A in L, we investigate the optimal set mapping $P: A \longrightarrow POW(\mathbb{R}^N)$, $b \longmapsto P_b$. We ask for continuity properties of this mapping. We denote by POW(.) the power set, i.e. the collection of all subsets, the empty set included.

A *submap* P^* *for* P *on* A is a set-valued mapping $P^*: A \longrightarrow POW(\mathbb{R}^N)$ such that $P_b^* \subset P_b$ for each b in A. A *selection* S *for* P *on* A is a mapping $S: A \longrightarrow \mathbb{R}^N$ such that $S_b \in P_b$ for each b in A. It can be considered as a submap where the images are singletons.

In the following, the term continuity of a selection is used in the usual sense. Upper- resp. lower-semicontinuity (usc resp. lsc) is used in the sense of Hahn [6] and Michael [7]. This, the definition of an extremal subset and closedness of a set-valued mapping can be found in Brosowski [2].

In this paper we are concerned with the existence of a continuous selection for the optimal-set-mapping P. In two earlier works, lsc submaps have been considered. Firstly there is a theorem of Deutsch, Indumathi and Schnatz [4,Theorem 2.4], which states that P admits a continuous selection if and only if P admits a unique maximal lsc submap. (A lsc submap G_o is called maximal if G is a lsc submap and $G_o \subset G$ implies $G_o = G$.) Secondly there is a uniqueness result for continuous selections, in a special case, of Blatter and Schumaker [1]. Given a continuous selection, they characterized the largest lsc submap for P. (A lsc submap G_o is called the largest lsc submap if G is a lsc submap implies $G \subset G_o$.) Now the problem is, to give a definition of a subset of the optimal set, such that the corresponding set-valued mapping is the largest lsc submap of the optimal-set-mapping. We will carry out this similar to Blatter and Schumaker, but without assuming the existence of a continuous selection.

2. The Kernel of the Optimal-Set-Mapping

For an element b in L and $v \in Z_b$, we define the set of active constraints as

$$M_{b,v} := \{t\in T \mid \langle B(t),v\rangle = b(t)\}.$$

Given a nonempty set H in Z_b,

$$E_{b,H} := \bigcap_{v\in H} M_{b,v}$$

denotes the set of common active constraints of elements in H. Moreover lin H denotes the affine hull of H and relint H the relative interior of

H, i.e. the interior of H w.r.t. the affine hull of H.

We state the following lemma without proof. It is a slight modification of IV Lemma 24 in Brosowski [2].

Lemma 2.1: Let $b \in \mathcal{L}$, $H \subset Z_b$ and $v_o \in \operatorname{relint} H$. Then $M_{b,v_o} = E_{b,H}$.

For $b \in \mathcal{L}$ and a nonempty set $H \subset Z_b$, we set

$$\mathcal{N}_{b,H} := \bigcap_{v \in H} \{t \in T \mid \langle B(t), v - v_o \rangle = 0\} \quad , \ v_o \in H$$

and $L(H) := \operatorname{lin} H - v_o$, $v_o \in H$. Then both definitions are independent of the choice of $v_o \in H$.

Let $Q = Q_H$ be the orthogonal projection of $\mathbb{R}^N$ onto $L(H)$. For each element t in $T \setminus \mathcal{N}_{b,H}$, we have $QB(t) \neq \theta$, hence $\frac{QB(t)}{\|QB(t)\|}$ is well-defined. Here and in the sequel the index $_2$ has been omitted. Let

$$D_H := \{q \in \mathbb{R}^N \mid \text{there exists a net } (t_\lambda) \text{ in } T \setminus \mathcal{N}_{b,H} \text{ which converges to } t_o \text{ in } E_{b,H} \text{ and the net } \left(\frac{QB(t_\lambda)}{\|QB(t_\lambda)\|}\right) \text{ converges to } q\}.$$

For consistency, we set $D_\emptyset := \emptyset$.

Lemma 2.2: Let $b \in \mathcal{L}$ and $H \subset Z_b$. Then D_H is compact.

Proof: It suffices to show that D_H is closed. Let $(q_n) \subset D_H$ a sequence converging to $q_o \in \mathbb{R}^N$. For each $n \in \mathbb{N}$ resp. q_n, there exists a net $(t_{\lambda,n})$ in $T \setminus \mathcal{N}_{b,H}$, which converges to $\bar{t}_n \in E_{b,H}$ and the net

$$\left(\frac{QB(t_{\lambda,n})}{\|QB(t_{\lambda,n})\|}\right)$$

converges to q_n. T is compact, hence the sequence $(\bar{t}_n)$ has a clusterpoint $\bar{t}_o \in E_{b,H}$. For each open neighborhood U of $\bar{t}_o$ and each open neighborhood V of q_o there exists $n = n(U,V) \in \mathbb{N}$ such that $\bar{t}_n \in U$ and $q_n \in V$. Hence there exists $\lambda = \lambda(U,V)$ such that $t_{\lambda,n} \in U$ and

$$\frac{QB(t_{\lambda,n})}{\|QB(t_{\lambda,n})\|} \in V,$$

where $n = n(U,V)$. Then the net $(t_{\lambda,n})$, indexed by the open neighborhoods U and V, converges to $\bar{t}_o$ and the net

$$\left(\frac{QB(t_{\lambda,n})}{\|QB(t_{\lambda,n})\|}\right),$$

again indexed by U and V, converges to q_o. This shows $q_o \in D_H$. □

Let $(\mathbb{R}^N)^* := \cup\{(\mathbb{R}^N)^k \mid k \in \mathbb{N}_o\}$, where $\mathbb{N}_o := \mathbb{N} \cup \{0\}$ and $(\mathbb{R}^N)^o := \{\theta\}$. Each element $q_k \in (\mathbb{R}^N)^k$, $k \in \mathbb{N}$ has a unique representation $q_k = (q_{k-1}, q)$, where $q \in \mathbb{R}^N$ and $q_{k-1} \in (\mathbb{R}^N)^{k-1}$. We identify the elements q_k and (θ, q_k).

<u>Definition 2.3:</u> Let $b\in L$. We set $H(\theta):=P_b$. For $k\in\mathbb{N}$ and $q_k=(q_{k-1},q)\in(\mathbb{R}^N)^{k-1}\times\mathbb{R}^N$, we define inductively: if $q\in D_{H(q_{k-1})}$, then

$$H(q_k) := \{v_o\in H(q_{k-1}) \mid \langle q,v_o-v\rangle \le 0 \text{ for each } v\in H(q_{k-1})\},$$

otherwise $H(q_k):=H(q_{k-1})$. The *kernel* P_b^* *of* P_b is defined as follows:

$$P_b^* := \bigcap_{q\in(\mathbb{R}^N)^*} H(q).$$

We next show some basic facts about the kernel P_b^* of P_b, $b\in L$.

<u>Lemma 2.4:</u> Let $b\in L$ and $q\in(\mathbb{R}^N)^*$. Then $H(q)$ is a closed, convex and extremal subset of P_b. If P_b is compact, then $H(q)$ is nonempty.

Proof: We show by induction on k that the claim is true for each $q_k\in(\mathbb{R}^N)^k$, $k\in\mathbb{N}_o$. The case k=0 is clear, $P_b=H(\theta)$ has all these properties. Let $q_k=(q_{k-1},q)\in(\mathbb{R}^N)^k$, $k\ge1$, then, by induction hypothesis $H(q_{k-1})$ is a closed, convex, extremal subset of P_b and nonempty if P_b is compact. If $q\notin D_{H(q_{k-1})}$, then $H(q_k)=H(q_{k-1})$ and we are done. Otherwise $H(q_k)$ is the set of minimal points of the following minimization problem:

$$\text{minimize} \quad \langle q,x\rangle \qquad \text{subject to} \quad x\in H(q_{k-1}).$$

W.l.o.g. we may assume $H(q_k)\neq\emptyset$. Now $\langle q,.\rangle$ is linear and continuous. $H(q_{k-1})$ is closed and convex. Hence $H(q_k)$ can be written as the intersection of two closed and convex sets and is an extremal subset of $H(q_{k-1})$. By induction hypothesis $H(q_{k-1})$ is an extremal subset of P_b, hence $H(q_k)$ is an extremal subset of P_b. If P_b is compact, then $H(q_{k-1})\neq\emptyset$ is compact, hence $H(q_k)$ is nonempty. □

<u>Theorem 2.5:</u> Let $b\in L$. Then P_b^* is a closed, convex and extremal subset of P_b.

Proof: Follows from Lemma 2.4. □

Our next result, in section 4, shows that the kernel mapping $b \longmapsto P_b^*$ is lower-semicontinuous. With Michael's selection theorem [7] this yields a sufficient condition for the existence of a continuous selection. For the proof, we need some lemmas, which we collect in section 3.

3. Preliminaries

We define $(\tilde{\mathbb{R}}^N)^o:=\{\theta\}$, for $k\in\mathbb{N}$
$(\tilde{\mathbb{R}}^N)^k := \{q\in(\mathbb{R}^N)^k \mid q=(q_o,\dots,q_k),\ q_o=\theta \text{ and } q_i\in D_{H(q_o,\dots,q_{i-1})},\ i=1,\dots,k\}$
and $\tilde{\mathbb{R}}^N:=\cup\{(\tilde{\mathbb{R}}^N)^k \mid k\in\mathbb{N}_o\}$.

<u>Definition 3.1:</u> An element $q\in(\mathbb{R}^N)^*$ is called a *leaf* if $q\in\tilde{\mathbb{R}}^N$ and $D_{H(q)}=\emptyset$.

<u>Lemma 3.2:</u> Let $b\in\mathcal{L}$. Then

(i) for each $q\in(\tilde{\mathbb{R}}^N)^k, k\in\mathbb{N}$, we have $\dim H(q) \le N-k$ and $k\le N$,

(ii) for each $q\in(\mathbb{R}^N)^*$ there exists $q^*\in\tilde{\mathbb{R}}^N$ such that $H(q)=H(q^*)$,

(iii) for each $q\in(\mathbb{R}^N)^*$ there exists a leaf $q^*\in\tilde{\mathbb{R}}^N$ such that $H(q^*)\subset H(q)$,

(iv) for each $q\in(\mathbb{R}^N)^*$ and each leaf $q^*\in\tilde{\mathbb{R}}^N$, we have $H(q^*)\subset H(q)$ or $H(q^*)\cap H(q) = \emptyset$,

(v) if $P_b^*\neq\emptyset$, then $H(q)=P_b^*$ for each leaf $q\in\tilde{\mathbb{R}}^N$ and

(vi) for each $q\in(\mathbb{R}^N)^*$ and $\bar{v}\in H(q)$ there exists $q^*\in\tilde{\mathbb{R}}^N$ such that $\bar{v}\in H(q^*)$, $H(q^*)\subset H(q)$ and $\bar{v}\notin H(q^*,\bar{q})$ for each $\bar{q}\in D_{H(q^*)}$.

Proof: (i) We have $q=(q_o,q_1,\dots,q_k)$, where $q_o=\theta$ and $q_i\in D_{H(q_o,\dots,q_{i-1})}$, $i=1,\dots,k$. Since $\dim H(q_o,\dots,q_i) \le \dim H(q_o,\dots,q_{i-1}) - 1$, $i=1,\dots,k$ and $\dim H(q_o) \le N$, we have $\dim H(q) = \dim H(q_o,\dots,q_k) \le N-k$. If $\dim H(q) \ge 0$, then $k\le N$, otherwise $1 \le \dim H(q_o,\dots,q_{k-1}) \le N-(k-1) = N-k+1$, hence $k\le N$.

(ii) We have $q=(q_o,q_1,\dots,q_k)$, where $q_o=\theta$ and $q_i\in\mathbb{R}^N$, $i=1,\dots,k$, $k\in\mathbb{N}$. We obtain q^* by deleting those q_i, $1\le i\le k$ such that $q_i\notin D_{H(q_o,\dots,q_{i-1})}$. Then $q^*\in\tilde{\mathbb{R}}^N$ and $H(q)=H(q^*)$.

(iii) Suppose there exists $q=(q_o,\dots,q_k)$, $q_o=\theta$, $q_i\in\mathbb{R}^N$, $i=1,\dots,k$, $k\in\mathbb{N}$ and there does not exist a leaf $q^*\in\tilde{\mathbb{R}}^N$ such that $H(q^*)\subset H(q)$. Then, by (ii), we may assume that $q\in\tilde{\mathbb{R}}^N$. Since q is no leaf, there exists an element $q_{k+1}\in D_{H(q)}$, again $(q_o,\dots,q_{k+1})$ is no leaf and there is an element $q_{k+2}\in D_{H(q_o,\dots,q_{k+1})}$. In that way, we construct a sequence of sets $(H(q_o,\dots,q_i))$ with strictly decreasing dimension. Since $q_i\in D_{H(q_o,\dots,q_{i-1})}$, $i\in\mathbb{N}$, we know $H(q_o,\dots,q_{i-1})\neq\emptyset$, $i\in\mathbb{N}$, hence $\dim H(q_o,\dots,q_{i-1}) \ge 0$, $i\in\mathbb{N}$. This is a contradiction.

(iv) We show by induction on k, that for each $q\in(\mathbb{R}^N)^k$, $k\in\mathbb{N}_o$ and each leaf $q^*\in\tilde{\mathbb{R}}^N$, we have $H(q^*)\subset H(q)$ or $H(q^*)\cap H(q)=\emptyset$. This is true for $k=0$, since $H(\theta)=P_b$. Let $q\in(\mathbb{R}^N)^k$, $k\ge1$ and a leaf $q^*\in\tilde{\mathbb{R}}^N$ be given. Assume $H(q^*)\cap H(q)\neq\emptyset$. By (ii), we may assume that $q\in(\tilde{\mathbb{R}}^N)^k$. Since the intersection is nonempty there exists $v_1\in H(q^*)\cap H(q)$ and $q=(q_o,\dots,q_k)$, $q_o=\theta$, $q_i\in D_{H(q_o,\dots,q_{i-1})}$, $i=1,\dots,k$. Since $H(q)\cap H(q^*)\neq\emptyset$, we have $H(q_o,\dots,q_{k-1})\cap H(q^*)\neq\emptyset$ and by induction hypothesis $H(q^*)\subset H(q_o,\dots,q_{k-1})=:H$. We know that $q_k\in D_H$, hence there exists a net $(t_\lambda)\subset T\setminus N_{b,H}$, which converges to $t_o\in E_{b,H}$ and

$$\left(\frac{QB(t_\lambda)}{||QB(t_\lambda)||}\right)$$

converges to q_k. Here $Q=Q_H$ denotes the orthogonal projection onto $L(H)$. Since $H(q^*)\subset H$, we have $t_o\in E_{b,H(q^*)}$ and

$$\left(\frac{Q_{H(q^*)}B(t_\lambda)}{||Q_HB(t_\lambda)||}\right) = \left(Q_{H(q^*)}\frac{Q_HB(t_\lambda)}{||Q_HB(t_\lambda)||}\right) \quad \text{converges to } Q_{H(q^*)}q_k.$$

Suppose there exists $v_o\in H(q^*)$ such that $\langle q_k,v_o-v_1\rangle\neq0$. Then $Q_{H(q^*)}q_k\neq\theta$, hence there exists a subnet of (t_λ), again denoted by (t_λ), such that $Q_{H(q^*)}B(t_\lambda)\neq\theta$ and

$$\left(\frac{Q_{H(q^*)}B(t_\lambda)}{||Q_{H(q^*)}B(t_\lambda)||}\right) = \left(\frac{\frac{Q_{H(q^*)}B(t_\lambda)}{||Q_H B(t_\lambda)||}}{\left|\left|\frac{Q_{H(q^*)}B(t_\lambda)}{||Q_H B(t_\lambda)||}\right|\right|}\right) \quad \text{converges to}$$

$\frac{Q_{H(q^*)}q_k}{||Q_{H(q^*)}q_k||}$. Since $Q_{H(q^*)}B(t_\lambda)\neq\theta$, we have $(t_\lambda)\subset T\setminus N_{b,H(q^*)}$. That is there exists a net $(t_\lambda)\subset T\setminus N_{b,H(q^*)}$, converging to $t_o\in E_{b,H(q^*)}$, and

$$\left(\frac{Q_{H(q^*)}B(t_\lambda)}{||Q_{H(q^*)}B(t_\lambda)||}\right) \quad \text{converges to } \bar{q} := \frac{Q_{H(q^*)}q_k}{||Q_{H(q^*)}q_k||}.$$

Then $\bar{q}\in D_{H(q^*)}$. But q^* is a leaf and this is a contradiction. Hence $<q_k,v_o-v_1>=0$ for each $v_o\in H(q^*)$ and $<q_k,v_o-v>=<q_k,v_o-v_1>+<q_k,v_1-v>\leq 0$ for each $v\in H$ and $v_o\in H(q^*)$. This shows $H(q^*)\subset H(q)=H(q_o,\dots,q_k)$ and the induction is complete.
(v) If $P_b^*\neq\emptyset$, then for each $q\in(\mathbb{R}^N)^*$ and each leaf $q^*\in\mathbb{R}^N$, we have $H(q)\cap H(q^*)\neq\emptyset$. Hence by (iv), $H(q^*)\subset H(q)$ and $H(q^*)\subset P_b^*$. The converse $P_b^*\subset H(q^*)$ is obvious.
(vi) Let $q\in(\mathbb{R}^N)^*$ and $\bar{v}\in H(q)$. By (ii), we may assume $q\in\mathbb{R}^N$. Suppose the claim is not true, i.e. there does not exist $q^*\in\mathbb{R}^N$ such that $\bar{v}\in H(q^*)$, $H(q^*)\subset H(q)$ and $\bar{v}\notin H(q^*,\bar{q})$ for each $\bar{q}\in D_{H(q^*)}$. Then $q=(q_o,q_1,\dots,q_k)$, $k\in\mathbb{N}$, $q_o=\theta$, $q_i\in D_{H(q_o,\dots,q_{i-1})}$, $i=1,\dots,k$ and in particular there is $q_{k+1}\in D_{H(q_o,\dots,q_k)}$ such that $\bar{v}\in H(q_o,\dots,q_{k+1})$. Again by our supposition, there exists $q_{k+2}\in D_{H(q_o,\dots,q_{k+1})}$ such that $\bar{v}\in H(q_o,\dots,q_{k+2})$. In the same way as in (iii), we obtain a sequence of sets $(H(q_o,\dots,q_i))$ with strictly decreasing dimension. Since $q_i\in D_{H(q_o,\dots,q_{i-1})}$, $i\in\mathbb{N}$, we know $H(q_o,\dots,q_{i-1})\neq\emptyset$, $i\in\mathbb{N}$, hence $\dim H(q_o,\dots,q_i)\geq 0$, $i\in\mathbb{N}$. This is a contradiction. □

It follows from Lemma 3.2 (iii), that $(\mathbb{R}^N)^*$ contains always at least one leaf and that in the definition of the kernel it is enough to take the intersection over all leaves of $(\mathbb{R}^N)^*$ instead of all elements of $(\mathbb{R}^N)^*$.

<u>Theorem 3.3:</u> Let $b\in L$ and $P_b^*\neq\emptyset$. Then there is a neighborhood W of E_{b,P_b^*} such that $W\subset N_{b,P_b^*}$.
Proof: By Lemma 3.2 (iii) and (v), there exists a leaf $q\in\mathbb{R}^N$ such that $H(q)=P_b^*$. Suppose for each neighborhood W of $E_{b,P_b^*}=E_{b,H(q)}$ there exists $t_W\in W\setminus N_{b,P_b}$. Then $t_W\notin N_{b,H(q)}$ and $QB(t_W)\neq\theta$, where Q is the orthogonal projection onto $L(H(q))$. Hence

$$\frac{QB(t_W)}{||QB(t_W)||}$$

is well-defined. T is compact, hence the net (t_W), where W runs over all neighborhoods of $E_{b,H(q)}$, has a convergent subnet. We denote it by (t_W)

again, it converges to $t_o \in E_{b,H(q)}$. W.l.o.g.

$$\left(\frac{QB(t_W)}{\|QB(t_W)\|}\right)$$

converges to a point $\bar{q} \in R^N$. Then $\bar{q} \in D_{H(q)}$. But q is a leaf, hence $D_{H(q)} = \emptyset$. This is a contradiction. □

<u>Lemma 3.4:</u> Let $b \in L$, $P_b^* \neq \emptyset$, $q \in (\mathbb{R}^N)^*$ and $\bar{v} \in H(q)$. Then there is a neighborhood W of $E_{b,H(q)}$ such that

$$\langle B(t), v^* - \bar{v} \rangle \leq 0$$

for each $t \in W$ and $v^* \in P_b^*$.

Proof: By Lemma 3.2 (vi), there exists $q^* \in \tilde{R}^N$ such that $\bar{v} \in H(q^*)$, $H(q^*) \subset H(q)$ and $\bar{v} \notin H(q^*, \bar{q})$ for each $\bar{q} \in D_{H(q^*)}$. Suppose for each neighborhood W of $E_{b,H(q^*)}$ there exists $t_W \in W$ and $v_W^* \in P_b^*$ such that

$$\langle B(t_W), v_W^* - \bar{v} \rangle > 0.$$

Using Theorem 3.3, we can choose $v^* \in P_b^*$ such that

$$\langle B(t_W), v^* - \bar{v} \rangle > 0$$

for each t_W. Then there is a subnet of (t_W), again denoted by (t_W), which converges to a point $t_o \in E_{b,H(q^*)}$ and

$$\left(\frac{QB(t_W)}{\|QB(t_W)\|}\right)$$

converges to $\bar{q}$. Here Q denotes the orthogonal projection onto $L(H(q^*))$. This shows $\bar{q} \in D_{H(q^*)}$. We have $\langle \bar{q}, v^* - \bar{v} \rangle \geq 0$, since $v^* - \bar{v}$ $L(H(q^*))$, hence $\langle \bar{q}, v^* - \bar{v} \rangle = 0$ and $\bar{v} \in H(q^*, \bar{q})$. This is a contradiction, hence there exists a neighborhood W of $E_{b,H(q^*)}$ such that

$$\langle B(t), v^* - \bar{v} \rangle \leq 0$$

for each $t \in W$ and $v^* \in P_b^*$. The claim follows easily from the inclusion $E_{b,H(q)} \subset E_{b,H(q^*)}$. □

<u>Lemma 3.5:</u> Let $b \in L$, $P_b^* \neq \emptyset$ and $\bar{v} \in P_b$. Then there exists $v_o \in P_b$ and a neighborhood W of M_{b,v_o} such that

$$\langle B(t), v_o - \bar{v} \rangle \leq 0$$

and

$$\langle B(t), v^* - \bar{v} \rangle \leq 0$$

for each $t \in W$ and $v^* \in P_b^*$.

Proof: By Lemma 3.2 (vi), there exists $q^*\in\tilde{\mathbb{R}}^N$ such that $\bar{v}\in H(q^*)$ and $\bar{v}\notin H(q^*,q)$ for each $q\in D_{H(q^*)}$. We define

$$H' := \{v_o\in H(q^*) \mid \langle q,v_o-\bar{v}\rangle \le 0 \quad \text{for each } q\in D_{H(q^*)}\}.$$

Since $P_b^*\subset H'$ and $P_b^*\neq\emptyset$, we have $H'\neq\emptyset$.
Claim: $\dim H' = \dim H(q^*)$.
Choose $v_1\in P_b^*\subset H'\subset H(q^*)$ and $v\in\mathbb{R}^N$ with $v_1+v\in H(q^*)$. For each $q\in D_{H(q^*)}$, we have $\bar{v}\notin H(q^*,q)$, hence

$$\langle q,v_1-\bar{v}\rangle < 0.$$

Then there exists $0<\varepsilon_q<1$ such that

$$\langle q,v_1+\varepsilon_q v-\bar{v}\rangle \le \langle q,v_1-\bar{v}\rangle + \varepsilon_q|\langle q,v\rangle| < -\varepsilon_q.$$

There exists an open neighborhood U_q of q such that

$$\langle\tilde{q},v_1-\bar{v}\rangle + \varepsilon_q|\langle\tilde{q},v\rangle| < 0$$

for each $\tilde{q}\in U_q$. The collection of open sets $\{U_q|q\in D_{H(q^*)}\}$ covers the compact set $D_{H(q^*)}$, hence there exist points $q_1,\dots,q_l$ in $D_{H(q^*)}$, $l\in\mathbb{N}$, such that $D_{H(q^*)}\subset\cup\{U_{q_i}\,|i=1,\dots,l\}$. We define $\varepsilon:=\min\{\varepsilon_{q_i}\,|i=1,\dots,l\}$. For each $q\in D_{H(q^*)}$ there exists $1\le i\le l$ with $q\in U_{q_i}$ and

$$\begin{aligned}\langle q,v_1+\varepsilon v-\bar{v}\rangle &\le \langle q,v_1-\bar{v}\rangle + \varepsilon|\langle q,v\rangle|\\ &\le \langle q,v_1-\bar{v}\rangle + \varepsilon_{q_i}|\langle q,v\rangle|\\ &< 0.\end{aligned}$$

We obtain $v_1+\varepsilon v\in H'$. Since $H(q^*)$ is finitedimensional and $H'\subset H(q^*)$, we have $\dim H' = \dim H(q^*)$. This proves the claim.

H' is a nonempty, convex and finitedimensional set, hence relint $H'\neq\emptyset$. Choose $v_o\in$relint H', then $M_{b,v_o}=E_{b,H'}=E_{b,H(q^*)}$, since relint $H'\subset$ relint $H(q^*)$. For each $q\in D_{H(q^*)}$, we have $q\in L(H(q^*))=L(H')$, hence there exists $\varepsilon>0$ such that $v_o+\varepsilon q\in H'$ and

$$\langle q,v_o+\varepsilon q-\bar{v}\rangle \le 0$$

resp. $$\langle q,v_o-\bar{v}\rangle \le -\varepsilon||q||^2 < 0.$$

Suppose for each neighborhood W of $E_{b,H(q^*)}$, there exists $t_W\in W$ such that

$$\langle B(t_W),v_o-\bar{v}\rangle > 0.$$

Then the net (t_W), which is indexed by the neighborhoods W of $E_{b,H(q^*)}$, contains a subnet, again denoted by (t_W), which converges to a point t_o

in $E_{b,H(q^*)}$. We have $(t_W)\subset T\setminus N_{b,H(q^*)}$, since $v_o-\bar{v}\in L(H(q^*))$. We may assume w.l.o.g. that

$$\left(\frac{QB(t_W)}{||QB(t_W)||}\right)$$

converges to an element $\bar{q}\in\mathbb{R}^N$, hence $\bar{q}\in D_{H(q^*)}$. Here Q denotes the orthogonal projection onto $L(H(q^*))$. We have $<\bar{q},v_o-\bar{v}>\geq 0$ and this is a contradiction. Hence there exists a neighborhood W of $M_{b,v_o}=E_{b,H(q^*)}$ such that

$$<B(t),v_o-\bar{v}> \leq 0$$

for each $t\in W$. By Lemma 3.4, we may assume that

$$<B(t),v^*-\bar{v}> \leq 0$$

for each $t\in W$ and $v^*\in P_b^*$. □

4. Main Results

Definition 4.1: Let $b\in A\subset L$. A mapping $P^*:A \longrightarrow POW(\mathbb{R}^N)$ admits *locally a bounded selection at* $b\in A$, if there exists a neighborhood U of b, a mapping $\pi:U \longrightarrow \mathbb{R}^N$ and a constant $C\geq 0$ such that $\pi(b')\in P_{b'}^*$, and $||\pi(b')||\leq C$ for each $b'\in U$.

Theorem 4.2: Let $b\in A\subset L$, P be closed at b and P^* admit locally a bounded selection at b. Then the kernel mapping $P^*:A \longrightarrow POW(\mathbb{R}^N)$ is lsc at b.
Proof: Suppose P^* is not lsc at b. Then there exists a sequence $(b_n)\subset A$, converging to b, $v^*\in\mathbb{R}^N$ and an open set $U\subset\mathbb{R}^N$ such that $v^*\in U\cap P_b^*\neq\emptyset$ and $U\cap P_{b_n}^*=\emptyset$ for each $n\in\mathbb{N}$. Since P^* admits locally a bounded selection at b, there exists a subsequence of (b_n), again denoted by (b_n), and points $v_n\in P_{b_n}^*$, $n\in\mathbb{N}$, such that (v_n) converges to $\bar{v}\in\mathbb{R}^N$. P is closed at b, hence $\bar{v}\in P_b$. By Lemma 3.5 there exists a point $v_o\in P_b$ and a neighborhood W of M_{b,v_o} such that

$$<B(t),v_o-\bar{v}> \leq 0$$

and

$$<B(t),v^*-\bar{v}> \leq 0$$

for each $t\in W$. We choose $0<\lambda<1$ so large that $v_\lambda:=\lambda v^*+(1-\lambda)v_o\in U$, then we have $M_{b,v_\lambda}\subset M_{b,v_o}$. W is a neighborhood of M_{b,v_λ}, hence there exists $\varepsilon>0$ such that

$$<B(t),v_\lambda> \leq b(t) - \varepsilon$$

for each $t\in T\setminus W$. We define for $n\in\mathbb{N}$

$$\tilde{v}_n := v_n+v_\lambda-\bar{v},$$

and $(\tilde{v}_n)$ converges to v_λ. Choose $n_o \in \mathbb{N}$ so large that

$$||\langle B(.),\tilde{v}_n - v_\lambda\rangle|| < \frac{\varepsilon}{2} \quad \text{and} \quad ||b-b_n|| < \frac{\varepsilon}{2} \quad \text{for } n \geq n_o.$$

We obtain for $n \geq n_o$ and $t \in T\backslash W$

$$\begin{aligned}\langle B(t),\tilde{v}_n\rangle &< \langle B(t),v_\lambda\rangle + \frac{\varepsilon}{2}\\ &\leq b(t) - \varepsilon + \frac{\varepsilon}{2}\\ &= b(t) - \frac{\varepsilon}{2}\\ &< b_n(t)\end{aligned}$$

and for $t \in W$

$$\begin{aligned}\langle B(t),\tilde{v}_n\rangle &= \langle B(t),v_n\rangle + \langle B(t),v_\lambda - \overline{v}\rangle\\ &\leq b_n(t) + \lambda\langle B(t),v^*\rangle + (1-\lambda)\langle B(t),v_o\rangle - \langle B(t),\overline{v}\rangle\\ &= b_n(t) + \lambda\langle B(t),v^* - \overline{v}\rangle + (1-\lambda)\langle B(t),v_o - \overline{v}\rangle\\ &\leq b_n(t),\end{aligned}$$

hence $\tilde{v}_n \in Z_{b_n}$, $n \geq n_o$. Moreover, we have

$$\begin{aligned}\langle p,\tilde{v}_n\rangle &= \langle p,v_n\rangle + \langle p,v_\lambda\rangle - \langle p,\overline{v}\rangle\\ &= \langle p,v_n\rangle \quad , n \in \mathbb{N},\end{aligned}$$

hence $\tilde{v}_n \in P_{b_n}$, $n \geq n_o$.
Claim: $\tilde{v}_n \in P^*_{b_n}$, $n \geq n_o$.
We fix $n \geq n_o$ and show by induction on k that $\tilde{v}_n \in H_n(q_k)$ for each $q_k \in (\mathbb{R}^N)^k$, $k \in \mathbb{N}_o$. Here H_n denotes the set H in the definition of the kernel which belongs to the parameter b_n. The case k=0 is clear since $H_n(\Theta) = P_{b_n}$. Let $q_k \in (\mathbb{R}^N)^k$, $k \geq 1$, then $q_k = (q_{k-1},q) \in (\mathbb{R}^N)^{k-1} \times \mathbb{R}^N$. By induction hypothesis $\tilde{v}_n \in H_n(q_{k-1})$. If $q \notin D_{H_n(q_{k-1})}$, then there is nothing to prove. Otherwise

$$\langle B(t),\tilde{v}_n\rangle < b_n(t)$$

for each $t \in T\backslash W$, hence $E_{b_n,H_n(q_{k-1})} \subset M_{b_n,\tilde{v}_n} \subset W$ and

$$\langle B(t),v_\lambda - \overline{v}\rangle \leq 0$$

for each $t \in W$. Since $v_\lambda - \overline{v} = \tilde{v}_n - v_n \in L(H_n(q_{k-1}))$, we have

$$\langle q,v_n - v_n\rangle = \langle q,v_\lambda - \overline{v}\rangle \leq 0.$$

Since $v_n \in P^*_{b_n} \subset H_n(q_k)$, we have $\tilde{v}_n \in H_n(q_k)$. This completes the induction and shows the claim.
We found a sequence $(\tilde{v}_n)$ converging to $v_\lambda \in U$ such that $\tilde{v}_n \in P^*_{b_n}$, $n \geq n_o$ and $U \cap P^*_{b_n} = \emptyset$, $n \in \mathbb{N}$. This is a contradiction. □

Note that the kernel mapping is not upper-semicontinuous, in general. The assumption that P^* admits locally a bounded selection is satisfied e.g. if P_b is nonempty in a neighborhood of b and P is upper-semicontinuous and compact-valued. In the latter case each local selection of P is bounded.

Definition 4.3: A subset $A \subset \mathsf{L}$ has *property (I)*, if for each $b \in A$, $\bar{v} \in P_b$ and $\varepsilon > 0$, there exist $b_\varepsilon \in A$ and a neighborhood W of $M_{b,\bar{v}}$ such that (i) $\|b - b_\varepsilon\| < \varepsilon$, (ii) $\langle B(t), \bar{v}\rangle = b_\varepsilon(t)$ for each $t \in W$ and (iii) $\langle B(t), \bar{v}\rangle \leq b_\varepsilon(t)$ for each $t \in T$.

The next theorem provides an example for such a subset.

Theorem 4.4: The solvability set L has property (I).

Proof: Let $b \in \mathsf{L}$, $\bar{v} \in P_b$ and $\varepsilon > 0$. Then

$$V_\varepsilon := \{t \in T \mid |\langle B(t), \bar{v}\rangle - b(t)| < \varepsilon\}$$

is an open neighborhood of $M_{b,\bar{v}}$. Since T is normal, there exists a closed neighborhood W of $M_{b,\bar{v}}$ with $W \subset V_\varepsilon$. By Urysohn's lemma there exists a continuous function $\tau : T \longrightarrow [0,1]$ such that $\tau(t) = 1$ for $t \in W$ and $\tau(t) = 0$ for $t \in T \setminus V_\varepsilon$. We define for $t \in T$

$$b_\varepsilon(t) := (1 - \tau(t)) b(t) + \tau(t) \langle B(t), \bar{v}\rangle$$

and we have for each $t \in T$

$$|b_\varepsilon(t) - b(t)| = \tau(t) |\langle B(t), \bar{v}\rangle - b(t)|.$$

The last term is less than ε for $t \in V_\varepsilon$ and is equal to 0 for $t \notin V_\varepsilon$, hence $\|b - b_\varepsilon\| < \varepsilon$. For each $t \in W$, we have $\tau(t) = 1$, hence $\langle B(t), \bar{v}\rangle = b_\varepsilon(t)$ and obviously $\langle B(t), \bar{v}\rangle \leq b_\varepsilon(t)$ for each $t \in T$. □

Theorem 4.5: Let $b \in \mathsf{L}$, $b_\varepsilon \in C(T)$, $\bar{v} \in Z_b$ and W be a neighborhood of E_{b,P_b} such that $W \subset M_{b_\varepsilon,\bar{v}}$. Then $b_\varepsilon \in \mathsf{L}$ and $\bar{v} \in P_{b_\varepsilon}^{\varepsilon}$.

Proof: By Brosowski and Fischer [3, Theorem 4.6], we have for each neighborhood W of E_{b,P_b}

$$-p \in \overline{\mathrm{CON}}\{B(t) \in \mathbb{R}^N \mid t \in W\},$$

where CON{...} denotes the convex cone generated by {...}. From our assumption, it follows

$$-p \in \overline{\mathrm{CON}}\{B(t) \in \mathbb{R}^N \mid t \in M_{b_\varepsilon,\bar{v}}\}.$$

There exists a sequence (p_n), converging to -p, with the following representation

$$p_n = \sum_{i=1}^{k_n} \rho_{i,n} B(t_{i,n}),$$

where $\rho_{i,n} \geq 0$ and $t_{i,n} \in M_{b_\varepsilon,\bar{v}}$, $i=1,\dots,k_n, n \in \mathbb{N}$. We obtain for each $v \in Z_{b_\varepsilon}$

$$\begin{aligned}
<p,\bar{v}> &= <-\lim p_n,\bar{v}> \\
&= -\lim <p_n,\bar{v}> \\
&= -\lim < \sum_{i=1}^{k_n} \rho_{i,n} B(t_{i,n}),\bar{v}> \\
&= -\lim \sum_{i=1}^{k_n} \rho_{i,n} <B(t_{i,n}),\bar{v}> \\
&= -\lim \sum_{i=1}^{k_n} \rho_{i,n} b_\varepsilon(t_{i,n}) \\
&\leq -\lim \sum_{i=1}^{k_n} \rho_{i,n} <B(t_{i,n}),v> \\
&= <-\lim \sum_{i=1}^{k_n} \rho_{i,n} B(t_{i,n}),v> \\
&= <p,v>
\end{aligned}$$

and since $\bar{v} \in Z_{b_\varepsilon}$, we have $\bar{v} \in P_{b_\varepsilon}$. □

<u>Theorem 4.6:</u> Let $A \subset \mathbb{L}$ with property (I). Then for each $b \in A$, $\bar{v} \in P_b$ and $\varepsilon > 0$ there exist $b_\varepsilon \in A$ and a neighborhood W of $M_{b,\bar{v}}$ such that (i) $||b-b_\varepsilon|| < \varepsilon$, (ii) $W \subset M_{b_\varepsilon,\bar{v}}$ and (iii) $\bar{v} \in P_{b_\varepsilon}$.

Proof: Follows immediately from Theorem 4.5. □

<u>Definition 4.7:</u> Let $(P_n) \subset POW(\mathbb{R}^N)$ a sequence of sets. The *upper* or *outer limit* is defined as

$$\overline{\lim}\, P_n := \{y \in \mathbb{R}^N \mid \text{there exists a sequence } (v_n) \subset \mathbb{R}^N \text{ with clusterpoint } y \text{ and } v_n \in P_n,\ n \in \mathbb{N}\}.$$

<u>Lemma 4.8:</u> Let $A \subset \mathbb{L}$ with property (I), $b \in A$, P be closed at b and $q \in (\mathbb{R}^N)^*$ such that $H(q) \neq \emptyset$. Then there exists a sequence $(b_n) \subset A$, converging to b and $\overline{\lim}\, P_{b_n} \subset H(q)$.

Proof: Since $H(q) \neq \emptyset$, there exists $\bar{v} \in H(q) \subset P_b$. By property (I) of A, there exist for $b \in A$, $\bar{v} \in P_b$ and $\varepsilon := \frac{1}{n} > 0$ a $b_n \in A$ and a neighborhood W_n of $M_{b,\bar{v}}$ such that (i) $||b-b_n|| < \frac{1}{n}$, (ii) $W_n \subset M_{b_n,\bar{v}}$ and (iii) $\bar{v} \in P_{b_n}$. Then $(b_n) \subset A$ converges to b.

We fix $n \in \mathbb{N}$ and let $v \in P_{b_n}$. We have for each $t \in W_n$

$$<B(t),v-\bar{v}> = <B(t),v> - <B(t),\bar{v}> \leq b_n(t) - b_n(t) = 0.$$

W_n is a neighborhood of $E_{b,P_b} \subset M_{b,\bar{v}}$, hence by Brosowski and Fischer [3, Lemma 4.5] there exists $\lambda>0$ such that $\bar{v}+\lambda(v-\bar{v})\in Z_b$, hence $\bar{v}+\lambda(v-\bar{v})\in P_b$.

W.l.o.g. we may assume $q\in\tilde{\mathbb{R}}^N$ and $q=(q_0,\ldots,q_k)$, $k\in\mathbb{N}_0$, $q_0=\theta$, $q_i\in D_{H(q_0,\ldots,q_{i-1})}$, $i=1,\ldots,k$. We show by induction on i that $\bar{v}+\lambda(v-\bar{v})\in H(q_0,\ldots,q_i)$, $i=0,\ldots,k$. This is true for i=0, since $\bar{v}+\lambda(v-\bar{v})\in P_b=H(q_0)$. By induction hypothesis $\bar{v}+\lambda(v-\bar{v})\in H(q_0,\ldots,q_{i-1})$, $1\le i\le k$. Since $\bar{v}\in H(q_0,\ldots,q_{i-1})$, we have $E_{b,H(q_0,\ldots,q_{i-1})}\subset M_{b,\bar{v}}$, hence W_n is a neighborhood of $E_{b,H(q_0,\ldots,q_{i-1})}$. For each $t\in W_n$, we have $\langle B(t),v-\bar{v}\rangle\le 0$ and $v-\bar{v}\in L(H(q_0,\ldots,q_{i-1}))$, hence

$$\langle q_i,\bar{v}+\lambda(v-\bar{v})-\bar{v}\rangle = \lambda\langle q_i,v-\bar{v}\rangle \le 0$$

and $\bar{v}+\lambda(v-\bar{v})\in H(q_0,\ldots,q_i)$. This completes the induction, hence $\bar{v}+\lambda(v-\bar{v})\in H(q)$. It follows $v\in \mathrm{lin}\, H(q)$ and since P_{b_n} is finitedimensional, we have $P_{b_n}\subset \mathrm{lin}\, H(q)$, hence $\overline{\lim}\, P_{b_n} \subset \mathrm{lin}\, H(q)$.

P is closed at b, hence $\overline{\lim}\, P_{b_n} \subset P_b$. By Lemma 2.4, H(q) is an extremal subset of P_b, hence the following formula holds:

$$H(q) = \mathrm{lin}\, H(q) \cap P_b$$

and this completes the proof. □

<u>Theorem 4.9:</u> Let $A\subset\mathbb{L}$ with property (I), $b\in A$, P_b be compact and S be a continuous selection for P on A. Then $S_b\in P_b^*$ and in particular $P_b^*\neq\emptyset$.

Proof: Let $q\in(\mathbb{R}^N)^*$ arbitrary. P_b is compact, hence $H(q)\neq\emptyset$ by Lemma 2.4. By Lemma 4.8 there exists a sequence $(b_n)\subset A$, converging to b, such that $\overline{\lim}\, P_{b_n} \subset H(q)$. Since $S_{b_n}\in P_{b_n}$, $n\in\mathbb{N}$, we have $S_b\in H(q)$ and since $q\in(\mathbb{R}^N)^*$ was arbitrary, we have $S_b\in P_b^*$ and $P_b^*\neq\emptyset$. □

The Theorems 4.9 and 3.3 show that the kernel P_b^* is a generalization of the lower-semicontinuous kernel of Blatter and Schumaker [1].

<u>Theorem 4.10:</u> Let $A\subset\mathbb{L}$ with property (I), P be closed on A and P_b be compact for each $b\in A$. Then P admits a continuous selection on A if and only if $P_b^*\neq\emptyset$ for each $b\in A$.

Proof: If P admits a continuous selection, then $P_b^*\neq\emptyset$ for each $b\in A$ by Theorem 4.9. If P is closed at b and P_b is compact, then, by Brosowski and Fischer [3, Theorem 6.6], P is usc at b, hence each selection for P is locally bounded. Assume $P_b^*\neq\emptyset$ for each $b\in A$, then, by Theorem 4.2, the kernel mapping is lsc at each $b\in A$. By Michael's selection theorem [7, Theorem 1], P^*, hence P, admits a continuous selection on A. □

<u>Theorem 4.11:</u> Let $A\subset\mathbb{L}$ with property (I), P_b be compact for each $b\in A$ and P admit a continuous selection on A. Then we have for each $b\in A$

$$P_b^* = \{S_b \in P_b \mid S \text{ is a continuous selection for } P \text{ on } A\} =: \tilde{P}_b$$

and P^* is the largest lsc submap for P on A.

Proof: Obviously $\tilde{P}:A \longrightarrow POW(\mathbb{R}^N)$ is a lsc mapping and using Michael's selection theorem [7, Theorem 1], we see that $\tilde{P}$ is the largest lsc submap for P. By Theorem 4.2, P^* is lsc, hence $P_b^* \subset \tilde{P}_b$ for each $b \in A$. By Theorem 4.9, we have $\tilde{P}_b \subset P_b^*$ for each $b \in A$. □

5. Two-Lower-Semicontinuity

Lemma 5.1: Let $A \subset \mathbb{L}$ with property (I), $b \in A$, P be closed at b, P_b be compact and $P_b^* = \emptyset$. Then there exist disjoint, compact sets G and G^* in P_b and two sequences (b_n) resp. (b_n^*) in A, converging to b, with $\overline{\lim}\, P_{b_n} \subset G$ resp. $\overline{\lim}\, P_{b_n^*} \subset G^*$.

Proof: Choose a leaf $q^* \in \tilde{\mathbb{R}}^N$ arbitrary. By Lemma 2.4, $H(q^*) \neq \emptyset$ and by Lemma 3.2 (iv), there exists $q \in (\mathbb{R}^N)^*$ such that $H(q) \cap H(q^*) = \emptyset$. We define $G := H(q)$ and $G^* := H(q^*)$, then G and G^* are disjoint and compact subsets of P_b. By Lemma 4.8, there exist sequences (b_n) resp. (b_n^*) in A, both converging to b, such that $\overline{\lim}\, P_{b_n} \subset G$ resp. $\overline{\lim}\, P_{b_n^*} \subset G^*$. □

The ε-neighborhood of a nonempty set $P \subset \mathbb{R}^N$ is given by

$$B_\varepsilon(P) := \{x \in \mathbb{R}^N \mid d(x,P) < \varepsilon\},$$

where

$$d(x,P) := \inf\{\,\|x-v\| \mid v \in P\}.$$

Definition 5.2: Let $A \subset \mathbb{L}$. A mapping $P:A \longrightarrow POW(\mathbb{R}^N)$ is called *two-lower-semicontinuous* (two-lsc) *at* $b \in A$, if for each $\varepsilon > 0$ there exists a neighborhood U of b where

$$B_\varepsilon(P_{b_1}) \cap B_\varepsilon(P_{b_2}) \neq \emptyset$$

for each choice of points b_1 and b_2 in U. P is called *two-lower-semicontinuous on* A, if it is two-lsc at each $b \in A$.

The definition of two-lsc has been introduced by Deutsch and Kenderov [5]. They showed that, in some cases, this is the essential property for the existence of a continuous selection. But their results did not include Theorem 5.4.

Theorem 5.3: Let $A \subset \mathbb{L}$ with property (I), $b \in A$, P_b be compact. If P is two-lsc at b, then $P_b^* \neq \emptyset$.

Proof: Suppose $P_b^*=\emptyset$, then, by Lemma 5.1, there exist disjoint and compact sets G resp. G^* in P_b and two sequences (b_n) resp. (b_n^*) in A, both converging to b, with $\overline{\lim}\, P_{b_n} \subset G$ resp. $\overline{\lim}\, P_{b_n^*} \subset G^*$. Since P is two-lsc at b, there exist subsequences (b_{n_i}) resp. $(b_{n_i}^*)$ and a sequence (y_i) with

$$y_i \in B_{\frac{1}{i}}(P_{b_{n_i}}) \cap B_{\frac{1}{i}}(P_{b^*_{n_i}}) \quad , i \in \mathbb{N}.$$

Then there exist sequences (x_i) resp. (x_i^*) such that

$$x_i \in P_{b_{n_i}} \quad \text{resp.} \quad x_i^* \in P_{b^*_{n_i}} \quad , i \in \mathbb{N}$$

and $(x_i-x_i^*)$ converges to θ. P_b is compact, P is two-lsc at b, hence P is usc at b and (x_i) has a clusterpoint $x_o \in G$. Since $(x_i-x_i^*)$ converges to θ, x_o is a clusterpoint of (x_i^*) and $x_o \in G$. This is a contradiction to $G \cap G^*=\emptyset$. □

<u>Theorem 5.4:</u> Let $A \subset \mathbb{L}$ with property (I), P be closed on A and P_b be compact for each $b \in A$. Then the following statements are equivalent:
(i) P is two-lsc on A,
(ii) $P_b^* \neq \emptyset$ for each $b \in A$ and
(iii) P admits a continuous selection on A.
Proof: It follows from Theorem 5.3 that (i) implies (ii) and from Theorem 4.10 that (ii) implies (iii). The implication (iii) to (i) is trivial. □

6. Chebyshev Approximation

Let S be a compact Hausdorff space, $M=\text{span}\{m_1,\ldots,m_n\} \subset C(S)$ a subspace with dim M = n. For each $x \in C(S)$ let

$$d(x,M) := \inf\{||x-m|| \mid m \in M\}$$

denote the distance to x from M and

$$P_M(x) := \{m \in M \mid \; ||x-m|| = d(x,M)\}$$

denote the set of best Chebyshev-approximations to x from M.

We consider the metric projection $P_M: C(S) \longrightarrow \text{POW}(M)$ and ask for continuous selections for this mapping.

It is well-known, that this approximation problem is equivalent to a semi-infinite optimization problem. (See Brosowski [2, I Example 19].) Both problems are connected in a special way.

An element $m_o = \sum_{i=1}^{n} \bar{v}_i m_i$, where $\bar{v}_i \in \mathbb{R}$, $i=1,\dots,n$, is a best approximation to x from M with distance $d(x,M)=z_o$ if and only if $(\bar{v}_1,\dots,\bar{v}_n,z_o)$ is a minimal point of the following optimization problem:

$$\text{minimize} \quad z$$

$$\text{subject to} \quad z\in\mathbb{R},\ v_i\in\mathbb{R},\ i=1,\dots,n,$$

$$\text{and} \quad \sum_{i=1}^{n} v_i \eta m_i(s) - z \le \eta x(s)$$

$$\text{for each } (\eta,s) \in \{+1,-1\}\times S.$$

We define $N:=n+1$, $T:=\{+1,-1\}\times S$, $p:=(0,\dots,0,1)\in\mathbb{R}^N$, $v:=(v_1,\dots,v_n,z)$, $B(t):=B(\eta,s):=(\eta m_1(s),\dots,\eta m_n(s),-1)$ and $b(t):=b(\eta,s):=\eta x(s)$, $t=(\eta,s)\in T$. Equivalent to the metric projection we consider the optimal-set-mapping $P:A \longrightarrow POW(\mathbb{R}^N)$, where

$$A := \{b\in C(T) \mid b(\eta,s) = -b(-\eta,s) \text{ for each } (\eta,s)\in T\}.$$

In order to apply our results, we have to show that A has property (I).

Theorem 6.1: Let N,T,p,B and A given as above. Then A has property (I).

Proof: It is well-known that A is a subset of $\mathbb{L}$. Let $b\in A$, $\bar{v}=(\bar{v}_1,\dots,\bar{v}_N)\in P_b$ and $\varepsilon>0$. Obviously $\bar{v}_N\ge 0$ and we distinguish two cases.

Case 1: $\bar{v}_N=0$.

Set $W:=T$ and $b_\varepsilon:=b$, then we are through.

Case 2: $\bar{v}_N>0$.

For a subset $R\subset T$ let $R^\eta := R \cap \{\eta\}\times S$, $\eta\in\{+1,-1\}$ and $\dot{R} := \{(\eta,s)\in T \mid (-\eta,s)\in R\}$. Let

$$U := \{t\in T \mid |\langle B(t),\bar{v}\rangle - b(t)| \le \varepsilon\}.$$

Then U^+ is a closed neighborhood of $M^+_{b,\bar{v}}$ resp. $\dot{U}^-$ is a closed neighborhood of $\dot{M}^-_{b,\bar{v}}$ and w.l.o.g. $U^+\cap\dot{U}^-=\emptyset$. Since T is normal, there exist neighborhoods W^η in $\{\eta\}\times S$ of $M^\eta_{b,\bar{v}}$ with $W^\eta\subset U^\eta$, $\eta\in\{+1,-1\}$. Then $W:=W^+\cup W^-$ is a neighborhood of $M_{b,\bar{v}}$.

By Urysohn's lemma there exists a continuous function $\tilde{\tau}:\{+1\}\times S \longrightarrow [0,1]$, where $\tilde{\tau}(t)=1$ for $t\in W^+\cup\dot{W}^-$ and $\tilde{\tau}(t)=0$ for $t\notin U^+\cup\dot{U}^-$. We define for $t=(\eta,s)\in T$

$$\tau(t) := \tau(\eta,s) := \tilde{\tau}(+1,s),$$

hence $\tau:T \longrightarrow [0,1]$ is a continuous mapping.

Again by Urysohn's lemma there exists a continuous function $\tilde{\rho}:\{+1\}\times S \longrightarrow [-1,+1]$, where $\tilde{\rho}(t)=-1$ for $t\in U^+$ and $\tilde{\rho}(t)=+1$ for $t\in\dot{U}^-$. We define a continuous mapping by setting

$$\rho(t) := \rho(\eta,s) := \eta\tilde{\rho}(+1,s)$$

for each $t=(\eta,s)\in T$. Now we define

$$b_\varepsilon(t) := (1-\tau(t))b(t) + \tau(t)(<B(t),\overline{v}> + (\rho(t)+1)\overline{v}_N), \ t\in T.$$

We have $b\in A$, $<B(.),\overline{v}>+\overline{v}_N \in A$ and $\rho\in A$. It follows from the special structure of τ that $b_\varepsilon\in A$.

From the definition of b_ε it follows

$$|b_\varepsilon(t) - b(t)| = \tau(t)|<B(t),\overline{v}> + (\rho(t)+1)\overline{v}_N - b(t)|, \ t\in T.$$

We have for $t\notin U\cup U$ that $\tau(t)=0$, hence $|b_\varepsilon(t) - b(t)| = 0$. For $t\in U$, we have $\rho(t)=-1$, hence

$$|b_\varepsilon(t) - b(t)| = \tau(t)|<B(t),\overline{v}> - b(t)| < \varepsilon.$$

Finally, for $t=(\eta,s)\in U$, there is $(-\eta,s)\in U$, i.e.

$$\begin{aligned} |b_\varepsilon(t) - b(t)| &= |b_\varepsilon(\eta,s) - b(\eta,s)| \\ &= |-b_\varepsilon(-\eta,s) + b(-\eta,s)| \\ &= |b_\varepsilon(-\eta,s) - b(-\eta,s)| \\ &< \varepsilon. \end{aligned}$$

All three cases together show $||b-b_\varepsilon||<\varepsilon$.

For $t\in W$, we have $\tau(t)=1$ and $\rho(t)=-1$, hence

$$<B(t),\overline{v}> = b_\varepsilon(t).$$

For each $t\in T$

$$\begin{aligned} <B(t),\overline{v}> &\le (1-\tau(t))b(t) + \tau(t)<B(t),\overline{v}> + \tau(t)(\rho(t)+1)\overline{v}_N \\ &= b_\varepsilon(t). \end{aligned}$$

This completes the proof. □

The approximation problem is a special case of the optimization problem. For each $x\in C(S)$ let $P_M^*(x)$ denote this subset of $P_M(x)$ which corresponds to the kernel of the corresponding optimal set.

<u>Corollary 6.2:</u> The following statements are equivalent:

(i) P_M is two-lsc on $C(S)$,

(ii) $P_M^*(x)\neq\emptyset$ for each $x\in C(S)$ and

(iii) P_M admits a continuous selection on $C(S)$.

Proof: Follows from Theorem 5.4 and Theorem 6.1. □

REFERENCES

[1] J.Blatter and L.Schumaker, The set of continuous selections of a metric projection in C(X), J. Approx. Th., 36(1982), 141-155.

[2] B.Brosowski, Parametric semi-infinite optimization, Verlag Peter Lang, Frankfurt am Main und Bern, 1982.

[3] B.Brosowski and T.Fischer, Relations between approximation theory and parametric optimization, lecture at the conference "Parametric optimization and related topics", Plaue, GDR, 1985.

[4] F.Deutsch, V.Indumathi and K.Schnatz, Lower semicontinuity, almost lower semicontinuity and continuous selections for set-valued mappings, preprint.

[5] F.Deutsch and P.Kenderov, Continuous selections and approximate selections for set-valued mappings and applications to metric projections, SIAM J. Math. Anal., 14(1983), 185-194.

[6] H.Hahn, Reelle Funktionen, Chelsea Publishing Company, New York, 1948.

[7] E.Michael, Selected selection theorems, Amer. Math. Monthly, 63(1956), 230-237.

HÖLDER CONTINUITY OF SOLUTIONS OF PERTURBED OPTIMIZATION PROBLEMS UNDER MANGASARIAN-FROMOVITZ CONSTRAINT QUALIFICATION

H.Gfrerer *)

ABSTRACT:

We prove that if a strong second order sufficient condition and the Mangasarian-Fromovitz constraint qualification hold at a local minimizer for a nonlinear programming problem, then for sufficiently smooth perturbations of the constraints and the objective the minimizers - regarded as a function of the perturbations - satisfy a Hölder condition with exponent 1/2. An example shows that 1/2 is the best possible exponent. The result is obtained by using a twice continuously differentiable augmented Lagrangian.

1. Introduction

We consider in this paper the stability of solutions of nonlinear programming problems, when the data of the problem are subjected to small perturbations. Precisely, we deal with the perturbed optimization problem

$$\begin{array}{lll} \text{minimize} & f(x,p) & \\ \text{subject to} & h_j(x,p) = 0 & j = 1,\dots,l \\ & g_j(x,p) \leq 0 & j = 1,\dots,m \end{array} \qquad P(p)$$

where p is a perturbation parameter belonging to an open subset P of a normed space and $x \in \mathbb{R}^n$ is the variable in which the minimization is done.

Our interest here will be in investigating stability properties of solutions of P(p), when p varies near some fixed $p_0 \in P$ and at a local minimizer x_0 of $P(p_0)$

1.) appropriate differentiability and continuity assumptions on the problem functions

2.) the Mangasarian-Fromovitz constraint qualification

3.) a strong second-order sufficient condition

hold.

Many authors have already investigated various stability questions in mathematical programming. We mention here only a few which are in direct connection with our paper.

---

*) Universität Linz, Institut für Mathematik, A-4o4o Linz, Altenbergerstraße 69

Fiacco and Mc Cormick [3] proved the existence of a differentiable function of p that locally solves a particular form of P(p). Then Robinson [9] and Fiacco [4] extended this result to programs in which the perturbations appear generally, as in P(p).
In all three cases the results were established by applying the standard implicit-function theorem under the assumptions of strict complementary slackness, linear independence of the gradients of the binding constraints, and the second-order sufficient condition.

If the assumption of strict complementary slackness is dropped, then the standard implicit-function theorem can no longer be used. However, if the linear independence assumption and a strengthened form of the second-order sufficient condition hold, it was shown by Robinson [1o], that the minimizers and associated multipliers are (locally) single-valued functions of the perturbations, and that they are locally Lipschitzian. This was shown by using a general implicit-function theorem for generalized equations. Kojima [7] established similar results; his main tool has been the theory of piecewise continuously differentiable maps.

If one replaces the assumption of linear independence of the binding gradients by the Mangasarian-Fromovitz constraint qualification, then one can still show the existence of a locally unique minimizer for P(p) near p_0 which is continuous provided that an appropriate form of the strengthened second-order sufficient condition previously mentioned hold. This result was also established by Kojima [7]. Robinson [1] proved under a weaker sufficient second order condition, that the sets of stationary points near x_0 and related multipliers, regarded as multifunctions of the perturbations, are nonempty and locally upper Lipschitzian at p_0. Further, Robinson has given an example which shows that under the Mangasarian-Fromovitz constraint qualification and Kojima's second order condition one cannot establish some stronger property such as local Lipschitz continuity.
However, we will show that the solutions are Hölder continuous with exponent 1/2 in some neighbourhood of p_0.

Robinson's example shows that 1/2 is the best possible constant.
Using a somewhat weakened form of the second order sufficient condition, Alt [1] proved upper Hölder continuity with exponent 1/2 for the sets of local minimizers, i.e. the distance between local minimizers of P(p) and the local minimizer of the original problem $P(p_0)$ may be bounded by $O(\|p-p_0\|^{1/2})$. Auslender [2] obtained this result for nondifferentiable problems.
To prove our results we use a special, twice continuously differentiable augmented Lagrangian similar to that proposed by Tapia [14].

Throughout this paper we employ the following assumptions:

<u>Assumption 1.1:</u>

There is an open convex set $\Omega \subset \mathbb{R}^n$ and an open subset P of a normed space, so that for all $p \in P$ the functions $f(.,p)$, $h(.,p)$ and $g(.,p)$ - where $h := (h_1,\dots,h_l)^T$, $g := (g_1,\dots,g_m)^T$ - are twice continuously Frêchet differentiable on Ω, that f, h, g,

f', h' and g' are Lipschitzian on $\Omega \times P$ and f'', h'' and g'' are continuous on $\Omega \times P$. (Here and in what follows, the prime symbol will denote partial differentiation with respect to the variable x).

Now let $x_0 \in \Omega$ be a local minimizer to $P(p_0)$ for some fixed $p_0 \in P$.

Assumption 1.2: (Mangasarian and Fromovitz [8]).

a) $\{h'_j(x_0,p_0): j \in \{1,\dots,l\}\}$ is linearly independent.

b) There exists a $z \in \mathbb{R}^n$ such that

$$h_j'(x_0,p_0)z = 0 \quad \text{for } j \in \{1,\dots,l\}$$

$$g_j'(x_0,p_0)z < 0 \quad \text{if } j \in \{1,\dots,m\} \text{ and } g_j(x_0,p_0) = 0$$

It is well known that under Assumption 1.2 there exist one or more Lagrange multipliers $(\lambda_0,\mu_0) \in \mathbb{R}^l \times \mathbb{R}^m_+$ ($\mathbb{R}^m_+$ denotes the nonnegative orthant of $\mathbb{R}^m$) such that $(x_0,\lambda_0,\mu_0,p_0)$ satisfy the Karush-Kuhn-Tucker conditions

$$f'(x,p) + \sum_{j=1}^{l} \lambda_j h_j'(x,p) + \sum_{j=1}^{m} \mu_j g_j'(x,p) = 0$$

$$\begin{aligned} h_j(x,p) &= 0 \quad && \text{for } j \in \{1,\dots,l\} \\ g_j(x,p) &\le 0 && \text{for } j \in \{1,\dots,m\} \\ \mu_j g_j(x,p) &= 0 && \text{for } j \in \{1,\dots,m\} \end{aligned} \tag{1.1}$$

and that the set of Lagrange multipliers forms a bounded polyhedron.
Before starting our second-order sufficient condition we first employ the following notations:

$I(x,p):= \{j \in \{1,\dots,m\}: g_j(x,p) = 0\}$ for each $x \in \Omega$ and $p \in P$

$I^+(\mu):= \{j \in \{1,\dots,m\}: \mu_j > 0\}$ for each $\mu \in \mathbb{R}^m_+$

$U(x,p):= \{(\lambda,\mu) \in \mathbb{R}^l \times \mathbb{R}^m_+: (x,\lambda,\mu,p) \text{ satisfies } (1.1)\}$ for each $x \in \Omega$ and $p \in P$

$N(x,p,\mu):= \{z \in \mathbb{R}^n: h'(x,p)z = 0,\ g_j'(x,p)z = 0,\ j \in I^+(\mu)\}$ for each $x \in \Omega$, $p \in P$ and $\mu \in \mathbb{R}^m_+$

$L(x,p,\lambda,\mu):= f(x,p) + \lambda^T h(x,p) + \mu^T g(x,p)$ for each $x \in \Omega$, $p \in P$, $(\lambda,\mu) \in \mathbb{R}^l \times \mathbb{R}^m$.

All Euclidian spaces occuring in the context are equipped with the usual Euclidian norm. We denote the distance from a point b to a set A by $d(b,A):= \inf\{\|b-a\|: a \in A\}$ ($+\infty$ if $A = \emptyset$), the distance between two sets A,B by $d(B,A):= \inf\{d(b,A): b \in B\}$, and we write $V_\delta(A)$ for the set $\{b \mid d(b,A) \le \delta\}$, the space and the norm involved should be clear from the context.

Assumption 1.3: (Kojima [7], Condition 7.3)

$L''(x_0,p_0,\lambda,\mu)$ is positive definite on the subspace $N(x_0,p_0,\mu)$ of $\mathbb{R}^n$ for every $(\lambda,\mu) \in U(x_0,p_0)$.

We recall also, in our notation, the second-order sufficient condition used by Robinson [11] to prove the upper Lipschitz continuity.

Condition 1.4:

For each $(\lambda,\mu) \in U(x_0,p_0)$ there is some scalar $\alpha > 0$, so that for all $z \in \mathbb{R}^n$ with

$$h'(x_0,p_0)z = 0, \quad g_j'(x_0,p_0)z \leq 0, \quad j \in I(x_0,p_0), \quad f'(x_0,p_0)z = 0$$

one has $z^T L''(x_0,p_0,\lambda,\mu)z \geq \alpha\|z\|^2$

If Assumption 1.3 holds then also Condition 1.4 is satisfied. Hence by combination of the results established by Kojima [7, Theorem 7.2] and Robinson [11, Theorem 3.2, Corollary 4.3] we get the following Lemma:

Lemma 1.5:

Suppose that the Assumption 1.1 and for some fixed $p_0 \in P$ and a local minimizer x_0 of $P(p_0)$ the Assumptions 1.2 and 1.3 are satisfied.

Then there are positive scalars δ, ε and β such that for each $p \in V_\delta(p_0)$

(i) the ball $V_\varepsilon(x_0)$ contains a unique local minimizer $x(p)$ of $P(p)$ and the associated set of multipliers $U(x(p),p)$ is not empty;

(ii) for each $(\lambda,\mu) \in U(x(p),p)$ the inequality
$d((x(p),\lambda,\mu), \; \{x_0\} \times U(x_0,p_0)) \leq \beta|p-p_0|$
holds.

2. A Twice Continuously Differentiable Augmented Lagrangian

Augmented Lagrangian are usually formed by adding a penalty term to the usual Lagrangians to obtain certain desirable properties (e.g. convexity or strict convexity) that the usual Lagrangian does not possess. But the augmented Lagrangians most frequently used (Rockafellar [12], Fletcher [5]) in the presence of inequalities have the disadvantage that they are not twice continuously differentiable if the strict complementary slackness condition is not satisfied.

To avoid some difficulties resulting from this fact we define for each $x \in \Omega$, $(\lambda,\mu) \in \mathbb{R}^1 \times \mathbb{R}^m$, $p \in P$ and $c \in \mathbb{R}_+$

$$L(x,p,\lambda,\mu,c) := L(x,p,\lambda,\mu) + \frac{c}{2} \cdot [\sum_{j=1}^{1} h_j(x,p)^2 + \sum_{j \in I(x_o,p_o)} \mu_j g_j(x,p)^2]$$

where $(x_o,p_o) \in \Omega \times P$ is fixed. (To simplify the notation we drop the dependence of L on (x_o,p_o)).

Observe that the multipliers are also part of the penalty term.

Lemma 2.1:

Assume the hypothesis of Lemma 1.5. Then there exist scalars $\alpha > 0$, $c^* \geq 0$ such that for all $c \geq c^*$, all $z \in \mathbb{R}^n$ and all $(\lambda,\mu) \in U(x_o,p_o)$

$$z^T L''(x_o,p_o,\lambda,\mu,c) z \geq \alpha |z|^2,$$

i.e. $L''(x_o,p_o,\lambda,\mu,c)$ is positive definite.

Proof:

For any symmetric n x n matrix A we denote by e(A) the smallest eigenvalue of A, i.e. $e(A) := \min\{z^T A z: |z| = 1\}$

Then it is obvious, that

$$e(\lambda A) = \lambda e(A) \text{ for } \lambda \geq 0, \quad e(A+B) \geq e(A) + e(B)$$

and hence e(.) is concave. This implies that e(.) is continuous, if we regard the space of symmetric n x n matrices as an Euclidian space (see Rockafellar [12, Corollary 1o.1.1.]).

Now, let $\{(\lambda^i,\mu^i): i = 1,\ldots,k\}$ be the set of extreme points of the bounded polyhedron $U(x_o,p_o)$. Then for each (λ^i,μ^i), $i \in \{1,\ldots,k\}$ the matrix $L''(x_o,p_o,\lambda^i,\mu^i)$ is positive definite on the subspace $N(x_o,p_o,\mu^i)$ by Assumption 1.3. and hence (see Hestenes [6, p. 428] there exist scalars $\alpha^i > 0$, $c^i \geq 0$ such that for all $c \geq c^i$

$$e(L''(x_o,p_o,\lambda^i,\mu^i) + c.[\sum_{j=1}^{1} h_j'(x_o,p_o)^T h_j'(x_o,p_o) +$$

$$+ \sum_{j \in I^+(\mu^i)} g_j'(x_o,p_o)^T g_j'(x_o,p_o)]) \geq \alpha^i$$

Setting $\bar{c}^i = \max\{c^i, \max\{c^i/\mu_j^i: j \in I^+(\mu^i)\}\}$ and taking $h(x_o,p_o) = 0$, $g_j(x_o,p_o) = 0$ for $j \in I(x_o,p_o) \supset I^+(\mu^i)$ into account, we obtain

$$e(L''(x_0,p_0,\lambda^i,\mu^i,c) \geq \alpha^i$$

for each $c \geq \bar{c}^i$.

Next, set $\alpha := \min\{\alpha^i : i \in \{1,\dots,k\}\}$, $c^* := \max\{\bar{c}^i := i \in \{1,\dots,k\}\}$ and choose any $(\lambda,\mu) \in U(x_0,p_0)$. Then (λ,μ) has a representation as a convex combination of the extrem points of the bounded polyhedron $U(x_0,p_0)$, or in other words

$$(\lambda,\mu) = \sum_{i=1}^{k} t^i(\lambda^i,\mu^i) \quad \text{with} \quad t^i \geq 0, \quad \sum_{k=1}^{k} t^i = 1$$

Using the fact that e(.) is concave we get

$$e(L''(x_0,p_0,\lambda,\mu,c)) = e(\sum_{i=1}^{k} t^i L''(x_0,p_0,\lambda^i,\mu^i,c)) \geq$$

$$\geq \sum_{i=1}^{k} t^i \, e(L''(x_0,p_0,\lambda^i,\mu^i,c)) \geq \sum_{i=1}^{k} t^i \alpha^i \geq \alpha$$

and this completes the proof of our Lemma.

In the following we develope a saddle-point theory for our augmented Lagrangian. By Assumption 1.1 and using the fact that e(.) is continuous we can assume without loss of generality under the hypotheses of Lemma 1.5, that the scalars δ, ε and β from Lemma 1.5 resp. α and c^* from Lemma 2.1 satisfy:

<u>Condition 2.2:</u>

(i) $e(L''(x,p,\lambda,\mu,c^*)) \geq \frac{\alpha}{2}$ for each $p \in V_\delta(p_0)$, $x \in V_\varepsilon(x_0)$ and $(\lambda,\mu) \in V_\varepsilon(U(x_0,p_0))$

(ii) $g_j(x,p) < 0$, $j \notin I(x_0,p_0)$ resp. $g_j(x,p) > -\frac{2}{c^*}$, $j \in I(x_0,p_0)$ for each $x \in V_\varepsilon(x_0)$, $p \in V_\delta(p_0)$

(iii) $\beta.\delta < \varepsilon$

Now we define our dual functional by

$$\varphi(p,\lambda,\mu) := \min \{L(x,p,\lambda,\mu,c^*) : x \in V_\varepsilon(x_0)\} \tag{2.1}$$

for each $p \in V_\delta(p_0)$ and $(\lambda,\mu) \in V_\varepsilon(U(x_0,p_0))$ with $\mu \geq 0$. Observe that the optimization problem defining this functional has an unique solution $x(p,\lambda,\mu)$ since the feasible region is a compact convex set and the objective is continuous and strictly convex by Condition 2.2 (i).

We establish now the fundamental results of our saddle-point theory needed in the next section in the following theorem.

Theorem 2.3:

Assume the hypotheses of Lemma 1.5 and suppose also that Condition 2.2 is satisfied. Then for all $p \in V_\delta(p_0)$ the following results hold:

(i) $x(p) \in \text{int } V_\varepsilon(x_0)$ and $U(x(p),p) \subset \text{int } V_\varepsilon(U(x_0,p_0))$

(ii) $\varphi(p,.,.)$ is concave on $V_\varepsilon(U(x_0,p_0))$

(iii) If $(\bar\lambda,\bar\mu) \in U(x(p),p)$ then $x(p) = x(p,\bar\lambda,\bar\mu)$ and

$\varphi(p,\bar\lambda,\bar\mu) = \max\{\varphi(p,\lambda,\mu): (\lambda,\mu) \in V_\varepsilon(U(x_0,p_0)),\ \mu \geq 0\}$.

Proof:

(i) This follows immediately from Lemma 1.5 and Condition 2.2 (iii).

(ii) For all $t \in [0,1]$ there holds

$$\varphi(p,t(\lambda_1,\mu_1)+(1-t)(\lambda_2,\mu_2)) = \min\{t\ L(x,p,\lambda_1,\mu_1,c^*) +$$

$$+ (1-t)\ L(x,p,\lambda_2,\mu_2,c^*): x \in V_\varepsilon(x_0)\} \geq \min\{t\ L(x,p,\lambda_1,\mu_1,c^*): x \in V_\varepsilon(x_0)\} +$$

$$+ \min\{(1-t)\ L(x,p,\lambda_2,\mu_2,c^*): x \in V_\delta\} = t\ \varphi(p,\lambda_1,\mu_1) + (1-t)\ \varphi(p,\lambda_2,\mu_2)$$

(iii) First we remind, that if $(\bar\lambda,\bar\mu) \in U(x(p),p)$ then $(x(p),\bar\lambda,\bar\mu,p)$ satisfy (1.1). Now a straightforward calculation gives

$$L'(x(p),p,\bar\lambda,\bar\mu,c^*) = 0$$

and this proves together with (i) and the strict convexity of $L(.,p,\bar\lambda,\bar\mu,c)$ on $V_\varepsilon(x_0)$ the first part of our proposition.

To prove the second part choose any $(\lambda,\mu) \in V_\varepsilon(U(x_0),p_0)$ with $\mu \geq 0$. By definition of φ resp. L there holds

$$\varphi(p,\lambda,\mu) \leq L(x(p),p,\lambda,\mu,c^*) = L(x(p),p,\bar\lambda,\bar\mu,c^*) +$$

$$+ \sum_{j=1}^{l} h_j(x(p),p).[\lambda_j-\bar\lambda_j + (c^*/2)\ h_j(x(p),p)] +$$

$$+ \sum_{j\in I(x_0,p_0)} \mu_j g_j(x(p),p).[1+(c^*/2)\ g_j(x(p),p)] + \sum_{j\notin I(x_0,p_0)} \mu_j g_j(x(p),p) -$$

$$- \sum_{j=1}^{m} \bar\mu_j g_j(x(p),p) - (c^*/2) \sum_{j\in I(x_0,p_0)} \bar\mu_j g_j(x(p),p)^2$$

Now, observe that by (1.1) and Condition 2.2 (ii) we have:

a) $h_j(x(p),p) = 0$ for each $j \in \{1,\dots,l\}$

b) $\mu_j(g_j(x(p),p).[1+(c^*/2)\ g_j(x(p),p)] \leq 0$ for each $j \in I(x_0,p_0)$

c) $\mu_j g_j(x(p),p) \leq 0$ for each $j \notin I(x_o,p_o)$

d) $\bar{\mu}_j\ g_j(x(p),p) = 0$ for each $j \in \{1,\dots,m\} \supset I(x_o,p_o)$

Thus

$\varphi(p,\lambda,\mu) \leq L(x(p),p,\bar{\lambda},\bar{\mu},c^*) = \varphi(p,\bar{\lambda},\bar{\mu})$

and this completes our proof.

3. Hölder Continuity of Solutions of Perturbed Optimization Problems

In this section we use the results of Theorem 2.3 to establish the Hölder continuity of solutions of P(p). Our main result on continuity of local minimizers is given in the next theorem.

Theorem 3.1:

Assume the hypotheses of Theorem 2.3. Then there exist some scalars γ_1, γ_2 such that for all p_1, $p_2 \in V_\varepsilon(p_o)$

$$|x(p_1)-x(p_2)| \leq \gamma_1 \cdot \sqrt{d(U(x(p_1),p_1),\ U(x(p_2),p_2)) \cdot \| p_1-p_2\|} + \gamma_2 |p_1-p_2|$$

Proof:

By compactness of $U(x(p_1),p_1)$ resp. $U(x(p_2),p_2)$ we can choose $(\lambda^1,\mu^1) \in U(x(p_1),p_1)$ resp. $(\lambda^2,\mu^2) \in U(x(p_2),p_2)$ with $\|(\lambda^1,\mu^1)-(\lambda^2,\mu^2)\| = d(U(x(p_1),p_1),\ U(x(p_2),p_2))$.

If we set $\bar{x} := x(p_2,\lambda^1,\mu^1)$ then obviously $t = 1$ is solution of the one-dimensional optimization problem

$$\text{minimize}\quad L(x(p_1) + t((\bar{x}-x(p_1)),\ p_2,\ \lambda^1,\mu^1,c^*)$$

$$\text{s.t.}\quad t \leq t^* := \sup\{t:\ x(p_1)+ t(\bar{x}-x(p_1)) \in V_\varepsilon(x_o)\}$$

Now we consider for this problem the Karush-Kuhn-Tucker conditions:

$$L'(\bar{x},p_2,\lambda^1,\mu^1,c^*)(\bar{x}-x(p_1)) + u = 0$$

$$u \geq 0 \text{ and } u(1-t^*) = 0$$

By the mean-value theorem there exists a number $\xi \in [0,1]$ with

$$L'(\bar{x},p_2,\lambda^1,\mu^1,c^*)(\bar{x}-x(p_1)) = L'(x(p_1),p_2,\lambda^1,\mu^1,c^*)(\bar{x}-x(p_1)) +$$

$$+ (\bar{x}-x(p_1))^T \cdot (L''(x(p_1) + \xi(\bar{x}-x(p_1)),p_2,\lambda^1,\mu^1,c^*)(\bar{x}-x(p_1))$$

and hence by Condition 2.2 (i) and $u \geq 0$ we obtain

$$(\alpha/2).\ \|\bar{x}-x(p_1)\|^2 + L'(x(p_1),p_2,\lambda^1,\mu^1,c^*).(\bar{x}-x(p_1)) \le 0.$$

This implies

$$\|\bar{x}-x(p_1)\| \le (2/\alpha)\ \|L(x(p_1),p_2,\lambda^1,\mu^1,c^*)\| =$$

$$= (2/\alpha)\|L'(x(p_1),p_2,\lambda^1,\mu^1,c^*) - L'(x(p_1),p_1,\lambda^1,\mu^1,c^*)\|$$

and hence for some scalar γ_2 we get by Assumption 1.1

$$\|\bar{x}-x(p_1)\| \le \gamma_2.\|p_1-p_2\| \tag{3.2}$$

We will now estimate the distance between $\bar{x}$ and $x(p_2)$. First note that by Theorem 2.3 (iii) the inequality

$$\varphi(p_2,\lambda^1,\mu^1) = L(\bar{x},p_2,\lambda^1,\mu^1,c^*) \le L(x(p_2),p_2,\lambda^2,\mu^2,c^*) = \varphi(p_2,\lambda^2,\mu^2)$$

holds. This is equivalent to

$$L(\bar{x},p_2,\lambda^2,\mu^2,c^*)- L(x(p_2),p_2,\lambda^2,\mu^2,c^*) \le L(\bar{x},p_2,\lambda^2,\mu^2,c^*) - L(\bar{x},p_2,\lambda^1,\mu^1,c^*)$$

Using Taylor's formula, Condition 2.2 (i) and observing that $L'(x(p_2),p_2,\lambda^2,\mu^2,c^*) = 0$, the term on the left-hand side can be bounded below by $\frac{\alpha}{4}\ \|\bar{x}-x(p_2)\|^2$. The term on the right-hand side is equal to

$$\sum_{j=1}^{l}(\lambda_j^2-\lambda_j^1).h_j(\bar{x},p_2) + \sum_{j=1}^{m}(\mu_j^2-\mu_j^1).g_j(\bar{x},p_2) + (c^*/2)\sum_{j\in I(x_o,p_o)}(\mu_j^2-\mu_j^1).g_j(\bar{x},p_2)^2 =$$

$$= \sum_{j=1}^{l}(\lambda_j^2-\lambda_j^1).(h_j(\bar{x},p_2)-h_j(x(p_1),p_1)) + \sum_{j\notin I(x_o,p_o)}(\mu_j^2-\mu_j^1)(g_j(\bar{x},p_2)-g_j(x(p_1),p_1)) +$$

$$+ \sum_{j\in I(x_o,p_o)}(\mu_j^2-\mu_j^1).(g_j(\bar{x},p_2)-g_j(x(p_1),p_1)).(1+(c^*/2)(g_j(\bar{x},p_2) + g_j(x(p_1),p_1)))+$$

$$+ [\sum_{j=1}^{l}(\lambda_j^2-\lambda_j^1)h_j(x(p_1),p_1) - \sum_{j=1}^{m}\mu_j^1\ g_j(x(p_1),p_1) - (c^*/2)\sum_{j\in I(x_o,p_o)}\mu_j^1 g_j(x(p_1),p_1)^2 +$$

$$+ \sum_{j\in I(x_o,p_o)}\mu_j^2\ g_j(x(p_1),p_1).(1+(c^*/2)g_j(x(p_1),p_1)) + \sum_{j\notin I(x_o,p_o)}\mu_j^2\ g_j(x(p_1),p_1)]$$

Using a similar argument as in the proof of Theorem 2.3 (iii) we see that the term in brackets is not positive.
Hence we obtain the inequality

$$\frac{\alpha}{4}\|\bar{x}-x(p_2)\|^2 \le \sum_{j=1}^{l}(\lambda_j^2-\lambda_j^1)(h_j(\bar{x},p_2)-h_j(x(p_1),p_1)) +$$

$$+ \sum_{j\notin I(x_o,p_o)}(\mu_j^2-\mu_j^1)\cdot(g_j(\bar{x},p_2)-g_j(x(p_1),p_1)) +$$

$$+ \sum_{j\in I(x_o,p_o)}(\mu_j^2-\mu_j^1)(g_j(\bar{x},p_2)-g_j(x(p_1),p_1))(1+(c^*/2)(g_j(\bar{x},p_2)+g_j(x(p_1),p_1)))$$

Therefore, we get for some scalar γ_1 by (3.2) and Assumption 1.1

$$\|\bar{x}-x(p_2)\| \le \gamma_1\sqrt{\|(\lambda^2,\mu^2)-(\lambda^1,\mu^1)\|\cdot|p_1-p_2|}$$

which completes together with (3.2) the proof.

Corollary 3.2:

Assume the hypotheses of Theorem 3.1. Then for some constant γ the bound (3.1) can be replaced by

$$|x(p_1)-x(p_2)| \le \gamma\cdot|p_1-p_2|^{1/2} \tag{3.3}$$

Proof:

As $(\lambda^i,\mu^i) \in V_\varepsilon(U(x_o,p_o))$, $i = 1,2$ and this set is bounded, (3.3) follows from (3.1).

One might ask for whether the exponent (1/2) in (3.3) could be enlarged by means of an estimate for the distance between the set of multipliers in (3.1).
Such conjecture is encouraged by the existence of some known results concerning continuity properties of $U(x(p),p)$, regarded as a multifunction of the perturbation parameter p. So one can proof for a sufficient small neighborhood V of p_o that for each fixed $p_1 \in V$ there exists a constant $\gamma(p_1)$ such that for each $p \in V$

$$d(U(x(p),p),\ U(x(p_1),p_1)) \le \gamma(p_1)\cdot\ \|p-p_1\| \tag{3.4}$$

(see Robinson [11, Corollary 4.3], Kojima [7, Lemma 7.4, 7.5].
We shall give here an example, originally discussed by Robinson [11], to show that (1/2) is in fact the best possible exponent in (3.3). The reason is that the functional $d(U(x(p_1),p_1),\ U(x(p_2),p_2))$ has not to be continuous on $V \times V$ although (3.4) is valid.

For example consider the family of convex polytopes $S(b,c)$ defined for each pair (b,c) of real numbers by the linear system $A(b)x \ge a(c)$, where

$$A(b): = \begin{bmatrix} 0 & -1 & 1 & 0 \\ 0 & 1 & 1 & 0 \\ -1 & 0 & 1 & 0 \\ 1 & 0 & 1 & b \end{bmatrix}, \qquad a(c): = \begin{bmatrix} 1 \\ 1 \\ 1 \\ 1+c \end{bmatrix}$$

Now the projection of the origin on S(b,c) is the unique solution of the convex quadratic programming problem

$$(QP(b,c)) \qquad \begin{array}{ll} \text{minimize} & \frac{1}{2} \|x\|^2 \\ \text{subject to} & A(b)x \geq a(c) \end{array}$$

It is easy to check, that the vector x(0,0): = (0, 0, 1, 0) is the unique solution of QP(0,0) and that all assumptions are satisfied. However, if one consider the vectors

$$x(b,c): = (0, 0, 1, 0)^T + cb^{-1}(0, 0, 0, 1)^T$$

$$u(b,c): = (0.5, 0.5, 0, 0,)^T + cb^{-2}(-1, -1, 1, 1)^T$$

one finds that for $b > 0$ and $0 \leq c \leq \frac{1}{2} b^2$, x(b,c) is the unique solution of QP(b,c) and $U(x(b,c),(b,c)) = \{-u(b,c)\}$. For any positive b and $c_1: = 0$, $c_2: = \frac{1}{2} b^2$ we get

$$\|x(b,c_1)-x(b,c_2)\| = b^{-1}.|c_1-c_2| = \frac{1}{2} b = \frac{1}{2} b = \sqrt{\frac{1}{2} \cdot |c_2-c_1|} =$$

$$= \sqrt{\frac{1}{2}} \ \|(b,c_1)-(b,c_2)\| \text{ and } \|u(b,c_1)-u(b,c_2)\| = \frac{1}{2} \|(-1, -1, 1, 1)\| = 1$$

where we have normed the space of perturbation parameters with the Euclidian norm. This means, that the exponent (1/2) in bound (3.3) cannot be enlarged and that the functional defined by the distance of two sets of multipliers cannot be continuous.

References:

[1] Alt, W.: Lipschitzian Perturbations of Infinite Optimization Problems, in: Fiacco, A.V., ed.: Mathematical Programming with Data Perturbations II, Dekker, New York, 1983, 7-22.

[2] Auslender, A.: Stability in Mathematical Programming with Nondifferentiable Data. SIAM Journal of Control and Optimization 22 (1984), 239-253.

[3] Fiacco, A.V. and G.P.Mc Cormick: Nonlinear Programming: Sequential Unconstrained Minimization Techniques, Wiley, New York, 1968.

[4] Fiacco, A.V.: Sensitivity Analysis for Nonlinear Programming Using Penalty Methods. Mathematical Programming 1o (1976), 287-311.

[5] Fletcher, R.: An Ideal Penalty Function for Constrained Optimization. Journal of the Inst.of Mathematics and Applications 15 (1975), 319-342.

[6] Hestenes, M.R.: Augmentability in Optimization Theory. Journal of Optimization Theory and Applications 32 (198o), 427-44o.

[7] Kojima, M.: Strongly Stable Stationary Solution in Nonlinear Programs, in: Robinson, S.M., ed.: Analysis and Computation of Fixed Points, Academic Press, New York, 198o, 93-138.

[8] Mangasarian, O.L. and S. Fromovitz: The Fritz John Necessary Optimality Conditions in the Presence of Equality and Inequality Constraints. Journal of Mathematical Analysis and Applications 17 (1967), 37-47

[9] Robinson, S.M.: Perturbed Kuhn-Tucker points and rates of Convergence for a Class of Nonlinear Programming Algorithms, Mathematical Programming 7 (1974), 1-16

[1o] Robinson, S.M.: Strongly Regular Generalized Equations. Mathematics of Operations Research 5 (198o), 43-62

[11] Robinson, S.M.: Generalized Equations and Their Solutions, Part II: Applications to Nonlinear Programming. Mathematical Programming Study 19 (1982), 2oo-221

[12] Rockafellar, R.T.: Convex Analysis, Princeton University Press, Princeton, New Jersey 197o.

[13] Rockafellar, R.T.: A Dual Approach to Solving Nonlinear Programming Problems by Unconstrained Optimization, Mathematical Programming 5 (1973), 354-372

[14] Tapia, R.A.: Diagonalized Multiplier Methods and Quasi-Newton Methods for Constrained Optimization. Journal of Optimization Theory and Application 2 (1977), 135-194.

PARAMETRIC OPTIMIZATION: PIVOTING AND PREDICTOR-CORRECTOR CONTINUATION, A SURVEY

Jürgen Guddat [1)]

Abstract.

This article includes a survey on solution algorithms for parametric optimization problems based on pivot technique and predictor-correction continuation, respectively. Mainly we consider single-parametric optimization problems and distinguish two techniques that depend on the given class. Linear optimization problems with one parameter at the objective and/or at the right-hand side of the constraints and some extensions can be solved in an efficient manner using pivot technique. Predictor-corrector continuation methods have proved their value for more general smooth single-parametric optimization problems.
Some applications (e. g. solving naturally given parametric optimization problems, globalization of locally convergent algorithms, multi-objective optimization, stochastic optimization, multi-level optimization, semi-infinite optimization) show the utility of such solution algorithms.

1. Introduction

We consider the following parametric optimization problem

$$P(t):\ \min\ \{f(x,t)\mid x\in M(t)\},\quad t\in T\subseteq R^q\quad (q\geq 1) \qquad (1.1)$$

where

$$M(t) := \{x\in R^n \mid h_i(x,t)=0,\ i\in I,\ g_j(x,t)\leq 0,\ j\in J\}$$

$$I := \{1,\dots,m\},\ m<n,\ J := \{1,\dots,s\}\ .$$

By a solution algorithm for $P(t)$, $t\in T$, we understand the solution of the following problem:

---

1) Humboldt University, Department of Mathematics,
GDR-1086 Berlin, PO 1297

(A) find a solution $x(t)$ of $P(t)$ for all $t \in T$ if such a solution exists. Of course, the notion of a solution depends on the class we consider.

There are many motivations to develop solution algorithms for $P(t)$, $t \in T$, e. g.:

(i) Solving naturally given parametric optimization problems,

(ii) globalization of locally convergent algorithms,

(iii) multiobjective optimization based on parametric optimization,

(iv) stochastic optimization based on parametric optimization (see e. g. J. Guddat et al. [39]),

(v) multi-level optimization problems which especially appear in connection with decomposition methods (see e. g. K. Beer [11], B. Bank et al.[10], A. M. Geoffrion [36], K. Tammer [113]),

(vi) semi-infinite optimization in connection with (finite dimensional) multiparametric optimization (see e. g. R. Hettich and H. Th. Jongen [49], [50], R. Hettich and G. Still [51], R. Hettich and P. Zencke [52], H. Th. Jongen et al. [63], H. Th. Jongen and G. Zwier [65], [66], [67]).

Chapter 4 includes an example to (i) and a brief survey to (ii) and (iii).

Since there exist more efficient algorithms for single-parametric optimization problems, we restrict ourselves mainly to this case, i. e. we consider the problem $P(t)$ defined in (1.1) for $t \in T \subseteq R$. Without loss of generality we choose $T := [0,1]$. We give also some hints for the multi-parametric case.

Chapter 2 gives a brief survey for some special classes (linear optimization problems with one parameter in the objective and in the right-hand sides, convex quadratic optimization problems with one parameter in the linear part of the objective and in the right-hand sides, a special complementarity problem) for which it is possible to compute a vectorfunction $x(t)$ defined on $[0,1]$ if we assume that a solution $x(t)$ exists for each $t \in [0,1]$. More precisely, we compute a discretization of $[0,1]$:

$$0 = t_o < \ldots < t_i < t_{i+1} < \ldots < t_N = 1 \tag{1.2}$$

and linear functions $x^i(t)$ corresponding to $[t_i, t_{i+1}]$, $i=0,\ldots,N-1$, such that

$$x(t) = \begin{cases} x^o(t), & [0,t_1] \\ \vdots & \\ x^i(t), & [t_i,t_{i+1}] \\ \vdots & \\ x^{N-1}(t), & [t_{N-1},0] \,. \end{cases} \tag{1.3}$$

We can use modified simplex algorithms for the above special classes, which are described in several books (e. g. B. Bank et al. [9], W. Dinkelbach [19], T. Gal [29], [30], J. Guddat et al. [40], F. Nožička et al. [86], C. van de Panne [87]). We note that the computation of $x(t)$ is really efficient.

These procedures were, however, previously founded on the inner rules of the simplex method. In particular, the basic idea for the methods for convex quadratic optimization problems with one parameter in the linear part of the objective and one parameter in the right-hand side (see (2.4)) is originally due to Ph. Wolfe [117], who, in the third phase of his method for solving convex quadratic optimization problems, found it necessary to solve a special single-parametric quadratic auxiliary problem. The same basic idea underlies the well-known procedure of C. E. Lemke [78] for solving linear complementarity problems, in which a special single-parametric linear complementarity problem is involved in a subsidiary rule. Other authors have shown that this procedure of Lemke and, consequently, as its central point, the procedure for solving the auxiliary parametric problem always give a solution in a larger class of matrices than that considered by Lemke (see e. g. R. Chandrasekaran [16], B. C. Eaves [24], R. Saigal [105]). Detailed expositions are given by H. J. Lüthi [80] and C. van de Panne [87]. What can we say about other classes of single-parametric optimization problems? The monograph of C. van de Panne [87] contains a description of a procedure for solving single-parametric complementarity problems in general form. We refer also to works of L. Grygerova [47], R. Gollmer [38] and the articles cited there to linear optimization problems with parameter-dependent constraint matrices. A procedure which is based on the pivot-technique and designed to solve certain single-parametric piece-wise linear complementarity problems is that of M. Kojima et al. [74], which also applies to a procedure of M. Kojima [71], which determines a piece-wise linear approximation for a continuously parameter-dependent Kuhn-Tucker point of a not necessarily convex nonlinear single-parametric optimization problem. Proposals for dealing with single-parametric nonlinear complementarity problems have been given by N. Meggido [81]. We further note a procedure of I. Zsigmond [122] for linear problems with one parameter in the constraint matrix and for which a computer program exists, as well as an algorithm of A. M Geoffrion [35], for solving strictly convex single-parametric problems entailing analytic functions. In case of quadratic problems in which the parameter also occurs in the objective function matrix, there also exist proposals for solution procedures (see e. g. J. C. G. Boot [12], K. Ritter [93], J. Guddat and K. Tammer [44], K. Tammer [110], [112], H. Välialo [114]). Studies by F. Guerra [46] confirm that most of the procedures

mentioned last are hardly suitable for computing a discretization (1.2) and a corresponding vectorfunction $x(t)$ of (1.3) for larger problems. On the other hand there are many theoretical arguments (see e. g. Chapter 5 in B. Bank et al. [9]) that there is a clear qualitative distinction between the special classes considered in Chapter 2 and more general classes considered in Chapter 3.
That is the reason why we propose another technique for more general classes, namely the so-called continuation methods (also called homotopy methods or path-following methods). We consider the problem $P(t)$ under the assumption

(V1), $f, h_i, g_j \in C^2(R^{n+1}, R)$, $i \in I$, $j \in J$,

and we use the following notations:

$(x(t), \lambda(t), \mu(t))$ as a solution of the Kuhn-Tucker-system corresponding to $P(t)$ is denoted by a stationary solution, $x(t)$ is called Kuhn-Tucker point and $(\lambda(t), \mu(t))$ the corresponding Lagrange-vector.

The set of all stationary solutions of $P(t)$ is denoted by $\psi_{stat}(t)$, the set of all Kuhn-Tucker points by $\psi_{KT}(t)$, the set of all local minimizers by $\psi_{loc}(t)$, and the set of all global minimizers by $\psi(t)$. We note that the notations stationary solutions and Kuhn-Tucker point are not unique in the literature.

Now we can give a concretization of problem (A) under the assumption that $\psi_{stat}(t) \neq \emptyset$ for all $t \in [0,1]$, namely

(A') find a suitable fine discretization (1.2) of $[0,1]$ and corresponding points $(\tilde{x}(t_i), \tilde{\lambda}(t_i), \tilde{\mu}(t_i))$ $(i=1,\dots,N)$ which are sufficiently good approximations of $(x(t_i), \lambda(t_i), \mu(t_i))$ where $(x(t_i), \lambda(t_i), \mu(t_i)) \in \psi_{stat}(t_i)$.

We have assumed:

(E0) A sufficiently good approximation $(\tilde{x}(t_o), \tilde{\lambda}(t_o), \tilde{\mu}(t_o))$ of $(x(t_o), \lambda(t_o), \mu(t_o)) \in \psi_{stat}(t_o)$ is known.

Continuation methods for parameter-dependent nonlinear equations are well-known and successful (see e. g. J. C. Alexander et al. [1], E. Allgower and K. Georg [2], [3], J. W. Avila [8], C. B. Garcia and W. I. Zangwill [32], K. Georg [37], A. Reinoza [88], S. M. Robinson [99], H. Schwetlick [109] and the works cited in all these articles and books, respectively).

There are two approaches for continuation methods (see e. g. E. Allgower and K. Georg [3] and the cited references there):

1. Predictor-corrector methods (where an implicitly given continuous path P of solutions will be numerically approximated),

2. simplicial methods (where a piece-wise linear approximation of the path P will be followed numerically exactly).

We restrict ourselves to predictor-corrector continuation methods for single-parametric optimization problems and call them briefly continuation methods. Now there are several proposals for such continuation methods which are developed directly for the optimization problem (e. g. H. Gfrerer et al. [33], [34], J. Guddat et al. [45], J. Hackl [48], R. Lehmann [75], [76], P. Meravý [83], [84], C. Richter [89], [90], [91], A. Ruske [103], [104]). On the other hand, we can reformulate the Kuhn-Tucker system of P(t) as an equation system or a generalized equation and can use the predictor-corrector continuation methods cited above for parameter-dependent equations or generalized equations. We may point out that there exists also a program package for parametric optimization problems (see A. B. Jadikin [56]).
To solve the problem (A') by using continuation methods only, it is necessary to have a continuous connection between a stationary solution (x^0, λ^0, μ^0) of P(0) and a stationary solution (x^1, λ^1, μ^1) of P(1).

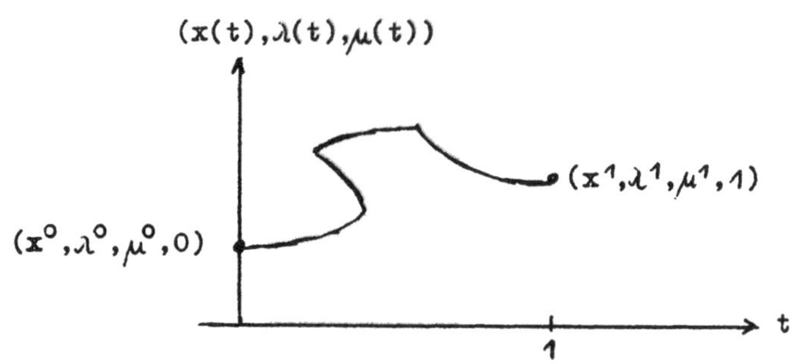

Fig. 1.1

The general existence of continuous Kuhn-Tucker curves is investigated by A. L. Dontchev and H. Th. Jongen [22], H. Th. Jongen et al. [57], [59], [60], [61], M. Kojima [72], M. Kojima and P. Hirabayashi [73], of course we have Kuhn-Tucker manifolds for multi-parametric optimization problems under certain assumptions (see H. Th. Jongen et al. [62], S. Schecter [107]. Continuous properties of a Kuhn-Tucker function are related to

a) implicit function theorem results (see e. g. A. V. Fiacco [26], [27], S. M. Robinson [97], [98], [100], [101] and the cited references there),
b) upper and lower semicontinuity of the optimal-set-mapping (see e. g.

B. Bank et al. [9], B. Brosowski [13], [15], A. V. Fiacco [27], D. Klatte and B. Kummer [69], S. Zlobec and A. Ben-Israel [121] and the articles cited there),

c) continuous selections (see e. g. J. P. Aubin and A. Cellina [8], F. Deutsch and P. Kenderov [18], G. Dommisch [20], T. Fischer [28] and the articles cited there).

Furthermore, the following works from our bibliography, which have not been cited so far, belong to the theoretical background of solution procedures for parametric optimization problems, R. L. Armacost and A. V. Fiacco [6], B. Brosowski [15], A. L. Dontchev [21], J. Guddat et al. [41], H. Th. Jongen et al. [58], [64], H. Th. Jongen and G. Zwier [65], [66], [67], E. S. Levitin [78].

We may point out that the works of A. V. Fiacco [26], [27], R. L. Armacost and A. V. Fiacco [6], S. M. Robinson [98] on the one hand and the article of E. S. Levitin [78] on the other hand give the background for different notions of approximate selection functions (see B. Bank et al. [9], Chapter 6.3) and their computations.
Under certain conditions, in particular differentiability assumptions, the computation of a so-called L-function (in the sense of Levitin, see B. Bank et al. [9]) is reduced to the calculation of a selection function for a particular linear parametric optimization problem. This problem is obtained by linearization of the initial problem.
R. L. Armacost and A. V. Fiacco [6], A. V. Fiacco [27] use certain constraints qualifications and convexity conditions and construct so-called F-functions (in the sense of Fiacco, see B. Bank et al. [9]) with the aid of a linearization of a differentiable selection function $x(t) \in \psi(t)$ in the following manner:

$$x(t) = \bar{x}(t^o) + D_\lambda \bar{x}(\lambda^o)^T(\lambda - \lambda^o) \; ;$$

formulas for calculating $D_\lambda \bar{x}(\lambda^o)$ are also derived.
We introduce the following notion: For $\Phi \in C^2(R^{n+1},R)$ we denote by $D_x \Phi(x,t)$ the vector of the first partial derivatives and by $D_x^2 \Phi(x,t)$ the $n \times n$-matrix of the second derivatives of Φ with respect to x. We can consider an algorithm to compute an approximate selection function to be a method which solves problem (A') partially.
Finally we note that the articles of H. Th. Jongen et al. [59], [61] contain a valuable discussion of various degenerating points on a Kuhn-Tucker curve. These investigations show the difficulties to solve problem (A') if a "connecting" curve like in Fig. 1.1 does not exist.
Chapter 3 includes a brief survey on different approaches for (predictor-corrector) continuation methods. It will be shown that these methods are a good tool to solve problem (A') in an efficient way if

we have the situation as in Figure 1.1.
As noted above, Chapter 4 includes some applications with respect to (i), (ii), (iii). The problem of optimal dispatch for electric power stations is modelled by a single-parametric optimization problem (Section 4.1). Section 4.2 includes some proposals for special imbeddings, which are useful to globalize locally convergent algorithms under certain conditions. In this connection we note that we can consider some algorithms for solving quadratic optimization problems and linear complementarity, respectively (e. g. methods of Ph. Wolfe [117], S. Houthakker [54], C. E. Lemke [77], K. Ritter [93] and K. Tammer [112]) as an imbedding algorithm. From this point of view we can interprete the pivoting approach for solving certain linear parametric optimization problems, quadratic parametric optimization problems and parametric linear complementarity problems (see Chapter 2) as a special continuation method.

2. On linear single-parametric optimization and some extensions

We consider the problem

$$P(t): \min \{(c+tc')^T x \mid Ax=b+tb',\ x \geq 0\},\ t \in [0,1] \tag{2.1}$$

under the following assumptions

(A1) A is an $(m \times n)$-matrix with the rank $m < n$,
(A2) rank $(b,b') = 2$,
(A3) rank $(c,c') = 2$,
(A4) $P(0)$ is solvable.

The assumptions (A1), (A2) and (A3) do not restrict the problem $P(t)$. Assumption (A4) simplifies the development of the algorithm and ensures a normal start.
We denote by $M(t)$ the feasible set, by $\mathfrak{A} := \{t \in R \mid \psi(t) \neq \emptyset\}$ the so-called solubility set, and by $\varphi(t) := \min \{(c+tc')^T x \mid x \in M(t)\}$ the so-called extreme value function.
The following theorem is the theoretical basis for the algorithm to solve problem (A).

Theorem 2.1 (see e. g. B. Bank et al. [9] and F. Nožička et al. [86]):

a) (i) $\mathfrak{A}$ is a convex set
   (ii) $\psi|_{\mathfrak{A}}$ is upper-semicontinuous (in the sense of Hausdorff) on $\mathfrak{A}$,
   (iii) $\varphi|_{\mathfrak{A}}$ is continuous on $\mathfrak{A}$,

b) Let $\mathfrak{A}$ be a non-trivial interval. Then there is a unique partitioning of $\mathfrak{A}$ into closed intervals $\mathfrak{A}^k = [t_k, t_{k+1}]$, $t_k < t_{k+1}$, $k = 1, \ldots, s$, $s \geq 1$ (where t_1 and t_{s+1} can be $-\infty$ or $+\infty$, respectively) with the following properties:

(i) $\dim \psi(t) = \text{const.}$, $t \in (t_k, t_{k+1})$

(ii) ψ is Hausdorff-continuous on (t_k, t_{k+1})

(iii) There exists a vector function $x^k(t) \in \psi(t)$ which is affin-linear on $[t_k, t_{k+1}]$.

(iv) φ is a quadratic function on $[t_k, t_{k+1}]$.

For an arbitrary feasible basic point $x(t)$ of problem $P(t)$ for fixed $t \in R$, we use the following notations.
Let x_{i_r}, $r = 1, \ldots, m$, be the corresponding basic variables and x_{i_s}, $s = 1, \ldots, n-m$, be non-basic variables. Then we obtain the unique representation

$$x_{i_r}(t) = d_{r,-1} + td_{r,o} - \sum_{s=1}^{n-m} d_{r,s} x_{j_s}, \quad r = 1, \ldots, m .$$

The objective function z can be written as

$$z(t) = d_{-1,-1} + t(d_{o,-1} + d_{-1,o}) + t^2 d_{o,o} - \sum_{s=1}^{n-m} (d_{-1,s} + td_{o,s}) x_{j_s} .$$

Again, we form the table $T(t)$ with the coefficients of the last two formulas.

	x_{j_1}	...	x_{j_s}	...	$x_{j_{n-m}}$	1	t
x_{i_1}	$d_{1,1}$	...	$d_{1,s}$	...	$d_{1,n-m}$	$d_{1,-1}$	$d_{1,o}$
...	...		...		...	...	...
x_{i_r}	$d_{r,1}$	...	$d_{r,s}$	...	$d_{r,n-m}$	$d_{r,-1}$	$d_{r,o}$
...	...		...		...	...	...
x_{i_m}	$d_{m,1}$	...	$d_{m,s}$	...	$d_{m,n-m}$	$d_{m,-1}$	$d_{m,o}$
1	$d_{-1,1}$	...	$d_{-1,s}$	...	$d_{-1,n-m}$	$d_{-1,-1}$	$d_{-1,o}$
t	$d_{o,1}$	...	$d_{o,s}$	...	$d_{o,n-m}$	$d_{o,-1}$	$d_{o,o}$

$T(t)$

The basic point is feasible if

$$d_{r,-1} + td_{r,o} \geq 0, \quad r = 1, \ldots, m ,$$

and fulfils the sufficient condition for optimality if moreover

$$d_{-1,s} + td_{o,s} \leq 0, \quad s = 1,\dots,n .$$

Let

$$\underline{t} = \max \left\{0, \sup \left\{- \frac{d_{-1,s}}{d_{o,s}} \mid d_{o,s} < 0\right\}, \sup \left\{- \frac{d_{r,-1}}{d_{r,o}} \mid d_{r,o} > 0\right\}\right\},$$

$$\overline{t} = \min \left\{1, \inf \left\{- \frac{d_{-1,s}}{d_{o,s}} \mid d_{o,s} > 0\right\}, \inf \left\{- \frac{d_{r,-1}}{d_{r,o}} \mid d_{r,o} < 0\right\}\right\}.$$

The basic point $x(t)$ with $x_{i_r}(t) = d_{r,-1} + td_{r,o}$, $r=1,\dots,m$, $x_{j_s}(t) = 0$, $s=1,\dots,n$, remains feasible and fulfils the criterion for optimality for $\underline{t} \leq t \leq \overline{t}$. If there exists an index $s_o \in \{1,\dots,n-m\}$ with

$$d_{-1,s_o} + td_{o,s_o} > 0$$

such that

$$I^{s_o} := \{r \in \{1,\dots,m\} \mid d_{r,s_o} > 0\} = \emptyset ,$$

then $P(t)$ is unsolvable for

$$t < t^* = - \frac{d_{-1,s_o}}{d_{o,s_o}} \quad \text{if} \quad d_{o,s_o} < 0$$

or

$$t > t^* = - \frac{d_{-1,s_o}}{d_{o,s_o}} \quad \text{if} \quad d_{o,s_o} > 0$$

or for each t if $d_{o,s_o} = 0$, respectively.

If $I^{s_o} \neq \emptyset$, we can carry out a primal simplex step for fixed t with the pivot element d_{r_o,s_o} where a pivot row r_o is determined by

$$\frac{d_{r_o,-1} + td_{r_o,o}}{d_{r_o,s_o}} = \min \left\{\frac{d_{r,-1} + td_{r,o}}{d_{r,s_o}} \mid d_{r,s_o} > 0\right\}. \tag{2.2}$$

If there exists an index $r_o \in \{1,\dots,m\}$ with

$$d_{r_o,-1} + td_{r_o,o} < 0.$$

such that

$$J^{r_o} := \{s \in \{1,\dots,n-m\} \mid d_{r_o,s} < 0\} = \emptyset ,$$

then $P(t)$ is unsolvable for

$$t > t^* = -\frac{d_{r_o,-1}}{d_{r_o,o}} \quad \text{if} \quad d_{r_o,o} < 0$$

or

$$t < t^* = -\frac{d_{r_o,-1}}{d_{r_o,o}} \quad \text{if} \quad d_{r_o,o} > 0$$

or for each t if $d_{r_o,o} = 0$, respectively.

If $J^{r_o} \neq \emptyset$, we can carry out a dual simplex step for fixed t with the pivot element d_{r_o,s_o} where a pivot column s_o is determined by

$$\frac{d_{-1,s_o} + td_{o,s_o}}{d_{r_o,s_o}} = \min \left\{ \frac{d_{-1,s} + td_{o,s}}{d_{r_o,s}} \;\middle|\; d_{r_o,s} < 0 \right\}. \tag{2.3}$$

The following lemma gives a sufficient condition for the fact that a fixed value $\bar{t}$ is the upper boundary point of the solubility set $\mathfrak{M}$ of $P(t)$.

Lemma 2.1 (see J. Guddat et al. [39]):

We assume that $P(\bar{t})$ is solvable.

If there exists an index

$$s_o \in J^{\bar{t}} := \{ s \in \{1,\dots,n-m\} \mid d_{o,s} > 0,\ d_{-1,s} + \bar{t}d_{o,s} = 0 \}$$

with $I^{s_o} = \emptyset$ or

if there exists an index

$$r_o \in I^{\bar{t}} := \{ r \in \{1,\dots,m\} \mid d_{r,o} < 0,\ d_{r,-1} + \bar{t}d_{r,o} = 0 \}$$

with $J^{r_o} = \emptyset$, then $P(t)$ is unsolvable for $t > \bar{t}$.

We obtain the following algorithm:

Al(t)

Step 0: $k = 1$, $t_o = 0$, $T = T(0)$ is a simplex table which fulfils the criterion of feasibility and optimality for $P(0)$:

Step 1: Calculate

$$\bar{t} = \min \left\{ 1, \inf \left\{ -\frac{d_{-1,s}}{d_{o,s}} \;\middle|\; d_{o,s} > 0 \right\}, \inf \left\{ -\frac{d_{r,-1}}{d_{r,o}} \;\middle|\; d_{r,o} < 0 \right\} \right\}$$

if $\bar{t} = t_{k-1}$, go to Step 4.

Else $t_k = \bar{t}$,

print (t_{k-1}, t_k), x^k with $x^k_{i_r}(t) = d_{r,-1} + td_{r,o}$, $r = 1, \ldots, m$;
$x^k_{j_s}(t) = 0$, $s = 1, \ldots, n-m$.
If $t_k = 1$, go to Step 5,
else $k = k+1$.

Step 2: If $J^{\bar{t}} = \emptyset$, go to Step 3.
Else if there exists an index $s_o \in J^{\bar{t}}$ with $I^{s_o} = \emptyset$, go to Step 5.
Else carry out a primal simplex step for T with a pivot column $s_o \in J^{\bar{t}}$ and a pivot row r_o according to (2.2). Let T be the resulting simplex table and go to Step 1.

Step 3: If there exists an index $r_o \in I^{\bar{t}}$ with $J^{r_o} = \emptyset$, go to Step 5.
Else carry out a dual simplex step for T with a pivot row $r_o \in I^{\bar{t}}$ and a pivot column s_o according to (2.3). Let T be the resulting simplex table and go to Step 1.

Step 4: $t = \bar{t} + \varepsilon$.
Solve the problem $P(t)$ using table T as the initial table.
If there exists an index $s_o \in \{1, \ldots, n-m\}$ with

$$d_{-1,s_o} + td_{o,s_o} > 0$$

and $I^{s_o} = \emptyset$, go to Step 5.
If there exists an index $r_o \in \{1, \ldots, m\}$ with

$$d_{r_o,-1} + td_{r_o,o} < 0$$

and $J^{r_o} = \emptyset$, go to Step 5.
Else go to Step 1.

Step 5: Stop.

<u>Remark 2.1:</u>

(i) Such algorithms are proposed in several books (see the references in Chapter 1).

(ii) We have to choose ε sufficiently small (especially, smaller than the smallest length of the existing partitioning intervals in $[0,1]$).

(iii) In case of non-degeneracy, Step 4 is omitted and we obtain the partitioning described in Theorem 2.1. In particular, we obtain the affin-linear vectorfunction $x^k(t)$ after one simplex step. In case of degeneracy it is possible that we obtain a finer partitioning.

(iv) Should the case of degeneracy arise, additional rules are necessary in order to avoid cycles (for example, the lexicographical simplex method).

Now we introduce two more general classes:

$$P(t):\ \min\{x^TCx + (p+tp')^Tx \mid Ax = b+tb',\ x \geq 0\},\quad t\in[0,1] \tag{2.4}$$

where C is positive-definite, and

$$\psi(t):=\{(x,y)\in R^{2n} \mid -Kx+y = q+tq',\ x^Ty = 0,\ x \geq 0,\ y \geq 0\} \tag{2.5}$$

where K is positive-semidefinite. Of course, $P(t)$ from (2.1) or (2.4) can be considered as a special case of (2.4). However, the problems (2.4) and (2.5) are analyzed in chapter 5 of Bank et al.[9]. We obtain similar results as in Theorem 2.1.

Remark 2.2:

(i) We can also use the pivot technique to compute a vectorfunction $x(t)$, which solves the problem (2.4) or (2.5) for each t. Such methods are similar to Al(t) (see e. g. B. Bank et al. [9], Section 6.2).

(ii) If we have parametric optimization problems which are not included in (2.5), then we propose in general the procedures described in Chapter 3 (see the discussion in Chapter 1).

Now we consider a multi-parametric problem

$$\psi(\lambda):=\{(x,y)\in R^{2n} \mid -Kx+y=q+\sum_{j=1}^{L}\lambda_j q^j\},\ x\geq 0, y\geq 0, x^Ty=0\},\quad \lambda\in R^L \tag{2.6}$$

where K is positive-definite and $L \geq 2$.

Remark 2.3:

(i) There are also some methods to solve this problem by pivot technique (see e. g. books mentioned above). Using these procedures it is principally possible to calculate the complete partitioning of the solubility set or a given subset of the solubility set, however multi-parametric problems of large dimension will probably prove intractable. The computation of the partitioning is not necessary in many applications, thus in multiobjective optimization and decision problems of stochastic optimization referred to in J. Guddat et al. [39]. One is not so much interested in a partitioning but in finding solutions for particular parameter points in an efficient manner. We therefore propose to reduce the analysis of multi-parametric optimization problems to that of a sequence of single-parametric problems. This idea has been realized in U. Kaussmann [68] for linear multi-parametric optimization problems, and now we give a brief description of the principle. We consider, in the parameter space, a finite number of given lines of the form

$$g(\lambda^0,\lambda^1)= \{\lambda \mid \lambda = \lambda^0 + t(\lambda^1-\lambda^0), t \in R\} , \qquad (2.7)$$

where λ^0 and λ^1, $\lambda^0 \neq \lambda^1$, are fixed. Substituting (2.7) in the initial problem $\psi(\lambda)$ we obtain a single-parametric problem. The actual choice of the lines $g(\lambda^0,\lambda^1)$ in (2.7) depends on the concrete application (see e. g. Section 4.3).

(ii) There is one special case of (2.6) for which the numerical expense is not so large, namely the two-parametric optimization problem

$$\min \{(c+\eta c')^T x \mid Ax = b+\nu b', \ x \geq 0\} \ (\eta,\nu) \in R^2$$

(see e. g. F. Nožička et al. [86], J. Guddat et al. [39]).

3. On nonlinear parametric optimization

We consider the problem $P(t)$ from (1.1) with $T := [0,1]$ under the differentiability assumption (V1).
The Kuhn-Tucker system corresponding to $P(t)$ can be described as follows:

$$D_x f(x,t) + \sum_{i \in I} \lambda_i D_x h_i(x,t) + \sum_{j \in J} \mu_j D_x g_j(x,t) = 0 , \qquad (3.1)$$

$$h_i(x,t) = 0, \quad i \in I, \quad g_j(x,t) \leq 0, \quad \mu_j \geq 0, \quad j \in J , \qquad (3.2)$$

$$\mu_j \cdot g_j(x,t) = 0 , \quad j \in J . \qquad (3.3)$$

If we assume the problem $P(t)$ to be a convex problem and $\psi(t)$ to be non-empty for all $t \in [0,1]$, then there is a very simple procedure to solve problem (A'):
For any discretization of $[0,1]$: $0 = t_0 < \ldots < t_i < t_{i+1} < \ldots < t_N = 1$ the problem $P(t_i)$ can be solved by several globally convergent algorithms (e. g. penalty methods). But, in the non-convex case, we get real numerical difficulties to solve the problems $P(t_i)$ separately. In particular, we cannot use the globally convergent feasible direction methods since we do not know a feasible point for $P(t_i)$ in general. That is the reason for proposing continuation methods where we use information about $x(t_i)$ to compute $x(t_{i+1})$. Furthermore, we get more numerical efficiency by using this approach also in the convex case.
As noted in Chapter 1 we need a continuous connection between a

stationary solution $z^o := (x^o, \lambda^o, \mu^o)$ of $P(0)$ and a stationary solution $z^1 := (x^1, \lambda^1, \mu^1) \in P(1)$. Therefore, we introduce the following conditions:

(E1) There exists a continuous function $x: [0,1] \longrightarrow R^n$ such that $x(t) \in \psi_{loc}(t)$ for all $t \in [0,1]$.

(E1') For given $z^o \in \psi_{stat}(0)$ there exists a curve

$$K := \{v \in R^{n+m+s+1} \mid \Theta = w(s) := (x(s), \lambda(s), \mu(s), t(s)), s \in [0,1]\}$$

where Θ is continuous on $[0,1]$ with $(x(0), \lambda(0), \mu(0)) = z^o$, $t(0) = 0$, $t(1) = 1$ and $z(s) := (x(s), \lambda(s), \mu(s)) \in \psi_{stat}(t(s))$ for all $s \in [0,1]$.

Now we are going to describe the general idea of a predictor-corrector continuation method. For simplicity we shall restrict ourselves to (E1). Under the assumptions (E0) and (E1) we can formulate problem (A') in a more concrete way:

(A'') find a suitably fine discretization (1.2) of $[0,1]$ and for a given $\varepsilon > 0$ a finite sequence of points $\tilde{x}^i$ $(i=1,\dots,N)$ with $\|\tilde{x}^i - x(t_i)\| \leq \varepsilon$.

To solve this problem we consider such computation methods which generate $\tilde{x}^{i+1}$ in a small number of iteration steps with an efficient numerical effort using the previous point $\tilde{x}^i$ as a starting point. We apply here locally convergent algorithms with at least superlinear or quadratic rate of convergence (see e. g. the methods of S. M. Robinson [96], R. B. Wilson [120]). Methods of this type proved their value on the computer, too (see W. Hock and K. Schittkowski [53], K. Schittkowski [108]).
These methods require the important assumption for convergence that the chosen starting point must belong to a certain neighbourhood of the unknown solution searched for. Thus, the idea described above can be realized by using the known point $\tilde{x}^{i-1}$ where $\|\tilde{x}^{i-1} - x(t_{i-1})\| \leq \varepsilon$ (see Fig. 3.1) if $|t_i - t_{i-1}|$ is chosen sufficiently small.
We can distinguish three general conceptions to realize this idea in the literature.

1. Conception (using locally convergent algorithms for $P(t)$):

We follow here R. Lehmann [75], and assume (E1), (V1) and furthermore for $t \in [0,1]$:

(V2) $D_x h_i(x(t),t), i \in I$, $D_x g_j(x(t),t), j \in J(t) := \{j \in J \mid g_j(x(t),t) = 0\}$ are linearly independent.

Then there is a unique Lagrange-vector $(\lambda(t), \mu(t))$.
Let $J^+(t) := \{j \in J(t) \mid \mu_j(t) > 0\}$, $J^o(t) := \{j \in J(t) \mid \mu_j(t) = 0\}$.

(V3) (strongly sufficient second order condition):

For each $y \neq 0$ with $D_x h_i(x(t),t)^T y = 0,\ i \in I$, $D_x g_j(x(t),t)^T y = 0,\ j \in J^+(t)$ we have

$$y^T D_x^2 L(x(t), \lambda(t), \mu(t), t) y > 0$$

where $L = f(x,t) + \sum_{i \in I} \lambda_i h_i(x,t) + \sum_{j \in J} \mu_j g_j(x,t)$ is the Lagrangian.

Now we consider a locally convergent algorithm denoted by (AL) for solving the problem P(t). Let $\alpha \in (0,1)$. The radius of convergence of the algorithm (AL) for solving P(t) in a neighbourhood of $z(t) = (x(t), \lambda(t), \mu(t))$ is defined by

$$r_\alpha(t) := \sup \{ r \mid \text{The algorithm (AL) generates, for all } z^0 \in R^{n+m+s} \text{ with } \|z^0 - z(t)\| \leq r \text{ as a starting point, a sequence } \{z^i\},\ i=1,2,\ldots \text{ with the property } \|z^{i+1} - z(t)\| \leq \alpha \|x^i - x(t)\|,\ i=0,1,2,\ldots \}.$$

This implies

$$\|x^{i+1} - x(t)\| \leq \|z^{i+1} - z(t)\| \leq \alpha \|x^i - x(t)\| \leq \alpha \|z^i - z(t)\|,\ i=0,1,\ldots,$$

i. e. we have at least Q-linear convergence for $\{x^i\}$ and $\{z^i\}$. The following theorem for a general algorithm is proved (including the algorithms of S. M. Robinson [96], R. B. Wilson [120] and Wilson's method with a consistent approximation of the Hessian of the Lagrangian (see R. Lehmann [75])).

Theorem 3.1 (R. Lehmann [75]):
Assume (E1), (V1), (V2), (V3). Then there is a positive number r such that

$$r_\alpha(t) \geq r \quad \text{for all} \quad t \in [0,1].$$

Now one can find a suitably fine discretization by using the predictor-corrector scheme to solve the problem (A'') (see Figure 3.1) in a finite number of steps (the so-called numerical feasibility in the sense of J. W. Avila [8], that means implementability in the strongest sense).

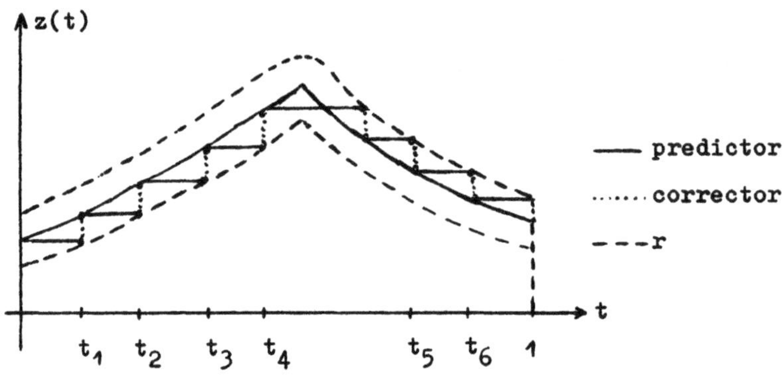

Fig. 3.1

The so-called trivial predictor (P_r) is used in Fig. 3.1 (see also Fig. 3.2 for details) and, as a predictor, one of the local convergent algorithms mentioned above can be used.

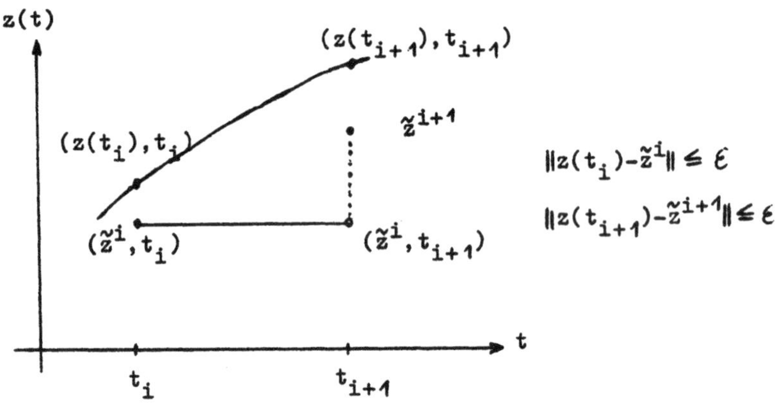

Fig. 3.2

We refer also to works by C. Richter [89], [90], [91], who follows this conception, too.

<u>2. Conception</u> (Transformation of the Kuhn-Tucker system to Kuhn-Tucker equations or a generalized equation and using continuation methods for nonlinear equations or generalized equations):

The transformation of the Kuhn-Tucker system to Kuhn-Tucker equations is well-known and standard in nonlinear complementarity theory and also in nonlinear optimization (see e. g. B. C. Eaves and H. Scarf

[24] for the linear case; A. Charnes et al. [17], C. B. Garcia and W. I. Zangwill [32], M. Kojima [71], M. Kojima and P. Hirabayashi [73], N. Meggido and M. Kojima [82]).

Of course, the Kuhn-Tucker system (3.1) - (3.3) is a mixture of equations and inequalities. The reformulation to a system of equations is quite simple. Let k be a given positive integer, and given $y \in R$. We define

$$y^+ := \max\{y_j, 0\}^k, \quad y^- := \min\{y_j, 0\}^k.$$

Then y^+ and y^- are $(k-1)\times$ continuously differentiable. Thus if e. g. $k = 1$, then y^+ and y^- are continuous, and if $k = 2$, then y^+ and y^- are continuously differentiable (C^1).

Then we can formulate the Kühn-Tucker system as an equivalent system of equations:

$$H(x,y,t) := \begin{bmatrix} D_x f(x,t) + \sum_{i\in I} y_i D_x h_i(x,t) + \sum_{j\in J} y_j^+ D_x g_j(x,t) \\ -h_i(x,t), \quad i \in I \\ y_j^- - g_j(x,t), \quad j \in J \end{bmatrix} = 0 \tag{3.4}$$

More precisely, if (x, λ, μ) satisfies the Kuhn-Tucker system (3.1) - (3.3), then (x,y) solves (3.4) where $y_i = \lambda_i$, $i \in I$ and for $j \in J$

$$y_j = \begin{cases} (\mu_j)^{1/k} & \text{if } g_j(x,t) = 0 \\ +(g_j(x,t))^{1/k} & \text{if } g_j(x,t) < 0 \end{cases}.$$

Conversely, if (x,y) satisfies (3.4), then (x, λ, μ) fulfils the Kuhn-Tucker system where

$$\lambda_i = y_j,\ i \in I, \quad \mu_j = y_j^+,\ j \in J \quad \text{and} \quad h_i(x,t) = y_j^-,\ j \in J .$$

Now we have to distinguish two cases:

<u>Case 1:</u> $k > 1$. Then we can use standard continuation methods for differentiable nonlinear parameter-dependent equations (see the references in chapter 1).

<u>Case 2:</u> $k = 1$. Then (see M. Kojima and P. Hirabayashi [73]) the mapping H is piecewise continuously differentiable with respect to the subdivision $\{\tau(K) \mid K \subseteq J\}$ of $R^{n+m+s+1}$, where the so-called cells $\tau(K)$ are defined by

$$\tau(K) = R^n \times \{y \in R^{m+s} \mid y_j \geq 0,\ j \in K,\ y_j \leq 0,\ j \in J \setminus K\} \times R,$$

i. e. for each $\tau(K)$ there exists an open set $U \supseteq \tau(K)$ and a continuously differentiable function $G: U \longrightarrow R^{n+m+s}$ with $G|_{\tau(K)} = H|_{\tau(K)}$. Such a mapping H is called a PC^1-mapping.

In this case we can use continuation methods for the PC^1-mapping (see e. g. J. C. Alexander et al. [1], A. Reinoza [88]). As noted in Chapter 1 we can also consider the Kuhn-Tucker system to be a generalized equation (see e. g. S. M Robinson [98], [100], [101] and use corresponding continuation methods (see e. g. A. Reinoza [88], C. Richter [91]).

These two conceptions have the disadvantage that we have to use the full Kuhn-Tucker system or the equivalent full Kuhn-Tucker equations. This disadvantage is omitted in the next conception.

3. Conception (Active index-set strategy):
We consider instead of the original problem $P(t)$, $t \in T$, a finite number of auxiliary parametric optimization problems with equality constraints of the following type only:

$P^{J_s}(t)$: $\min \{ f(x,t) \mid h_i(x,t) = 0,\ i \in I,\ g_j(x,t) = 0,\ j \in J_s \}$, $t \in [t_s, t_{s+1}]$,

where $J_s \subseteq J$, $t_s < t_{s+1}$, $s = 0, 1, \ldots, N$, $t_o = 0$, $t_N = 1$.

This has the advantage of reducing the size of the problem since the sets I_s are certain active index sets. Moreover, since $P^{I_s}(t)$ has only equality constraints, we can use the predictor-corrector schemes for solving the nonlinear equations mentioned above (the corresponding Kuhn-Tucker system is a nonlinear equation system). However, the points, where the active index sets will be changed, have to be overcome and the new index set of active constraints has to be computed. In the following we give a brief abstract of the articles by H. Gfrerer et al. [33], [34], J. Guddat et al. [45]. We refer the reader also to the article of R. Lehmann [76] in this volume.
Firstly we follow H. Gfrerer et al. [34]. The basic assumption is that 0 is regular value of H ($k = 1$, i. e. H is a PC^1-mapping), i. e. for each $K, K_o \subseteq J$, there is an open set $U \supseteq \sigma(K,K_o)$ and a continuously differentiable function $G: U \to R^{n+m+s+1}$, such that $G|_{\sigma(K,K_o)} = H|_{\sigma(K,K_o)}$ and the Jacobian of $G \circ i$ is of rank $n+m+s$ at each point w with $i(w) \in U$ and $G(i(w)) = 0$, where $\sigma(K,K_o) := \{(x,y,t) \in \tau(K) \mid y_j = 0,\ j \in K_o\}$ and i is the natural imbedding from $R^n \times R^{J \setminus K_o} \times R$ into R^{n+m+1}. Then it can be shown:

Theorem 3.2 (M. Kojima and P. Hirabayashi [73]):
Assume that 0 is regular value of H. Then we have:

(a) Each connected component S of $H^{-1}(0) := \{(x,y,t) \in R^{n+m+s+1} \mid H(x,y,t) = 0\}$ is either homeomorphic to R (path) or to the unit circle $\{(s_1,s_2) \in R^2 \mid s_1^2 + s_2^2 = 1\}$ (loop) and, for each $\tau(K)$, $K \subseteq J$, with $S \cap \tau(K) \neq \emptyset$, each connected component of $S \cap \tau(K)$ is diffeomorphic to an interval or to the unit circle (piecewise continuously differentiable).

(b) (transversality condition) If S intersects with a face σ (i. e. $K_o \neq \emptyset$) of some cell $\tau(K)$, $K \subseteq J$, then there is a unique index $j_o \in J$ with $y_{j_o} = 0$, and S traverses σ at (x,y,t), i. e. the one-sided tangent vectors at (x,y,t) are not parallel to σ;

(c) if the connected component S is a path, i. e. $S := \{\Theta \mid \Theta = \Theta(s),\ s \in R\}$, then $\|\Theta(s)\| \longrightarrow \infty$ if $s \longrightarrow \pm\infty$.

We note that the transversality condition implies the changing of the index-set of active constraint in exactly one index.

Furthermore we may point out that the character of the Kuhn-Tucker point can be changed at a turning point (for details see M. Kojima and P. Hirabayashi [73]).

Let S be a connected component of $H^{-1}(0)$. If 0 is regular value of H, then Theorem 3.2 implies that each compact connected subset of S can be described parametrically in the form

$S = \{\Theta = (x,y,t) \in R^{n+m+s+1} \mid \Theta = \Theta(s),\ s \in L\}$, L compact interval,
for some continuous function Θ.
The index set $J(x(s),t(s))$ of active constraints is denoted by $J(s)$.

<u>Theorem 3.3:</u> Assume that 0 is regular value of H. Then we have:
There exists a unique subdivision of L into a finite number of nontrivial intervals

$$L = \bigcup_{l=1}^{N} [s^{l-1}, s^{l}]$$

and a corresponding family of index sets K^l, $l=1,\dots,N$, such that

(i) $K^l \neq K^{l+1}$, $l=1,\dots,N-1$,

(ii) $\Theta(s) \in \operatorname{int} \tau(K^l)$ for all $s \in (s^{l-1}, s^l)$,

(iii) $J_o(s) = K^l$ for all $s \in (s^{l-1}, s^l)$.

The intervals $[s^{l-1}, s^l]$ are said to be local stability sets. On local stability sets, Theorem 3.3 allows to replace $P(t)$ by a reduced problem

$$P^K(t): \min \{f(x,t) \mid h_j(x,t) = 0,\ i \in I \cup K\}$$

with $K \subsetneq J$. The corresponding Kuhn-Tucker system is

$$H^K(x, \lambda, \mu, t) = 0 , \tag{3.5}$$

where $H^K: R^n \times R^{|I \cup K|} \times R \longrightarrow R^n \times R^{|I \cup K|}$ is defined by

$$H^K(x,\lambda,\mu,t) = \begin{bmatrix} D_x f(x,t) + \sum_{i \in I} \lambda_i D_x h_i(x,t) + \sum_{j \in K} \mu_j D_x g_j(x,t) \\ \\ h_j(x,t), \quad j \in I \cup K \end{bmatrix} .$$

The next theorem gives a computationally important characterization of local stability sets:

Theorem 3.4: Assume that 0 is regular value of H.
For each $l \in \{1,\dots,N\}$, there exists an open neighbourhood $U^l \supseteq [s^{l-1}, s^l]$ and a continuously differentiable function $\Theta^l: U^l \longrightarrow R^n \times R^{|I \cup K^l|} \times R$; $\Theta^l(s) = (x^l(s), \lambda^l(s), \mu^l(s), t^l(s))$ with

(i) $x^l(s) = x(s)$, $\lambda_i^l(s) = y_i$, $i \in I$, $\mu_j^l(s) = y_j(s)$, $j \in K^l$, $t^k(s) = t(s)$ for all $s \in [s^{l-1}, s^l]$.

(ii) $\Theta^k(s)$ solves (2.7) with $K = K^l$ for all $s \in U^l$.

(iii) $[s^{l-1}, s^l]$ is a connected component of $\{s \in U^k \mid h_j(x^l(s), t^l(s)) \leq 0,\ j \in J \setminus K^l,\ \mu_j^l \geq 0,\ j \in K^l\}$.

The next theorem describes how the index set K^{l+1} varies from the index set K^l and is the basis to compute the new index set of active constraints.

Theorem 3.5: Assume that 0 is regular value of H.
For each $l \in \{1,\dots,N-1\}$, there is a unique index set $j^l \in K^l$ with either $u_{jl}^l(s^l) = 0$ and $j^l \in K^l$ or $h_{jl}(x^l(s^l), t^l(s^l)) = 0$, $j^l \in J \setminus K^l$ and

$$K^{l+1} = \begin{cases} K^l \cup \{j^l\} & , \quad j^l \in J \setminus K^l \\ K^l \setminus \{j^l\} & , \quad j^l \in K^l \end{cases} .$$

On the local stability set $[s^{l-1}, s^l]$ the Kuhn-Tucker system of $P(t)$ is reduced to a system of equations $H^K(x, \lambda, \mu, t) = 0$ with $K = K^l$. Thus, all algorithms which numerically follow a path of solutions of such a system can be used (see the references in Chapter 1). If a boundary point s^l of the local stability set $[s^{l-1}, s^l]$ is passed, the index set K of system (3.5) changes from K^l to K^{l+1} (see Theorem 3.5).
So, for P(t), the standard path-following techniques have to be

extended by a procedure for determining the J^{k+1} after passing s^1. This procedure is given by checking whether the local stability set was left:

If $h_j(\bar{x}^p, \bar{t}^p) \leq 0$ for all $j \in J \setminus K^{p-1}$ and $\bar{\mu}_j^p \geq 0$ for all $j \in K^{p-1}$, set $\bar{x}^p = \bar{x}^p$, $\bar{\lambda}^p = \bar{\lambda}^p$, $\bar{\mu}_j^p = \bar{\mu}_j^p$, $j \in K^{1-1}$, $K^p = K^{p-1}$.

Otherwise, there is an index $j^p \in J$ with either $h_{j^p}(\bar{x}^p, \bar{t}^p) > 0$ and $j^p \in J \setminus K^{p-1}$ or $\bar{u}_{j^p}^p < 0$ and $j^p \in K^{p-1}$. Then the index set will be changed to

$$K^p = \begin{cases} K^{p-1} \cup \{j^p\}, & \text{if } j^p \in J \setminus K^{p-1} \\ K^{p-1} \setminus \{j^p\}, & \text{if } j^p \in K^{p-1} \end{cases}$$

The Theorems 3.3, 3.4 and 3.5 give the theoretical background for the numerically feasible algorithm described in H. Gfrerer et al. [34]. Of course we can use the trivial predictor described in Fig. 3.2. There is also the so-called Euler-predictor (P_E) (see Fig. 3.3).

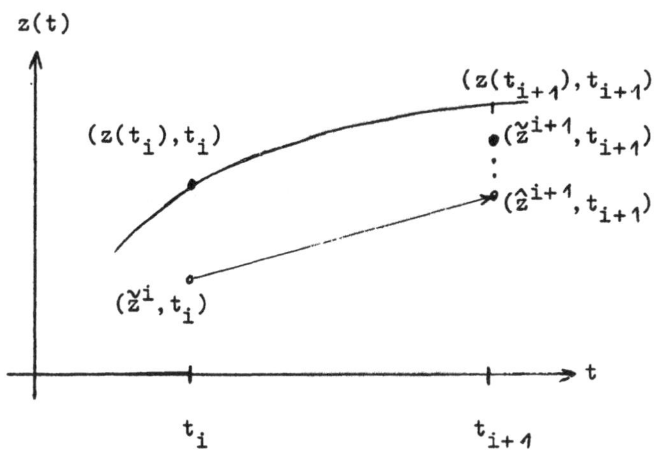

Fig. 3.3

$\hat{z}^{i+1} = \tilde{z}^i + (t_{i+1} - t_i)\tilde{\dot{z}}$ where $\tilde{\dot{z}}$ is an approximation of $\dot{z}(t_i)$.

The Euler-predictor does not demand essentially more computational effort than the trivial predictor.

Now we ask under which additional conditions the assumption "0 is regular value of H" implies (E1').

Theorem 3.6: Assume that

(i) 0 is regular value.

(ii) for each t in some neighbourhood of 0, there is a unique solution (x^t, y^t) of $H(\cdot,\cdot,t) = 0$,

(iii) $\emptyset \neq M(t) \subseteq C$, C compact, for each $t \in [0,1]$,

(iv) $P(t)$ satisfies the Mangasarian-Fromovitz constraint qualification (briefly MFCQ) for all $t \in [0,1]$.

Then (E1') holds.

Now we discuss the assumptions of Theorem 3.6. The assumption (i) is a so-called generic assumption, namely, if we consider instead of $P(t)$ the problem

$$P'(t): \min \{f'(x,t) \mid h_i'(x,t) = 0,\ i \in I,\ g_j'(x,t) \leq 0,\ j \in J\}$$

where

$$f'(x,t) := f(x,t) + c^T x,$$

$$h_i'(x,t) := h_i(x,t) + d_i,\ i \in I,$$

$$g_j(x,t) := g_j(x,t) + d_j,\ j \in J,$$

then almost every $(c,d) \in R^{n+m+s}$ is regular value of the mapping H (see M. Kojima and P. Hirabayashi [73]).
The assumptions (ii), (iii), (iv) realize (E1'). From (iii) it follows that there is no $\bar{t} \in [0,1]$ with $\lim_{t \to \bar{t}} x_j(t) = +\infty$. (iv) implies that the Langrange multipliers have a good behaviour (there exists no $\bar{t} \in [0,1]$ with $\lim_{t \to \bar{t}} \mu_j(t) = +\infty$); (ii) stops the following behaviour of the curve of stationary solutions:

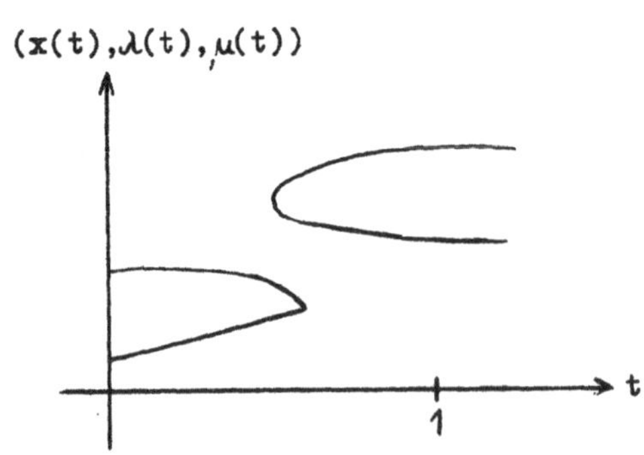

Fig. 3.4

Of course, the case described in Fig. 3.4 is possible without assumption (ii). Assumption (iv) together with (iii) is also very restrictive, namely, it follows that the sets $M(t)$ are homeomorphic for all $t \in [0,1]$ (see J. Guddat et al. [41]). Furthermore we have under the additional assumption (ii) that $M(1) = M$ is homeomorphic to a unit ball. The case that the MFCQ fails to hold for some $M(t)$ is possible in general. The behaviour of the Kuhn-Tucker curve and the corresponding Lagrange-vector is studied by H. Th. Jongen et al. [59], [61]. This study of four types of degeneration provides a very good analysis of the difficulties. In particular, if the MFCQ fails to hold, a Lagrange-multiplier tends to infinity and the Kuhn-Tucker path vanishes. From this it follows that we cannot solve problem (A'') in general by using continuation methods only. We can try to find another connected component of $H^{-1}(0)$ by using additional techniques (see J. Guddat et al. [43]).
Now we come back to the assumption (i).
As noted in Theorem 3.2,b) and Theorem 3.5 the index-set of active constraints changes exactly one index under this assumption.
On the other hand, there are naturally given parametric optimization problems where more than one index can be changed (see e. g. the model considered in Chapter 4.1). That is the reason why we ask ourselves how we can compute the new index set of active constraints in this case. At present there are three procedures under the assumptions (E1), (V1), (V2) and (V3).

1) Let $\bar{t}$ be the right boundary point of a local stability set. The article of H. Gfrerer et al. [33] includes a possibility to compute the right sided derivative $\dot{z}(t)|_{t=\bar{t}+} = (\dot{x}(t), \dot{\lambda}(t), \dot{\mu}(t))|_{t=\bar{t}+}$ and uses this information to calculate the new index set of active constraints. A numerically feasible continuation method is also proposed by using this approach (for details see H. Gfrerer et al. [33]). This procedure has the disadvantage that the numerical expense can be great.

2) We apply a locally convergent algorithm, for which the statement of Theorem 3.1 is fulfilled, to the "whole" problem $P(t)$ for $t \in [\bar{t}-\varepsilon, \bar{t}+\varepsilon]$, ε sufficiently small. Then we can compute the new index set of active constraints without any numerical effort.

3) We apply a locally convergent algorithm, for which the statement of Theorem 3.1 is fulfilled, to the "whole" problem $P(t_k)$, $k = 1,2,\ldots$, but solve an occurring subproblem $P^K(t)$ using the active index set strategy where K represents the "almost" active constraints. This interesting and successful procedure is proposed by R. Lehmann [76] in this volume.

Finally we note that bifurcation is impossible under the assumption (i). On the other hand, there are examples of Kuhn-Tucker points where bifurcation (see Fig. 3.5) occurs.

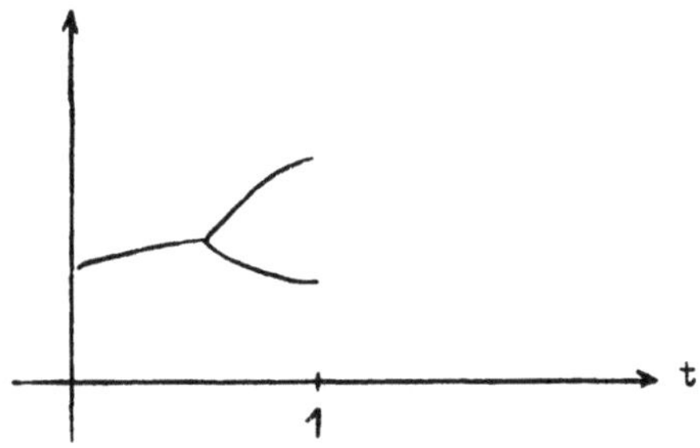

Fig. 3.5

Such bifurcation points can also be interesting for practical applications (see e. g. A. Ruske [104]). That is the reason why A. Ruske [103], [104] weakens the assumption (i) in such a way that a certain bifurcation occurs. An algorithm is proposed in order to compute all branches in the considered case of bifurcation.
The algorithms which are introduced in Chapter 3 and based on the active index set strategy as well as their modifications mentioned have been implemented on the computer at the Kepler-University Linz and/or the Humboldt-University Berlin with good success. Some of the first numerical results are included in H. Gfrerer et al. [33] and R. Lehmann [76].
Now we consider the multi-parametric optimization problem $P(t)$, $t \in T$, $T \subseteq R^q$ $(q > 1)$ defined in (1.1).

Remark 3.1:

(i) We do not know any predictor-corrector continuation method to solve problem (A) for multi-parametric optimization problems. At present, in our opinion we have only the following three procedures:

a) Computation of an approximate selection function as a partial solution of problem (A) (see the references mentioned in Chapter 1),
b) using simplicial continuation methods (see E. L. Allgower and P. H. Schmidt [4], E. L. Allgower and S. Gnutzmann [5]),
c) considering a sequence of single-parametric optimization problems instead of problem $P(t)$ (see Remark 2.3,(i) and Section 4.3).

4. Some applications

4.1. A simplified model of optimal economic dispatch of energy power stations

In different energy power stations let N blocks (energy generators) which are active for a fixed time $T = [\underline{t},\bar{t}]$ for instance $\underline{t} = 9^{oo}$, $\bar{t} = 10^{oo}$ be given. x_i denotes the, bounded above and below, unknown load for the i-th block, i. e. there exist numbers $\underline{x}_i, \bar{x}_i$, $i=1,\dots,N$, with $\underline{x}_i \leq x_i \leq \bar{x}_i$, $i=1,\dots,N$. $a_i(x_i)$ denotes the energy cost function for the block i $(i \in \{1,\dots,N\})$. Let $P(t)$ be the load at the time t. We assume $P(t)$ to be explicitly known and approximated as a continuous, piecewise linear function on T. Furthermore, we have the power balance equation

$$\sum_{i=1}^{N} x_i = P(t) .$$

Then we have to consider the following single-parametric optimization problem

$$P(t):\ \min \left\{ \sum_{i=1}^{N} a_i(x_i) \;\middle|\; \sum_{i=1}^{N} x_i = P(t),\ \underline{x}_i \leq x_i \leq \bar{x}_i,\ i=1,\dots,N \right\},\quad t \in T.$$

Such models are also introduced and investigated in M. Huneault et al. [55]. We propose the following procedure: Let t_l, $l=1,\dots,L$, be the points in $[\underline{t},\bar{t}]$ where $P(t)$ is not differentiable. Then we can consider the following $L+1$ parametric optimization problems

$$P(\Theta, t_l, t_{l+1}):\ \min \left\{ \sum_{i=1}^{N} a_i x_i \;\middle|\; x \in M(\Theta, t_l, t_{l+1}) \right\},\quad \Theta \in [0,1],$$

where

$$M(\Theta, t_l, t_{l+1}) := \Big\{ x \in R^n \;\Big|\; \sum_{i=1}^{N} x_i = P(t_l) + \Theta(P(t_{l+1}) - P(t_l)),$$
$$\underline{x}_i \leq x_i \leq \bar{x}_i,\ i=1,\dots,N \Big\}$$

$l=0,1,\dots,L+1$ with $t_o := \underline{t}$ and $t_{L+1} := \bar{t}$.

If we assume, as in the literature (see e. g. M. Huneault et al. [55]), that $a_i(x_i)$ is quadratic and strictly convex, then we obtain a convex quadratic optimization problem with one parameter in the right-hand side. Such problems can be solved by methods described in Chapter 2. Furthermore, it is possible to use the special structure of the problem $P(\Theta, t_l, t_{l+1})$ to develop more efficient algorithms. This is done in P. Kleinmann and R. Schultz [70]. The corresponding computer program calculates a piecewise linear continuous function $x(t)$ in a very short time. In case we assume $a_i(x_i)$ to be a more general nonlinear function, we can use the algorithms described in Chapter 3.

4.2. Globalization of locally convergent algorithms by imbedding

We consider the following nonlinear optimization problem

$$(P) \quad \min \{f(x) \mid h_i(x) = 0,\ i \in I,\ g_j(x) \leq 0,\ j \in J\} \tag{4.1}$$

where I and J are finite index sets and $f, h_i, g_i \in C^2(R^n, R)$, $i \in I$, $j \in J$; and a corresponding single-parametric optimization problem

$$P(t): \min \{f(x,t) \mid h_i(x,t) = 0,\ i \in I,\ g_j(x,t) \leq 0,\ j \in J\},\ t \in [0,1] \tag{4.2}$$

with the properties (E1) or (E1') (see Chapter 3) and (E0) (see Chapter 1). Furthermore, we assume for this application (E2) $P(1) \equiv (P)$.

We have the following standard imbeddings:

(i) $x^0 \in M := \{x \in R^n \mid h_i(x) = 0,\ i \in I,\ g_j(x) \leq 0,\ j \in J\}$ is known

$$f(x,t) := tf(x) + (1-t)\|x - x^0\|^2,$$
$$h_i(x,t) := h_i(x),\ i \in I,$$
$$g_j(x,t) := g_j(x),\ j \in J.$$

(ii) $x^0 \in R^n$ is arbitrarily chosen

$$f(x,t) := tf(x) + (1-t)\|x - x^0\|^2,$$
$$h_i(x,t) := h_i(x) + (t-1)|h_i(x^0)|,\ i \in I,$$
$$g_j(x,t) := g_j(x) + (t-1)|g_j(x^0)|,\ j \in J.$$

The articles of H. Gfrerer [33], [34] and J. Guddat et al. [45] include some special imbeddings, which have interesting properties. Here we represent some of them. First, without loss of generality we consider instead of the problem (P) the following problem

$$(P_1) \quad \min \{x_n \mid h_i(x) = 0,\ i \in I,\ g_j(x) \leq 0,\ j \in J\}$$

and the corresponding parametric optimization problem (4.2) where

$$f(x,t) := tx_n + (1-t)\|x - x^0\|^2$$
$$h_i(x,t) := h_i(x) + (t-1)|h_i(x^0)|,\ i \in I,$$
$$g_j(x,t) := g_j(x) + (t-1)|g_j(x^0)|,\ j \in J.$$

Then we have:

Theorem 4.1: Let $M := \{x \in R^n \mid h_i(x) = 0,\ i \in I,\ g_j(x) \leq 0,\ j \in J\} \neq \emptyset$.
Then it holds:

a) (i) $P(1) \equiv (P_1)$,

(ii) $\psi(t) \neq \emptyset$ for all $t \in [0,1)$,

(iii) $\psi(0) = \{x^o\}$.

b) Assume additionally: g_j, $j \in J$ are convex on R^n, h_i, $i \in I$ are affine-linear and M compact. Then

(i) $\psi|_{[0,1]}$ is upper-semicontinuous in the sense of Berge (see B. Bank et al. [9]) on $[0,1]$,

(ii) $\psi(t) = \{x(t)\}$ for all $t \in [0,1)$ and $x(t)$ is continuous on $[0,1)$.

The following imbedding secures condition (E1') for certain nonconvex optimization problems for almost all starting points: We consider the problem (P) defined by (4.1) and the corresponding parametric optimization problem

$\bar{P}(t)$: $\min \{f(\bar{x},t) \mid h_i(\bar{x},t) = 0,\ i \in I,\ g_j(\bar{x},t) \leq 0,\ j \in \bar{J}\},\ t \in [0,1]$

where $\bar{x} = (x,v) \in R^n \times R^m$, $\bar{J} = J \cup \{s+1, s+2\}$ and

$$f(\bar{x},t) := t(f(x) + \frac{1}{2}\|v\|^2) + (1-t)\frac{1}{2}(\|x-x^o\|^2 + \|v-v^o\|^2)$$

$$h_i(\bar{x},t) := th_i(x) + (1-t)(v_i - d_i),\quad i \in I,$$

$$g_j(\bar{x},t) := tg_j(x) + (1-t)(v_j - d_j),\quad j \in J,$$

$$g_{s+1}(\bar{x},t) := \|x\|^2 - r\ ,$$

$$g_{s+2}(\bar{x},t) := \|v\|^2 - q\ .$$

Let $\bar{H}(\bar{x},\bar{y},t,w)$ for $\bar{P}(t)$ be analogously defined as $H(x,y,t)$ from (3.4) for $P(t)$.
For problem (P) we assume that

(A1) M is non-empty and compact,

(A2) for each stationary solution (x,λ,μ) of (P)

(a) $D_x h_i(x)$, $i \in I$, $D_x g_j(x)$, $j \in J_o(x) := \{j \in J \mid g_j(x) = 0\}$ are linearly independent,

(b) the strict complementary slackness condition holds for (x,u),

(c) $y^T D_x^2 L(x,u) y \neq 0$ for each $y \neq 0$ with
$D_x h_i(x)^T y = 0,\ i \in I,\ D_x g_j(x)^T y = 0,\ j \in J_o(x)$
(L is the Lagrangian of (P)).

Next we assume that (P), extended by the constraint $\|x\|^2 \leq r$, satisfies a constraint qualification stronger than MFCQ:

(A3) For each $r \in [r_1, r_2]$ and for each $x \in C(r) := \{x \in R^n \mid \|x\|^2 \leq r\}$

(a) $D_x h_i(x)$, $i \in I$, are linearly independent,

(b) there is a vector $z \in R^n$ with

$$h_i(x) + D_x h_i(x)^T z = 0, \ i \in I,$$
$$g_j(x) + D_x g_j(x)^T z < 0, \ j \in J \text{ with } g_j(x) \geq 0,$$
$$2x^T z < 0 \text{ if } \|x\|^2 = r.$$

Finally,

(A4) $f, h_i, g_j \in C^3(R^n, R)$, $i \in I$, $j \in J$.

By using Theorem 3.6 the following theorem is proved in H. Gfrerer et al. [33].

Theorem 4.2: Assume that (P) satisfies (A1), (A2), (A3), (A4) and let $W := \{x^o, v^o, d, r, q) \in R^{n+2m+2} \mid \|x^o\|^2 < r, \ \|d\|^2 < q, \ v_j^o > d_j > 0, \ j \in J, \ r_1 < r < r_2, \ q > 0\}$.

Then, for almost all $w = (x^o, v^o, d, r, q) \in W$, the point $(\bar{x}^o, \bar{u}^o)$ with $\bar{x}^o = (x^o, d)$, $\bar{u}^o = (v^o - d, 0, 0)$ is the unique stationary solution of $\bar{P}(0)$, and the connected component $\bar{S}_o$ of $\bar{H}^{-1}(0)$ which contains $(\bar{x}^o, \bar{y}^o, 0)$ is a piecewise continuously differentiable path that connects $(\bar{x}^o, \bar{y}^o, 0)$ with some point $(\bar{x}^*, \bar{y}^*, 1)$, where $\bar{x}^* = (x^*, 0)$ and x^* is a Kuhn-Tucker point of (P).

Remark 4.1: It follows (see the discussion of the assumptions of Theorem 3.6) that the feasible set M is homeomorphic to a unit ball, i. e. the considered class of nonconvex optimization problems, for which we have a "globally convergent algorithm with probability 1", is restrictive.

Finally we note that we numerically follow piecewise continuously differentiable Kuhn-Tucker pathes by using the proposed imbeddings. We may point out that there is another approach given by P. Meravý [83], [84] with smooth homotopy pathes leading to a Kuhn-Tucker point of P by imbedding the equations appearing in the Kuhn-Tucker system.

4.3. Multiobjective optimization based on parametric optimization

We consider the following multiobjective optimization problem

$$\min \{(f_1(x),\dots,f_L(x)) \mid x \in M\}, \qquad (4.1)$$

where M is a given non-empty subset of R^n and $f_1,\dots,f_L$ are given functions defined on R^n. Let Φ^m (Φ^m_{loc}) be the set of all (locally) efficient (also called Pareto-optimal, admissible, non-dominated etc.) points of the problem (4.1). There are many possibilities to consider a parametric optimization problem $P(t)$, $t \in T \subseteq R^q$ corresponding to (4.1) in such a manner that $\psi(t)$ is related to Φ^m (see e. g. J. Guddat et al. [39], [40] and A. P. Wierzbicki [118], [119] and the references cited there). We follow here the approach given in J. Guddat et al. [39], [40] and consider the multi-parametric optimization problem

$$P(\lambda,\mu):\ \min \left\{\sum_{i=1}^{L} \lambda_i f_i(x) \mid x \in M(\mu)\right\},\ \lambda \in \Lambda,\ \mu \in \mathcal{M},$$

where

$$M(\mu) := \{x \in M \mid f_j(x) \leq \mu_j,\ j=1,\dots,L\},$$
$$\Lambda := \{\lambda \in R^L \mid \sum_{j=1}^{L} \lambda_j = 1,\ \lambda_j > 0,\ j=1,\dots,L\},$$
$$\mathcal{M} := \{\mu = (\mu_1,\dots,\mu_L)^T \mid \mu_j \in R \cup \{+\infty\}\}$$

($\mu_j = +\infty$ means that the corresponding j-th inequality does not appear).

Then we have

<u>Theorem 4.3:</u> $\Phi^m = \bigcup_{\mu \in \mathcal{M}} \psi(\lambda,\mu)$ for each $\lambda^0 \in \Lambda$ and

$$\Phi^m_{loc} = \bigcup_{\mu \in \mathcal{M}} \psi_{loc}(\lambda,\mu) \text{ for each } \lambda^0 \in \Lambda.$$

We note that this relation holds without any assumptions.

Taking into consideration Remark 2.3 and Remark 3.1 we solve a sequence of single-parametric optimization problems instead of the multi-parametric optimization problems $P(\lambda,\mu)$. These can be obtained by taking suitable points λ^0, λ^1 in Λ and μ^0, μ^1 in $\mathcal{M}$ and the straight lines connecting them. This implies:

$$P(t,\lambda^0,\lambda^1,\mu^0,\mu^1):\ \min \left\{\sum_{i=1}^{L}(\lambda_i^0+t(\lambda_i^1-\lambda_i^0))f_i(x) \mid x \in M(t,\mu^0,\mu^1)\right\},\ t \in [0,1]$$

where

$$M(t,\mu^0,\mu^1) := \{x \in M \mid f_j(x) \leq \mu_j^0+t(\mu_j^1-\mu_j^0),\ j=1,\dots,L\}$$

and $\lambda^0, \lambda^1 \in \Lambda$ and $\mu^0, \mu^1 \in \mathcal{M}$ are chosen such that $(\lambda^0, \mu^0) \neq (\lambda^1, \mu^1)$.

Assuming that $P(t, \lambda^0, \lambda^1, \mu^0, \mu^1)$ is solvable for all $t \in [0,1]$ we can find a discretization (1.2) in an efficient manner under the assumptions introduced in Chapter 2 and Chapter 3, and the corresponding computed points $x(t_i)$ or $\tilde{x}(t_i)$, $i=1,\dots,N$, belong to Φ^m and Φ^m_{loc}, respectively.

Moreover, this approach is suitable to develop dialogue procedures (see J. Guddat et al. [39], [40]). The essential dialogue control is based on the following telescreen picture:

$\underline{f}_1$	$f_1(x^k)$	$\frac{f_1(x^k)-\underline{f}_1}{\vert\underline{f}_1\vert} \cdot 100$
$\vdots$	$\vdots$	$\vdots$
$\underline{f}_j$	$f_j(x^k)$	$\frac{f_j(x^k)-\underline{f}_j}{\vert\underline{f}_j\vert} \cdot 100$
$\vdots$	$\vdots$	$\vdots$
$\underline{f}_L$	$f_L(x^k)$	$\frac{f_L(x^k)-\underline{f}_L}{\vert\underline{f}_L\vert} \cdot 100$

where $\underline{f}_j = \inf \{f_j(x) \mid x \in M\}$.

The third column contains the percentage derivations of the objective function value $f_j(x^k)$ (x^k is the current computed point belonging to Φ^m and Φ^m_{loc}, respectively) from the lower bounds.

If the decision maker cannot accept the point x^k as a "solution" of the problem (4.1), we should answer the following questions by using this telescreen picture:

a) Which f_j $(j \in \{1,\dots,L\})$ do you wish to improve? Let $K \subseteq \{1,\dots,L\}$ be the corresponding index set.

b) Which goals a_j do you wish for f_j, $j \in K$? ($a_j \geq \underline{f}_j$ is feasible).

c) Which upper bounds a_j can you accept for f_j, $j \in \{1,\dots,L\} \setminus K$? ($a_j \leq +\infty$ is feasible)

In answering these three questions on the basis of the telescreen picture the decision-maker can formulate his current wishes. The intended dialogue control in the sense of realizing these wishes can be achieved, above all, by choosing $\mu^1_j = a_j$ $(j=1,\dots,L)$.

Here $a_j = +\infty$ means that the corresponding inequality $f_j(x) \leq \mu_j$ must be deleted. The dialogue control with the aid of the parameter vector μ can be supported and even substituted partially by choosing a suitable λ^1 in comparison with λ^0.
However, the control with the aid of λ has the disadvantage that it is impossible, in general, to foresee exactly which objective function will be decreased and which not. It is of course also possible to fix λ (to set e. g. $\lambda^1 = \lambda^0$) such that the dialogue control can be achieved by changing μ only. This dialogue procedure is implemented at Humboldt-University Berlin.

Remark 4.2: We may point out that the feasible set $M(t)$ can be empty for $t \in (t_1, t_2)$ with $t_1, t_2 \in (0,1)$ and $t_1 < t_2$ in the non-convex case. The parametric approaches proposed by A. P. Wierzbicki (see e. g. [118], [119]) have some advantages from this point of view.

Remark 4.3: As in multiobjective optimization, there are also relations between decision problems of stochastic optimization and multi-parametric optimization. This approach is useful to develop dialogue procedures in a similar way for stochastic optimization problems (for details see J. Guddat et al. [39]).

Acknowledgement: I thank H. Gfrerer (Kepler-Universität Linz), H. Th. Jongen (Technische Hogeschool Twente), K. Tammer (Technische Hochschule Leipzig), W. Zulehner (Kepler-Universität Linz) for their helpful comments, criticisms and hints.

References:

[1] Alexander, J. C., T.-Y. Li and J. A. Yorke: Piecewise Smooth Homotopics, in: B. C. Eaves, F. J. Gould, H. O. Peitken, M. J. Todd (eds.): Homotopy Methods and Global Convergence, Plenum P. C. 1983.

[2] Allgower, E. L. and K. Georg: Simplicial and Continuation for Approximating Fixed Points and Solutions to Systems of Equations. SIAM Review 22 (1980), 28 - 85.

[3] Allgower, E. L. and K. Georg: Predictor-Corrector Methods for Approximating Fixed Points and Zero Points of Nonlinear Mappings. In: A. Bachem, M. Grötschel, B. Korte (eds.): Mathematical Programming The State of the Art Bonn 1982, Springer-Verlag Berlin-Heidelberg-New York-Tokyo 1983, 15 - 57.

[4] Allgower E. L. and P. H. Schmidt: An Algorithm for Piecewise-linear Approximation of an Implicitly Defined Manifold, SIAM J. Numer. Anal. 22 (1985), 322 - 346.

[5] Allgower, E. L. and S. Gnutzmann: An Algorithm for Piecewise Linear Approximation of Implicitly Defined 2-Dimensional Surfaces, Arbeitspapiere der Gesellschaft für Mathematik und Datenverarbeitung 137, Bonn 1985.

[6] Armacost, R. L. and A. V. Fiacco: Second-order Parametric Sensitivity Analysis in Nonlinear Programming and Estimates by Penalty Function Methods. Technical Paper T-324, The George Washington Univ., Inst. of Management, Sci. and Engin., Washington 1975.

[7] Aubin, J. P. and A. Cellina: Differential Inclusions. Springer-Verlag, Berlin 1984.

[8] Avila, J. W.: The Feasibility of Continuation Methods for Nonlinear Equations. SIAM J. Num. Anal. 11 (1974), 102 - 121.

[9] Bank, B., J. Guddat, D. Klatte, B. Kummer and K. Tammer: Non-linear Parametric Optimization, Akademie-Verlag, Berlin 1982.

[10] Bank, B., R. Mandel und K. Tammer: Parametrische Optimierung und Aufteilungsverfahren. In [79], 107 - 123.

[11] Beer, K.: Lösung großer linearer Optimierungsaufgaben, VEB Deutscher Verlag der Wissenschaften, Berlin 1977.

[12] Boot, J. C. G.: On Sensitivity Analysis in Convex Quadratic Programming Problems. Oper. Res. 11 (1963), 771 - 786.

[13] Brosowski, B.: Parametric Semi-Infinite Optimization. Verlag Peter Lang, Frankfurt/Main-Bern 1982.

[14] Brosowski, B. and A. Conci: On Vector Optimization and Parametric Programming. In: Proc. of the Conf. Segundas Jornadas Latino Americas de Matematica Aplicada, Rio de Janeiro 1983, Vol. II, 483 - 495.

[15] Brosowski, B.: On the Continuity of the Optimum Set in Parametric Semiinfinite Programming. In: A. V. Fiacco: Mathematical Programming with Data Perturbations II, Lecture Notes in Pure and Applied Mathematics 85, Marcel Dekker, Inc. New York-Basel 1983, 23 - 49.

[16] Chandrasekaran, R.: A Special Case of the Complementarity Pivot Problem, Op. Research 7(1970), 263 - 268.

[17] Charnes, A., C. B. Garcia and C. E. Lemke: Constructive Proofs of Theorems Relating to F(x) = y, with Applications, Math. Programming 12 (1977), 328 - 343.

[18] Deutsch, F. and P. Kenderov: Continuous Selections and Approximate Selections for Set-valued Mappings and Applications to Metric Projections, SIAM J. Math. Anal. 14 (1983), 185 - 194.

[19] Dinkelbach, W.: Sensitivitätsanalysen und parametrische Programmierung, Springer-Verlag, Berlin 1969.

[20] Dommisch, G.: Existence of Lipschitz-continuous and Differentiable Selections for Multifunctions. In: [42].

[21] Dontchev, A. L.: Perturbations, Approximations and Sensitivity Analysis of Optimal Control Systems. Lecture Notes Control Inform. Sci., Springer-Verlag, Berlin 1983.

[22] Dontchev, A. L. and H. Th. Jongen: On the Regularity of the Kuhn-Tucker Curve. Memorandum Nr. 460, Twente University of Technology, The Netherlands (1984)
To appear in: SIAM J. Control and Optimization.

[23] Eaves, B. C.: The Linear Complementarity Problem. Management Sci. 17 (1971), 612 - 634.

[24] Eaves, B. C. and H. Scarf: The Solution of Systems of Piecewise Linear Equations, Math. of Op. Research 1 (1976), 1 - 27.

[25] Eremin, I. I. and J. Guddat (eds.): Parametric Optimization and Ill-posed Problems in Mathematical Optimization, Seminarbericht Nr. 81, Sektion Mathematik der Humboldt-Universität zu Berlin 1986.

[26] Fiacco, A. V.: Sensitivity Analysis for Nonlinear Programming Using Penalty Functions. Math. Programming 10 (1976), 287 - 311.

[27] Fiacco, A. V.: Introduction to Sensitivity and Stability Analysis in Nonlinear Programming. Academic Press, New York-London, 1983.

[28] Fischer, T.: Continuous Selection for Semi-infinite Optimization. In: [42].

[29] Gal, T.: Betriebliche Entscheidungsprobleme, Sensitivitätsanalyse und parametrische Programmierung. Walter de Gruyter, Berlin-New York 1973.

[30] Gal, T.: Post optimal Analysis Parametric Programming Analysis and Related Topics. McGraw Hill, New York 1979.

[31] Garcia Polomares, U. M. and O. L. Mangasarian: Superlinearly Convergent Quasi-Newton Algorithms for Nonlinearly Constrained Optimization Problems, Math. Programming 11 (1976), 1 - 13.

[32] Garcia, C. B. and W. I. Zangwill: Pathways to Solutions, Fixed Points and Equilibria. Prentice-Hall, Inc., Englewood Cliffs. N. J. 07632, 1981.

[33] Gfrerer, H., J. Guddat and J. Wacker: A Globally Convergent Algorithm Based on Imbedding and Parametric Optimization, Computing 30 (1983), 3, 225 - 252.

[34] Gfrerer, H., J. Guddat, Hj. Wacker and W. Zulehner: Path-following Methods for Kuhn-Tucker Curves by an Active Index Set Strategy. In: Bagchi, A.; Jongen, H. Th. (eds.): System and Optimization, Proceedings of the Twente Workshop, Lecture Notes in Control and Information Sciences 66, Springer-Verlag Berlin-Heidelberg-New York-Tokyo 1985, 111 - 132.

[35] Geoffrion, A. M.: Strictly Concave Parametric Programming, Part I: Basic Theory. Management Sci., Ser. A, 13 (1966), 244 - 253. Part II: Additional Theory and Computational Considerations. Management Sci., Ser. A, 13 (1967), 359 - 370.

[36] Geoffrion, A. M.: Generalized Bender's Decomposition, JOTA 10 (1972), 237 - 259.

[37] Georg, K.: On Tracing an Implicitly Defined Curve by Quasi-Newton Steps and Calculating Bifurcation by Local Perturbation. SIAM J. Stat. Comp. 2 (1981) 1, 35 - 49.

[38] Gollmer, R.: Eine Klasse linearer Optimierungsprobleme mit mehreren Parametern in der Restriktionsmatrix, Dissertation, Humboldt-Universität zu Berlin 1980.

[39] Guddat, J., F. Guerra Vasquez, K. Tammer and K. Wendler: Multiobjective and Stochastic Optimization Based on Parametric Optimization, Akademie-Verlag, Berlin 1985.

[40] Guddat, J., F. Guerra Vasquez, K. Tammer and K. Wendler: On Dialogue Algorithms of Multiobjective Optimization Problems Based on Parametric Optimization. In: [25], 46 - 69.

[41] Guddat, J., H. Th. Jongen and J. Rückmann: On Stability and Stationary Points in Nonlinear Optimization. Memorandum Nr. 526, Twente University of Technology, The Netherlands (1985) To appear in: Journal Australian Math. Soc., Series B.

[42] Guddat, J., H. Th. Jongen, B. Kummer and F. Nožička (eds.): Parametric Optimization and Related Topics, to appear in Akademie-Verlag, Berlin 1986.

[43] Guddat, J., H. Th. Jongen and D. Nowack: A Conceptual Algorithm in Non-convex Parametric Optimization, Manuscript. Sektion Mathematik, Humboldt-Universität zu Berlin 1986.

[44] Guddat, J. und K. Tammer: Eine Modifikation der Methode von Theil und van de Panne zur Lösung einparametrischer quadratischer Optimierungsprobleme. Math. Operationsforsch. Statist. 1 (1970), 199 - 206.

[45] Guddat, J., Hj. Wacker and W. Zulehner: On Imbedding and Parametric Optimization - a Concept of a Globally Convergent Algorithm for Nonlinear Optimization Problems. Math. Programming Study 21 (1984), 79 - 96.

[46] Guerra Vasquez, F.: Aplicacion de la optimizacion parametrica el problema de optimizacion vectorial cuadratic. Diss. Univ. Habana 1980.

[47] Grygerova, L.: Lösungsbereich von Optimierungsproblemen mit Parametern in den Koeffizienten der Matrix der linearen Restriktionsbedingungen, Apl. Mat. 17 (1972), 388 - 400.

[48] Hackl, J.: Solution of Optimization Problems via Continuation Methods, in [115], 95 - 127.

[49] Hettich, R. and H. Th. Jongen: Semiinfinite Programming: Conditions of Optimality and Applications. In: Optimization Techniques, Part 2, Lecture Notes Control and Inf. Sciences, 7, Springer-Verlag (1978), 1 - 11.

[50] Hettich, R. and H. Th. Jongen: On the Local Continuity of the Chebyshev Operator, JOTA, 33 (1981),4, 296 - 307.

[51] Hettich, R. and G. Still: Local Aspects of a Method for Solving Membrane-Eigenvalue Problems by Parametric Semi-infinite Programming, in [42].

[52] Hettich, R. und P. Zencke: Numerische Methoden der Approximation und semi-infiniten Optimierung. Teubner Studienbücher, Stuttgart 1982.

[53] Hock, W. and K. Schittkowski: A Comparative Performance Evaluation of 27 Nonlinear Programming Codes, Computing 30 (1983),4, 335 - 358.

[54] Houthakker, S.: The Capacity Method of Quadratic Programming, Econometrica 28 (1960), 62 - 87.

[55] Huneault, M., R. Calderon and F. D. Galiana: Fast Secure Economic Dispatch Using Continuation Methods. Preprint Department of Electrical Engineering McGill University, Montreal, Quebec 1984.

[56] Jadikin, A. B. (Ядыкин, А. Б.): Не линейно параметрическая оптимизационная система омега большой размерности, Сборник трудов, ВНИ Системных Исследовании 13 (1984), 55 - 67.

[57] Jongen, H. Th., P. Jonker and F. Twilt: On Deformation in Optimization. Methods of Operations Research 37 (1980), 171 - 184.

[58] Jongen, H. Th., P. Jonker and F. Twilt: On One-parameter Families of Sets Defined by (In)Equality Constraints. Nieuw Archief v. Wiskunde (3), XXX, (1982), 307 - 322.

[59] Jongen, H. Th., P. Jonker and F. Twilt: One-parameter Families of Optimization Problems: Equality Constraints. Journal Optimization Theory and Applications 48, (1986), 141 - 161.

[60] Jongen, H. Th., P. Jonker and F. Twilt: On Index-sequence Realization in Parametric Optimization. Seminarbericht Nr. 50 der Sektion Mathematik der Humboldt-Universität zu Berlin 1983, 159 - 166.

[61] Jongen, H. Th., P. Jonker and F. Twilt: Critical Sets in Parametric Optimization, Mathematical Programming 35 (1986), 1 - 21.

[62] Jongen, H. Th., P. Jonker and F. Twilt: Parametric Optimization: The Kuhn-Tucker Set, in [42].

[63] Jongen, H. Th., P. Jonker and F. Twilt: Nonlinear Optimization in $\mathbb{R}^n$. I. Morse Theory, Chebyshev Approximation. Peter Lang Verlag, Frankfurt a. M. 1983.

[64] Jongen, H. Th., P. Jonker and F. Twilt: Nonlinear Optimization in $\mathbb{R}^n$, II. Transversality, Flows, Parametric Aspects. Peter Lang Verlag, Frankfurt a. M. (to appear).

[65] Jongen, H. Th. and G. Zwier: On the Local Structure of the Feasible Set in Semi-infinite Optimization. Int. Series Num. Math. 72 (1985), 185 - 202.

[66] Jongen, H. Th. and G. Zwier: Structural Analysis in Semi-infinite Optimization. In: Lemarechal, C. (ed.): Proceedings Third Franco-German Conference in Optimization, INRIA 1985, 56 - 67.

[67] Jongen, H. Th. and G. Zwier: On Regular Semi-infinite Optimization. In: Anderson, E.J., Philpott, A.B. (eds.): Infinite Programming, Lect. Notes Econ. Math. System, 259, Springer-Verlag 1985, 53 - 64.

[68] Kaussmann, U.: Mehrparametrische lineare Optimierung. Diss. Humboldt-Universität zu Berlin 1971.

[69] Klatte, D. and B. Kummer: Stability Properties of Infima and Optimal Solutions of Parametric Optimization Problems, in: Abstracts of the IIASA Workshop on Nondifferentiable Optimization: Motivations and Applications, Sept. 17 - 22, 1984, Sopron, Hungary, IIASA, Laxenburg, Austria, 1984.

[70] Kleinmann, P. and R. Schultz: An Approach to Optimal Dispatch of Electric Power by Convex Quadratic Parametric Optimization, Manuskript Sektion Mathematik der Humboldt-Universität Berlin 1986.

[71] Kojima, M.: A Complementarity Pivoting Approach to Parametric Nonlinear Programming, Meth. of Op. Research 4 (1979), 464 - 472.

[72] Kojima, M.: Strongly Stable Stationary Solutions in Nonlinear Programs. In: Analysis and Computing of Fixed Points. Robinson, S. M. (ed.) Academic Press New York 1980, 93 - 138.

[73] Kojima, M. and P. Hirabayashi: Continuation Deformation of Nonlinear Programs. Math. Programming Study 21 (1984), 150 - 198.

[74] Kojima, M., H. Nishino and T. Sekine: An Extension of Lemke's Method to the Piecewise Linear Complementarity Problem. SIAM J. Appl. Math. 31 (1976), 600 - 613.

[75] Lehmann, R.: On the Numerical Feasibility of Continuation Methods for Nonlinear Programming Problems. Math. Operationsforsch. Statist. Ser. Opt. 15 (1984) 4, 517 - 530.

[76] Lehmann, R.: An Algorithm for Solving One-parametric Optimization Problems Based on an Active-index Set Strategy, in [42].

[77] Lemke, C. E.: Bimatrix Equilibrium Points and Mathematical Programming, Management Sci.(1965), 681 - 689.

[78] Levitin, E. S. (Левитин, Е. С.): О коррекции решений задач нелинейного программирования с непольной информации. Всесоюзная летняя школа о методах оптимизации и их применении. Акад. наук СССР, сиб. отдел., Иркутск 1974.

[79] Lommatzsch, K. (ed.): Anwendungen der linearen parametrischen Optimierung, Akademie-Verlag, Berlin 1979.

[80] Lüthi, H. J.: Komplementaritäts- und Fixpunktalgorithmen in der mathematischen Programmierung, Springer L. N. in Economics and Math. Systems, 129, Springer-Verlag, Berlin-New York 1976.

[81] Megiddo, N.: On the Parametric Nonlinear Complementarity Problem. Math. Programming Study 7 (1978), 142 - 150.

[82] Megiddo, N. and M. Kojima: On the Existence and Uniqueness of Solutions in Nonlinear Complementarity Theory, Math. Programming 12 (1977), 110 - 130.

[83] Meravý, P.: A Note about some Relations between the Method of Parametric Transformation Functions on the Smooth Homotopy Methods for Solving Constrained Optimization Problems. Proc. of 8. Conf. Math. Meth. in Economy, Sellin, GDR 1984.

[84] Meravý, P.: Smooth Homotopies of Math. Programming, in [42].

[85] Michael, E.: Selected Selection Theorems, Amer. Math. Monthly 63 (1956), 230 - 237.

[86] Nožička, F., J. Guddat, H. Hollatz und B. Bank: Theorie der linearen parametrischen Optimierung. Akademie-Verlag, Berlin 1974.

[87] Panne, C. van de: Methods for Linear and Quadratic Programming. North-Holland Publ. Comp., 1975.

[88] Reinoza, A.: Solving generalized equations via homotopies. Mathematical Programming 31 (1985), 307 - 320.

[89] Richter, C.: Über die numerische Behandlung von Einbettungsverfahren in der nichtlinearen Optimierung, TU Dresden, Preprint 1980.

[90] Richter, C.: Ein implementierbares Einbettungsverfahren der nichtlinearen Optimierung, Math. Operationsforsch. u. Stat., ser. optimization 15 (1984) 4, 545 - 553.

[91] Richter, C.: Über die numerische Behandlung nichtlinearer Optimierungsprobleme mit Hilfe von verallgemeinerten Variationsungleichungen und von Nichtoptimalitätsmaßen, Diss. B, Technische Universität Dresden 1980.

[92] Ritter, K.: Ein Verfahren zur Lösung parameter-abhängiger, nichtlinearer Maximumprobleme. Unternehmensforschung 6 (1962), 149 - 166.

[93] Ritter, K.: A Method for Solving Maximum-problems with a Nonconcave Quadratic Objective Function, Z. Wahrsch. Verw. Gebiete 4 (1965). 340 - 351.

[94] Ritter, K.: A Method for Solving Nonlinear Maximum-problems Depending on Parameters, Naval Res. Logist. Quart. 14 (1967), 147 - 162.

[95] Ritter, K.: A Parametric Method for Solving Certain Nonconcave Maximum Problems, J. Computer System Sci. 1 (1967), 44 - 54.

[96] Robinson, S. M.: A Quadratically-convergent Algorithm for General Nonlinear Programming Problems. Math. Programming 3 (1972), 145 - 156.

[97] Robinson, S. M.: Perturbed Kuhn-Tucker Points and Rates of Convergence for a class of Nonlinear Programming Algorithms. Math. Programming 7 (1974), 1 - 16.

[98] Robinson, S. M.: Strongly Regular Generalized Equations. Math. Oper. Res. 5 (1980), 43 - 62.

[99] Robinson, S. M. (ed.): Analysis and Computation of Fixed Points. Academic Press New York 1980.

[100] Robinson, S. M.: Generalized Equations and their Solutions, Part I: Basic Theory. Math. Programming Study 10 (1979), 128 - 141.

[101] Robinson, S. M.: Generalized Equations and their Solutions, Part II: Applications to Nonlinear Programming. Math. Programming Study 19 (1982), 200 - 221.

[102] Rückmann, J.: Struktur- und Stabilitätsuntersuchungen des lokalen und globalen Verhaltens der Mengen der stationären und Kuhn-Tucker-Punkte eines ein-parametrischen Optimierungsproblems, Diplomarbeit, Sektion Mathematik der Humboldt-Universität zu Berlin 1984.

[103] Ruske, A.: Numerical Treatment of Bifurcation Problems in Case of One-parametric Nonlinear Optimization Problems, in [25], 109 - 130.

[104] Ruske, A.: Numerische Behandlung von Bifurkationsproblemen bei einparametrischen Optimierungsaufgaben, Dissertation, Sektion Mathematik der Humboldt-Universität zu Berlin 1986.

[105] Saigal, R.: A Note on a Special Linear Complementarity Problem, Op. Research 7 (1970), 179 - 183.

[106] Lüthi, H. J.: Komplementaritäts- und Fixpunktalgorithmen in der mathematischen Programmierung, Spieltheorie und Ökonomie. Lecture Notes in Economics and Math. Systems, 129, Springer-Verlag, Berlin-New York 1976.

[107] Schecter, S.: Structure of the First-order Solution Set for a Class of Nonlinear Programs with Parameters. Math. Programming (to appear). (Preprint, North Carolina State University, 1985).

[108]a Schittkowski, K.: Nonlinear Programming Codes-information, Tests, Performance. Lecture Notes Econ. Math. Systems 183, Springer-Verlag, Berlin-Heidelberg-New York 1980.

[108]b Schittkowski, K.: The Nonlinear Programming Method of Wilson, Han and Powell with an Augmented Lagrangian. Part 1: Convergence Analysis, Num. Math. 38 (1981), 83 - 114; Part 2: An Efficient Implementation with Linear Least Squares Subproblems. Num. Math. 38 (1981),1, 115 - 127.

[109] Schwetlick, H.: Numerische Lösungen nichtlinearer Gleichungen. VEB Deutscher Verlag der Wissenschaften, Berlin 1979.

[110] Tammer, K.: Die Abhängigkeit eines quadratischen Optimierungsproblems von einem Parameter in der Zielfunktion. Math. Operationsforsch. Statist. 5 (1974), 573 - 590.

[111] Tammer, K.: Theorie und Verfahren für parameter-abhängige lineare Komplementaritätsprobleme. Jahrestagung Math. Optim., Vitte (Hiddensee) 1979, 205 - 213.

[112] Tammer, K.: Ein Verfahren zur Lösung einer Klasse von indefiniten quadratischen Optimierungsproblemen. VII. IKM der Hochsch. Architektur Bauwesen Weimar-Berichte, Berlin 1976, 217 - 221.

[113] Tammer, K.: The Application of Parametric Optimization and Imbedding for the Foundation and Realization of a Generalized Primal Decomposition Approach, in [42].

[114] Välialo, H.: A unified approach to one-parametric general quadratic programming, Math. Programming 33 (1985), 318 - 338.

[115] Wacker, Hj. (ed.): Continuation Methods, Academic Press New York 1978.

[116] Wacker, Hj.: A Summary of the Developments on Imbedding Methods. In Wacker, Hj.: Continuation Methods. Proc. of a Symp. at the Univ. of Linz (Austria), Oct. 3 - 4, 1977, Academic Press New York 1978, 1 - 35.

[117] Wolfe, Ph.: The Simplex-method for Quadratic Programming. Econometrica 27 (1959), 382 - 398.

[118] Wierzbicki, A. P.: Basic Properties of Scalarizing Functional for Multiobjective Optimization. Math. Operationsforsch. Statistik, Optimization 8 (1977), 55 - 60.

[119] Wierzbicki, A. P.: On the Completeness and Constructiveness of Parametric Characterizations to Vector Optimization Problems, Manuscript IIASA-Inst. Laxenburg 1985.

[120] Wilson, R. B.: A Simplicial Method for Concave Programming. Cambridge Harvard Univ., Ph. D. Diss., 1963.

[121] Zlobec, S. and A. Ben-Israel: Perturbed Convex Programs: Continuity of Optimal Solutions and Optimal Values, Oper. Res. Verf. 31 (1979), 737 - 749.

[122] Zsigmond, I.: Parametrization of all the Coefficients of a Linear Programming Problem, Computing Center for Universities, ESZK4, Budapest 1976.

PARAMETRIC OPTIMIZATION IN CERTAIN LATTICE-ORDERED GROUPS: VARIABLE OBJECTIVE FUNCTION

Siegfried Helbig *)

Abstract

On a fully-ordered, complete, lattice-ordered group $(F,\leq,\circ)$ with zero-element a further binary operation $\oplus$ is introduced by $x \oplus y = y \iff x \leq y$ for $x, y \in F$. Such groups turn out to be extremal algebras as defined by Nedoma [10] and K. Zimmermann [11]. The aim of this paper is to consider optimization problems in $(F,\leq,\circ)$ (introduced by Helbig [8]) in dependence on the objective function. Some continuity properties of the optimal-set-mapping and the minimal-value-mapping are given. Finally, an application of these results is given.

I. Introduction

A set F with a binary relation "≤" is called a lattice, if

i) $(F,\leq)$ is a partial ordered set

ii) any pair of elements $x, y \in F$ have a least upper bound and a greatest lower bound.

Definition I.1.: Let F be a set with a binary relation "≤" and a binary operation "∘", such that $(F,\circ)$ is a group with neutral element $\bar{1}$. $(F,\leq,\circ)$ is called

a) a lattice-ordered-group (l.o. group), if

  i) $(F,\leq)$ is a lattice

  ii) ∘ is monotonous, that is $x \leq y$ implies $x\circ z \leq y\circ z$ and $z\circ x \leq z\circ y$ for all $z \in F$.

b) a fully-ordered group (f.o. group), if for all $x, y \in F$ either $x < y$, or $x = y$, or $y < x$.

c) Archimedian, if

  i) $(F,\leq)$ is a partial ordered set

  ii) $x^n < y$ for all $n \in \mathbb{Z}$ implies $x = \bar{1}$

d) (conditionally) complete, if every non-void subset of F bounded from above has a least upper bound in F.

The relationship between those groups are given in Fuchs [5].

The following theorem is important for topological properties of complete groups:

---

*) Johann-Wolfgang-Goethe-Universität, Fachbereich Mathematik, Robert-Mayer-Straße 6-10, D-6000 Frankfurt/Main

Theorem I.2.: A f.o. group is complete if and only if it is order-isomorphic to the naturally ordered additive group of real numbers. Thus all complete f.o. groups are commutative.

Remark: An order-isomorhpism is an isomorphism which preserves the order.

Proof: The proof (see for instance Kokorin and Kopytov [9], p. 110) uses a theorem of Hölder (see for instance Fuchs [5], p. 45, Theorem 1): "A f.o. group is Archimedian if and only if it is order-isomorphic to a subgroup of the additive group of real numbers with the natural ordering Thus all Archimedian f.o. groups are commutative." ∎

For the following we need two further properties of groups:

Definition I.3.: a) The f.o. group $(F,\leq,\circ)$ is called extremal, if a further binary operation $\oplus$ is introduced on F by

$$x \oplus y = y \iff x \leq y \qquad \text{for } x, y \in F.$$

b) $(F,\leq,\circ)$ is called a l.o. group with zero-element $\bar{0}$, if

i) $(F\setminus\{\bar{0}\},\leq,\circ)$ is a l.o. group

ii) $(F,\circ)$ is a semi-group

iii) $\bar{0} \leq x$ and $\bar{0}\circ x = x\circ\bar{0} = \bar{0}$ for all $x \in F$.

In section II we show that an extremal f.o. l.o. group with zero-element is an extremal algebra which was introduced by Nedoma [10]. So we can apply the results of [10], Cunninghame-Green [3], and K. Zimmermann [11], [12], [13].

Because our aim is to consider optimization problems in $(F,\leq,\circ)$ and to investigate the continuous dependence of such problems from a parameter, it is convenient to assume the completeness of F. Note that an Archimedian f.o. group can be embedded in a complete f.o. group under preserving the order (see Fuchs [5], p. 92ff). Thus we apply Theorem I.2. in section III to obtain topological properties in $(F,\leq,\circ)$. For this purpose we endow F (resp. F^n) with the so-called open-intervall-topology (see Fuchs [5], p. 33f).

In section IV we introduce optimization problems in $(F,\leq,\circ)$, which are linear with respect to $\oplus$ and $\circ$. In [8] the author gives characterization theorems of optimal points for such problems.

In this paper we consider optimization problems in dependence on the objective function and ask for continuity properties of the minimal-value-mapping and the optimal-set-mapping. Let L be the set of all objective functions, for which the optimization has a solution.

In the sections V and VI the main results are stated:

- the minimal-value-mapping is continuous over L
- the optimal-set-mapping is closed over L
- a necessary and sufficient condition for the upper-semi-continuity of the optimal-set-mapping is given.

Some of these results have pendants in semi-infinite linear optimization

as given by Brosowski [1].
In section VII these results are applied to bottleneck-assignment-problems.

II. Extremal algebras

A set F with two binary operations $\oplus$ and $\circ$ is called extremal algebra, if

(A1) $(F,\oplus)$ and $(F,\circ)$ are commutative semi-groups

(A2) $(x \oplus y)\circ z = x\circ z \oplus y\circ z$ for all x, y, z $\in$ F

(A3) There exist elements $\overline{0}$ and $\overline{1}$ in F with the properties:

$$x \circ \overline{1} = x$$
$$x \oplus \overline{0} = x \qquad \text{for all } x \in F$$
$$x \circ \overline{0} = \overline{0}$$

(A4) $x \oplus y = x$ or $x \oplus y = y$

In some papers (see [11], [12], [13]) K. Zimmermann requires a further property for extremal algebras:

(A5) For $x \in F$, $x \neq \overline{0}$, there exists $x^{-1} \in F$ with $x^{-1}\circ x = \overline{1}$.

This algebraic structure has various applications in practice, for instance in network- and scheduling-problems (see Cunninghame-Green [3], K. Zimmermann [14], and U. Zimmermann [15]).

Lemma II.1.: a) Let $(F,\oplus,\circ)$ be an extremal algebra. If an order on F is introduced by

(*) $\quad x \leq y \iff x \oplus y = y \quad$ for all x, y $\in$ F

then $(F,\leq,\circ)$ is an extremal f.o. l.o. semi-group with zero-element $\overline{0}$. With (A5) $(F\setminus\{\overline{0}\},\bullet)$ is a group.

b) Let $(F,\leq,\circ)$ be an extremal f.o. l.o. semi-group with zero-element $\overline{0}$ such that $(F,\circ)$ is a commutative semi-group. Then $(F,\oplus,\circ)$ is an extremal algebra. If $(F\setminus\{\overline{0}\},\circ)$ is a group, then (A5) holds.

Proof: a) By definition $(F,\circ)$ is a commutative semi-group (with (A5) $(F\setminus\{\overline{0}\},\circ)$ is a group). It is easy to see that the relation "$\leq$" defined by (*) is partial order on F (see Nedoma [10], p. 2f). Let x, y $\in$ F with $x\neq y$. Then according to (A4) either $x < y$ or $y < x$. Hence, $(F,\circ)$ is a fully-ordered semi-group.
$(F,\circ)$ is a lattice, because for x, y $\in$ F with $x \oplus y = y$

- y is the least upper bound since $x \leq y$ and $y \leq y$
- x is the greatest lower bound since $x \leq y$ and $x \leq x$.

To show the monotony of "$\circ$",let x, y, z be in F with $x \leq y$; it follows

$$x \circ z \oplus y \circ z = (x \oplus y) \circ z = y \circ z \quad \Longleftrightarrow \quad x \circ z \le y \circ z.$$

The commutativity of "$\circ$" implies $z \circ x \le z \circ y$ and hence the assertion.
Thus $(F, \le, \circ)$ is a f.o. l.o. (semi-)group.
According to (A4) and (A3) F is extremal and has a zero-element with the required properties.
b) (A1), (A3), (A4), and (A5) follow immediately by definition.
Apply a result mentioned in Fuchs [5](property A, p. 66):
the monotony of "$\circ$" implies the distributive law (A2). ∎

For the remaining part of this paper let $(F, \oplus, \circ)$ be an extremal f.o. l.o. group with zero-element $\bar{0}$.

Example II.2.: (see Nedoma [10], Helbig [8]):
Denote the real numbers by $\mathbb{R}$, the non-negative real numbers by $\mathbb{R}^+$, the positive real numbers by $\mathbb{R}^>$, and let a, b be in $\mathbb{R} \cup \{-\infty, \infty\}$.
Examples of $(F, \oplus, \circ)$ are:

$(F, \oplus, \circ)$	$\bar{0}$	$\bar{1}$	Properties	(A5)
$(\mathbb{R} \cup \{-\infty\}, \max, +)$	$-\infty$	0	extremal f.o. l.o. group with $\bar{0}$	$x^{-1} = -x$
$(\mathbb{R} \cup \{\infty\}, \min, +)$	∞	0	dito	$x^{-1} = -x$
$(\mathbb{R}^+, \max, \cdot)$	0	1	dito	$x^{-1} = 1/x$
$(\mathbb{R}^> \cup \{\infty\}, \min, \cdot)$	∞	1	dito	$x^{-1} = 1/x$
$([a,b], \max, \min)$	a	b	extremal f.o. l.o. semi-group with $\bar{0}$	does not exist for $x \ne \bar{1}$
$([a,b], \min, \max)$	b	a	dito	

In all examples $\mathbb{R}$ can be replaced by the rational numbers. □

In the cartesian product $F^n := F \times F \ldots\ldots \times F$ the relation "$\le$" induces a partial order. The operations $\oplus$ and $\circ$ can be extended to F^n as follows (see [8]):
let x, y be in F^n and $\alpha \in F$; then define

- $x \oplus y := (x_1 \oplus y_1, \ldots\ldots, x_n \oplus y_n) \in F^n$
- $\alpha \circ x := (\alpha \circ x_1, \ldots\ldots, \alpha \circ x_n) \in F^n$
- $x \circ y := (x,y) := \sum_{i=1}^{n}{}^{\oplus}\, x_i \circ y_i \in F$

The function (.,.) is called extremal-inner-product, because it satisfies all properties of a real inner product with respect to $\oplus$ and $\circ$.
According to (A4)

$$(x,y) = x_j \circ y_j$$

for at least one $j \in \{1, \ldots, n\}$.

In this way, we define the matrix operations $A \circ B$ and $A \oplus B$ with suitable matrices $A \in F^{n,k}$ and $B \in F^{k,m}$. For instance $C := A \circ B$ is defined by

$$c_{ij} := \sum_{l=1}^{k}{}^{\oplus} a_{il} \circ b_{lj} \qquad \text{for } i=1, \ldots, n \quad j=1, \ldots, m$$

III. Topological aspects

For the next chapters it is convenient to assume the completeness of the l.o. group F. Thus, let $(F, \oplus, \circ)$ be a

complete extremal f.o. l.o. group with zero-element.

Applying Theorem I.2. we obtain

- the commutativity of the operation "$\circ$"
- the isomorphie of $(F \setminus \{\overline{0}\}, \circ)$ to $(\mathbb{R}, +)$ with natural order

Let φ be the isomorphism; extend φ by defining

$$\varphi(\overline{0}) := -\infty$$

With regard to rule of addition $-\infty + x = -\infty$ for all $x \in \mathbb{R} \cup \{-\infty\}$, $(F, \circ)$ is isomorphic to $(\mathbb{R} \cup \{-\infty\}, +)$.

Example III.1.: Consider the examples in II.2.

$(F, \oplus, \circ)$	φ	order in F is the ...
$(\mathbb{R} \cup \{-\infty\}, \max, +)$	id	usual greater-equal-relation in $\mathbb{R}$
$(\mathbb{R} \cup \{\infty\}, \min, +)$	$-$ id	" smaller-equal- " in $\mathbb{R}$
$(\mathbb{R}^{+}, \max, \cdot)$	$\begin{cases} \ln(x) & \text{for } x \neq 0 \\ -\infty & \text{for } x = 0 \end{cases}$	" greater-equal- " in $\mathbb{R}$
$(\mathbb{R}^{>} \cup \{\infty\}, \min, \cdot)$	$\begin{cases} -\ln(x) & \text{for } x \neq \infty \\ -\infty & \text{for } x = \infty \end{cases}$	" smaller-equal- " in $\mathbb{R}$

□

Take the sets

$$U_a := \{x \in F \mid x < a\}$$

and $\qquad\qquad$ for $a, b \in F$

$$L_b := \{x \in F \mid b < x\}$$

and the set F as a subbasis of open sets of a topology $\mathcal{T}$ on F, which is called the open-interval -topology (see Fuchs [5], p. 32f, Kokorin and

Kopytov [9], p. 108).
With this topology $(F,\mathcal{T})$ is a normal topological space, which satisfies the Hausdorff-property (see Fuchs [5], p. 33, Proposition 12).
For sake of uniformity denote an element of the basis of neighbourhoods of some $x \in F$ by

$$U_{ab} := \{z \in F \mid a < z < b\} \qquad \text{if } x \neq \bar{0}$$

resp.

$$U_{\bar{0}b} := \{z \in F \mid \bar{0} \le z < b\} \qquad \text{if } x = \bar{0}$$

(see [8]).
To apply the topological properties of $\mathbb{R} \cup \{-\infty\}$ we have the useful

<u>Lemma III.2.:</u> The isomorphism

$$\varphi\colon (F,\circ) \longrightarrow (\mathbb{R}\cup\{-\infty\},+)$$

is a homeomorphism, i.e. φ is a topological mapping.
<u>Proof:</u> Let V be a neighbourhood of an element $z \in \mathbb{R} \cup \{-\infty\}$; there exist $x, y \in \mathbb{R} \cup \{-\infty\}$ with $x < y$ such that

$$]x,y[\;\subset V \qquad \text{if } z \neq -\infty$$

resp. $[x,y[\;\subset V$ otherwise, where $].,.[$ denotes the open and $[.,.[$ the half-open interval.
Since φ preserves the order (see Theorem I.2.) and F is complete

$$\varphi^{-1}(z) \subset U_{ab} := U_{\varphi^{-1}(x)\varphi^{-1}(y)} \subset \varphi^{-1}(V) \qquad \text{if } z \neq -\infty$$

resp.

$$\bar{0} = \varphi^{-1}(z) \subset U_{\bar{0}b} := U_{\bar{0}\varphi^{-1}(y)} \subset \varphi^{-1}(V) \qquad \text{if } z = -\infty$$

Thus $\varphi^{-1}(V)$ is a neighbourhood of $\varphi^{-1}(z)$.
To proof the continuity of φ in $c \in F$ $(c \neq \bar{0})$, let W be an open set in $\mathbb{R} \cup \{-\infty\}$ and let x, y in $\mathbb{R}$ such that

$$z := \varphi(c) \subset \;]x,y[\; \subset W.$$

With the above considerations

$$c \in \;]\varphi^{-1}(x),\varphi^{-1}(y)[\; \subset \varphi^{-1}(W),$$

which implies the assertion. The proof in the case $c = \bar{0}$ is analogous.
In the same way we proof the continuity of φ^{-1}.
Hence φ is a topological mapping. ∎

Let F^n be equipped with the product topology T^n. Then it follows:

1) (F^n, T^n) satisfies the first axiom of countability
2) The sets
$$Q_{ab} := \{x \in F^n \mid a_i \leq x_i \leq b_i \text{ for } i=1, \dots, n\}$$
for a, b $\in F^n$ are compact in (F^n, T^n).
3) Every bounded sequence in F^n has a cluster point.
4) The extremal-inner-product is a continuous function.

For abbreviation we write F^n instead of (F^n, T^n).
Finally, we proof

<u>Lemma III.3.:</u> Let a_i, b_i be in F for i=1, ... , k such that $a_i < b_i$; then there exists $\alpha \in F$ with $\overline{0} < \alpha < \overline{1}$ such that

$$a_i < \alpha \circ b_i < b_i \qquad \text{for } i=1, \dots, k$$

<u>Remark:</u> For k=1 the assertion is equivalent to a further property for an extremal algebra considered in some papers of K. Zimmermann (see [11], [12]), there denoted as (A6). For k≥1 this lemma is a conclusion of (A6) and was proved by K. Zimmermann in [11](Lemma 1h). In our case the assertion follows by reason of the completeness of F and the order-preserving isomorphism.

<u>Proof:</u> Since $b_i > \overline{0}$

$$a_i \circ b_i^{-1} < \overline{1} \qquad \text{for } i=1, \dots, k$$

According to (A4) it follows

$$z := \sum_{i=1}^{k}{}^{\oplus}\, a_i \circ b_i^{-1} < \overline{1}$$

Let a be a real number such that

$$\varphi(z) < a < \varphi(\overline{1})$$

Define $\alpha := \varphi^{-1}(a)$; therefore $\alpha \neq \overline{0}$ (since $a \neq -\infty$). Furthermore, since φ preserves the order

$$z < \alpha < \overline{1}$$

This implies $a_i \circ b_i^{-1} < \alpha < \overline{1}$ for i=1, ... , k , which is equivalent to

$$a_i < \alpha \circ b_i < b_i \qquad \text{for } i=1, \dots, k$$

If $\alpha \circ b_i = b_i$ for one $i \in \{1, \dots, k\}$, the equality

$$\alpha = \alpha \circ b_i \circ b_i^{-1} = b_i \circ b_i^{-1} = \overline{1}$$

implies a contradiction. ∎

It should be mentioned that K. Zimmermann introduces in [12] on an extremal algebra with the additional properties (A5) and (A6) (see remark above) a topology and calls it topological extremal algebra. An easy computation shows that $(F,\oplus,\circ)$ is a topological extremal algebra.

IV. The extremally-linear otpimization problem

In the remaining part of this paper let $(F,\oplus,\circ)$ be a complete, extremal, f.o. l.o. group with zero-element $\overline{0}$.
Furthermore, let T be a compact Hausdorff-space, $n \in \mathbb{N}$, $p \in F^n$, and $b: T \to F$ and $B: T \to F^n$ continuous mappings. Then consider the extremally-linear minimization problem

$$\text{ELP} \qquad \min\ (p,x)$$

$$\text{subject to: } (B(t),x) \geq b(t) \qquad \text{for all } t \in T,$$

where $(.,.)$ denotes the extremal-inner-product. Minimization of the function (p,x) means to minimize with respect to the order relation "$\leq$" on $(F,\oplus,\circ)$.
The problem ELP and a corresponding extremally-nonlinear problem were first discussed and presented by Helbig [8]. Moreover, a necessary and sufficient condition for optimal points of ELP and a Kuhn-Tucker-type theotheorem are given in [8]. It should be mentioned that U. Zimmermann [15] considers optimization problems in ordered algebraic structures with a finite number of restrictions.
In this paper, the problem ELP is considered in dependence on the objective function p, whereas B and b are fixed mappings. On account of this we write ELP(p) and describe the feasible set, minimal value, and optimal set as follows:

$$Z := \{x \in F^n \mid (B(t),x) \geq b(t) \text{ for all } t \in T\}$$

$$E_p := \inf_{x \in Z} (p,x)$$

$$P_p := \{x \in Z \mid (p,x) = E_p\}.$$

In the next chapters we study continuity properties of the mappings

$$E : L_{Bb} \longrightarrow F \qquad \underline{\text{minimal-value-mapping}}$$
$$p \longrightarrow E_p$$

$$P : L_{Bb} \longrightarrow POT(F^n) \qquad \underline{\text{optimal-set-mapping}}$$
$$p \longrightarrow P_p$$

where L_{Bb} denotes the solvability set, that is

$$L_{Bb} := \{p \in F^n \mid P_p \neq \emptyset\}$$

and $POT(F^n)$ denotes the power set of F^n, that is the set of all subsets of F^n including the empty set and the set F^n.

V. The minimal-value-mapping

It should be mentioned that a function $f: F^n \longrightarrow F$ is called extremal-concave (e-concave), if

$$f(\alpha \circ x \oplus \beta \circ y) \geq \alpha \circ f(x) \oplus \beta \circ f(y)$$

for all $x, y \in F^n$ and $\alpha, \beta \in F$ with $\alpha \oplus \beta = \overline{1}$.
A set $A \subset F^n$ is called extremal-convex (e-convex), if for all $x, y \in F^n$ and $\alpha, \beta \in F$ with $\alpha \oplus \beta = \overline{1}$

$$\alpha \circ x \oplus \beta \circ y \in A.$$

Then we have:

Theorem V.1.: The mapping $E: L_{Bb} \longrightarrow F$ has the following properties:
a) E is e-concave on every e-convex subset of L_{Bb}.
b) E is continuous on L_{Bb}.
Proof: a) Let p^1, p^2 be in L_{Bb} and $\alpha, \beta \in F$ with $\alpha \oplus \beta = \overline{1}$.
Furthermore let v be an arbitrary element of $P_q := P_{\alpha \circ p^1 \oplus \beta \circ p^2}$.
It follows:

$$E_q = (q,v) = \alpha \circ (p^1,v) \oplus \beta \circ (p^2,v) \geq \alpha \circ E_{p^1} \oplus \beta \circ E_{p^2}.$$

b) Suppose that E is not continuous in $p \in L_{Bb}$. Then there exists a neighbourhood U_{ac} from E_p and a sequence $(p^k)_{k \in \mathbb{N}}$ with $p^k \rightarrow P$ such that

$$E_{p^k} \notin U_{ac} \qquad \text{for all } k \in \mathbb{N}.$$

Consider two cases:
1.) $E_p < c \leq E_{p^k} \qquad$ for all $k \in \mathbb{N}' \subset \mathbb{N}$

2.) $E_{p^k} \leq c < E_p$ for all $k \in \mathbb{N}'' \subset \mathbb{N}$

<u>To 1.)</u> For arbitrary $v \in P_p$ there exists a $k_o \in \mathbb{N}$ such that

$$(p^k, v) \in U_{ac} \qquad \text{for all } k \geq k_o$$

This leads to a contradiction since

$$(p^k, v) < c \leq E_{p^k} \leq (p^k, v).$$

<u>To 2.)</u> Lemma III.3. implies the existence of $f \in F$ such that

$$E_{p^k} \leq a < f < E_p$$

for all $k \in \mathbb{N}''$. For abbreviation define the set $M := \{i \in \{1,...,n\} | p_i \neq \bar{0}\}$.
Define a neighbourhood U_{de} of p by

$$d_i := f \circ p_i \circ E_p^{-1} \qquad \text{for } i=1, \dots, n$$

$$e_i := \begin{cases} c \circ p_i \circ E_p^{-1} & \text{if } i \in M \\ \bar{1} & \text{if } i \notin M \end{cases}$$

Then $p^k \in U_{de}$ for all $k \geq k_1$, k_1 sufficiently large. Choose for each $k \geq k_1$ an element $v^k \in Z$ with

$$(p^k, v^k) < f.$$

For $i \in M$, we have

$$v_i^k < (p_i^k)^{-1} \circ f < d_i^{-1} \circ f,$$

which implies $v_i^k \longrightarrow v_i$ for $i \in M$ and $v_i \in F$ suitable.
Since $v^k \in Z$

$$f < E_p \leq (p, v^k) = \sum_{i \in M}^{\oplus} p_i \circ v_i^k.$$

Because of the convergence of

$$\sum_{i \in M}^{\oplus} p_i^k \circ v_i^k \longrightarrow \sum_{i \in M}^{\oplus} p_i \circ v_i,$$

a contradiction follows by

$$(p^k, v^k) < f < (p^k, v^k)$$

for sufficiently large $k \in \mathbb{N}''$. ∎

A similar result for semi-infinite linear optimization problems is due to Brosowski [1], p. 115.

VI. The optimal-set-mapping

The set-valued mapping $P: L_{Bb} \longrightarrow POT(F^n)$, which assigns to an objective function p the set of the optimal points of ELP(p), is said to be (see Brosowski [1], p. 26ff):

- closed at $p \in L_{Bb}$ if and only if for all sequences $(p^k)_{k\in\mathbb{N}}$ in L_{Bb} and $(v^k)_{k\in\mathbb{N}}$ in P_{p^k} satisfying $p^k \longrightarrow p$ and $v^k \longrightarrow v$, one has $v \in P_p$
- upper semicontinuous (u.s.c.) at $p \in L_{Bb}$ if and only if for each open set W with $P_p \subset W$, there exists an open set V with $p \in V$ such that $P_q \subset W$ for every $q \in V$.

Theorem VI.1.: The mapping $P : L_{Bb} \longrightarrow POT(F^n)$ is closed at each $p \in L_{Bb}$.
Proof: Let $(p^k)_{k\in\mathbb{N}}$ be a sequence in L_{Bb} and let $(v^k)_{k\in\mathbb{N}}$ be a sequence in P_{p^k} such that

$$p^k \longrightarrow p \quad \text{and} \quad v^k \longrightarrow v \qquad \text{for } k \to \infty.$$

Suppose there exist $w \in Z$ and $c \in F$ with

$$(p,w) < c < (p,v).$$

From the continuity of the extremal-inner-product it follows for $k \geq k_o$

$$(p^k,w) < c < (p^k,v^k),$$

which contradicts $v^k \in P_{p^k}$. ∎

A similar result holds in semi-infinite linear optimization (see Brosowski [1], p. 113, Theorem 9(1)).

For abbreviation define for $p \in F^n$ a set N(p) by

$$N(p) := \{i \in \{1, \ldots , n\} \mid p_i \neq \bar{0}\}$$

Lemma VI.2.: Let $Z \neq \emptyset$ and $p \in F^n$. If

$$\#N(p) = n,$$

then $P_p \neq \emptyset$.

Proof: If $(\overline{0}, \ldots, \overline{0}) \in Z$, then it follows immediately $P_p \neq \emptyset$, since $(\overline{0}, \ldots, \overline{0}) \in P_p$.
Otherwise $(p,x) > \overline{0}$ for all $x \in Z$.
For an element $z \in Z$ define

$$K := \{x \in Z \mid (p,x) \leq (p,z)\}.$$

This set is e-convex and compact, since

$$x_i \leq p_i^{-1} \circ (p,z) \qquad \text{for } i=1, \ldots, n$$

for all $x \in K$. The continuous mapping $(p,.)$ attains its minimum over the compact set K. Hence the assertion holds. ■

Since P is u.s.c. in the case $\#N(p) = 0$, we derive a necessary and sufficient condition for the u.s.c. of P under the assumption $\#N(p) \neq 0$.

Theorem VI.3.: Let p be an element of L_{Bb} such that $\#N(p) \neq 0$ and $E_p \neq \overline{0}$.
The mapping $P : L_{Bb} \longrightarrow POT(F^n)$ is u.s.c. in p if and only if $\#N(p) = n$ or $\#N(p) = 1$.

Proof: Let $N := N(p)$.

=>: Let $2 \leq \#N \leq n-1$, which implies $CN \neq \emptyset$, where CN denotes the complement of the set N. Without loss of generality let

$$N = \{1, \ldots, m\}$$
$$CN = \{m+1, \ldots, n\}$$

and $E_p = \overline{1}$. Otherwise introduce $q \in F^n$ with

$$q_i := p_i \circ E_p^{-1} \qquad \text{for } i=1, \ldots, n.$$

An easy computation shows that $E_q = \overline{1}$ and $P_q = P_p$.
Let $(s^k)_{k \in \mathbb{N}}$, resp. $(r^k)_{k \in \mathbb{N}}$ be strictly increasing, resp. decreasing sequences with $s^k \longrightarrow \overline{1}$, resp. $r^k \longrightarrow \overline{0}$. For $k \in \mathbb{N}$ define

$$p_i^k := \begin{cases} s^k \circ p_i & \text{if } i = 1 \\ p_i & \text{if } i \in N,\ i \neq 1 \\ r^k & \text{if } i \in CN \end{cases}$$

Lemma VI.2. implies $P_{p^k} \neq \emptyset$ for all $k \in \mathbb{N}$. For arbitrary $w^k \in P_{p^k}$ define

$$v_i^k := \begin{cases} (p_i^k)^{-1} \circ E_{p^k} & \text{if } i \in N \\ (p_i^k)^{-1} \circ E_{p^k} \oplus w_i^k & \text{if } i \in CN \end{cases}$$

This implies $w_i^k \leq v_i^k$ for $i \in \{1, \ldots , n\}$, since

$$(*) \qquad p_i^k \circ w_i^k \leq E_{p^k} = p_i^k \circ v_i^k \qquad \text{for } i \in N$$

and

$$(**) \qquad p_i^k \circ w_i^k \leq E_{p^k} \dot{\oplus} p_i^k \circ w_i^k = p_i^k \circ v_i^k \qquad \text{for } i \in CN.$$

v^k is a feasible point, because

$$(\nabla) \qquad b(t) \leq (B(t),w^k) = B(t)_i \circ w_i^k \leq B(t)_i \circ v_i^k \leq (B(t),v^k)$$

for all $t \in T$. Together with (*) and (**) we see:

$$v^k \in P_{p^k} \qquad \text{for all } k \in \mathbb{N}.$$

For the open set W defined by

$$W := \bigcup_{l \in \mathbb{N}} W_l := \bigcup_{l \in \mathbb{N}} \{x \in F^n \mid x_1 < (s^l \circ p_1)^{-1};\ x_n < (r^l)^{-1}\}$$

we conclude: i) $P_p \subset W$

ii) $P_{p^k} \cap CW \neq \emptyset$

which contradicts the u.s.c. of P.
To proof i) let u be in P_p. $((r^k)^{-1})_{k \in \mathbb{N}}$ is an unbounded sequence; hence, there exists a $l_o \in \mathbb{N}$ such that $u_n < (r^l)^{-1}$ for all $l \geq l_o$. Since $u_1 \leq p_1^{-1} \leq (s^l \circ p_1)^{-1}$ for all $l \in \mathbb{N}$ the assertion i) holds.
If $l \leq k$, it is

$$(r^l)^{-1} < (r^k)^{-1} \leq v_n^k$$

which implies $v^k \notin W_l$. If $l > k$, we have

$$(s^l \circ p_1)^{-1} < (s^k \circ p_1)^{-1} = v_1^k$$

which implies $v^k \notin W_l$.
Alltogether: $v^k \notin W_l$ for all $l \in N$. Hence, ii) is proved.

<u><=:</u> If P is not u.s.c. in p, then there exists an open set W with $P_p \subset W$ and a sequence $(p^k)_{k \in \mathbb{N}}$ with $p^k \longrightarrow p$ such that

$$P_{p^k} \cap CW \neq \emptyset.$$

<u>Case $\#N = n$</u>

Let v^k be in $P_{p^k} \cap CW$. Let U_{ac} and U_{de} be arbitrary neighbourhoods of E_p

resp. p. Then for sufficiently large $k \in \mathbb{N}$

$$(p^k, v^k) = E_{p^k} \in U_{ac}$$
$$p^k \in U_{de}.$$

Without loss of generality $d_i > \overline{O}$ for $i=1, \ldots, n$. Then the following inequality holds for $i=1, \ldots, n$:

$$v_i^k < (p_i^k)^{-1} \circ c < d_i^{-1} \circ c.$$

The bounded sequence $(v^k)_{k \in \mathbb{N}}$ has a convergent subsequence, which converges to an element $v \in F^n$. Since P is a closed mapping and CW is a closed set,

$$v \in P_p \cap CW$$

in contradiction to the assumption.

Case $\#N = 1$

Let $N = \{1\}$; for arbitrary $v \in P_p$ let $k_o \in \mathbb{N}$ be such that

$$\sum_{i=2}^{n} {}^{\oplus}\, p_i^k \circ v_i < p_1^k \circ v_1 \qquad \text{for all } k \geq k_o.$$

If $x \notin P_p$ we see that

$$x_1 > v_1 = p_1^{-1} \circ E_p$$

and

$$(p^k, v) = p_1^k \circ v_1 < p_1^k \circ x_1 \leq (p^k, x)$$

which implies $x \notin P_{p^k}$; therefore $P_{p^k} \subset P_p \subset W$ in contradiction to the assumption. ∎

Since we do not need the assumption $E_p \neq \overline{O}$ in the case $\#N = n$ in the sufficient part of the above theorem, we conclude:

Corollary VI.4.: Let $p \in L_{Bb}$. The mapping $P : L_{Bb} \longrightarrow POT(F^n)$ is u.s.c. in p, if P_p is compact.

Proof: It suffices to show the equivalence

$$\#N = n \quad \Longleftrightarrow \quad P_p \text{ is compact}$$

=>: If $\#N = n$ define a point $v \in P_p$ by

$$v_i := p_i^{-1} \circ E_p \qquad \text{for } i=1, \ldots , n.$$

Like (∇) in Theorem VI.3. we conclude $v \in Z$.
The optimality of v is obvious.
Thus, P_p is a closed subset of the compact set

$$K := \{x \in F^n \mid x_i \le v_i \text{ for } i=1, \ldots , n\}$$

and hence compact.
<u><=:</u> Since P_p is compact

$$P_p \subset K := \{x \in F^n \mid x_i \le c_i \text{ for } i=1, \ldots , n\}$$

with a suitable $c \in F^n$.
If there exists a $j \in \{1, \ldots , n\}$ with $p_j = \overline{0}$, then define for an arbitrary $v \in P_p$ and $s > \overline{1}$

$$u_i := \begin{cases} v_i & \text{if } i \neq j \\ v_i \oplus s \circ c_i \oplus s & \text{if } i = j \end{cases}$$

since $u_i \ge v_i$ $(i=1, \ldots , n)$ u is feasible (see (∇) in Theorem VI.3.).
u is an optimal point, because

$$(p,u) = \sum_{i \neq j}^{\oplus} p_i \circ u_i = \sum_{i \neq j}^{\oplus} p_i \circ v_i = (p,v) = E_p.$$

But $c_j < s \circ c_j \le u_j$, which immediately implies

$$u \notin K$$

in contradiction to the assumption. ∎

This corollary has a pendant in semi-infinite linear optimization (see Brosowski [1], p. 113, Theorem 9(2)).
The following example shows that we cannot drop the assumption $E_p > \overline{0}$ in the case $\#N = 1$.

<u>Example VI.5.:</u> Let $(F,\oplus,\circ) = (\mathbb{R}^+,\max,\cdot)$.
Furthermore, let $n = 3$, $T = \{1\}$, $B(1) = (0,0,1)$, $b(1) = 1$, and $p=(1,0,0)$.
In this example we have

$$\#N = 1; \; E_p = 0$$
$$P_p = \{x \in F^3 \mid x_1 = 0; \; x_3 \ge 1\}.$$

We define a sequence $(p^k)_{k\in\mathbb{N}}$ in L_{Bb} by

$$p^k := (1, 1/k^2, 1/k)$$

and an open set W by

$$W := \bigcup_{l\in\mathbb{N}} W_l := \bigcup_{l\in\mathbb{N}} \{x \in F^3 \mid x_1 < 1/l;\ x_2 < l\}$$

For $k \in \mathbb{N}$, the points

$$v^k := (1/k, k, 1)$$

are feasible. They are optimal since otherwise there exists a w^k in Z such that

$$(p^k, w^k) < (p^k, v^k) = 1/k.$$

In particular, it follows

$$w_3^k < 1/k \circ (p_3^k)^{-1} = 1.$$

Therefore $w^k \notin Z$ in contradiction to the last assumption.
We claim that i) $P_p \subset W$
ii) $P_{p^k} \cap CW \neq \emptyset$

To i): For arbitrary $u \in P_p$ let $l_o \in \mathbb{N}$ such that $u_2 < l_o$. Since $u_1 < 1/l$ for all $l \in \mathbb{N}$, we conclude

$$u \in W_l \subset W.$$

To ii): If $l \le k$, we have $l \le k = v_2^k$ and $v^k \notin W_l$.
If $l > k$, we have $1/l < 1/k = v_1^k$ and $v^k \notin W_l$.
Alltogether: $v^k \notin W$. ∎

VII. An application

Beside the wellknown asssignment problem some authors (see for instance Ducamp [4], Garfinkel [6], Gross [7]) consider the bottleneck-assignment problem BAP which differs only in the objective function and can be formulated as follows:

Let N be in $\mathbb{N}$ and c_{ij} real numbers for $i=1, \ldots, N$ and $j = 1, \ldots, N$.
Then consider:

$$\text{BAP} \quad \min \max_{i,j=1,\dots,N} c_{ij}x_{ij}$$

subject to:

$$\sum_{i=1}^{N} x_{ij} = 1 \qquad \text{for } j=1, \dots , N$$

$$\sum_{j=1}^{N} x_{ij} = 1 \qquad \text{for } i=1, \dots , N$$

$$x_{ij} \in \{0,1\} \qquad \text{for } i,j=1, \dots , N$$

One can interpret BAP like follows:

- suppose N jobs are to be done in parallel, each requiring one man
- let c_{ij} be the time man i needs to realize job j
- let x_{ij} be the indicator whether man i is assigned to job j or not
- the minimal value of BAP is the completition time, if all jobs start simultaneously.

In [2], Burkard, Hahn and U. Zimmermann study an algebraic approach to solve the usual assignment problem with the sum objective function as well as the bottleneck-assignment problem.

Consider the extremal f.o. l.o. group $(F,\oplus,\circ) = (\mathbb{R}^+,\max,\cdot)$ with zero-element $\bar{0}$ and define

$$n = N^2$$
$$T = \{1, \dots , 2N\}$$
$$p_k = c_{ij} \qquad \text{for } k = (i-1)N + j \quad i,j=1, \dots , N$$

For $t \in \{1, \dots , N\}$ let

$$B(t)_i := \begin{cases} \bar{1} & \text{if } i=(t-1)N+1, \dots , tN \\ \bar{0} & \text{otherwise} \end{cases}$$

and for $t \in \{N+1, \dots , 2N\}$

$$B(t)_i := \begin{cases} \bar{1} & \text{if } i=t-N,(t-N)+N, \dots , (t-N)+(N-1)N \\ \bar{0} & \text{otherwise} \end{cases}$$

Then the problem

$$\text{EAP}(p) \qquad \min\ (p,y)$$

subject to

$$(B(t),y) \geq b(t) \qquad \text{for all } t \in T$$

is a minimization problem of type ELP(p) with the following properties:

<u>Theorem VII.1.:</u> Consider the problems BAP and EAP. Let Z_B, E_B, P_B, Z_E, E_E, and P_E be the feasible set, minimal value, and optimal set of BAP, resp. EAP. Then

$$\text{i) } Z_B \subset Z_E$$
$$\text{ii) } E_B = E_E$$
$$\text{iii) } P_B \subset P_E.$$

<u>Proof:</u> Since i) is obvious, the inequality $E_B \geq E_E$ is fulfilled because of i), and iii) follows from i) and ii), it suffices to show that the assumption $E_E < E_B$ leads to a contradiction.
For sake of clearness let EAP be transformed like follows:

$$\text{EAP} \quad \min \max_{i,j=1,\dots,N} c_{ij}x_{ij}$$

subject to

$$\max_{i=1,\dots,N} x_{ij} \geq 1 \qquad \text{for } j=1, \dots , N$$
$$\max_{j=1,\dots,N} x_{ij} \geq 1 \qquad \text{for } i=1, \dots , N$$

In this way we identify a vector $x \in F^n = F^{N^2}$ with a matrix $x \in F^{N,N}$.
Let $x \in P_E$ be arbitrary; without loss of generality we can assume $x_{ij} \leq 1$ for $i,j=1, \dots , N$. Otherwise define a set K by

$$K := \{(i,j) \in \{1, \dots , N\} \times \{1, \dots , N\} \mid x_{ij} > 1\}$$

and put

$$\tilde{x}_{ij} := \begin{cases} x_{ij} & \text{if } (i,j) \notin K \\ 1 & \text{if } (i,j) \in K \end{cases}$$

If $\tilde{x}$ is not feasible for EAP, then there exists either i_o or j_o in $\{1, \dots , N\}$ with

$$\max_{j=1,\dots,N} \tilde{x}_{i_oj} < 1 \qquad \text{or} \qquad \max_{i=1,\dots,N} \tilde{x}_{ij_o} < 1.$$

This implies either $(i_o,j) \notin K$ for all $j \in \{1, \dots , N\}$ or $(i,j_o) \notin K$ for all $i \in \{1, \dots , N\}$, which contradicts $x \in Z_E$.
Since $\tilde{x}_{ij} \leq x_{ij}$ for all $i,j=1, \dots , N$, $\tilde{x} \in P_E$ with the above property.
The i-th row (resp. the j-th column) of x is called <u>critical</u>, if only one component of this row (resp. column) differs from zero. Since $x \in Z_E$ this component is equal to one. The following procedure determines a feasible point of BAP starting with $x \in P_E$ with $x_{ij} \leq 1$ for $i,j=1, \dots , N$:

S1) If i-th row is critical (let $x_{ij_o} = 1$), put

$$x_{lj_o} = 0 \qquad \text{for } l=1, \dots , N, \ l \neq i$$

S2) If j-th column is critical (let $x_{i_oj} = 1$), put

$$x_{i_ol} = 0 \qquad \text{for } l=1, \dots , N, \ l \neq j$$

S3) If neither a row nor a column is critical, let $(i_o,j_o) \in \{1, \dots , N\} \times \{1, \dots , N\}$ such that $x_{i_oj_o} = 1$ and put

$$x_{i_ol} = 0 \qquad \text{for } l=1, \dots , N, \ l \neq j_o$$
$$x_{lj_o} = 0 \qquad \text{for } l=1, \dots , N, \ l \neq i_o$$

S4) Repeat S1), S2), S3) while all rows and columns are critical

After finishing this procedure one gets a $x \in P_E$ which has exactly one component in each row and each column equal to one and zero otherwise; hence, x is feasible for BAP. Therefore

$$\max_{i,j=1,\dots,N} c_{ij}x_{ij} = E_E < E_B$$

which contradicts $x \in Z_B$. ■

In consequence of this theorem one can compute a solution of BAP by application of an algorithm for solving EAP (see for instance K. Zimmermann [13]) and of the little procedure above.
Moreover, the continuity investigations of the chapters V and VI can be applied to BAP. In this case we consider the coefficients of the objective function as parameters of the bottleneck-assignment problem.

References:

[1] B. BROSOWSKI: Parametric semi-infinite optimization
Peter D. Lang-Verlag, Frankfurt A.M. and Bern, 1982

[2] R.E. BURKARD, W. HAHN and U.ZIMMERMANN: An algebraic approach to assignment problems, Mathematical Programming 12, 1977, 318-327

[3] R. CUNNIGHAME-GREEN: Minimax Algebra
Lecture Notes in Economics and Mathematical Systems 166
Springer-Verlag, Berlin, Heidelberg, New York, 1979

[4] A. DUCAMP: Un problème d'affectation
Cahiers du Centre d'Etudes de Recherche Operationelle 8
1966, 69-72

[5] L. FUCHS: Partially ordered algebraic systems
Pergamon Press, Oxford, 1963

[6] R.S. GARFINKEL: An improved algorithm for the bottleneck-assignment problem, Operations Research 19, 1971, 1747-1751

[7] O. GROSS: The bottleneck-assignment problem Report P-1630, The Rand Corporation, Santa Monica, California, 1959

[8] S. HELBIG: Optimization problems on extremal algebras: necessary and sufficient conditions for optimal points in: B. Brosowski and F. Deutsch (Eds.), Parametric optimization and approximation, International Series of Numerical Mathematics 72, Birkhäuser-Verlag, Basel, Boston, Stuttgart, 1985, 166-184

[9] A.I. KOKORIN and V.M. KOPYTOV: Fully ordered groups John Wiley & sons, New York, Toronto, 1974

[10] J. NEDOMA: Řešeni optimalizačnich úloh na ohodnocenych orientovaných grafech pomoci extremâlni algebry Internal publication of EML EU CSAV, Praha 1974

[11] K. ZIMMERMANN: A general separation theorem in extremal algebras Ekonomicko-Matematický Obzor 13, 1977, 179-201

[12] K. ZIMMERMANN: Extremally convex functions Wissenschaftliche Zeitschrift der Pädagogischen Hochschule 'N.K. Krupskaja', Halle, XVII, 1977, 3-7

[13] K. ZIMMERMANN: Solution of some optimization problems on the extremal algebra, Studies on mathematical programming, Mathematical Methods of Operations Research 1, 1980, 179-185

[14] K. ZIMMERMANN: On some synchronization problems with extremal operations, Ekonomicko-Matematický Obzor 21, 1985, 196-209

[15] U. ZIMMERMANN: Linear and combinatorial optimization in ordered algebraic structures, Annals of discrete mathematics 10, North Holland, 1981

LOCAL ASPECTS OF A METHOD FOR SOLVING MEMBRANE-EIGENVALUE PROBLEMS BY PARAMETRIC SEMI-INFINITE PROGRAMMING

Rainer Hettich[1] and Georg Still[1]

Abstract In former papers [7,9] a method has been proposed to compute approximations to eigenvalues and eigenfunctions of the Laplacean by defect-minimization. This method leads to a parametric (linear) semi-infinite programming problem $P(\lambda)$, the parameter λ of which is to be determined such that a certain function ε of the value $v(\lambda)$ and the solution $z(\lambda)$ of $P(\lambda)$ has a local minimum. In general ε depends nondifferentiable on λ. A method of Mifflin is taken to find the local minima of ε. As this requires the knowledge of generalized gradients of ε differentiability of $v(\lambda)$ and $z(\lambda)$ is studied under appropriate regularity conditions. Some numerical examples are given.

1. Introduction

In [7,9] a method was presented for computing approximations to eigenvalues and eigenfunctions of the Laplace-operator by means of a parametric semi-infinite optimization problem $P(\lambda)$. To determine such approximations requires the computation of local minima of a function of this problem-parameter $\lambda \in \mathbb{R}$ given by the ratio of the value $v(\lambda)$ and a norm of a solution $z(\lambda)$ of problem $P(\lambda)$.

In [7] mainly the global search for local minima in some λ-interval was considered and it was shown that a sort of continuation method can be applied to speed up computation. In this paper local aspects are emphasized, instead. Nevertheless, to keep the presentation self-contained, in the next section we shortly present the method (slightly modified w.r.t. [7,9]) and the global aspects. In Section 3 an algorithm of Mifflin for minimizing univariate nondifferentiable functions is sketched which has been used by us to improve a given rough estimate of an eigenvalue. As this algorithm makes use of generalized gradients, in Section 4 the differentiability of the solution $z(\lambda)$ of problem $P(\lambda)$ is investigated under appropriate regularity conditions. Results are obtained which are quite similar to those given in [5] for finite parametric problems. In the final section a number of computational results are presented.

---

1) Prof. Dr. Rainer Hettich, Fachbereich IV - Mathematik
Universität Trier, P. O. Box 3825, D-5500 Trier, Germany

Dr. Georg Still, Fachbereich IV - Mathematik
Universität Trier, P. O. Box 3825, D-5500 Trier, Germany

2. Outline of the method

We consider the following problem:
Given an open, bounded domain $R \subset \mathbb{R}^2$ with boundary B. Let $\bar{R} = R \cup B$,

$$V = C^2(R) \cap C(\bar{R}),$$

and

$$V_o = \{f \in V \mid f(x) = 0 \quad \text{for all } x \in B\}.$$

Then eigenvalues λ and eigenfunctions $u \in V_o$ of the Laplace-operator are sought, i. e. pairs (λ, u), $\lambda \in \mathbb{R}$ and $u \in V_o$, $u \neq 0$, such that

$$\Delta u(x) + \lambda u(x) = 0 \quad \text{for all } x \in R \,. \tag{2.1}$$

In the sequel we use the norm

$$\| f \| = \left\{ \int_R f^2(x)dx \,/ \int_R dx \right\}^{\frac{1}{2}}$$

on V. We remark that very mild conditions on R imply the existence of a sequence λ_i of eigenvalues with no finite point of accumulation and a corresponding complete, orthonormal system of eigenfunctions in V_o (cf. [3] for instance).

Our approach is based on the following theorem given by Moler and Payne [12]:

Theorem 2.1 Given $\hat{\lambda} \geq 0$, $\hat{u} \in V$, $\| \hat{u} \| = 1$ such that

$$\Delta \hat{u}(x) + \hat{\lambda}\hat{u}(x) = 0 \quad \text{for all } x \in R$$

and

$$\varepsilon := \max_{x \in B} \mid \hat{u}(x) \mid < 1.$$

Then the following holds:

a) There exists an eigenvalue λ_k of (2.1) such that

$$\frac{\hat{\lambda}}{1+\varepsilon} \leq \lambda_k \leq \frac{\hat{\lambda}}{1-\varepsilon} \tag{2.2}$$

b) There exists an eigenfunction v_k corresponding to λ_k such that

$$\| \hat{u} - v_k \| \leq \frac{\varepsilon}{\alpha}\left(1 + \frac{\varepsilon^2}{\alpha^2}\right)^{\frac{1}{2}}$$

with

$$\alpha := \min \left\{ \frac{|\lambda_i - \hat{\lambda}|}{\lambda_i} \,, \quad \lambda_i \neq \lambda_k \right\} . \quad \blacksquare$$

Choosing N linearly independent trial functions $\varphi_1(\lambda;x),..,\varphi_N(\lambda;x) \in V$ satisfying for any $\lambda \in \mathbb{R}_+$

$$\Delta\varphi_i(\lambda;x) + \lambda\varphi_i(\lambda;x) = 0 \quad , \quad x \in R \quad , \quad i = 1,..,N \quad ,$$

(for examples cf. [4,6,12]) and setting

$$\phi_N(\lambda;\alpha;x) = \sum_{i=1}^{N} \alpha_i\varphi_i(\lambda;x) \quad , \quad \alpha \in \mathbb{R}^N \quad , \tag{2.3}$$

we consider the following linear semi-infinite programming problem:

$P_N(\lambda)$ Minimize $\ell(\alpha,\mu) = \mu$ subject to

$$\left.\begin{aligned} \phi_N(\lambda;\alpha;x) &\le \mu \\ -\phi_N(\lambda;\alpha;x) &\le \mu \end{aligned}\right\} \text{ for all } x \in B$$

$$\alpha_1 \ge 1 \quad .$$

The last constraint is added to exclude the zero solution. Other (linear) constraints such as $\sum_{i=1}^{N} \alpha_i \ge 1$ may be more appropriate in some cases to admit solutions with $\alpha_1 = 0$. With a solution $(\mu(\lambda),\alpha(\lambda))$ of $P_N(\lambda)$ the value

$$\tilde{\varepsilon} = \frac{\mu(\lambda)}{\| \phi_N(\lambda;\alpha(\lambda);.)\|}$$

gives an error bound

$$|\lambda_k - \lambda| \le \lambda \frac{\tilde{\varepsilon}}{1-\tilde{\varepsilon}} \tag{2.4}$$

according to (2.2) provided that $\tilde{\varepsilon} < 1$. We define

$$\tilde{\varepsilon}_N(\lambda) := \min \left\{ \frac{\mu}{\| \phi_N(\lambda;\alpha;.)\|} \;\middle|\; (\mu,\alpha) \text{ is a solution of } P_N(\lambda)\right\} \tag{2.5}$$

Then "good" error bounds may be expected when taking values λ for which the right-hand side of (2.4), i. e. the value

$$\varepsilon_N(\lambda) := \lambda \frac{\tilde{\varepsilon}_N(\lambda)}{1 - \tilde{\varepsilon}_N(\lambda)} \tag{2.6}$$

is (locally) minimized. For fixed $N \in \mathbb{N}$ (i. e. a fixed number of trial functions in (2.3)) this means that we have to find all local minima of $\varepsilon_N(\lambda)$ for instance in a given interval $[\underline{\lambda},\overline{\lambda}]$ where eigenvalues are expected.

We first determine the local minima of $\varepsilon_{N_0}(\lambda)$ (N_0 an initial number of trial functions) on a discrete set $S = \{\overline{\lambda}_i = \underline{\lambda} + i\, h \;,\; i = 0,1,..\}$ in the interval $[\underline{\lambda},\overline{\lambda}]$ ($h > 0$ a stepsize). For that purpose we have to compute solutions $(\alpha(\lambda), \mu(\lambda))$ of $P_{N_o}(\lambda)$ at all points $\bar{\lambda}_i$. This can be done very efficiently by applying a sort of continuation method as described in [7]:

Given a solution of $P_N(\lambda^o)$ (by means of a simplex method) we may expect that for λ "near" λ^o the basis indices corresponding to the solution of problem $P_N(\lambda^o)$ lead immediately to an optimal solution or at least to a "good" dual or primal starting vertex of $P_N(\lambda)$ (cf. [7, Theorem 2]).

3. A method of R. Mifflin for minimizing a univariate function

Being interested in small values $\varepsilon_N(\lambda)$, we perform two steps:

a) Starting with an inclusion

$$a \le \lambda^*_{N_o} \le b$$

for a local minimum $\lambda^*_{N_o}$ of $\varepsilon_{N_o}(\lambda)$ given by minimum search on the discrete set S, the local minimum $\lambda^*_{N_o}$ is computed.

b) If the minimal value $\varepsilon_{N_o}(\lambda^*_{N_o})$ (i. e. the bound (2.4) for an eigenvalue λ_k) is not as small as desired, the number N_o of the trial functions is increased and a local minimum λ^*_N of $\varepsilon_N(\lambda)$ "near" $\lambda^*_{N_o}$ is computed for a number $N > N_o$. A possible starting interval is given by

$$J_N = [\lambda^*_{N_o} - 2\varepsilon_N(\lambda^*_{N_o}) \quad , \quad \lambda^*_{N_o} + 2\varepsilon_N(\lambda^*_{N_o})]$$

since it follows from Theorem 2.1 that $\lambda^*_N, \lambda_k \in J_N$. Step b) is repeated until for some N a bound $\varepsilon_N(\lambda)$ is as small as desired or N exceeds a given bound. (In general it will be difficult to guarantee density of the trial functions).

Since the computation of the function values $\varepsilon_N(\lambda)$ is rather expensive it is of great interest to use a fastly convergent minimization algorithm. Therefore, to compute the local minima of $\varepsilon_N(\lambda)$ (N fixed), we make use of an algorithm of R. J. Mifflin [11] having favourable convergence properties. We shortly outline this algorithm for solving the problem:

$$\text{Minimize } \varepsilon(\lambda) \quad , \quad \lambda \in [a,b]$$

with a real valued continuous function ε. We further assume that for all $\lambda \in [a,b]$ a generalized gradient $\varepsilon'(\lambda)$ (in the sense of Clarke [2]) is available.

In each step the algorithm starts with two points $x_k, y_k \in [a,b]$ satisfying

$$\varepsilon(x_k) < \varepsilon(y_k)$$

$$\varepsilon'(x_k)\,(y_k - x_k) < 0$$

(this implies that a local minimum λ^* exists in the interval defined by x_k, y_k) and finite difference approximations $D^2\varepsilon(x_k)$, $D^2\varepsilon(y_k)$ for second derivatives $\varepsilon''(x_k)$, $\varepsilon''(y_k)$. (In the beginning we set $D^2\varepsilon(x_o) = D^2\varepsilon(y_o) = 0$).

Applying an interpolation method based on quadratic or linear approximations to $\varepsilon(\lambda)$ near x_k, y_k a new point z_k between x_k and y_k is computed. Depending on the values $\varepsilon(z_k)$, $\varepsilon'(z_k)$ the algorithm stops if $\varepsilon'(z_k) = 0$ or defines new inclusion points x_{k+1}, y_{k+1} for λ^* and corresponding difference approximations $D^2\varepsilon(x_{k+1})$, $D^2\varepsilon(y_{k+1})$.

The following convergence results hold for the minimization algorithm (cf. [11]):

If ε is a (locally) Lipschitz function then the numbers x_k converge to a stationary point $\lambda^* \in (a,b)$.

If, moreover,

either ε is convex on $[a,b]$

or

$\varepsilon \in C^2[a,\lambda^*) \cap C^2(\lambda^*,b]$ and $\lim_{x \mapsto \lambda^*_\pm} \varepsilon''(x)$ exists, then setting

$$r_k := \begin{cases} |y_k - \lambda^*|\;|x_k - \lambda^*| & \text{if } x_k \neq \lambda^* \\ |y_k - \lambda^*| & \text{if } x_k = \lambda^* \end{cases}$$

the algorithm is superlinearly convergent in the following sense

$$\lim_{k\to\infty} \frac{r_{k+1}}{r_k} = 0 .$$

The algorithm requires the computation of generalized gradients $\varepsilon'(\lambda)$. In practice these could be approximated by means of finite differences. However, with regard to efficiency and numerical stability it seems preferable to use analytic expressions obtained by assuming appropriate regularity conditions to hold. The next section is intended to provide such formulae for our problem.

4. Differentiability

In this section we assume that

- $\Lambda \subset \mathbb{R}$ is an open set
- $X \subset \mathbb{R}$ is the union of a finite number of pairwise disjoint compact intervals, non of them reducing to a point
- $a: \Lambda \times X \to \mathbb{R}^n$, $b: \Lambda \times X \to \mathbb{R}$, and $c: \Lambda \to \mathbb{R}^n$ are twice continuously differentiable functions.

For abbreviation let $g: \Lambda \times X \times \mathbb{R}^n \to \mathbb{R}$ be the function

$$g(\lambda,x,z) = a^T(\lambda,x)z - b(\lambda,x) \ . \tag{4.1}$$

Then we consider the following parametric semi-infinite programming problem

$$P(\lambda): \text{ Minimize } F(\lambda,z) := c^T(\lambda)z \text{ subject to } z \in Z(\lambda)$$

where the feasible set $Z(\lambda)$ is given by

$$Z(\lambda) = \{z \mid g(\lambda,x,z) \le 0 \quad \text{for all } x \in X\} \ . \tag{4.2}$$

Obviously problem $P_N(\lambda)$ of Section 2 can be easily put into this form. Our aim in this section is to derive sufficient conditions on $P(\bar{\lambda})$, $\bar{\lambda} \in \Lambda$, such that in a neighborhood of $\bar{\lambda}$ problem $P(\lambda)$ has a unique solution $z(\lambda)$ depending continuously differentiable on λ. In addition we are interested in explicit formulae to compute $z_\lambda(\lambda)$.

We proceed in two steps: First, we shall show that under appropriate assumptions in a neighborhood of $(\bar{\lambda},\bar{z})$, $\bar{z}$ a solution of $P(\bar{\lambda})$, the problem can be reduced to a differentiable, convex, parametric problem (with a finite number of constraints!). In a second step this finite problem is treated like in [5] to obtain the required results.

In the sequel let $\bar{\lambda} \in \Lambda$ and $\bar{z} \in Z(\bar{\lambda})$ be fixed. We define

$$\bar{E} = \{x \in X \mid g(\bar{\lambda},x,\bar{z}) = 0\} \subset X \ , \tag{4.3}$$

the set of x such that the constraint is active.

<u>Assumption 1</u> For every $x \in \bar{E}$ we have

$$g_x(\bar{\lambda},x,\bar{z}) \ne 0 \quad \text{if } x \in \partial X = X \setminus \operatorname{int}(X) \tag{4.4}$$

and

$$g_{xx}(\bar{\lambda},x,\bar{z}) < 0 \quad \text{if } x \in \operatorname{int}(X) \ . \ \blacksquare \tag{4.5}$$

Note, that $\bar{z}$ being feasible, $g(\bar{\lambda},x,\bar{z})$ attains maxima with value 0 in the points in $\bar{E}$. Thus, for $x \in \operatorname{int}(X)$, in addition to (4.5) we have

$$g_x(\bar{\lambda},x,\bar{z}) = 0 \quad , \quad x \in \operatorname{int}(X) \cap \bar{E}. \tag{4.6}$$

Assumption 1 together with the assumptions on X implies that $\bar{E}$ is a finite set

$$\bar{E} = \{\bar{x}^1,..,\bar{x}^r\} \ . \tag{4.7}$$

For brevity we often write g^i, g_x^i, a_x^i etc for $g(\bar{\lambda},\bar{x}^i,\bar{z})$, $g_x(\bar{\lambda},\bar{x}^i,\bar{z})$, $a_x(\bar{\lambda},\bar{x}^i)$ etc.

Our first theorem is an immediate consequence of the Implicit-Function-Theorem applied to the equation

$$g_x(\lambda,x,z) = 0 \ . \tag{4.8}$$

Theorem 4.1 Let Assumption 1 hold. Then there exist

- neighborhoods $U_{\bar\lambda}$, $U_{\bar z}$ of $\bar\lambda$, $\bar z$ resp.
- continuously differentiable functions $x^i: U_{\bar\lambda} \times U_{\bar z} \to X$, $i = 1,..,r$, with $x^i(\bar\lambda,\bar z) = \bar x^i$
- a constant $\varepsilon > 0$

such that:

(i) For every $(\lambda,z) \in U_{\bar\lambda} \times U_{\bar z}$ the $x^i(\lambda,z)$, $i = 1,..,r$, are the only local maxima of $g(\lambda,\ .\ ,z)$ on X with value of g greater than $-\varepsilon$.

(ii) If $\bar x^i \in \partial X$, then $x^i(\lambda,z) \equiv \bar x^i$.

(iii) If $\bar x^i \in \mathrm{int}(X)$ then the derivatives $\bar x^i_\lambda = x^i_\lambda(\bar\lambda,\bar z)$ and $\bar x^i_z = x^i_z(\bar\lambda,\bar z)$ are given by

$$\bar x^i_\lambda = -(g^i_{xx})^{-1}\, g^i_{\lambda x} \quad , \quad \bar x^i_z = -(g^i_{xx})^{-1}\, g^i_{zx} \ . \quad ■ \tag{4.9}$$

As an easy consequence of Theorem 4.1 we obtain

Theorem 4.2 Let Assumption 1 hold. Then, with the notations of Theorem 4.1, we have for every fixed $\lambda \in U_{\bar\lambda}$

(i) $U_{\bar z} \cap Z(\lambda) = U_{\bar z} \cap Z_{red}(\lambda)$ with

$$Z_{red}(\lambda) = \{z \mid g(\lambda,x^i(\lambda,z),z) \le 0 \quad , \quad i = 1,..,r\} \ . \tag{4.10}$$

(ii) Locally in $U_{\bar z}$ problem $P(\lambda)$ may be replaced by

$$P_{red}(\lambda):\quad \text{Minimize } F(\lambda,z) = c^T(\lambda)z \quad \text{subject to} \quad z \in U_{\bar z} \cap Z_{red}(\lambda) \ . \tag{4.11}$$

(iii) The functions $\varphi^i(\lambda,z) := g(\lambda,x^i(\lambda,z),z)$ are convex functions of z. ■

The convexity of the functions φ^i is an easy consequence of the definition of the functions x^i and the linearity of $g(\lambda,x,z)$ as a function of z (cf. [10, p. 90]). Theorems 4.1 and 4.2 generalize results for non-parametric problems given in [8].

To obtain the desired differentiability results, we consider the so-called Karush-Kuhn-Tucker equations for $P_{red}(\lambda)$. With the Lagrangean

$$L(\lambda,z,\mu) = F(\lambda,z) + \sum_{i=1}^{r} \mu_i g(\lambda,x^i(\lambda,z),z) \tag{4.12}$$

these may be written as

$$L_z(\lambda,z,\mu) = 0 \tag{4.13 a}$$

$$L_\mu(\lambda,z,\mu) = 0 . \tag{4.13 b}$$

We get

$$L_z(\lambda,z,\mu) = c(\lambda) + \sum_{i=1}^{r} \mu_i[a(\lambda,x^i(\lambda,z)) + g_x(\lambda,x^i(\lambda,z),z)x_z^i(\lambda,z)] .$$

From the definition of the functions x^i we obtain

$$L_z(\lambda,z,\mu) = c(\lambda) + \sum_{i=1}^{r} \mu_i a(\lambda,x^i(\lambda,z)) . \tag{4.14}$$

(4.13) is a system of n + r equations for the n + r + 1 unknowns z, μ,λ. Under the assumption of a constraint qualification (for instance Assumption 2 lateron) a necessary condition for $\bar{z}$ to solve $P(\bar{\lambda})$ is that there exists a $\bar{\mu} \in \mathbb{R}^r$, $\bar{\mu} \geq 0$, such that $(\bar{\lambda},\bar{z},\bar{\mu})$ solves (4.13). If the Jacobian matrix w.r.t. z,μ is regular in $(\bar{\lambda},\bar{z},\bar{\mu})$, then, by the Implicit-Function-Theorem, in a neighborhood of $\bar{\lambda}$ (4.13) defines continuously differentiable functions $z(\lambda)$, $\mu(\lambda)$ with $z(\bar{\lambda}) = \bar{z}$ and $\mu(\bar{\lambda}) = \bar{\mu}$.

The Jacobian of (4.13) w.r.t. z, in $(\bar{\lambda},\bar{z},\bar{\mu})$ is given by

$$J = \begin{bmatrix} \mathcal{J} & A \\ A^T & 0 \end{bmatrix} \tag{4.15 a}$$

with

$$\mathcal{J} = \sum_{i=1}^{r} \bar{\mu}_i \, a_x^i(\bar{x}_z^i)^T , \tag{4.15 b}$$

an n×n-matrix, and the n×r-matrix

$$A = (a^1,\ldots,a^r) \tag{4.15 c}$$

with columns $a^i = a(\bar{\lambda},\bar{x}^i)$. The matrix $\mathcal{J}$ in (4.15 b) is positive semidefinite. In fact, from Theorem 4.1 we obtain

$$\mathcal{J} = \sum_{i=1}^{r} \bar{\mu}_i m_i a_x^i (a_x^i)^T \tag{4.16}$$

with $m_i \geq 0$ defined by

$$m_i = \begin{cases} 0 & \text{if } \bar{x}^i \in \partial X \\ -(g_{xx}^i)^{-1} & \text{if } \bar{x}^i \in \operatorname{int}(X) \end{cases} . \tag{4.17}$$

<u>Assumption 2</u> A has full rank r, i. e. the vectors $a^i = a(\bar{\lambda},\bar{x}^i)$, $i = 1,..,r$, are linearly independent. Moreover, the a^i, $i = 1,..,r$, together with the vectors $m_i a_x^i$, $i = 1,..,r$, span the whole of $\mathbb{R}^n$ (i. e. we can select a basis from $\{a^i, m_i a_x^i \mid i = 1,..,r\}$). ■

Theorem 4.3 If Assumption 2 holds and $\bar{\mu}_i > 0$ for all i such that $\bar{x}^i \in \mathrm{int}(X)$, then J, given by (4.15), is regular. ∎

Note that $\bar{J}$ is positive definite on $\{a^1,..,a^r\}^{\perp}$. The assertion of the theorem then follows from [8, page 6,7].

Observing that for $\lambda \in U_{\bar{\lambda}}$ the $z(\lambda) \in U_{\bar{z}}$ are solutions of $P(\lambda)$ as long as $\mu(\lambda) \geq 0$, we have shown the following theorem.

Theorem 4.4 Suppose that Assumptions 1 and 2 hold together with the condition $\bar{\mu} > 0$ (strict complementarity). Then there is a neighborhood $U_{\bar{\lambda}}$ of $\bar{\lambda}$ such that the (unique) solution $z(\lambda)$ of $P(\lambda)$, $\lambda \in U_{\bar{\lambda}}$, depends continuously differentiable on λ. Furthermore, $z_\lambda(\bar{\lambda})$ can be computed from

$$J \begin{pmatrix} z_\lambda(\bar{\lambda}) \\ \mu_\lambda(\bar{\lambda}) \end{pmatrix} = - \begin{pmatrix} L_{z\lambda}(\bar{\lambda},\bar{z},\bar{\mu}) \\ L_{\mu\lambda}(\bar{\lambda},\bar{z},\bar{\mu}) \end{pmatrix} \quad . \quad \blacksquare \qquad (4.18)$$

Remark

(1) Note that Assumption 2 is a sufficient condition for $\bar{z}$ to be an extreme point of the feasible set of $P(\bar{\lambda})$ (cf. [1], [13]).

(2) Using the Karush-Kuhn-Tucker equations to derive results like Theorem 4.4 is a well-known technique in parametric optimization and is not restricted to linear problems. Cf. [5] for an extensive study in this area.

5. Numerical experiences

The approach outlined above for solving the eigenvalue problem (2.1) is illustrated by taking the L-shaped region R as shown in Figure 1. Appropriate trial functions in this case are given by

$$\phi_N(\lambda;\alpha;x) = \sum_{\ell=1}^{N} \alpha_\ell J_{2\ell/3}(\sqrt{\lambda}r)\sin(2\ell\varphi/3)$$

where J_α denotes the Bessel function of order α, and (r,φ) are polar coordinates (cf. [6], [7]).

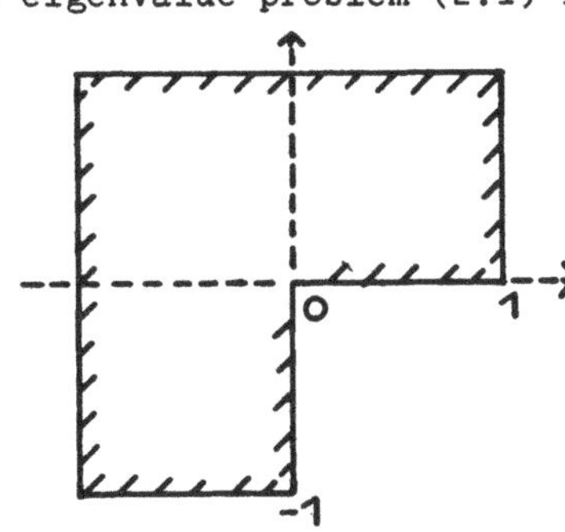

Figure 1

For computational purposes only one solution $(\bar{\alpha},\bar{\mu})$ of $P_N(\lambda)$ is calculated and $\tilde{\varepsilon}_N(\lambda)$ (cf. 2.5) is replaced by

$$\bar{\varepsilon}_N(\lambda) = \frac{\bar{\mu}(\lambda)}{\| \phi_N(\lambda;\bar{\alpha}(\lambda);.) \|}$$

to give bounds $\epsilon_N(\lambda)$ according to (2.6).

Note that $\bar{\epsilon}_N(\lambda) = \tilde{\epsilon}_N(\lambda)$, if the solution of $P_N(\lambda)$ is unique. The functions ϵ_N, $\tilde{\epsilon}_N$ are not differentiable and even not continuous in general. However we never met discontinuity during the computation.

To get first approximations for eigenvalues, local minima of $\epsilon_{N_o}(\lambda)$ have been computed for $N_o = 21$ on the discrete set $S = \{\bar{\lambda}_\nu = 8 + \nu, \nu = 0,1,..,92\}$ by taking a discretization of the boundary consisting of $M = 60$ points on B, which in our experiences may be considered as a good approximation to the semi-infinite problem. It would be preferable to deal in a final implementation of the algorithm with the semi-infinite problem.

To solve all these problems $P_{N_o}(\bar{\lambda}_\nu)$ by using the continuation method indicated in Section 2, 2359 simplex iterations were needed, i. e. on average ~26 per problem, which must be compared with ~100 simplex iterations for one independent solution of $P_{N_o}(\lambda)$. In some other computed examples the advantage of the continuation method is even more evident (cf. [7] for a more detailed discussion).

The computed values of $\epsilon_{N_o}(\bar{\lambda}_\nu)$, $\nu = 8,..,33$ indicated in Figure 2 provide approximations λ^* to 5 local minima $\lambda^*_{N_o}$ of $\epsilon_{N_o}(\lambda)$ and corresponding starting intervals for the minimization algorithm (for example $\lambda^* = 29$ with inclusion interval [28, 30] for the corresponding minimum $\lambda^*_{N_o}$).

On the whole set S inclusion intervals for 15 local minima of $\epsilon_{N_o}(\lambda)$ were found.

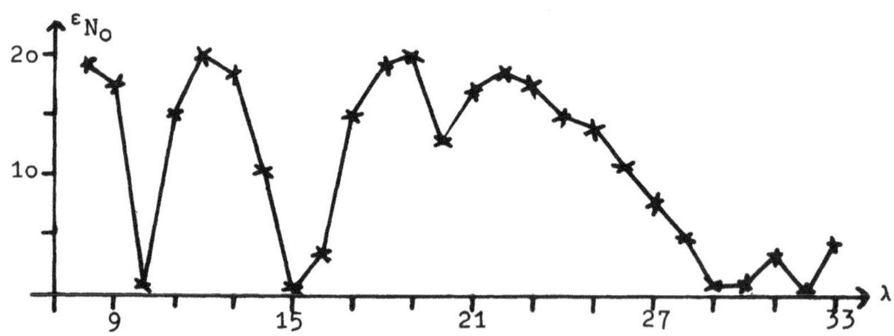

.Figure 2

For the minimization algorithm the derivatives $\epsilon'_N(\lambda)$ are needed. In first instance we computed approximations by divided differences

$$\epsilon'_N(\lambda) \sim \frac{\epsilon_N(\lambda+\delta) - \epsilon_N(\lambda)}{\delta} \qquad (\delta \text{ small}) . \tag{5.1}$$

But a choice of δ necessarily smaller than the length of the final inclusion interval may lead to a loss of accuracy in the values $\epsilon'_N(\lambda)$ and more seriously in the approximations for the second derivatives disturbing the favourable convergence of Mifflin's algorithm when dealing with small inclusions. We therefore compute the derivatives $\varepsilon_N'(\lambda)$ with the help of a representation (4.18) for the derivatives $(\alpha_\lambda,\mu_\lambda)$ of the solution (α,μ) of $P_N(\lambda)$.

Besides the improved numerical stability, such an evalutation of $\varepsilon_N'(\lambda)$ is more efficient: In our example an additional calculation of N^2 values of Bessel functions J_α is needed to compute $\varepsilon_N'(\lambda)$. This has to be compared with the evaluation of $N{\times}M$ function values of J_α and the solution of $P_N(\lambda+\delta)$ when $\varepsilon_N'(\lambda)$ is approximated by (5.1) instead.

The computation of an error bound for an eigenvalue is illustrated by a typical example:
The global search has led to an inclusion interval [29, 30] for an eigenvalue. The procedure b) in Section 3 has been accomplished for numbers N = 36,39,..,48 of trial functions. Table 1 contains the approximate local minima λ_N^* and the values $\varepsilon_N(\lambda_N^*)$. Furthermore, the length of the initial and the final interval is given for each minimization step (N fixed together with the number of iterations required to get this reduction with Mifflin's algorithm.
To compute the function values in the minimization procedure we again use the continuation idea above.

N	λ_N^*	$\varepsilon_N(\lambda_N^*)$	initial length of interval	final	# of iter.
36	29.521 481 116 8	$0.441\ 10^{-6}$	1	$0.181\ 10^{-8}$	9
39	29.521 481 114 7	$0.279\ 10^{-6}$	$0.282\ 10^{-6}$	$0.213\ 10^{-8}$	4
42	29.521 481 114 0	$0.270\ 10^{-7}$	$0.274\ 10^{-7}$	$0.758\ 10^{-11}$	7
45	29.521 481 114 2	$0.189\ 10^{-7}$	$0.192\ 10^{-7}$	$0.115\ 10^{-9}$	5
48	29.521 481 114 1	$0.187\ 10^{-8}$	$0.194\ 10^{-8}$	$0.577\ 10^{-9}$	5

Table 1

The computed approximations λ^* of eigenvalues of the L-shaped membrane are given in Table 2 together with error bounds and the number of trial functions. For completeness we have added the eigenvalues $2\pi^2$, $5\pi^2$, $8\pi^2$ which are also the first eigenvalues of the unit square.

λ^*	$\epsilon_N(\lambda^*)$	N
9.639 723 844	0.119 10^{-7}	42
15.197 251 926	0.110 10^{-7}	42
$2\pi^2$		
29.521 481 111 4	0.187 10^{-8}	48
31.912 635 956 7	0.764 10^{-8}	48
41.474 509 890 6	0.757 10^{-9}	54
44.948 487 781 3	0.316 10^{-8}	48
$5\pi^2$		
56.709 609 888	0.186 10^{-7}	48
65.376 535 709	0.245 10^{-7}	48
71.057 755 648	0.601 10^{-7}	48
$8\pi^2$		
89.301 668 352	0.182 10^{-6}	48
92.306 906 763	0.971 10^{-7}	48
97.380 722 646	0.985 10^{-7}	48

Table 2

Acknowledgement: We thank the unknown referee for many valuable suggestions to improve the presentation.

References

[1] E. J. Anderson: A new primal algorithm for semi-infinite linear programming. In: E. J. Anderson and A. B. Philpott (eds.), Infinite Programming, Springer Lecture Notes in Economics and Mathematical Sciences, 1985, 108-122

[2] F. H. Clarke: Generalized gradient and applications, Transactions of the AMS 205 (1975), 247-262

[3] R. Courant and D. Hilbert: Methoden der mathematischen Physik, Bd. I, Springer, 1968

[4] J. D. P. Donnelly: Eigenvalues of membranes with reentrant corners, SIAM J. Numer. Anal. 6 (1969), 47-61

[5] A. V. Fiacco: Introduction to sensitivity and stability analysis in nonlinear programming, Academic Press, New York etc, 1983

[6] L. Fox, P. Henrici and C. Moler: Approximations and bounds for eigenvalues of elliptic operators, SIAM J. Numer. Anal. 4 (1967), 89-102

[7] R. Hettich: On the computation of membrane-eigenvalues by semi-infinite programming. In: E. J. Anderson and A. B. Philpott (eds.), Infinite Programming, Springer Lecture Notes in Economics and Mathematical Sciences, 1985, 79-89

[8] R. Hettich and H. Th. Jongen: Semi-infinite programming: conditions of optimality and applications. In: J. Stoer (ed.), Optimization Techniques, Part 2, Springer Lecture Notes in Control and Information Sciences, 1978, 1-11

[9] R. Hettich and P. Zencke: Two case-studies in parametric semi-infinite programming. In: Bagchi and H. Th. Jongen (eds.), Systems and Optimization, Springer Lecture Notes in Control and Information Sciences, 1984, 132-155

[10] R. Hettich and P. Zencke: Numerische Methoden der Approximation und semi-infiniten Optimierung, Teubner Studienbücher Mathematik, Stuttgart, 1982

[11] R. Mifflin: Stationarity and superlinear convergence of an algorithm for univariate locally Lipschitz constrained minimization, Mathem. Programming 28 (1984), 50-71

[12] C. Moler and L. F. Payne: Bounds for eigenvalues and eigenvectors of symmetric operators, SIAM J. Numer. Anal. 4 (1967), 483-488

[13] P. Nash: Algebraic fundamentals of infinite dimensional linear programming. In: E. J. Anderson and A. B. Philpott (eds.), Infinite Programming, Springer Lecture Notes in Economics and Mathematical Sciences, 1985, 37-52

PARAMETRIC OPTIMIZATION: THE KUHN-TUCKER SET

Hubertus Th. Jongen, Peter Jonker and Frank Twilt[1)]

Abstract. We consider finite dimensional smooth optimization problems depending on a finite number of real parameters. Our main interest lies in the structure of the Kuhn-Tucker set Σ_{KT}. If the linear independence constraint qualification everywhere holds, we show that Σ_{KT} is in general a creased manifold (in the sense of S. Schecter). If the latter constraint qualification does not hold, the set Σ_{KT} need not to be closed. A special case (only one equality constraint) is treated in detail. By means of an illustrative example we show that the closure $\bar{\Sigma}_{KT}$ need not even to be a topological manifold. The structure of this example is topologically stable under perturbations of the problem data. Finally, we indicate a possible way of extending the ideas of the special case to the general case.

1. Introduction

Let I and J be finite index sets and $f, h_i, g_j \in C^\infty(\mathbb{R}^n \times \mathbb{R}^r, \mathbb{R})$, $i \in I$, $j \in J$; here, $C^k(\mathbb{R}^\ell, \mathbb{R})$ stands for the linear space of real valued functions on the Euclidean space $\mathbb{R}^\ell$ which are k-times continuously differentiable.

The following parametric optimization problem $P(t)$, $t \in \mathbb{R}^r$, will be the subject of our investigation:

$$P(t) : \text{Minimize } f(x,t) \text{ on } M(t), \tag{1.1}$$

where

$$M(t) = \{x \in \mathbb{R}^n \mid h_i(x,t) = 0,\ g_j(x,t) \geq 0,\ i \in I,\ j \in J\}. \tag{1.2}$$

Since our main goal is a structural analysis of certain stable properties, we restrict to smooth functions, noting that C^k-functions can be arbitrarily well approximated by C^∞-functions (cf. [6]).

The variable $t \in \mathbb{R}^r$ denotes the parameter(s). So, for each fixed parameter t, we are dealing with an optimization problem in the variable $x (\in \mathbb{R}^n)$.

For $\phi \in C^\infty(\mathbb{R}^n \times \mathbb{R}^r, \mathbb{R})$ we denote by $D\phi$ (resp. $D_x\phi$, $D_t\phi$) the row vector of the first partial derivatives with respect to x and t (resp. x, t). By z we denote the general point in $\mathbb{R}^n \times \mathbb{R}^r$ and z will always be partitioned as follows:

$$z = (x,t),\ x \in \mathbb{R}^n \text{ and } t \in \mathbb{R}^r. \tag{1.3}$$

1) Twente University of Technology, Department of Applied Mathematics, P.O. Box 217, 7500 AE Enschede, The Netherlands.

Moreover, $J_o(z)$ represents the active inequality constraints:

$$J_o(z) = \{j \in J \mid g_j(z) = o\}. \tag{1.4}$$

Let $\bar{x} \in M(\bar{t})$ and suppose that $\bar{x}$ is a local minimum for $f(\cdot,\bar{t})_{|M(\bar{t})}$. Then it is well-known ([2]) that there exist

λ, λ_i, μ_j, $i \in I$, $j \in J_o(\bar{z})$, not all vanishing, such that:

$$\left.\begin{aligned} &\lambda D_x f = \sum_{i\in I} \lambda_i D_x h_i + \sum_{j\in J_o(\bar{z})} \mu_j D_x g_j|_{\bar{z}} = (\bar{x},\bar{t}) \\ &\lambda \geq o,\ \mu_j \geq o,\ j \in J_o(\bar{z}). \end{aligned}\right\} \tag{1.5}$$

A point $\bar{x} \in M(\bar{t})$ at which (1.5) holds with $\lambda > o$ (and hence with $\lambda = 1$) is called a Kuhn-Tucker point. If some kind of constraint qualification holds, then a local minimum is also a Kuhn-Tucker point.
The Kuhn-Tucker set Σ_{KT} for $P(t)$ is defined by

$$\Sigma_{KT} = \left\{ z \in R^n x\ R^r \;\middle|\; \begin{array}{l} x \in M(t) \text{ and } x \text{ is a Kuhn-Tucker point} \\ \text{for } f(\cdot,t)_{|M(t)}. \end{array} \right\}. \tag{1.6}$$

The aim of this paper is a study of the local structure of the Kuhn-Tucker set Σ_{KT}, resp. of its closure $\bar{\Sigma}_{KT}$.

The special case $r = 1$ (one-parameter families) is nowadays well understood. We distinguish three typical ways of tackling the problem in case $r = 1$. In all of them a regularization is central; the main tools are "regular values" of certain associated mappings, resp. "transversal" approximations:

I. The approach via generalized equations (see [16]) (concept of regular value of multifunctions) by A. Reinoza (cf. [15]).

II. The approach via piecewise smooth mappings (concept of regular value of piecewise smooth mappings) by M. Kojima and R. Hirabyashi (cf. [13]).

III. The approach via transversal approximations (jet-transversality w.r.t. the defining differentiable functions f, h_i, g_j) by H.Th. Jongen, P. Jonker and F. Twilt (cf. [9], [10]).

It should be emphasized that (in contradistinction to I and II) in approach III <u>no</u> constraint qualification is assumed a priori. In fact if certain constraint qualifications are violated, then generic conditions take their place. The resulting Kuhn-Tucker set Σ_{KT} in I, II is a topological 1-dimensional manifold (without boundary) and closed as a set, whereas is III the closure $\bar{\Sigma}_{KT}$ may have boundary points (due to the lack of some constraint qualification).

If the dimension of the parameter space (r) exceeds one, the situation (with constraints) becomes much more complicated. If a Kuhn-Tucker point $\bar{x} \in M(\bar{t})$ happens to be strongly stable (in the sense of Kojima), then, locally around $(\bar{x},\bar{t})$, the Kuhn-Tucker set Σ_{KT} is a topological manifold of dimension r, parametrizable by means of the parameter t (cf. [12]). Apart from the latter result, there is a very interesting, recent study of S. Schecter on multiparametric nonlinear programming problems within the context of Pareto theory (cf. [18]). Under certain rank assumptions on the set of gradients corresponding to the active constraints, he proved that the Kuhn-Tucker set is a manifold with boundary (again boundary points indicate the lack of some constraint qualification). The approach of Schecter contains features of the approaches II, III as mentioned above.

The present paper is closely related to Schecter's study and it is organized as follows. In Section 2 we show, under the assumption of the linear independence constraint qualification, that Σ_{KT} is generically a creased manifold (in the sense of Schecter; for a precise definition see [18]); roughly speaking, an n-dimensional creased manifold is a topological manifold, locally pieced together (nonlinearly) from orthants of $\mathbb{R}^n$, where the manifold is smooth corresponding to points from the interior of an orthant and where nondifferentiability (creases) may occur along orthant-boundaries. This "creased manifold"-result was proved by Schecter in [18] in the case of a special type of parametric problems. We will use a rather different proof technique which will give at least some more clarification of some of Schecter's work. In Section 3 we treat the special case of one equality constraint only and we give a condition which guarantees that het closure $\bar{\Sigma}_{KT}$ is a manifold. Then for n = 2 and r = 3 we give an illustrative example which indicates that $\bar{\Sigma}_{KT}$ need not to be a topological manifold; it is shown that the structure of this example is topologically stable under perturbations of the problem data. So, this destroys any hope for $\bar{\Sigma}_{KT}$ to be a topological manifold in general. Finally, in Section 4 we make some final comments on the way to proceed a structural analysis in the general constrained case, based on the ideas developed in this paper.

The space $C^\infty(\mathbb{R}^\ell, \mathbb{R})$ will be topologized by means of the strong C^2-topology denoted by C^2_S (cf. [6]). In fact, a (typical) neighborhood $\mathcal{N}_{\varepsilon(\cdot)}$ of the zero function is generated by means of a strictly positive and continuous function $\varepsilon : \mathbb{R}^n \to \mathbb{R}$:

$$\mathcal{N}_{\varepsilon(\cdot)} = \left\{\phi \in C^\infty(\mathbb{R}^\ell, \mathbb{R}) \;\middle|\; |\phi(x)| + \sum_{i=1}^{\ell} \left|\frac{\partial \phi}{\partial x_i}(x)\right| + \sum_{i,j=1}^{\ell} \left|\frac{\partial^2 \phi}{\partial x_i \partial x_j}(x)\right| < \varepsilon(x) \right.$$

$$\left. \text{for all } x \in \mathbb{R}^n \right\}.$$

A base-neighborhood of any other function is obtained by translating $\mathcal{N}_{\varepsilon(\cdot)}$. Finally, a finite number of copies of $C^\infty(\mathbb{R}^\ell, \mathbb{R})$ will be topologized by means of the product-topology, generated by C^2_S.

2. The linear independence case

From now on we assume that the index sets I, J are fixed and that $|I| < n$, $|\cdot|$ denoting the cardinality.

The linear independence constraint qualification (shortly LICQ) is said to hold at $\bar{x} \in M(\bar{t})$ if the set of vectors

$$\{D_x h_i, D_x g_j, i \in I, j \in J_0(\bar{x},\bar{t})\}_{|(\bar{x},\bar{t})} \tag{2.1}$$

is linearly independent.
As an abbreviation we put $H = (h_i)_{i\in I}$ and $G = (g_j)_{j\in J}$.

A pair $(H,G) \in C^\infty(\mathbb{R}^n \times \mathbb{R}^r, \mathbb{R})^{|I|+|J|}$ is called regular if LICQ holds at every point $x \in M(t)$ for all $t \in \mathbb{R}^r$; here, $M(t)$ is the feasible set corresponding to (H,G) according to (1.2).

Let R denote the set consisting of all regular pairs (H,G). Then, it is not difficult to see that R is a C^1_S-open subset of $C^\infty(\mathbb{R}^n \times \mathbb{R}^r, \mathbb{R})^{|I|+|J|}$ and, moreover, R is nonempty since $|I| < n$.

In the remaining part of this section the constraints $H = (h_i)_{i\in I}$ and $G = (g_j)_{j\in J}$ are assumed to form a regular pair.

The LICQ implies that $M(t)$ is a manifold with boundary and corners (or a regular constraint set in the sense of [7]) and the concept of a critical point makes sense ([7]). A point $\bar{x} \in M(\bar{t})$ is called a critical point if there exist reals λ_i, μ_j, $i \in I$, $j \in J_0(\bar{z})$ such that

$$D_x f = \sum_{i\in I} \lambda_i D_x h_i + \sum_{j\in J_0(\bar{z})} \mu_j D_x g_j |_{\bar{z} = (\bar{x},\bar{t})}. \tag{2.2}$$

The numbers λ_i, μ_j in (2.2) are called Lagrange parameters, and they are unique since the set of vectors in (2.1) is linearly independent. Observe that $\bar{x}$ is a Kuhn-Tucker point iff $\mu_j \geq 0$, $j \in J_0(\bar{z})$. In particular, since LICQ holds, a local minimum $\bar{x}$ for $f(\cdot,\bar{t})_{|M(\bar{t})}$ is necessarily a Kuhn-Tucker point (cf. [2]). Note that LICQ implies in addition that Σ_{KT} is a closed set. We will show the following theorem, where the concept of creased manifold was introduced by S. Schecter in [18].
As we pointed out in the introduction, a creased manifold (of dimension n) is a special type of a piecewise smooth manifold: each piece is locally diffeomorphic with an orthant of $\mathbb{R}^n$ and they are fitted together, combinatorically, in the same way as the orthants of $\mathbb{R}^n$. In particular, a creased manifold is a topological manifold.

Theorem A. There exists a C^2_S-open and dense subset R^* of $C^\infty(\mathbb{R}^n \times \mathbb{R}^r, \mathbb{R}) \times R$ with the property:
If $(f,H,G) \in R^*$, then the corresponding Kuhn-Tucker set Σ_{KT} is either empty or a creased manifold of dimension r. □

The proof of Theorem A follows basically from two observations. Firstly (Step 1) we prove that Theorem A holds in a simplified situation with linear constraints, not depending on the parameter t; secondly (Step 2) we show that the situation in Step 1 can always be achieved by means of specially structured local coordinate transformations. The details for globalizing local approximations are standard and will be omitted (see for example [6],[8],[11]).

As a preparation we consider a special subset of $\mathbb{R}^{2n}$.
Let $(u_1,v_1,u_2,v_2,\cdots,u_n,v_n)$ denote the general point in $\mathbb{R}^{2n}$ and define the set $K(n) \subset \mathbb{R}^{2n}$:

$$K(n) = \{(u_1,v_1,\cdots,u_n,v_n) \mid u_i v_i = o, \quad i = 1,\cdots,n\}. \tag{2.3}$$

So, we see that $K(n)$ is nothing else but the product set $K(1) \times \ldots \times K(1)$. The set $K(1) = \{(u,v) \mid uv = o\}$ is depicted in Fig. 2.1.a. We stratify (= partition into manifolds) the set $K(1)$ by taking as strata the origin and the four components of the set $K(1)\backslash\{0\}$; see Fig. 2.1.b. This stratification is obviously Whitney-regular and the stratification of $K(n) = K(1) \times \ldots \times K(1)$ will then be the product-stratification (cf. [3]).

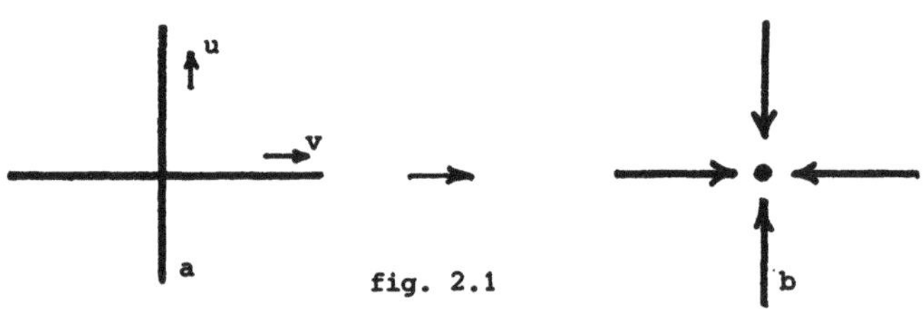

fig. 2.1

In $K(n)$ lies the subset $K^+(n)$, consisting of those points for which (in view of (2.3)) also $u_i \geq o$, $v_i \geq o$, $i = 1,\cdots,n$ holds. The set $K^+(n)$ is also the n-fold product of $K^+(1)$. Obviously, $K^+(n)$ is a creased manifold of dimension n, where the "creases" occur exactly along the strata of dimension less than n; see Fig. 2.2.

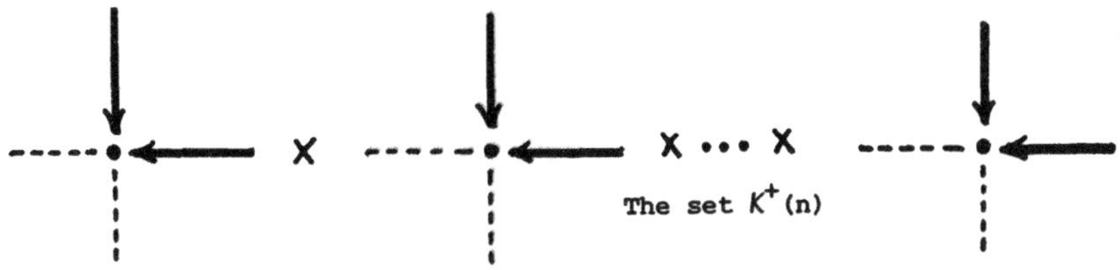

Fig. 2.2

<u>Step 1.</u> Put $n = p + q$ and consider the <u>constant</u> feasible set $\mathbb{H}^p \times \mathbb{R}^q$, where $\mathbb{H}^p = \{y \in \mathbb{R}^p | y_i \geq o,\ i = 1,\cdots,p\}$ (the nonnegative orthant of $\mathbb{R}^p$).
So we are dealing with the case $I = \emptyset$, $J = \{1,\cdots,p\}$ and $g_j(x,t) = x_j$, $j = 1,\ldots,p$ (cf. (1.2)).
For $f \in C^\infty(\mathbb{R}^n \times \mathbb{R}^r, \mathbb{R})$ we define the following (reduced) 1-jet extension $\hat{j}^1 f$ with $f_i := \frac{\partial}{\partial x_i} f(x,t)$:

$$\left.\begin{aligned} \hat{j}^1 f : (x_1,\cdots,x_p,\ x_{p+1},\cdots,x_n,\ t_1,\cdots,t_r) \longrightarrow \\ (x_1, f_1,\cdots,\ x_p, f_p,\ f_{p+1},\cdots,f_n). \end{aligned}\right\} \qquad (2.4)$$

In the image space of $\hat{j}^1 f$ we consider the set $K^+(p)\times\{o_q\}$ (o_q = origin in $\mathbb{R}^q$). The stratification of $K^+(p)\times\{o_q\}$ is in an obvious way induced by the stratification of $K^+(p)$. The set $K^+(p)\times\{o_q\}$ plays the following crucial role:

$$\boxed{(x,t) \in \Sigma_{KT} \text{ iff } \hat{j}^1 f(x,t) \in K^+(p)\times\{o_q\}.} \qquad (2.5)$$

Let F denote the subset of $C^\infty(\mathbb{R}^n \times \mathbb{R}^r, \mathbb{R})$ consisting of those f with the property that $\hat{j}^1 f$ is transversal to every stratum of $K^+(p)\times\{o_q\}$. Since $K^+(p)\times\{o_q\}$ is a closed set and since its stratification is Whitney-regular it follows from the jet-transversality theorem (cf. [5],[8],[11]) that F is C^2_S-open and dense in $C^\infty(\mathbb{R}^n \times \mathbb{R}^r, \mathbb{R})$. If $f \in F$, then Σ_{KT} is easily seen to be a creased manifold; in fact, recall that $\Sigma_{KT} = (\hat{j}^1 f)^{-1}(K^+(p)\times\{o_q\})$.

<u>Step 2.</u> Choose $\bar{x} \in M(\bar{t})$ and consider the special local coordinate transformation

$$(y,u) = \Phi(x,t), \qquad (2.6)$$

where Φ in (2.6) is specified by:

$$\left.\begin{aligned} y_i &= h_i(x,t) &&, i \in I \\ y_j &= g_j(x,t) &&, j \in J_o(\bar{x},\bar{t}) \\ y &= \xi_\ell^T(x-\bar{x}) &&, \ell \in L_o \\ u &= t\ , && \end{aligned}\right\} \qquad (2.7)$$

and where $|L_o| = n - |I| - |J_o(\bar{x},\bar{t})|$, the vectors $\xi_\ell \in \mathbb{R}^n$, $\ell \in L_o$, being chosen such that they form together with the vectors $D_x^T h_i(\bar{x},\bar{t})$, $D_x^T g_j(\bar{x},\bar{t})$, $i \in I$, $j \in J_o(\bar{x},\bar{t})$ a basis for $\mathbb{R}^n$.

Note that the set M(t) under Φ transforms locally into the following constant set for t near $\bar{t}$:

$$\{0_{|I|}\} \times \mathbb{H}^{|J_o(\bar{x},\bar{t})|} \times \mathbb{R}^{|L_o|} \tag{2.8}$$

So, M(t) takes essentially the form $\mathbb{H}^p \times \mathbb{R}^q$ as used in Step 1.

Finally, let η denote the function f in the new coordinates:

$$\eta(y,u) = f \circ \Phi^{-1}(y,u) \quad , \tag{2.9}$$

and let $\tilde{x} \in M(\tilde{t})$ be a critical point for $f_{|M(\tilde{t})}$ with $(\tilde{x},\tilde{t})$ near $(\bar{x},\bar{t})$; hence we have:

$$D_x f = \sum_{i\in I} \tilde{\lambda}_i D_x h_i + \sum_{j\in J_o(\tilde{z})} \tilde{\mu}_j D_x g_j \Big|_{\tilde{z} = (\tilde{x},\tilde{t})} \tag{2.10}$$

Note that $J_o(\tilde{z}) \subset J_o(\bar{z})$ and a short calculation shows:

$$\frac{\partial \eta}{\partial y_j}(\Phi(\tilde{z})) = \tilde{\mu}_j \ , \quad j \in J_o(\tilde{z}). \tag{2.11}$$

Hence, with the aid of the formulas (2.8) - (2.11) we are (locally) in the situation of Step 1.

3. The special case: one equality constraint

In this section we consider the case of only one constraint which, in addition, is an equality constraint; so $|I| = 1$ and $J = \emptyset$. Let $h(x,t)$ (resp. $f(x,t)$) denote the equality constraint function (resp. the objective function).

We call a point $z = (x,t)$ a generalized critical point (gc-point) if both $h(z) = 0$ and if $\{D_x h, D_x f\}$ is linearly dependent. Furthermore, Σ_{gc} stands for the set of all gc-points. Note that we obviously have the inclusion $\Sigma_{KT} \subset \Sigma_{gc}$.

The linear dependence of $\{D_x h, D_x f\}$ can occur in the following three ways, each of them giving rise to a smooth manifold M_i in $\mathbb{R}^n \times \mathbb{R}^n$:

$$\left.\begin{array}{lll} D_x h = 0 \text{ and } D_x f = 0 & , & M_0 = \{0_n\}\times\{0_n\} \\ D_x h = 0 \text{ and } D_x f \neq 0 & , & M_1 = \{0_n\}\times(\mathbb{R}^n\setminus\{0_n\} \\ D_x h \neq 0 \text{ and } D_x f = \lambda D_x h & , & M_2 = \bigcup_{u\in \mathbb{R}^n\setminus\{0\}} \{u\} \times \text{span}\{u\} \end{array}\right\} \tag{3.1}$$

If M denotes the subset of $\mathbb{R}^n \times \mathbb{R}^n$ consisting precisely of linearly dependent pairs (u,v), then we see that $\{M_o, M_1, M_2\}$ is a stratification of M. Furthermore, we have the contact properties:

$$\bar{M}_2 = M_2 \cup M_1 \cup M_o , \qquad \bar{M}_1 = M_1 \cup M_o , \tag{3.2}$$

and it is not hard to verify that the stratification $\{M_o, M_1, M_2\}$ of the set M is Whitney-regular (in fact, check the socalled regularity conditions (a) and (b') as explained in [14, chapter 5]).

We define the following (reduced) 1-jet extension $\hat{j}^1(h,f)$ associated with (h,f):

$$\left.\begin{aligned} \hat{j}^1(h,f) \; : \; \mathbb{R}^n \times \mathbb{R}^r &\longrightarrow \mathbb{R} \times \mathbb{R}^n \times \mathbb{R}^n \\ (x,t) &\longmapsto (h, D_x h, D_x f)_{|(x,t)} \end{aligned}\right\} \tag{3.3}$$

In the image space of $\hat{j}^1(h,f)$ we consider the manifolds $\mathcal{M}_i$:

$$\mathcal{M}_i = \{0_1\} \times M_i , \quad i = 0,1,2. \tag{3.4}$$

For the codimension of $\mathcal{M}_i$ in $\mathbb{R} \times \mathbb{R}^n \times \mathbb{R}^n$ we obtain:
codim $\mathcal{M}_o$ = 2n+1, codim $\mathcal{M}_1$ = n+1, codim $\mathcal{M}_2$ = n .

Now, it is crucial to note (compare (3.1) and (3.4)):

$$\Sigma_{KT} = \hat{j}^1(h,f)^{-1}[\mathcal{M}_o \cup \bar{\mathcal{M}}_2], \qquad \Sigma_{gc} = \hat{j}^1(h,f)^{-1}[\mathcal{M}_o \cup \mathcal{M}_1 \cup \mathcal{M}_2]. \tag{3.5}$$

Let $F(n,r)$ denote the "transversal set", i.e.

$$F(n,r) = \{(h,f) \in C^\infty(\mathbb{R}^n \times \mathbb{R}^r, \mathbb{R})^2 \,|\, \hat{j}^1(h,f) \text{ transversal to } \mathcal{M}_i, \; i = 0,1,2\}. \tag{3.6}$$

<u>Theorem B.</u>

a. The set $F(n,r)$ is C^2_S-open and dense in $C^\infty(\mathbb{R}^n \times \mathbb{R}^r, \mathbb{R})^2$.

b. If $(h,f) \in F(n,r)$, then $\Sigma_{gc} = \bar{\Sigma}_{KT}$.

c. If $r \le n$ and $(h,f) \in F(n,r)$, then both Σ_{KT} and $\bar{\Sigma}_{KT}$ are either empty or smooth manifolds or dimension r.

<u>Proof.</u> Statement a follows from the jet-transversality theorem (density) and the fact that $\{\mathcal{M}_o, \mathcal{M}_1, \mathcal{M}_2\}$ is a Whitney-regular stratification for the closed set $\bigcup_{i=o}^{2} \mathcal{M}_i$ (cf. [5],[8],[11]). The former Whitney-regularity and the contact conditions (3.2) show that every smooth manifold in $\mathbb{R}^n \times \mathbb{R}^n$ having a nonempty transversal intersection

with M_1 also intersects M_2 in every neighborhood of an intersection point with M_1. From this Statement b follows taking (3.5), (3.6) into account. So it remains to check Statement c. From the transversality assumption on $\hat{j}^1(h,f)$ (see (3.6)) it follows that $\hat{j}^1(h,f)$ does not meet $\mathcal{M}_0$ if the codimension of $\mathcal{M}_0$ is greater than the number of variables in (x,t), i.e. $n+r$. Recall that codim $\mathcal{M}_0 = 2n+1$. Hence, if $r \le n$ then $\hat{j}^1(h,f)$ does not meet $\mathcal{M}_0$. In the latter case we see that $\Sigma_{KT} = \hat{j}^1(h,f)^{-1}(\mathcal{M}_2)$. Consequently, Σ_{KT} is either empty or a smooth manifold of dimension ($n+r$ - codim $\mathcal{M}_2$) $= r$. From Statement b and (3.5) we see (if $r \le n$) that $\overline{\Sigma}_{KT} = \hat{j}^1(h,f)^{-1}(\mathcal{M}_1 \cup \mathcal{M}_2)$. Now, $\mathcal{M}_1 \cup \mathcal{M}_2 = \{0_1\} \times (M_1 \cup M_2)$ (cf. (3.4)) and $M_1 \cup M_2$ is the set of vector pairs $(u,v) \in \mathbb{R}^n \times \mathbb{R}^n$ with rank $(u \vdots v) = 1$. But the latter set is a smooth manifold of codimension $n-1$ in $\mathbb{R}^n \times \mathbb{R}^n$ (see for example [1],[8],[11]) and hence, $\mathcal{M}_1 \cup \mathcal{M}_2$ is a smooth manifold of codimension n in $\mathbb{R} \times \mathbb{R}^n \times \mathbb{R}^n$. But then the inequality $r \le n$ and the transversality assumption on $\hat{j}^1(h,f)$ show that $\overline{\Sigma}_{KT}$ is a smooth manifold of dimension $(n+r)-n = r$. □

Unfortunately, if $r > n$ and $(h,f) \in \mathcal{F}(n,r)$, then $\overline{\Sigma}_{KT}$ need even not to be a topological manifold. This will be shown in the following example ($n = 2$, $r = 3$), the structure of which is topologically stable under C^2_S-perturbations of the problem data (h,f).

Example. Let $x = (x_1,x_2)$, $t = (t_1,t_2,t_3)$ and put $h(x,t) = x_1^2 - x_2^2 - t_3$, $f(x,t) = (x_1 - t_1)^2 + (x_2 - t_2)^2$. A short calculation shows that $(h,f) \in \mathcal{F}(2,3)$ and, moreover, Σ_{gc} is determined by the equations (3.7), (3.8):

$$-2x_1x_2 + t_1x_2 + t_2x_1 = o , \tag{3.7}$$

$$x_1^2 - x_2^2 - t_3 = 0. \tag{3.8}$$

So, Σ_{gc} is the graph of the function $(x_1^2 - x_2^2)$ on the set defined by (3.7). Hence, Σ_{gc} is diffeomorphic with the set defined by (3.7). Put $\phi(x_1,x_2,t_1,t_2) = -2x_1x_2 + t_1x_2 + t_2x_1$ (cf. (3.7)). Then, $\phi(o) = o$, $D\phi(o) = o$ and the Hessian matrix $D^2\phi(o)$ is nonsingular and has two negative eigenvalues. From this we see that $D\phi(x) = o$ iff $x = o$, which implies that the zero set of ϕ is a smooth manifold outside the origin; moreover, by means of a linear change of coordinates we obtain that the zero set of ϕ, and hence Σ_{gc}, is diffeomorphic with the set Σ:

$$\Sigma = \{(y_1,y_2,y_3,y_4) \mid y_1^2 + y_2^2 - y_3^2 - y_4^2 = o\} . \tag{3.9}$$

The set Σ is not a topological manifold at the origin. To see this, we use the following argument from algebraic topology: if we delete one point from a topological manifold of dimension greater or equal to three, then this does not affect the fundamental group Π_1 (cf. [17]).

The set Σ in (3.9) is contractible and hence its fundamental group is trivial. On the other hand, $\Sigma\setminus\{o\}$ can be parametrized by putting $y_1^2 + y_2^2 = \varepsilon = y_3^2 + y_4^2$, $\varepsilon > o$; hence, the homotopy-type ([17]) of $\Sigma\setminus\{o\}$ coincides with the type of the set $\{y \mid y_1^2 + y_2^2 = 1, y_3^2 + y_4^2 = 1\}$, which is the product of two circles. Consequently, the fundamental group of $\Sigma\setminus\{o\}$ has two generators ([17]), which shows that $\Pi_1(\Sigma) \neq \Pi_1(\Sigma\setminus\{o\})$ and hence, Σ is not a topological manifold at the origin.

The latter phenomenon is stable under small C_S^2-perturbations of the data (h,f). In fact, choose a convex C_S^2-open neighborhood V of the given pair (h,f) in the transversal set F (2,3) (recall Statement a in Theorem B). Take an arbitrary $(\tilde{h},\tilde{f})$ in V and define the "linear combinations" connecting (h,f) with $(\tilde{h},\tilde{f})$:

$$H(\cdot,\cdot,u) = (1-u)h + u\,\tilde{h} \;, \quad F(\cdot,\cdot,u) = (1-u)f + u\tilde{f} \;. \tag{3.10}$$

Let $\Sigma_{gc}^* \subset \mathbb{R}^n \times \mathbb{R}^r \times \mathbb{R}$ denote the set of gc-points for (H,F), considering u as an additional parameter. The set Σ_{gc}^* could be viewed as an "unfolding" of the original set Σ_{gc}. Note that $\hat{J}^1(H(\cdot,\cdot,o),F(\cdot,\cdot,o))_{|(o,o)} \in M_o$. For each $u \in [o,1]$ we see that $\hat{J}^1(H(\cdot,\cdot,u),F(\cdot,\cdot,u))$ meets the manifolds M_o, M_1, M_2 (defined in (3.4)) transversally. From the implicit function theorem we obtain a unique smooth curve $\psi : [o,1] \to \mathbb{R}^n \times \mathbb{R}^r \times \mathbb{R}$, $u \mapsto (x(u),t(u),u)$, with $\psi(o) = o$ and $\hat{J}^1(H(\cdot,\cdot,u),F(\cdot,\cdot,u))_{|\psi(u)} \in M_o$.

Next, we take an open neighborhood $O^* \subset \mathbb{R}^n \times \mathbb{R}^r \times \mathbb{R}$ of the image $\psi([o,1])$ and consider especially the "height"-function $\eta(x,t,u) = u$ on the set $\Sigma_{gc}^* \cap O^*$. From the transversality condition it follows that $\Sigma_{gc}^* \cap O^*$ has also a Whitney-regular stratification, each stratum corresponding to and being one dimension higher than the corresponding stratum in Σ_{gc}. Finally, take O^* sufficiently small such that Thom's "first isotopy theorem" (cf. [3],[14]) is applicable. The latter theorem states, within this context, that locally in $\mathbb{R}^n \times \mathbb{R}^r \times \mathbb{R}$ around the point $\psi(u)$ the homeomorphy-type of the u-sections of the set $\Sigma_{gc}^* \cap O^*$ remains constant for $u \in [o,1]$. But, $u = o$, resp. $u = 1$, corresponds to the pair (h,f), resp. $(\tilde{h},\tilde{f})$ and hence, the required topological stability is proved.

4. Final remarks

The ideas in the foregoing analysis can be used in order to treat te general case of many (in)equality constraints. In fact, let us discuss briefly some aspects separately on the equality constrained case and the inequality constrained one. The ideas below will form the key for tackling the general case.

Equality constraints only. In that case we have $J = \emptyset$.
We assume that $I = \{1,\cdots,m\}$ and $m < n$. The equality constraint functions are taken together in the vector function $H = (h_1,\cdots,h_m)^T$. Instead of the reduced 1-jet extension taken in (3.3) we now consider $\hat{j}^1(H,f)$:

$$\hat{j}^1(H,f) : R^n \times R^r \to R^m \times R^{nm} \times R^n, \qquad (4.1)$$
$$(x,t) \mapsto (H, D_xH, D_xf)_{|(x,t)}.$$

In the image space of $\hat{j}^1(H,f)$ we consider the sets $\{0_m\} \times M_i$, $i = 0,1,\ldots,m$, where

$$M_i = \{(u_1,\cdots,u_m,v) \in R^{n(m+1)} | \mathrm{rank}(u_1 \vdots \cdots \vdots u_m \vdots v) = i\}. \qquad (4.2)$$

Here, u_i^T refers to the corresponding row in D_xH, whereas v^T refers to D_xf. The sets M_i are smooth manifolds in $R^{n(m+1)}$ (cf. [1],[8],[11]). Note that the manifolds M_i, $i = 0,1,\ldots,m$ form a stratification of the set of real $n\times(m+1)$ matrices of rank $\le m$. This stratification is known to be Whitney-regular (see e.g. [3],[4]). However, this stratification is not yet fine enough for our purpose. In fact, we have to refine each M_i by means of the smooth manifolds M_i^* and M_i^{**},

$$\left.\begin{array}{l} M_i^* = \{(u_1,\cdots,u_m,v) \in M_i | v \in \mathrm{Span}\{u_1,\cdots,u_m\}\} \\ M_i^{**} = M_i \setminus M_i^* \end{array}\right\} \quad i = 0,1,\cdots,m \qquad (4.3)$$

Note that $\Sigma_{KT} = \hat{j}^1(H,f)^{-1} \left(\bigcup_{i=0}^{m} \{0_m\} \times M_i^* \right)$.

So, a further study of the sets $\{0_m\} \times M_i^*, \{0_m\} \times M_i^{**}$ and the corresponding transversal set for the reduced 1-jet extensions $\hat{j}^1(H,f)$ gives the required information about general equality-constrained problems.

Inequality constraints only. Now, we have $I = \emptyset$ and $J = \{1,\cdots,s\}$, where s is some natural number (which certainly may exceed n). Now, we have to deal with the following reduced 1-jet extension:

$$\hat{j}^1(g_1,\cdots,g_s,f) : (x,t) \mapsto (g_1,\cdots,g_s,D_xg_1,\cdots,D_xg_s,D_xf) \qquad (4.4)$$

Apart from the fact that s might exceed n, there are two basic differences with the case of equality constraints only:

Difference 1. Concerning the image space of $\hat{j}^1(\cdot)$ we have to consider w.r.t. the $(g_1,\cdots,g_s)$-part the nonnegative orthant H^s of R^s. The stratification for H^s which corresponds to active/nonactive constraints is generated by means of all subsets of $\{1,\cdots,s\}$. In fact, for $\tilde{J} \subset \{1,\cdots,s\}$ the corresponding stratum of H^s is the subset $\{(y_1,\cdots,y_s) \in H^s | y_i = o,\ i \in \tilde{J}\}$.

Difference 2. For each $\tilde{J} \subset \{1,\cdots,s\}$ we have to consider a special set in the image space of $\hat{j}^1(\cdot)$ corresponding to the entries of $\{D_x g_i, D_x f,\ i \in \tilde{J}\}$. In fact, let $\tilde{J} = \{1,\cdots,p\}$. Then, the following subsets K_i are important (concerning Kuhn-Tucker relations), $i = 0,1\ldots,\min(n,p)$:

$$K_i = \{(u_1,\cdots,u_p,v) \in R^{n(p+1)} | \operatorname{rank}(u_1 \vdots \cdots \vdots u_p) = i \text{ and } v = \sum_{i=1}^{p} \alpha_i u_i, \quad \text{all } \alpha_i \geq o\}. \tag{4.5}$$

The stratification of the sets K_i has to be taken carefully since the polyhedral structure of the nonnegative convex cone generated by the vectors $u_1,\ldots,u_p$ may change its structure. The main idea is illustrated by means of the following example in R^3 with four vectors $u_1,\ldots,u_4$. With respect to arbitrarily small perturbations of the vector u_4, the cone can be turned into the "stable" situations as depicted in Fig. 4.1.b and Fig. 4.1.c respectively.

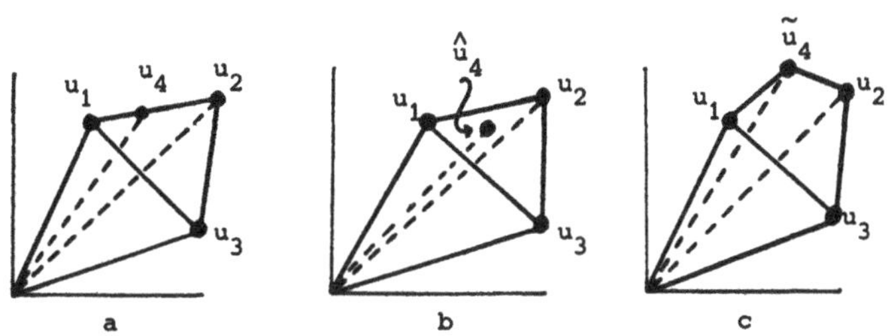

Fig. 4.1

References

[1] Bröcker, Th., and Lander, L.: Differential germs and catastrophes. Londen Math. Soc. Lect. Notes, Vol. 17. Cambridge University Press (1975).

[2] Fiacco, A.V., and McCormick, G.P.: Nonlinear programming: sequential unconstrained minimization techniques. Wiley, New York (1968).

[3] Gibson, C.G., Wirthmüller, K., du Plessis, A.A., and Looijenga, E.J.N.: Topological stability of smooth mappings. Lect. Notes in Mathematics, Vol. 552, Springer Verlag (1976).

[4] Gibson, C.G.: Singular points of smooth mappings. Research Notes in Mathematics, Vol. 25, Pitman (1979).

[5] Golubitsky, M., and Guillemin, V.: Stable mappings and their singularities. Graduate Texts in Mathematics. Vol. 14, Springer Verlag (1973).

[6] Hirsch, M.W.: Differential topology. Graduate Texts in Mathematics, Vol. 33, Springer Verlag (1976).

[7] Jongen, H.Th., Jonker, P., and Twilt, F.: Nonlinear optimization in $\mathbb{R}^n$, I. Morse theory, Chebyshev approximation. Peter Lang Verlag, Frankfurt a.M. (1983).

[8] Jongen, H.Th., Jonker, P., and Twilt, F.: Nonlinear optimization in $\mathbb{R}^n$, II. Transversality, flows, parametric aspects, Peter Lang Verlag, Frankfurt a.M. (to appear).

[9] Jongen, H.Th., Jonker, P., and Twilt, F.: One-parameter families of optimization problems: equality constraints. Journal of Optimization Theory and Applications 48 (198ᴗ), pp. 141-161.

[10] Jongen, H.Th., Jonker, P., and Twilt, F.: Critical sets in parametric optimization. Mathematical Programming 35 (1986), pp. 1-21.

[11] Jongen, H.Th., Jonker, P., and Twilt, F.: Nonlinear optimization theory in $\mathbb{R}^n$ from a global point of view VI. Memorandum Nr. 506, Twente University of Technology (1985).

[12] Kojima, M.: Strongly stable stationary solutions in nonlinear programs. In: Analysis and computation of fixed points (S.M. Robinson, ed.). Academic Press, New York, pp. 93-138 (1980).

[13] Kojima, M., and Hirabayashi, R.: Continuous deformations of nonlinear programs. Mathematical Programming Study 21 (1984), pp. 150-198.

[14] Lu, Y.-C.: Singularity theory and an introduction to catastrophe theory. Universitext, Springer Verlag (1976).

[15] Reinoza, A.: Solving generalized equations via homotopies. Mathematical Programming 31 (1985), pp. 307-320.

[16] Robinson, S.M.: Strongly regular generalized equations. Mathematics of Operations Research. Vol. 5, No. 1 (1980), pp. 43-62.

[17] Spanier, E.H.: Algebraic topology. McGraw-Hill (1966).

[18] Schecter, S.: Structure of the first-order solution set for a class of nonlinear programs with parameters. Mathematical Programming 34 (1986), pp. 84-110.

A NOTE ON BRANIN'S METHOD FOR FINDING THE CRITICAL POINTS OF SMOOTH FUNCTIONS

by

Hubertus Th. Jongen, Peter Jonker, and Frank Twilt[1)]

Abstract. For a generic subclass of smooth functions f on $\mathbb{R}^n$, we study a globally defined autonomous differential equation on $\mathbb{R}^n$, underlying the Newton-iteration method for finding critical points of f. We describe the set of all equilibria of the differential equation and discuss the local phase portraits around these points.

1. Introduction

Let f be a smooth function on $\mathbb{R}^n$, i.e. $f \in C^\infty(\mathbb{R}^n,\mathbb{R})$. By $Df(x)$, resp. $D^2f(x)$, we denote the $1{\times}n$-matrix of the first order partial derivatives, resp. the symmetric $n{\times}n$-matrix of the second order partial derivatives, of f at x. The adjoint matrix ([8]) of $D^2f(x)$ will be denoted by $\widehat{D^2f}(x)$. Apparently, we have: (I_n = $n{\times}n$-unit-matrix)

$$\widehat{D^2f}(x)\cdot D^2f(x) = \det D^2f(x)\cdot I_n \qquad (1.1)$$

We consider the following autonomous differential equation on $\mathbb{R}^n$:

$$\frac{dx}{dt} = -\widehat{D^2f}(x)\cdot D^Tf(x) \qquad (1.2)$$

Euler's discretization to (1.2) yields

$$x_{k+1} = x_k - h_k\cdot\widehat{D^2f}(x_k)\cdot D^Tf(x_k)\ , \qquad (1.3)$$

where the step lengths may be suitably chosen. If $\det D^2f(x) \neq 0$, then, from (1.1), it follows that $\widehat{D^2f}(x) = \det D^2f(x)\cdot(D^2f(x))^{-1}$. Hence, in this case, formula (1.3) is just the damped Newton iteration method for finding the zeros for Df (i.e. the critical points of f). In order to get more insight in the local and global features of Newton's iteration method, especially for small values of the step lengths, it is reasonable to study the underlying "continuous" version (1.2). This idea goes back to F.H. Branin ([1]). System (1.2) is referred to as to the Branin-method, or globalized, continuous Newton-method.

A critical point of f is an equilibrium for (1.2) and will be referred to as to an essential singularity. The set of all such equilibria is denoted by $Ess(f)$. Apart from

1) Twente University of Technology, Department of Applied Mathematics, P.O. Box 217, 7500 AE Enschede, The Netherlands

these essential singularities, there are - in general - still other equilibria for (1.2), namely those points $\tilde{x}$ for which

$$D^T f(\tilde{x}) \neq 0 \quad \text{and} \quad \widetilde{D^2 f}(\tilde{x}) \cdot D^T f(\tilde{x}) = 0 \ . \tag{1.4}$$

The point $\tilde{x}$ is called an <u>extraneous singularity</u> and the set of all such equilibria is denoted by Ext(f).
One easily verifies that det $D^2 f(x) = 0$ if $x \in \text{Ext}(f)$, whereas this needs not to be true if $x \in \text{Ess}(f)$.

The target of this work is the investigation of the possible local structures of the phase-portrait of (1.2) around its equilibria. To this aim, we firstly have to describe the (local) structure of the sets Ess(f) and Ext(f).
However, we do not deal with these questions for <u>all</u> functions f: we restrict ourselves to a "generic" subclass of $C^\infty(R^n,R)$. Roughly speaking, this means restriction to a subset F of $C^\infty(R^n,R)$, such that, w.r.t. "small" perturbations of the partial derivatives, F is "stable" and each $f \in C^\infty(R^n,R)$ can be approximated by elements of F. Here, we treat as an introduction the first (and most simple) restriction step.

Let F_* be the set of all functions $f \in C^\infty(R^n,R)$ which fulfil

<u>Condition (*)</u> $\quad \|Df(x)\| + |\det D^2 f(x)| > 0$, for all $x \in R^n$.

Let $f \in F_*$ and $\bar{x} \in \text{Ess}(f)$, thus det $D^2 f(\bar{x}) \neq 0$. Then, $\bar{x}$ is <u>isolated</u> from the other equilibria of (1.2). Moreover, the first order Taylor expansion of the r.h.s. of (1.2) around $\bar{x}$ is

$$-\widetilde{D^2 f}(\bar{x}) \cdot D^2 f(\bar{x}) \cdot (x-\bar{x}) = -\det D^2 f(\bar{x}) \cdot (x-\bar{x}) \tag{1.5}$$

Hence, the trajectory through any x ($x \neq \bar{x}$ and x sufficiently close to $\bar{x}$) tends to $\bar{x}$ (if det $D^2 f(\bar{x}) > 0$) or emanates from $\bar{x}$ (if det $D^2 f(\bar{x}) < 0$). So, we have proved:

<u>Lemma 1.1.</u> If $f \in F_*$, then Ess(f) is a discrete subset of R^n. Moreover, a point $\bar{x} \in \text{Ess}(f)$ is either an attractor or a repellor for (1.2). □

Intuitively it will be clear: if $f \in F_*$, then this remains true under (sufficiently) small perturbations of Df and $D^2 f$. In order to make this more precise, we need the concept of "Whitney-topology" on $C^\infty(R^n,R)$. On the other hand, an $f \in C^\infty(R^n,R)$ can be approximated arbitrarily well by means of elements from F_*. The main idea for the proof will be a (jet)-transversality argument.

This paper is organized as follows. In Section 2, we briefly explain our genericity concept; to this aim we recall some basic facts on the strong Whitney topology and jet-transversality. In Section 3, we describe the structure of the set of all extraneous singularities. Section 4 is the nucleus of this study: it is dedicated to the investigation of the possible local phase portraits of (1.2) around extraneous singularities. After a short introduction to the two dimensional case, we work out in full detail the case $n = 3$, the main tool - apart from jet-transversality - being the Malgrange-Mather theorem on universal unfoldings. Finally, in Section 5, we give a concise outline of what may be expected in the case where $n \geq 4$.

This work extends some of the results due to J. Gomulka [3] and to S. Smale [10]. As an overall reference, we refer to Chapter 9 of our recent treatise [6].

2. Genericity aspects and (jet-)transversality.

Let $k \geq 0$ be an integer. The space $C^{\infty}(\mathbb{R}^n,\mathbb{R})$ will be topologized by means of the strong C^k-topology denoted by C^k (cf. [6]). In fact, a (typical) neighborhood $N^k_{\varepsilon(\cdot)}$ of the zero function is generated by means of a strictly positive and continuous function $\varepsilon: \mathbb{R}^n \to \mathbb{R}$:

$$N^k_{\varepsilon(\cdot)} = \Big\{\phi \in C^{\infty}(\mathbb{R}^n,\mathbb{R}) \Big| \; |\phi(x)| + \sum_{i=1}^{n} \Big|\frac{\partial \phi}{\partial x_i}(x)\Big| + \ldots + \sum_{i_1+\ldots+i_k=k} \Big|\frac{\partial^k \phi}{\partial x_{i_1} \ldots \partial x_{i_k}}(x)\Big| < \varepsilon(x)$$
$$\text{for all } x \in \mathbb{R}^n\Big\} .$$

A base-neighborhood of any other function is obtained by translating $N^k_{\varepsilon(\cdot)}$. (So roughly speaking, two functions are "C^k-close" if all their corresponding partial derivatives of order 0 through k, are "close" at any $x \in \mathbb{R}^n$).

The topology on $C^{\infty}(\mathbb{R}^n,\mathbb{R})$, generated by <u>all</u> sets $N^k_{\varepsilon(\cdot)}$, $k = 0,1,2,\ldots$, (as base neighborhoods of the zero function), is called the C^{∞}-topolgy.

A subset of $C^{\infty}(\mathbb{R}^n,\mathbb{R})$ is called <u>generic</u> if it contains the intersection of a countable number of C^{∞}-open and -dense subsets of $C^{\infty}(\mathbb{R}^n,\mathbb{R})$. We emphasize that $C^{\infty}(\mathbb{R}^n,\mathbb{R})$, endowed with the C^{∞}-topology, is a so called <u>Baire</u> space (cf. [9]). In particular, a generic subset of $C^{\infty}(\mathbb{R}^n,\mathbb{R})$ is (C^{∞}-) <u>dense</u> (but not necessarily open). Note that in a Baire space, the concept "generic" is much stronger than "dense".

In the sequel, we shall select generic (or, if possible, C^k-open and -dense) subsets of $C^{\infty}(\mathbb{R}^n,\mathbb{R})$. The selection criteria will be certain (jet-)transversality conditions on the functions f in $C^{\infty}(\mathbb{R}^n,\mathbb{R})$.

Let f be a function in $C^{\infty}(\mathbb{R}^n,\mathbb{R})$ and let $\ell \geq 1$ be an integer. For the partial derivatives of f at x we introduce the following notation:

$$\frac{\partial f}{\partial x_i}(x) = \partial_i(x) \ , \ \frac{\partial^2 f}{\partial x_i \partial x_j}(x) = \partial_{ij}(x) \ , \text{ etc.}$$

We write down - in a definite, but arbitrary sequence - all partial derivatives of f at x of order 1 through ℓ, thereby cancelling symmetry. Let $N(n,\ell)$ be the length of this list. For instance, in case $n = \ell = 2$, we consider the list $\partial_1(x),\partial_2(x),\partial_{11}(x),\partial_{12}(x),\partial_{22}(x)$; apparently $N(2,2) = 5$. The vector in $\mathbb{R}^{N(n,\ell)}$ with entries according to this (list of) partial derivatives is denoted by $j^\ell f(x)$. The <u>reduced</u> ℓ-jet extension (shortly: <u>ℓ-jet</u>) $j^\ell f$ of f is defined to be the mapping

$$j^\ell f\colon \mathbb{R}^n \to \mathbb{R}^{N(n,\ell)}\colon x \mapsto j^\ell f(x) \ .$$

The Euclidean space $\mathbb{R}^{N(n,\ell)}$ is called the ℓ-jet space (of f) and will be denoted by $J(n,\ell)$. Note that $j^\ell f$ is a smooth mapping. Let $X \subset J(n,\ell)$ be a smooth submanifold of codimension m. We say that <u>$j^\ell f$ meets X transversally</u> ($j^\ell f \pitchfork X$) if for every $\bar{x} \in \mathbb{R}^n$:

$$\left.\begin{array}{ll} \text{Either} & j^\ell f(\bar{x}) \notin X \ , \\ \text{or} & Dj^\ell f(\bar{x})[\mathbb{R}^n] + T_{\bar{z}}X = J(n,\ell), \end{array}\right\} \qquad (2.1)$$

where $T_{\bar{z}}X$ stands for the tangent space to X at $\bar{z} = j^\ell f(\bar{x})$ and where $Dj^\ell f(\bar{x})$ is the derivative of $j^\ell f$ at $\bar{x}$.

Let $\{h_1,\dots,h_m\}$ be a <u>defining system of functions for X around $\bar{z}$</u> (cf. [6]). So, the smooth functions h_i, $i = 1,\dots,m$, are defined on an open $J(n,\ell)$- neighborhood, say $U_{\bar{z}}$, of $\bar{z}$, the set $\{Dh_i(z),\ i = 1,\dots,m\}$ is linearly independent for all $z \in U_{\bar{z}} \cap X$, and $U_{\bar{z}} \cap X = \{z \in U_{\bar{z}} | h_i(z) = 0,\ i = 1,\dots,m\}$. Then, the following useful criterion for "$j^\ell f \pitchfork X$" holds:

$$\left.\begin{array}{l} j^\ell f \pitchfork X \text{ iff for every } \bar{x} \in \mathbb{R}^n \text{ with } \bar{z} = j^\ell f(\bar{x}) \in X \text{ we have:} \\ \text{If } \{h_1,\dots,h_m\} \text{ is a defining system of functions for X around } \bar{z}, \\ \text{then the set } \{D[h_i \circ j^\ell f](\bar{x}),\ i = 1,\dots,m\} \text{ is linearly independent} \end{array}\right\} \qquad (2.2)$$

Let $\pitchfork^\ell X = \{f \in C^\infty(\mathbb{R}^n,\mathbb{R}) | j^\ell f \pitchfork X\}$. Then, the following theorem, due to R. Thom, will play a crucial role (for a proof, see e.g. [4], or [6]).

<u>Theorem 2.1.</u> <u>(Jet-transversality theorem)</u>

- The set $\pitchfork^\ell X$ is <u>generic</u> in $C^\infty(\mathbb{R}^n,\mathbb{R})$.
- Let $f \in \pitchfork^\ell X$, then $(j^\ell f)^{-1}(X)$ is a smooth submanifold of X in $J(n,\ell)$, or empty.

If, in addition, X is closed in $J(n,\ell)$, then $\pitchfork^\ell X$ is C^k-open and -dense for $k \geq \ell+1$. □

As a corollary, we have (compare also Section 1):

Corollary 2.1. The set F_* is C^k-open and -dense in $C^\infty(R^n,R)$, $k \geq 2$. If $f \in F_*$, then $Ess(f)$ is either empty or a discrete subset of R^n.

Proof. Put $\ell = 1$ and choose $X = \{o_n\}$, o_n being the origin in $J(n,1)(= R^n)$. Then, the transversality Condition (2.1) is equivalent with Condition(*). Moreover if $f \in F_*$, then $(j^1f)^{-1}(o_n)$ is the set of essential singularities. Application of Theorem 2.1 yields the desired result. □

A subset of R^N is called stratified, if it is the disjoint, locally finite, union of smooth submanifolds ("strata") of R^N. The codimension of a stratified set is defined as the codimension of its highest dimensional strata.

Now, we have the following slight extension of the previous theorem.

Theorem 2.2. Let X be a stratified subset of $J(n,\ell)$, with strata X_j, $j \in J$. Then, $\bar{\pi}^\ell X := \bigcap_{j \in J} \bar{\pi}^\ell X_j$ is generic in $C^\infty(R^n,R)$.

If $f \in \bar{\pi}^\ell X$, then the inverse image $(j^\ell f)^{-1}(X)$ is either empty, or a stratified subset of R^n, with strata $(j^\ell f)^{-1}(X_j)$, $j \in J$, and the same codimension as X in $J(n,\ell)$. □

3. The set of extraneous singularities.

The "augmented matrix" $[D^Tf(x) \vdots D^2f(x)]$ is the $n\times(n+1)$-matrix which is obtained from $D^2f(x)$ by adding $D^Tf(x)$ as a first column. We have the following characterization of the points in $Ext(f)$, cf. [6].

Lemma 3.1. Let $f \in F_*$ and let $\tilde{x} \in R^n\backslash Ess(f)$ be given.
Then $\tilde{x} \in Ext(f)$ iff $\operatorname{rank}[D^Tf(x) \vdots D^2f(x)] \leq n-1$. □

The so-called trajectory map $Tf: R^n\backslash Ess(f) \to S^{n-1}$ is given by $Tf(x) = f(x)/\|Df(x)\|$. Here, S^{n-1} is the (n-1)dimensional unit sphere in R^n. A point $\tilde{x} \in R^n\backslash Ess(f)$ is called critical point for Tf if the Jacobi-matrix $(DTf(\tilde{x}))$ of Tf at $\tilde{x}$ (w.r.t. any local coordinate systems in R^n and S^{n-1}) has rank less than n-1.

For $\tilde{x} \in R^n\backslash Ess(f)$, we may represent (no loss of generality) Tf by $Tf(\tilde{x}) = (g_1(\tilde{x}),\ldots,g_{n-1}(\tilde{x}))^T$, with $g_i = \frac{\partial_i}{\partial_n}$. We put $g = (g_1,\ldots,g_{n-1})^T$. Note that:

$$\operatorname{rank} DTf(\tilde{x}) = \operatorname{rank} Dg(\tilde{x}) \tag{3.1}$$

By elementary matrix manipulations one easily proves (cf. [6]):

$$\operatorname{rank} Dg(\tilde{x}) = \operatorname{rank}[D^Tf(\tilde{x}) \vdots D^2f(\tilde{x})] - 1 \tag{3.2}$$

As a corollary of (3.1), (3.2) and Lemma 3.1, we find:

Lemma 3.2. Let $f \in F_*$ and let $\tilde{x} \in R^n \setminus Ess(f)$ be given. Then, $\tilde{x} \in Ext(f)$ iff $\tilde{x}$ is a critical point for Tf. □

Remark 3.1. Let $x_0 \notin Ext(f) \cup Ess(f)$. The trajectory of (1.2) through x_0 is represented by $x(t)$, $x(0) = x_0$. Then, in view of (1.1) and (1.2), we have:

$$\frac{d}{dt} D^T f(x(t)) = D^2 f(x(t)) \cdot \frac{dx}{dt} = -D^2 f(x(t)) \cdot \widetilde{D^2 f}(x(t)) \cdot D^T f(x(t)) =$$
$$-\det D^2 f(x(t)) \cdot D^T f(x(t).$$

Hence, on a trajectory of (1.2), the map Tf remains constant. □

Let A be a symmetric $n \times n$-matrix and let $b \in R^n$. The augmented $n \times (n+1)$-matrix $[b \vdots A]$ may be considered as an element of R^m, $m = \frac{1}{2}n(n+3)$. We define:

$$V_i = \{[b \vdots A] \in R^m \mid \operatorname{rank}[b \vdots A] = i\} .$$

Lemma 3.3. The sets V_i constitute smooth manifolds in R^m; $\operatorname{codim} V_i = \frac{1}{2}(n-i)(n-i+3)$. □

For the proof of this lemma which must take the symmetry of A into account, we refer to [6].

We consider our next transversality condition on functions $f \in C^\infty(R^n, R)$.

Condition (**). $\qquad j^2 f \pitchfork V_i \, , \; i = 0,1,\dots,n-1$.

The subset of $C^\infty(R^n, R)$ consisting of all functions which fulfil both the Conditions (*) and (**) will be denoted by F_{**}.

Together with Lemma 1.1 and Corollary 2.1, the following theorem describes the structure of the set of equilibria of (1.2) for a generic subset of $C^\infty(R^n, R)$.

Theorem 3.1. The set F_{**} is generic in $C^\infty(R^n, R)$. Let $f \in F_{**}$, then:

(i) $Ext(f)$ is a closed stratified set in R^n of codimension ≥ 2, with strata $(j^2 f)^{-1}(V_i)$, $i = 0,1,\dots,n-1$.

(ii) In the special case where $n = 2,3$ or 4, we have: $Ext(f)$ is a submanifold of R^n of codimension two (or empty).

Proof. The genericity of F_{**} is a direct consequence of Corollary 2.1, Lemma 3.3 and Theorem 2.2.

(i) In view of Lemma 3.1, we have $\mathrm{Ext}(f) = (j^2f)^{-1}(\bigcup_{i=0}^{n-1} V_i)$. Hence, the result follows from Theorem 2.2.

(ii) From Lemma 3.3 we conclude that codim $V_{n-1} = 2$ and codim $V_i \geq 5$ if $i \leq n-2$. So a dimension argument based on Condition (2.1) and the fact that $n \leq 4$ yields: $x \in \mathrm{Ext}(f)$ iff $j^2f(x) \in V_{n-1}$. Now, the assertion is a direct consequence of Part(i) of this theorem. □

<u>Remark 3.1</u>. In the case where n = 2,3 or 4, the set F_{**} is <u>C^3-open</u> in $C^\infty(R^n R)$. The proof of this statement requires a more subtile approach than merely applying Theorem 2.2, cf. [6].

<u>4. The local phase-portraits around extraneous singularities (n = 2,3).</u>

Throughout this section, we assume that $f \in F_{**}$ and $\tilde{x} \in \mathrm{Ext}(f)$.

We briefly treat the case where n = 2; for details, we refer to [6].

We may assume (no loss of generality) that $\partial_2(\tilde{x}) \neq 0$. Hence, around $\tilde{x}$, the trajectory map Tf can be represented as the <u>function</u> $Tf = \frac{\partial_1}{\partial_2}$. By writing down Condition (**) explicitly, one proves that $\tilde{x}$ is not merely a critical point (Lemma 3.2) but even a <u>non-degenerate</u> critical point for Tf. So, in view of the <u>Morse-lemma</u> (cf. [5]), there exist local coordinates of R^2 around $\tilde{x}$ - say y_1,y_2 - such that Tf takes the form $\pm y_1^2 \pm y_2^2$. From Remark 3.1 it follows that the trajectories of (1.2) around $\tilde{x}$ are given by the equations $\pm y_1^2 \pm y_2^2$ = constant. Hence, the equilibrium $\tilde{x}$ is either a "saddlepoint" or a "center", see Fig. 4.1. As a matter of fact, this conclusion requires <u>only</u> Condition(**), since in case n = 2, the Condition(*) is already implied by Condition(**).

$\bar{x} \in \mathrm{Ess}(f)$	$\tilde{x} \in \mathrm{Ext}(f)$	$\tilde{x} \in \mathrm{Ext}(f)$
$\det D^2f(\bar{x}) > 0$	$\det D^2Tf(\tilde{x}) < 0$	$\det D^2Tf(\tilde{x}) > 0$

Fig. 4.1.

Now, we pass on to the 3-dimensional case.
An element of the 2-jet space $J(3,2)$ ($= R^9$) is denoted by $z = (z_1,\cdots,z_9)^T$.
The 2-jet, j^2f, of f may be represented as:

$$(D^Tf \vdots D^2f) = \begin{bmatrix} \partial_1 & \vdots & \partial_{11} & \partial_{12} & \partial_{13} \\ \partial_2 & \vdots & \partial_{21} & \partial_{22} & \partial_{23} \\ \partial_3 & \vdots & \partial_{31} & \partial_{32} & \partial_{33} \end{bmatrix} = (\partial_1,\partial_2,\partial_3,\partial_{11},\partial_{21},\partial_{31},\partial_{22},\partial_{23},\partial_{33})^T.$$

The matrix $H(z)$ is defined as:

$$H(z) = \begin{bmatrix} z_1 & \vdots & z_4 & z_5 & z_6 \\ z_2 & \vdots & z_5 & z_7 & z_8 \\ z_3 & \vdots & z_6 & z_8 & z_9 \end{bmatrix}.$$

The 3×3-matrix $H_i(z)$, $i = 1,\cdots,4$, is obtained from $H(z)$ by deleting the $i^{\underline{th}}$ column of this latter matrix. On $J(3,2)$, the function h_i is given by $h_i(z) = \det H_i(z)$, $i = 1,\cdots,4$, whereas $\hat{h}_i := h_i \circ j^2f$ are functions on R^3.

In view of Conditions (*) and (**) we have (see also Lemma 3.1 and Theorem 3.1):

$$\left.\begin{array}{l} \operatorname{rank} H(j^2f(x)) \geq 2 \quad , \quad \text{all } x \in R^3 \\ \operatorname{Ext}(f) = \{x \in R^3 \mid \operatorname{rank} H(j^2f(x)) = 2\} \end{array}\right\} \qquad (4.1)$$

So, the set Ext(f) is just the set of common zeros of $\hat{h}_2, \hat{h}_3, \hat{h}_4$ on the <u>open</u> set $R^3\setminus$Ess(f). However, in view of Theorem 3.1 (ii), the triple $\{\hat{h}_2,\hat{h}_3,\hat{h}_4\}$ does not form a defining system for Ext(f). In order to select such a defining system around the <u>extraneous singularity</u> $\tilde{x}$, we assume, that $\partial_3(\tilde{x}) \neq 0$. Hence, the functions $\phi = \frac{\partial_1}{\partial_3}$ and $\psi = \frac{\partial_2}{\partial_3}$ are well defined in a neighborhood U of $\tilde{x}$. In view of Relation (3.2) and (4.1) a point $x \in U$ belongs to Ext(f) iff the following condition holds:

$$\operatorname{rank} \binom{D\phi}{D\psi}_x = 1 \qquad (4.2)$$

So, locally around $\tilde{x}$, we may assume that $D\psi \neq 0$, and thus for instance $\psi_{x_1} \neq 0$. (Here, ψ_{x_1} stands for the partial derivative of ψ w.r.t. x_1). We introduce the following notation:

$$\phi_i = \det\begin{bmatrix} \partial_{1i} & \partial_1 \\ \partial_{3i} & \partial_3 \end{bmatrix}, \quad \psi_i = \det\begin{bmatrix} \partial_{2i} & \partial_2 \\ \partial_{3i} & \partial_3 \end{bmatrix}, \quad \chi_i = \det\begin{bmatrix} \partial_{1i} & \partial_1 \\ \partial_{2i} & \partial_2 \end{bmatrix}, \quad i = 1,2,3.$$

With this notation, the components ϕ_{x_i}, ψ_{x_i} of resp. $D\phi$ and $D\psi$ may be written as $\phi_{x_i} = \frac{1}{\partial_3^2}\phi_i$ and $\psi_{x_i} = \frac{1}{\partial_3^2}\psi_i$, $i = 1,2,3$. Locally around $\tilde{x}$, the set Ext(f) is just the set of common zeros of $\hat{h}_3$ and $\hat{h}_4$. This follows from the assumption $\psi_1(\tilde{x}) \neq 0$ and from (4.1). Now, under the assumptions $\partial_3(\tilde{x}) \neq 0$ and $\psi_{x_1}(\tilde{x}) \neq 0$, the following result holds:

Lemma 4.1. The pair $\{\hat{h}_3, \hat{h}_4\}$ constitutes a defining system of functions for Ext(f) around $\tilde{x}$.

Proof. Note that, locally around $\tilde{z} = j^2f(\tilde{x})$, the set of augmented matrices V_2 is given by $h_3 = h_4 = 0$. So, cf. Condition (**) and (2.2), we are done if we can prove that $\{Dh_3, Dh_4\}_{\tilde{z}}$ is linearly independent.

Now, a straightforward calculation shows that the adjoint matrices $\tilde{H}_i(z)$, $i = 3,4$, are of the following form

$$\tilde{H}_3(\tilde{z}) = \begin{bmatrix} \cdot & \cdot & \cdot \\ \psi_3 & -\phi_3 & \chi_3 \\ -\psi_1 & \phi_1 & -\chi_1 \end{bmatrix}_{\tilde{x}} \quad \text{and} \quad \tilde{H}_4(\tilde{z}) = \begin{bmatrix} \cdot & \cdot & \cdot \\ \psi_2 & -\phi_2 & \chi_2 \\ -\psi_1 & \phi_1 & -\chi_1 \end{bmatrix}_{\tilde{x}}$$

By an elementary property of adjoint matrices (cf. [8]), it follows that rank $H_i(\tilde{z}) = 2$ implies rank $\tilde{H}_i(\tilde{z}) = 1$, for $i = 3,4$. Consequently, we have:

$$\det\begin{bmatrix} \psi_3 & -\phi_3 \\ -\psi_1 & \phi_1 \end{bmatrix}_{\tilde{x}} = \det\begin{bmatrix} \psi_3 & \chi_3 \\ -\psi_1 & -\chi_1 \end{bmatrix}_{\tilde{x}} = \det\begin{bmatrix} \psi_2 & -\phi_2 \\ -\psi_1 & \phi_1 \end{bmatrix}_{\tilde{x}} = \det\begin{bmatrix} \psi_2 & \chi_2 \\ -\psi_1 & -\chi_1 \end{bmatrix}_{\tilde{x}} = 0 \qquad (4.3)$$

For the 2x6-submatrix of $\begin{bmatrix} Dh_3 \\ Dh_4 \end{bmatrix}_{\tilde{z}}$ containing the last six columns, we find by inspection:

$$\begin{bmatrix} \psi_3 & -\phi_3 & +\chi_3-\psi_1 & 0 & \phi_1 & -\chi_1 \\ \psi_2 & -\phi_2-\psi_1 & \chi_2 & \phi_1 & -\chi_1 & 0 \end{bmatrix}_{\tilde{x}} \qquad (4.4)$$

If $\chi_1(\tilde{x}) \neq 0$ or $\phi_1(\tilde{x}) \neq 0$, then matrix (4.4) is obviously of full rank. So, let us assume that $\chi_1(\tilde{x}) = \phi_1(\tilde{x}) = 0$. In that case, Relations (4.3) together with the assumption $\psi_1(\tilde{x}) \neq 0$, yield $\phi_2 = \chi_2 = \phi_3 = \chi_3 = 0$ at $\tilde{x}$. It follows that matrix (4.4) has full rank. □

Remark 4.1. The result above depends on the assumptions $\partial_3(\tilde{x}) \neq 0$ (and $D\psi(\tilde{x}) \neq 0$). However, under the assumption $\partial_1(\tilde{x}) \neq 0$ or $\partial_2(\tilde{x}) \neq 0$ it is also possible to select from $\{h_2, h_3, h_4\}$ a pair of functions which forms a defining system of Ext(f) around $\tilde{x}$. This is easily verified. □

On U, the functions $\bar{h}_3$ and $\bar{h}_4$ are defined as follows:

$$\bar{h}_3 = \det \begin{bmatrix} \phi_{x_1} & \phi_{x_3} \\ \psi_{x_1} & \psi_{x_3} \end{bmatrix} \qquad \bar{h}_4 = \det \begin{bmatrix} \phi_{x_1} & \phi_{x_2} \\ \psi_{x_1} & \psi_{x_2} \end{bmatrix}$$

Straightforwardly, we find $\hat{h}_i = \partial_3^3 \cdot \bar{h}_i$, $i = 3,4$. Now, the following lemma is an obvious corollary of Lemma 4.1.

<u>Lemma 4.2.</u> The pair $\{\bar{h}_3,\bar{h}_4\}$ constitutes a defining system of functions for Ext(f), around $\tilde{x}$. □

Locally around $\tilde{x}$ we consider a smooth coordinate transformation $y = \Phi(x)$, with $\Phi(\tilde{x}) = 0$ such that $y_3 = \psi(x)-\psi(\tilde{x})$ and $y = (y_1,y_2,y_3)^T$. (Note that in view of the assumption $D\psi(\tilde{x}) \neq 0$, such Φ always exists). From Lemma 4.2, it follows that

$$\text{rank}\begin{bmatrix} D_y & \bar{h}_3 \circ \Phi^{-1} \\ D_y & \bar{h}_4 \circ \Phi^{-1} \end{bmatrix}_{y=0} = 2 , \tag{4.5}$$

where the subscript y in D_y stands for differentiation w.r.t. the y-coordinates.
By ϕ_{y_i} and ψ_{y_i}, we denote the partial derivatives of respectively $\phi \circ \Phi^{-1}$ and $\psi \circ \Phi^{-1}$ w.r.t. the new coordinates y_i , $i = 1,2,3$.
From (4.2) it follows that, in the y-coordinates, the set Ext(f) is given (locally around 0) by

$$\phi_{y_1} = \phi_{y_2} = 0 \tag{4.6}$$

From the chain rule (and by use of (4.6)) it follows that:

$$\begin{bmatrix} D_y & \bar{h}_3 \circ \Phi^{-1} \\ D_y & \bar{h}_4 \circ \Phi^{-1} \end{bmatrix}_{y=0} = \begin{bmatrix} \Delta_{22} & -\Delta_{21} \\ -\Delta_{32} & \Delta_{31} \end{bmatrix}_{\tilde{x}} \cdot \begin{bmatrix} D_y \phi_{y_1} \\ D_y \phi_{y_2} \end{bmatrix}_{y=0} , \tag{4.7}$$

where Δ_{ij} stands for the $(i,j)^{th}$element of the adjoint matrix $\widetilde{D\Phi}$ of $D\Phi$. From (4.5) and (4.7) a rank-argument w.r.t. the product of matrices yields:

$$\text{rank}\begin{bmatrix} \phi_{y_1y_1} & \phi_{y_1y_2} & \phi_{y_1y_3} \\ \phi_{y_2y_1} & \phi_{y_2y_2} & \phi_{y_2y_3} \end{bmatrix}_{y=0} = 2 \tag{4.8}$$

Altogether, we have proved:

Lemma 4.3. The pair $\{\phi_{y_1}, \phi_{y_2}\}$ constitutes (w.r.t. the y-coordinates) a defining system of functions for Ext(f) around 0. □

In view of Theorem 3.1(ii), the set Ext(f) is a smooth curve in R^3. Let a connected component of this curve be parametrized by $x(s) = (x_1(s), x_2(s), x_3(s))$, $x(0) = \tilde{x}$, where s = arc length. In view of (4.2), and since $\psi_{x_1}(\tilde{x}) \neq 0$, a real $\varepsilon > 0$ exists, such that if $|s| < \varepsilon$: $D\phi(x(s)) = \lambda(s) \cdot D\psi(x(s))$, for some unique, smooth function $\lambda(s)$. From now on, we assume that $|s| < \varepsilon$. For fixed s, the local manifold M_s, through x(s) is given by the equation $\psi(x) = \psi(x(s))$. (Instead of M_s, we sometimes write $M_{x(s)}$ or M_x) Now, we may characterize the point x(s) ($\in$ Ext(f)) as a critical point of the "objective function" ϕ w.r.t. the "feasible set" M_s. Let us consider the corresponding one-parameter family of Lagrange functions L_s given by:

$$L_s(x) = \phi(x) - \lambda(s) \cdot \psi(x).$$

Given a local parametrization $u = x(s;\xi)$, $\xi = (\xi_1, \xi_2)^T$ of M_s around x(s), with $x(s;0) = x(s)$, the Hessian matrix at x(s) of the restriction $L_s|_{M_s}$ will be denoted by $D^2_\xi L_s|_{M_s}$. A direct calculation, using the fact that $D_x L_s(x(s))$ vanishes, shows:

$$D^2_\xi L_s|_{M_s} = (D^T_\xi u)_{(s;0)} \cdot (D^2_x L_s)_{x(s)} \cdot (D_\xi u)_{(s;0)}.$$

The 4x4-matrices A_s and B_s are defined as

$$A_s := \left[\begin{array}{c|c|c} D_\xi u|_{(s;0)} & D^T\psi(x(s)) & \begin{matrix}0\\0\\0\end{matrix} \\ \hline 0 \quad 0 & 0 & 1 \end{array}\right]; \quad B_s := \left[\begin{array}{c|c} D^2_x L_s & D^T_x \psi \\ \hline D_x \psi & 0 \end{array}\right]_{x(s)}.$$

By inspection, one proves: (use: $D^T\psi(x(s)) \perp M_s$)

$$\det(D^2_\xi L_s|_{M_s}) = \frac{1}{|D_x\psi(x(s))|^4} \cdot \det(A_s^T \cdot B_s \cdot A_s).$$

Obviously, the matrix A_s is regular. So, we have

$$\det D^2_\xi L_s|_{M_s} = 0 \quad \text{iff} \quad \det B_s = 0 \tag{4.9}$$

For a real symmetric matrix K, the inertia In(K) is defined as the triple (ρ_1, ρ_2, ρ_3), where ρ_1, ρ_2 and ρ_3 are respectively the numbers of positive, negative and vanishing eigenvalues of K, with multiplicity counted. We emphasize that $\text{In}(D^2_\xi L_s|_{M_s})$ is independent of the incidental choice of the parametrization ξ. In fact, it can be shown (cf. [5]) that $\text{In}(D^2_\xi L_s|_{M_s}) = \text{In}(V^T \cdot D^2_x L_s \cdot V)$, where V is a 3x2 matrix with as columns an arbitrary pair of vectors spanning the complement of $D^T\psi(x(s))$. Now, one easily derives (use also (4.8)) that:

$$\operatorname{rank} D^2_{\psi} L_s|_{M_s} = \operatorname{rank} \begin{bmatrix} \phi_{y_1 y_1} & \phi_{y_1 y_2} \\ \phi_{y_2 y_1} & \phi_{y_2 y_2} \end{bmatrix}_{y(s)} \geq 1 , \tag{4.10}$$

where $y(s)$ stands for the parametrization $x(s)$ expressed in the y-coordinates. Since, on M_s we have y_3 = constant, the curve Ext(f) is transversal to M_s at $y(s)$ iff $\dot{y}_3(s) \neq 0$. Here, $\dot{y}_3(\cdot)$ stands for the derivative of $y_3(\cdot)$ w.r.t. the parameter s. From (4.6), it follows that:

$$\begin{bmatrix} \phi_{y_1 y_1} & \phi_{y_1 y_2} & \phi_{y_1 y_3} \\ \phi_{y_2 y_1} & \phi_{y_2 y_2} & \phi_{y_2 y_3} \end{bmatrix}_{y(s)} \cdot \begin{bmatrix} \dot{y}_1 \\ \dot{y}_2 \\ \dot{y}_3 \end{bmatrix}_s = 0 \tag{4.11}$$

As a consequence of (4.8), (4.9), (4.10) and (4.11) we have:

Lemma 4.4. Ext(f) is transversal to M_s at $x(s)$ iff $\det D^2_{\xi} L_s|_{M_s} \neq 0$.

The results, we have obtained up till now, enable us to describe the qualitative features of the phase-portrait of system (1.2) around an extraneous singularity $x(s_0)$ at which Ext(f) is transversal to M_{s_0}. To this aim, we note that, around $x(s_0)$, the trajectory map Tf can be represented as $Tf = (\phi,\psi)^T$. Hence, cf. Remark 3.1 and the definition of M_s, in a neighborhood of $x(s_0)$, the trajectories of (1.2) are just the level lines of $\phi|_{M_s}$. Consequently, the local phase-portrait of (1.2) around $x(s_0)$ is as depicted in Fig. 4.2. Here Fig. 4.2.a stands for the case where $\det D^2_{\xi} L_{s_0}|_{M_{s_0}} > 0$, "center type", and Fig. 4.2.b. stands for the case where $\det D^2_{\xi} L_{s_0}|_{M_{s_0}} < 0$, "saddle-point type".

(a) (b)

Fig. 4.2.

Now, we focus our attention to the case where the curve Ext(f) is non-transversal to M_0 at $\bar{x}$ (= x(0)). On R^3, the functions α_{ij} are given by

$$\alpha_{ij} = \det \begin{bmatrix} \partial_{1ij} & \partial_{11} & \partial_1 \\ \partial_{2ij} & \partial_{21} & \partial_2 \\ \partial_{3ij} & \partial_{31} & \partial_3 \end{bmatrix}, \quad i,j = 1,2,3.$$

Under the assumptions $\partial_3(x(s)) \neq 0$ and $\psi_1(x(s)) \neq 0$, we can find the following expression for the matrix $D_x^2 L_s(x(s))$:

$$D_x^2 L_s(x(s)) = \frac{1}{\partial_3(x(s))\cdot\psi_1(x(s))} \cdot (\alpha_{ij})_{x(s)} \tag{4.12}$$

This relation is derived by inspection (use: $\lambda(s) = \left.\frac{\phi_1}{\psi_1}\right|_{x(s)}$, and the fact that all 2×2 submatrices of $\hat{H}_3(z(s))$ and $\hat{H}_4(z(s))$ are singular).

Relation (4.12) is crucial, since it enables us to express the criterion in Lemma 4.4 explicitly in terms of partial derivatives of f at x(s), w.r.t. the x-coordinates, cf. (4.9).

We introduce an additional transversality condition. To this aim, note that the 3-jet space J(3,3) (= R^{19}) contains J(3,2) (= R^9) as a linear subspace. Hence, an element of J(3,3) may be represented by (z^T,v^T), where $z \in J(3,2)$ and $v \in R^{10}$. The 3-jet of f will be represented by $j^3f(x) = (j^2f(x),\hat{j}^3f(x))$, where
$\hat{j}^3f = (\partial_{111},\partial_{112},\partial_{113},\partial_{122},\partial_{123},\partial_{133},\partial_{222},\partial_{223},\partial_{233},\partial_{333})$. On the <u>open</u> subset $M_1 = \{(z^T,v^T) | z_3 \neq 0,\ z_2z_6-z_3z_5 \neq 0\}$ we define the matrix $H_5(z,v)$ from B_0 by the "formal substitution" $(z^T,v^T) := j^3f(\bar{x})$ and by <u>using (4.12)</u>. We define $h_5(z,v) = \det H_5(z,v)$ and $\bar{h}_5 = h_5 \circ j^3f$. Note that, on $(j^3f)^{-1}(M_1)$, we have: $\partial_3 \neq 0$, the functions ϕ and ψ are well-defined, and ψ_{x_1} does not vanish.
Let V be the 3×2 matrix, with $(-\psi_2(\bar{x}),\psi_1(\bar{x}),0)^T$ as its first and $(0,-\psi_3(\bar{x}),\psi_1(\bar{x}))^T$ as its last column. On M_1, the matrix $L_1(z,v)$ is defined from $V^T \cdot D_x^2L_0(\bar{x})\cdot V$ by the "formal substitution" $(z^T,v^T) := j^3f(\bar{x})$, and by using (4.12). The <u>open</u> subset $\hat{M}_1$ of M_1 is obtained by deleting all points (z^T,v^T) for which rank $L_1(z,v) = 0$. Now, we introduce the following subset of J(3,3):

$$A_1 = \{(z^T,v^T) \in \hat{M}_1 | h_3(z) = h_4(z) = h_5(z,v) = 0\}.$$

<u>Lemma 4.5</u>. The set A_1 is a 16-dimensional manifold, with $\{h_3,h_4,h_5\}$ as a defining system of functions around each of its points.

<u>Proof</u>. Let $(\bar{z},\bar{v}) \in A_1$, and thus $\tilde{z} \in V_2$ According to the same reasoning as used in the proof of Lemma 4.1, one shows that $\{D_zh_3,D_zh_4\}_{\bar{z}}$ is linearly independent. So, we are

done, if we can prove that $D_v h_5(\tilde{z},\tilde{v}) \neq 0$. To this aim, let us assume that $D_v h_5(\tilde{z},\tilde{v}) = 0$. By inspection and using $(\tilde{z}_2\tilde{z}_6-\tilde{z}_3\tilde{z}_5) \neq 0$, one verifies that this implies: the adjoint matrix $\tilde{H}_5(\tilde{z},\tilde{v})$ is of the form:

$$\tilde{H}_5(\tilde{z},\tilde{v}) = \begin{bmatrix} 0 & 0 & 0 & \cdot \\ 0 & 0 & 0 & \cdot \\ 0 & 0 & 0 & \cdot \\ \cdot & \cdot & \cdot & \cdot \end{bmatrix} .$$

Using the relation $H_5(\tilde{z},\tilde{v})\cdot\tilde{H}_5(\tilde{z},\tilde{v}) = 0$ and - again - the inequality $\tilde{z}_2\tilde{z}_6-\tilde{z}_3\tilde{z}_5 \neq 0$, we conclude that $\tilde{H}_5(\tilde{z},\tilde{v}) = 0$. Thus, we have rank $H_5(\tilde{z},\tilde{v}) \leq 2$.
From the construction of $\hat{A}_1$, it follows that rank $L_1(\tilde{z},\tilde{v}) > 0$. On the other hand, we have: det $L_1(z,v) = 0$ iff det $H_5(z,v) = 0$; see [5] and compare also (4.9). Since $h_5(\tilde{z},\tilde{v}) = 0$, we may conclude that rank $L_1(\tilde{z},\tilde{v}) = 1$, or equivalently, $\mathrm{In}(L_1(\tilde{z},\tilde{v}))$ equals either (1,0,1) or (0,1,1). Finally, we apply the Inertia Theorem, cf. [7]. We find:

$$\mathrm{In}(H_5(\tilde{z},\tilde{v})) = \mathrm{In}(L_1(\tilde{z},\tilde{v})) + (1,1,0) ,$$

and thus rank $H_5(\tilde{z},\tilde{v}) = 3$. Contradiction. □

Apparently, the manifold A_1 depends heavily on the three inequalities: $z_3 \neq 0$, $z_2z_6-z_3z_5 \neq 0$ and rank $L_1(z,v) \neq 0$, which of course "reflect" - on the level of the 3-jet space - our assumptions $\partial_3(\bar{x}) \neq 0$, $\psi_1(\bar{x}) \neq 0$ and the fact that (for $f \in F_{**}$) we have: rank $D^2_\xi L_s|_{M_s} > 0$, (cf. (4.10)). If we start with another extraneous singularity, say x with $\partial_2(x) \neq 0$ and $\chi_2(x) \neq 0$, then we may construct a 16-dimensional submanifold A_j, similar to A_1. Taking into account all possible initial assumptions, we find the manifolds $A_1,\ldots,A_{18}$. Now, the following transversality condition on the functions $f \in C^\infty(R^3,R)$ makes sense.

<u>Condition (***)</u>. $\quad j^3f \pitchfork A_j$, $j = 1,\ldots,18$.

The subset of $C^\infty(R^3,R)$ consisting of all functions which fulfil the Conditions (*), (**) and (***) will be denoted by F_{***}. From now on, we assume that $f \in F_{***}$. By $\mathrm{Ext}_d(f)$ we denote the subset of Ext(f) consisting of all those extraneous singularities at which Ext(f) and M_x are non transversal. Let $x \in \mathrm{Ext}_d(f)$ be arbitrary and let A_j be a corresponding 16-dimensional manifold. In view of Lemma 4.4 and (4.10) we have: locally around x, the set $\mathrm{Ext}_d(f)$ in just the inverse image of A_j under j^3f. Consequently, Condition (***) and Theorem 2.1 yield: $\mathrm{Ext}_d f$ is a <u>discrete</u> subset of R^3. Moreover, the set F_{***} is <u>generic</u> in $C^\infty(R^3,R)$, cf. Theorem 2.1. and 3.1.

Now, we consider - as before - the extraneous singularity $\bar{x}$. Let $\bar{x} \in Ext_d(f)$. Then from (2.2) and Lemma 4.5, it follows that

$$\{D\bar{h}_3, D\bar{h}_4, D\bar{h}_5\}_{\bar{x}} \text{ is linearly independent} \tag{4.13}$$

The function $\det \phi := \det \begin{bmatrix} \phi_{y_1y_1} & \phi_{y_1y_2} \\ \phi_{y_2y_1} & \phi_{y_2y_2} \end{bmatrix}$ is - locally around $y(0) = 0$ - well-defined. Put $\det(s) = \det \phi|_{y(s)}$.

Lemma 4.6. The triple $\{\phi_{y_1}, \phi_{y_2}, \det \phi\}_{y=0}$ is linearly independent.

Proof. Obviously, we have

$$\left[\begin{array}{c|c} D_y^2\phi & \begin{matrix}0\\0\\1\end{matrix} \\ \hline 0\ 0\ 1 & 0 \end{array}\right]_{y(s)} = \left[\begin{array}{c|c} D_y^T\phi^{-1} & \begin{matrix}0\\0\\0\end{matrix} \\ \hline 0\ 0\ 0 & 1 \end{array}\right]_{y(s)} \cdot \left[\begin{array}{c|c} D_x^2L_s & D_x^T\phi \\ \hline D_x\phi & 0 \end{array}\right]_{x(s)} \cdot \left[\begin{array}{c|c} D_y\phi^{-1} & \begin{matrix}0\\0\\0\end{matrix} \\ \hline 0\ 0\ 0 & 1 \end{array}\right]_{y(s)},$$

and hence, $\det(s) = \det^2 D_y\phi^{-1}(y(s)) \cdot \bar{h}_5(x(s))$. Consequently,

$$\frac{d}{ds}\det(0) \neq 0 \quad \text{iff} \quad \frac{d}{ds}\bar{h}_5(\bar{x}) \neq 0 \tag{4.14}$$

On the other hand, from (4.8) and (4.11) it follows that the vectors $D^T\phi_{y_1}(0)$ and $D^T\phi_{y_2}(0)$ span the orthogonal complement $(\dot{y}^\perp(0))$ of $\dot{y}(0)$. So, we are done if we can prove that $D_y^T\det\phi(0) \notin \dot{y}^\perp(0)$, or - equivalently (cf. (4.14)) - that $\frac{d}{ds}\bar{h}_5(\bar{x}) \neq 0$. Let us assume that $\frac{d}{ds}\bar{h}_5(\bar{x}) = 0$. We already know that $\bar{h}_3(x(s)) = \bar{h}_4(x(s)) \equiv 0$, cf. Lemma 4.2, and thus $\frac{d}{ds}\bar{h}_3(\bar{x}) = \frac{d}{ds}\bar{h}_4(\bar{x}) = 0$. This however, is impossible in view of (4.13), and the fact that $\dot{x}(0) \neq 0$. □

The following lemma shows that the curve Ext(f) is quadratically tangent to $M_{\bar{x}}$; moreover, if we pass $\bar{x}$, when traversing Ext(f), the sign of det(s) changes.

Lemma 4.7. $\quad \ddot{y}_3(0) \cdot \frac{d}{ds}\det(0) \neq 0.$

Proof. From (4.8) and (4.11) it follows that $\dot{y}_3(s) = \kappa(s)\cdot\det(s)$, where $\kappa(s)$ is a non-vanishing, smooth function in s. We also have: $\det(0) = 0$, cf. (4.10) and Lemma 4.4. Thus, under the assumption that our assertion is not true, we have:

$$\ddot{y}_3(0) = \kappa(0)\cdot\frac{d}{ds}\det(0) = \kappa(0)\cdot D_y \det \phi_{|y=0}\cdot\dot{y}(0) = 0.$$

However, this is impossible in view of (4.11) and Lemma 4.6. □

Remark 4.2. The extraneous singularities around $\tilde{x}$ may be characterized as the critical points for the one-parameter family of functions $\hat{\phi}_{y_3}(y_1,y_2) := \phi(y_1,y_2,y_3)$, the points in $Ext_d(f)$ being the degenerate ones. From this point of view, Lemma 4.7 is a direct consequence of our results on one-parameter optimization problems, cf. [6], especially Theorem 10.2.2. □

In order to describe the local phase portrait of (1.2) around $\tilde{x} \in Ext_d(f)$, we derive a normal form for the trajectory map Tf around $\tilde{x}$. W.r.t. the y-coordinates, we have: $Tf(y) = (\phi(y),y_3)^T$. Put $\eta(y_1,y_2) := \phi(y_1,y_2,0)$. Then the Splitting lemma (cf. [6]) can be applied to η at 0, cf. (4.6), (4.10) and Lemma 4.4. It follows that local C^∞-coordinates around 0 exist, again denoted by y_1,y_2, sending (0,0) to (0,0) and such that η takes the form

$$\eta(y_1,y_2) = \pm y_1^2 + g(y_2)\ ;\ g'(0) = g''(0) = 0\ , \tag{4.15}$$

where the + sign, resp. - sign occurs if the nonvanishing eigenvalue of the matrix $(\phi_{y_i y_j}(0))_{i,j=1,2}$ is positive, resp. negative.

The new y_1,y_2-coordinates, together with y_3 $(= \psi(x)-\psi(\tilde{x}))$ constitute a local C^∞-coordinate system of R^3. One easily verifies that w.r.t. these coordinates, (4.6), (4.10) and Lemma 4.6 do hold. Together with (4.9) and (4.15), this yields:

$$\phi_{y_1y_1}(0) \neq 0\ ,\quad \phi_{y_1y_2}(0) = \phi_{y_2y_2}(0) = 0 \tag{4.16}$$

$$\phi_{y_2y_3}(0) \neq 0 \tag{4.17.a}$$

$$\phi_{y_2y_2y_2}(0) \neq 0 \tag{4.17.b}$$

In view of (4.15) and (4.17.b) a local C^∞-coordinate transformation in the "y_2-direction", preserving (4.6), (4.10) and Lemma 4.6, exists such that - if the new coordinates are again denoted by y_1,y_2,y_3 - the function η takes the form

$$\eta(y_1,y_2) = \pm y_1^2 + y_2^3. \tag{4.18}$$

From now on, we consider η and ϕ as germs ([2]) at the origin of R^2 resp. R^3. The germ ϕ is a one-parameter unfolding of η. The ring of all C^∞-germs on R^2 at the origin will be denoted by E. For k = 1,2,..., the E-module m^k consists of all germs in E with vanishing (k-1)-jet, whereas Δ is the E-module generated by η_{y_1} and η_{y_2}.

So, cf. (4.18), we have $\Delta = \langle y_1, y_2^2 \rangle_E$. It follows that $m^3 \subset m \cdot \Delta$. Thus, cf. [2], the germ η is <u>3-determined</u>. On the other hand, if $\langle \phi_{y_3}(y_1,y_2,0) - \phi_{y_3}(0,0,0)\rangle_R$ stands for the R-vectorspace generated by $\phi_{y_3}(y_1,y_3,0) - \phi_{y_3}(0)$, it follows from (4.17.a) that $m = \Delta + \langle \phi_{y_3}(y_1,y_2,0) - \phi_{y_3}(0)\rangle_R + m^4$, i.e. the unfolding ϕ is 3-transversal (cf. [2]). Altogether, we have proved that ϕ is a <u>versal unfolding</u> of η. From (4.18) it is also clear, (cf. [2]), that $\pm y_1^2 + y_2^3 + y_2 t$ is a <u>universal unfolding</u> of η. Now, application of the Mather-Malgrange theorem on (universal) unfoldings, cf. [2], yields the existence of a smooth local R^3- coordinate system around 0, say $\zeta = \theta(y)$, and a local smooth function α around $y_3 = 0$, such that

$$\phi(y_1,y_2,y_3) = \pm \zeta_1^2(y) + \zeta_2^3(y) + \zeta_2(y)\cdot\zeta_3(y) + \alpha(y_3),$$

and which, in addition, fulfil the following properties:

$\zeta_i(y_1,y_2,0) = y_i$, 1,2,

$\zeta_3(y) = \zeta_3(y_3)$, and $\zeta_3(0) = 0$

$\alpha(0) = 0$.

Consequently, under the transformation θ, the planes y_3 = constant and ζ_3 = constant, do correspond; this is especially true for the planes $y_3 = 0$ and $\zeta_3 = 0$. Since $\alpha(0) = 0$, we finally conclude that the level lines of Tf (i.e. trajectories of (1.2)) around $\tilde{x}$ are - w.r.t. the ζ-coordinates - given by

$$\pm \zeta_1^2 + \zeta_2^3 + \zeta_2\zeta_3 = \text{constant} \; ; \; \zeta_3 = \text{constant}. \tag{4.19}$$

From (4.19) it follows that the local phase portrait of (1.2) is as depicted in Fig. 4.3. (In this figure we have chosen for the <u>positive</u> square ζ_1^2).

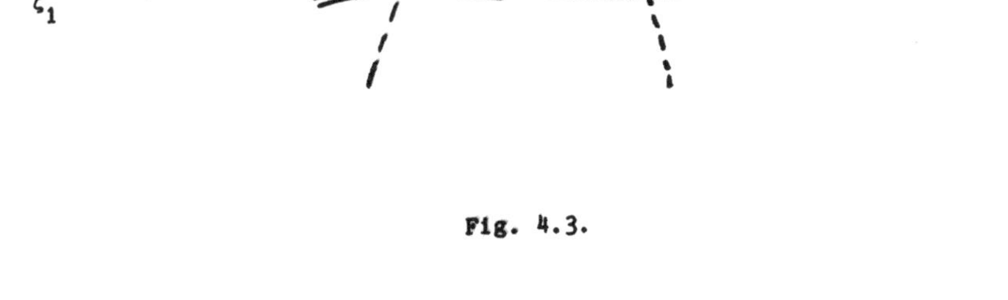

Fig. 4.3.

Altogether we have proved our main result:

<u>Theorem 4.1.</u> There exists a generic subset F_{***} of $C^{\infty}(R^3,R)$ such that, for $f \in F_{***}$, we have: the set Ext(f) - if not empty - is a closed, one-dimensional manifold. Moreover, with any $x \in$ Ext(f), a local two-dimensional manifold M_x through x can be associated, such that the local trajectory of (1.2) through a point of M_x remains in M_x. The curve Ext(f) intersects M_x non-transversally iff x is contained in a discrete subset $Ext_d(f)$. On a connected component of $Ext(f)\setminus Ext_d(f)$, the local phase portrait is either of a "center type" or of a "saddle-point type", whereas at a point in $Ext_d(f)$ such local phase portrait is of a "cusp type", cf. Fig. 4.3.

5. Final remarks: the case $n \geq 4$.

Let $f \in C^{\infty}(R^n,R)$, $n \geq 4$, be an element of F_{**}. We consider a point $\bar{x} \in$ Ext(f) and assume (no loss of generality) that $\partial_n(\bar{x}) \neq 0$. Hence, the function $\phi = \frac{\partial_1}{\partial_n}$ and the map $\psi = (\frac{\partial_2}{\partial_n},\ldots,\frac{\partial_{n-1}}{\partial_n})^T$ are well-defined in a neighborhood of x. Now, we distinguish between two cases:

$$\text{I: } J^2f(\bar{x}) \in V_{n-1}\ , \quad \text{and} \quad \text{II: } J^2f(\bar{x}) \in \bigcup_{k=0}^{n-2} V_k .$$

<u>We firstly consider the Case I.</u> In this case the set Ext(f) is - locally around $\bar{x}$ - a (n-2)-dimensional manifold (to be parametrized as x(s), $x(0) = \bar{x}$) In view of (3.1) and (3.2) we have $\text{rank}[D^T\phi \vdots D\psi]_{\bar{x}} = n-2$, and hence, we may assume that rank $D\psi(\bar{x}) = n-2$. Thus, locally around $\bar{x}$, we can associate with x(s) a(local) 2-dimensional manifold - say M_s - which is defined by $\psi(x) = \psi(x(s))$. As in the case n = 3, we may characterize x(s) as the critical points of ϕ w.r.t. the feasible set M_s, the degenerate ones being just those points for which Ext(f) and M_s are non-transversal. The proof of this statements is a straightforward generalization of the proof of Lemma 4.4; in fact, it is based on the introduction of local coordinates $(y_1,\ldots,y_n)$ such that $(y_3,\ldots,y_n)^T = \psi(x)-\psi(\bar{x})$ and $y(\bar{x}) = 0$. The set of all points of "non-transversal intersection" between M_s and Ext(f) is denoted by $Ext_d(f;\bar{x})$. In the y-coordinates, the trajectory map Tf is represented by $Tf(y) = (\phi(y),y_3,\ldots,y_n)^T$. We consider the 2x2-matrix $\Gamma(s)$:

$$\Gamma(s) = \begin{bmatrix} \phi_{y_1y_1} & \phi_{y_1y_2} \\ \phi_{y_2y_1} & \phi_{y_2y_2} \end{bmatrix}_{y(s)} ,$$

and distinguish between the following subcases:

I_a: rank $\Gamma(0) = 2$, I_b: rank $\Gamma(0) = 1$, and I_c: rank $\Gamma(0) = 0$.

In case I_a we have: $\bar{x} \in \text{Ext}(f)\backslash\text{Ext}_d(f;\bar{x})$, whereas in both other cases: $\bar{x} \in \text{Ext}_d(f;\bar{x})$. Now, we claim (under appropriate-transversality conditions):

In case I_a, a local trajectory of (1.2) through a point M_0 remains in M_0. Moreover, the restriction to M_0 of the local phase portrait of (1.2) around $\bar{x}$, is either of a "center type" (if det $\Gamma(0) > 0$), or of a "saddle point type" (if det $\Gamma(0) < 0$).

In case I_b, the Mather-Malgrange theorem on universal unfoldings yields - in the worst case - the existence of a local C^∞-coordinate system $\zeta = \theta(y)$ around $0 \in \mathbb{R}^n$, such that in the ζ-coordinates the level-lines of $Tf_{|M_s}$ are given by

$$\left.\begin{array}{l} \pm\, \zeta_1^2 + \delta\zeta_2^n + \zeta_3\zeta_2^{n-2} + \zeta_4\zeta_2^{n-3} + \ldots + \zeta_n\zeta_2 = \text{constant} \\ (\zeta_3,\ldots,\zeta_n) = \text{constant} \end{array}\right\} \tag{5.1}$$

Here, $\delta = 1$ if n is odd, and $\delta = \pm 1$ if n is even. In view of Remark 3 1, the local phase portrait of (1.2) around $\bar{x}$ is described by (5.1).

Case I_c does not occur if n = 4, but will generically occur if $n \geq 5$, see also [6, especially Theorem 10.1.2].

On the other hand, if n = 4, the Case II cannot occur (cf. Theorem 3.1 (ii)).

So, combining our comments on both Cases I and II, we conclude that - under an appropriate transversality condition - Theorem 4.1 remains true for n = 4, if we put: dim Ext(f) = 2, dim $\text{Ext}_d(f)$ = 1 and if we replace the entry "cusp type" by

"has a local phase-portrait, described by

$\pm\, \zeta_1^2 \pm \zeta_2^4 + \zeta_3\zeta_2^2 + \zeta_4\zeta_2$ = constant, (ζ_3,ζ_4) = constant".

However, if n > 4, then the Cases I_c and II cannot be excluded. The analysis of these cases will be the subject of further investigations.

References

[1] Branin, F.H.: A widely convergent method for finding multiple solutions of simultaneous non-linear equations. IBM J. Res. Develop., pp. 504-522 (1972).

[2] Bröcker, Th.: Differential germs and catastrophes, London Math. Soc. Lect. Notes Series 17, Cambridge Un. Press (1975).

[3] Gomulka, J.: Remarks on Branin's method for solving non-linear equations. In: Towards global optimization (L.C.W. Dixon and G.P. Szegö, eds.), Acad. Press (1976).

[4] Hirsch, M.W.: Differential topology. Springer Verlag (1976).

[5] Jongen, H.Th., Jonker, P. and Twilt, F.: Nonlinear optimization in R^n, I: Morse theory, Chebyshev approximation, Peter Lang Verlag, Frankfurt a.M. (1983).

[6] Jongen, H.Th., Jonker, P. and Twilt, F.: Nonlinear optimization in R^n, II: Transversality, Flows, Parametric Aspects, Peter Lang Verlag, Frankfurt a.M. (1986), to appear.

[7] Jongen, H.Th., Möbert, T., Rückmann, J and Tammer, K.: On inertia and Schur complement in optimization. Twente University of Technology, Memorandum nr. 552 (1986).

[8] Koecher, M.: Lineare Algebra und analytische Geometrie, Grundwissen Mathematik II, Springer Verlag (1983).

[9] Ljusternik, L.A., Sobolew, W.I.: Elemente der Funktionalanalysis. Akademie-Verlag, Berlin (1968).

[10] Smale, S.: A convergent process of price adjustment and global Newton-methods. J. Math. Economics, Vol. 3, pp. 107-120 (1976).

LIPSCHITZ CONTINUITY OF INFIMA AND OPTIMAL SOLUTIONS IN PARAMETRIC OPTIMIZATION: THE POLYHEDRAL CASE

Diethard Klatte [1)]

Abstract: In this paper we shall point out that various Lipschitz stability results known for parametric linear programs can be extended to more general parametric programs which have a polyhedral structure. This extension applies, in particular, to parametric (generally nonconvex) quadratic optimization problems.

1. Introduction

The purpose of this paper is to give a survey of results on Lipschitz properties of optimal solutions, stationary points and infima in parametric programs with polyhedral structure. It will be shown that many Lipschitz stability results known for parametric linear programs can be "saved" also for parametric problems of quadratic optimization and linear complementarity.

First we shall introduce some notations which will appear in the following. For simplicity throughout the paper we shall use the Euclidean norm $\|\cdot\|$ (and the corresponding induced norm for matrices). By $R^{m\times n}$ we denote the Euclidean space of (m,n)-matrices. The symbol dist(x,Z) stands for the Euclidean point-to-set distance from a point x to a set Z. The closed unit ball in the n-dimensional Euclidean space R^n has the standard symbol B_n , R^n_+ is the non-negative orthant of R^n. For a set $X\subset R^n$, int X and cl X denote its interior and its closure, respectively. If $X,Y\subset R^n$ and $\alpha,\beta\in R$, then we write $\alpha X+\beta Y := \{\alpha x+\beta y \,/\, x\in X,\ y\in Y\}$. Let $U_\varepsilon(x) := \{x\} + \varepsilon \operatorname{int} B_n$ denote the open ε-neighborhood of $x\in R^n$.

Given a parametric optimization problem

$$\min_x \{ f(x,t) \,/\, x\in M(t)\} \ , \ t\in T\subset R^r, \tag{1}$$

(where $f\colon R^s\times R^r \longrightarrow R$, $M\colon R^r \longrightarrow 2^{R^s}$, and t is the parameter vector), we define the <u>infimum function</u> φ of (1) by

$$\varphi(t) := \inf_x \{ f(x,t) \,/\, x\in M(t)\} \ , \ t\in T,$$

and the <u>(global) optimal set mapping</u> ψ of (1) by

$$\psi(t) := \{x\in M(t) \,/\, f(x,t) = \varphi(t)\} \, , \ t\in T.$$

A subset of an Euclidean space is said to be polyhedral if it is a

1) Pädagogische Hochschule Halle, Kröllwitzer Straße 44, DDR - 4020 Halle/Saale

union of finitely many polyhedral convex sets. Let F be a multivalued mapping (also called <u>multifunction</u>) from $T \subset R^r$ to R^s. Then we write

$$\text{dom } F \;:=\; \{ t \in T \,/\, F(t) \neq \emptyset \}$$

(<u>effective domain of F</u>) and

$$\text{graph } F := \{ (t,x) \in T \times R^s \,/\, x \in F(t) \}$$

(<u>graph of F</u>). We say that F is <u>polyhedral</u> if graph F is a polyhedral set, and F is <u>convex</u> (or <u>closed</u>) if graph F is convex (or closed).
F is <u>upper Lipschitzian at $t^o \in T$</u> (in the sense of [23]) if there are real numbers $L > 0$ and $\beta > 0$ such that

$$F(t) \subset F(t^o) + L\,\|t-t^o\|\,B_s \qquad (\forall t \in U_\beta(t^o) \cap T) \tag{2}$$

holds. F is <u>locally upper Lipschitzian on T</u> if for each $t^o \in T$, F is upper Lipschitzian at t^o. We shall say that F is <u>Lipschitzian on T</u> if there is some constant $L > 0$ such that

$$F(t'') \subset F(t') + L\,\|t''-t'\|\,B_s \qquad (\forall t',t'' \in T). \tag{3}$$

If (2) (or (3)) is satsfied, we also say that F is upper Lipschitzian (or Lipschitzian) <u>with modulus L</u>. A real-valued function g (or a vector-valued function g) is said to be Lipschitzian on T if the mapping $t \longrightarrow F(t) := \{g(t)\}$ has this property.
In order to motivate our studies, we recall typical Lipschitz continuity results in parametric linear optimization. Let us consider the linearly perturbed linear program

$$\min_x \{ p^T x \,/\, x \in M(b) \}, \quad t=(p,b) \in T \subset R^n \times R^m, \tag{4}$$

where p and b are regarded as parameters, A is an (m,n)-matrix, and $M(b) := \{ x \in R^n \,/\, Ax \leq b \}$ (we could also have added equality constraints and constrained variables; generally, we had to assume that M is a polyhedral convex multifunction). Then the following properties hold:

dom ψ is a polyhedral convex set, cf. [19], (4a)

φ is Lipschitzian on bounded subsets of dom ψ, cf. [26,19,23,28], (4b)

ψ is locally upper Lipschitzian on $R^n \times R^m$ with some uniform modulus, cf. [20,16,23]. (4c)

It is worth noting that, in this case, the optimal set mapping ψ is a polyhedral multifunction.
A very simple example demonstrates that it is not possible to strengthen the Lipschitz continuity results (4b) and (4c).

<u>Example 1</u>: Consider $px \longrightarrow \min$ s.t. $x \in [0,b] := \{ x \in R / 0 \leq x \leq b \}$, where $p \in R$ and $b \in R$ are parameters. Obviously, we have

$$\psi(p,b) = \begin{cases} \{0\} & \text{if } b \geq 0, p>0 \\ [0,b] & \text{if } b \geq 0, p=0 \\ \{b\} & \text{if } b \geq 0, p<0 \end{cases} \quad \text{and} \quad \varphi(p,b) = \begin{cases} 0 & \text{if } b \geq 0, p \geq 0 \\ pb & \text{if } b \geq 0, p \leq 0, \end{cases}$$

hence, for this parametric linear program, neither ψ nor φ are Lipschitzian on dom $\psi = R \times [0,+\infty)$.

For the arbitrarily perturbed linear program

$$\min_x \left\{ p^T x \;/\; Ax \leq b \right\}, \quad (p,A,b) \in T, \tag{5}$$

where $t=(p,A,b) \in T := R^n \times R^{m \times n} \times R^m$ is regarded as parameter, it is known that the following holds: Let $(p^o,A^o,b^o) \in T$, and suppose that Slater's condition is satisfied, i.e. $\{x \in R^n \;/\; A^o x < b^o\} \neq \emptyset$, and the optimal set $\psi(p^o,A^o,b^o)$ is nonempty and bounded. Then one has

$\psi(p,A,b) \neq \emptyset$ for (p,A,b) near (p^o,A^o,b^o), cf. [8,21], (5a)

ψ is upper Lipschitzian at (p^o,A^o,b^o), cf. [21], (5b)

φ is Lipschitzian on some neighborhood of (p^o,A^o,b^o), cf., e.g., [8]. (5c)

The paper is organized as follows. In §2 we shall recall Lipschitz continuity results for polyhedral multifunctions and linear generalized equations. To point out the essence of several main ideas, we give the proofs also for results which are known from the literature. In §3 we present sufficient conditions for the persistence and for some kind of Lipschitz stability of Kuhn-Tucker solutions and optimal points in arbitrarily perturbed quadratic (non-convex) programs. As by-products we also obtain continuity results for stationary solutions, local minimizing sets and infima. Finally, §4 surveys Lipschitz continuity properties of global optimal solutions and infima for linearly perturbed quadratic (non-convex) programs and some related structural properties.

2. Stability of linear generalized equations

Following Robinson [22,23] we shall use the formalisms of polyhedral multifunctions and linear generalized equations: this allows to analyze, in a unifying way, the stability of systems of linear equations and linear inequalities, of solution sets of linear complementarity problems, of optimal sets for linear and quadratic programs, etc.. First we recall a fundamental theorem of Hoffman [11] on the approximate solution of systems of linear inequalities and some related results.

Proposition 1 ([11,21]): Let H be an (r,s)-matrix, and define the multifunction $F(h) := \{x \in R^s \;/\; Hx \leq h\}$, $h \in R^r$. Then there exists some constant $L > 0$ such that for each $h \in \text{dom } F$ and for each $z \in R^s$, one has

$$\text{dist}(z,F(h)) \leq L \cdot \|(Hz-h)_+\| ,$$

where for $a=(a_1,\dots,a_r)$, $a_+ := (\max\{a_1,0\},\dots,\max\{a_r,0\})$.

Remark 1: We emphasize that the Lipschitz constant L only depends on H, we refer to the proofs, e.g., in [11,12,27].

The following proposition is a direct consequence of Hoffman's theorem.

Proposition 2 ([27]): If $G: R^r \longrightarrow 2^{R^s}$ is a polyhedral convex multifunction, then G is Lipschitzian on its effective domain dom G.

This proposition immediately applies to the solution set mapping G of a system of linear inequalities parameterized in the right-hand side. For this case, Mangasarian [17] and Mangasarian/Shiau [18] propose another approach (via a condition number of linear systems) which allows to compute explicitly Lipschitz constants.
A second classical application of Proposition 2 is provided by a parametric linear optimization problem

$$c^T x \longrightarrow \min \qquad \text{s.t.} \quad Ax \leq b , \tag{6}$$

where $c \in R^n$ and $A \in R^{m \times n}$ are given, and $b \in R^m$ is regarded as parameter vector. The multifunction

$$(b,t) \longrightarrow G(b,t) := \{ x \in R^n \;/\; Ax \leq b,\; c^T x \leq t \}$$

is convex and polyhedral, and so, G is Lipschitzian on dom G. The optimal set mapping ψ of problem (6) has, obviously, the representation

$$\psi(b) = G(b, \varphi(b)).$$

Since the infimum function of (6) is continuous and piecewise linear, cf. [19,28], Proposition 2 then yields that ψ is Lipschitzian on dom ψ. For possibilities of computing the Lipschitz constant in this situation, we refer to Kleinmann [16] and Mangasarian/Shiau [18].

Robinson [20,23] has proposed a technique that allows to carry over the Lipschitz continuity result for polyhedral convex multifunctions to the general case of polyhedral (but non-convex) multifunctions. Of course, classical examples from linear parametric optimization (cf. [19]) tell us that, in the non-convex case, the continuity of such multifunctions fails. Indeed, one can only expect some kind of upper semicontinuity. The technique in [20,23] (cf. also [16,12,13]) uses the idea of splitting a polyhedral multifunction into convex components. In Lemma 1 and Lemma 2 we present a rather general form of this "trick" in view of an application in §4.
As usual, the intersection of two multifunctions G and F from $T \subset R^r$ to R^s is defined by $(G \cap F)(t) := G(t) \cap F(t) \quad (\forall t \in T)$.

Lemma 1: Let $t^o \in T$, and let $G, G_1, \ldots, G_N$ be multifunctions from a closed set $T \subset R^r$ to R^s such that $G(t^o)$ is nonempty and

$$G(t) \subset \bigcup_{i=1}^{N} G_i(t) \qquad (\forall t \in T).$$

Suppose that the sets dom $(G \cap G_i)$ $(\subset T)$, $i=1,\dots,N$, are closed. Then there is a neighborhood U of t^o such that

$$G(t) \subset \bigcup_{i \in I} G_i(t) \quad \text{for all } t \in U \cap T, \tag{7}$$

where $I := \{ i \in \{1,\dots,N\} \ / \ G(t^o) \cap G_i(t^o) \neq \emptyset \}$.

Proof: We assume, on the contrary, that there are a sequence $\{t^k\} \subset T$ converging to t^o and some $j \in \{1,\dots,N\} \setminus I$ (without loss of generality j is independent of k) such that

$$G(t^k) \cap G_j(t^k) \neq \emptyset \quad \text{for all } k.$$

By hypothesis, dom $(G \cap G_j)$ is a closed set. Since for all k, $t^k \in$ dom $(G \cap G_j)$, and since $t^k \longrightarrow t^o$, this implies that t^o belongs to dom $(G \cap G_j)$. Hence, $G(t^o) \cap G_j(t^o)$ would be nonempty, contradicting $j \notin I$. //

Lemma 2: Assume the hypotheses of Lemma 1. If $G_1,\dots,G_N$ are upper Lipschitzian at t^o with moduli $L_1,\dots,L_N$, and if

$$G(t^o) = \bigcup_{i \in I} G_i(t^o),$$

where $I := \{ i \in \{1,\dots,N\} \ / \ G(t^o) \cap G_i(t^o) \neq \emptyset \}$, then G is upper Lipschitzian at t^o with modulus $L = \max_{i \in I} L_i$.

Proof: The result follows immediately from the definition of upper Lipschitzian multifunctions and Lemma 1. //

Now these lemmas will be applied to polyhedral multifunctions. If G is a polyhedral multifunction from R^r to R^s, then there exist nonempty polyhedral convex sets $S_1,\dots,S_N \subset R^r \times R^s$ such that

$$\text{graph } G = \bigcup_{i=1}^{N} S_i \quad . \tag{8}$$

Hence, the multifunction G can be splitted into multifunctions $G_1,\dots,G_N$, defined by

$$G_i(t) := \{ z \in R^s \ / \ (t,z) \in S_i \} \quad (i=1,\dots,N). \tag{9}$$

Obviously, the multifunctions $G_1,\dots,G_N$ are polyhedral and convex.

Proposition 3 ([20,23]): If G is a polyhedral multifunction from R^r to R^s, then there exists a real number $L > 0$ such that for each $t^o \in R^r$, G is upper Lipschitzian at t^o with modulus L, i.e., G is locally upper Lipschitzian on R^r with a uniform modulus.

Proof: Choose any $t^o \in R^r$. If $G(t^o) = \emptyset$ then $G(t) = \emptyset$ for all t belonging to some neighborhood of t^o, since dom G is closed. Hence, in this case, the assertion is trivially fulfilled. Suppose now $G(t^o) \neq \emptyset$, and let G be splitted, by (8) and (9), into polyhedral convex components $G_1,\dots,G_N$. By Proposition 2, G_i is Lipschitzian on dom G_i with modulus L_i $(i=1,\dots,N)$. Let $L := \max\{L_1,\dots,L_N\}$. In particular, the multifunctions $G_1,\dots,G_N$ are then upper Lipschitzian at t^o with modulus L (independent of t^o). Obviously,

$$G(t) \quad = \quad \bigcup_{i=1}^{N} G_i(t) \qquad (t\in R^r).$$

Thus, for each $i\in\{1,\dots,N\}$, dom $(G\cap G_i)$ = dom G_i. Recalling that dom G_i, $i\in\{1,\dots,N\}$, is closed, we hence observe that all assumptions of Lemma 2 are satisfied. This implies the desired result. //

Remark 2: Given a matrix $H\in R^{n\times n}$, the solution set mapping of a linear complementarity problem parameterized in the right-hand side,

$$\Gamma(h) := \{z\in R^n \;/\; Hz\geq h,\; z\geqslant 0,\; z^T(Hz-h)=0\},\quad h\in R^n,$$

is polyhedral, which is easy to verify. Hence, Γ is locally upper Lipschitzian on R^n. This result is helpful in analyzing Kuhn-Tucker sets and stationary solutions of perturbed quadratic programs, cf. §4.

Now we are going to study a more general situation: we consider the linear generalized equation

$$0 \in Hz - h + G(z)\;, \qquad z\in Z, \tag{10}$$

where G is a polyhedral multifunction from R^r to R^s, $Z\subset R^r$ is a polyhedral convex set, 0 is the origin in R^s, and $H\in R^{s\times r}$ as well as $h\in R^s$ will be regarded as parameters; for simplicity, we have written $Hz-h+G(z)$ instead of $\{Hz-h\}+G(z)$.
Let F denote the solution set mapping of the parameterized linear generalized equation (10)

$$F(H,h) := \{z\in Z \;/\; 0\in Hz - h + G(z)\}, \tag{11}$$

where H and h vary in $R^{s\times r}$ and R^s, respectively. Important special cases of (10) are the following ones:

(i) If $H\in R^{s\times r}$, $h\in R^s$, $G(z)\equiv R^r_+$ and $Z=R^r$ then we have a system of s linear inequalities in r variables: $Hz\leq h$.

(ii) $H\in R^{n\times n}$, $h\in R^n$, $Z = R^n_+$ and $G(z) := \{y\in -R^n_+ \;/\; y^Tz = 0\}$ give the linear complementarity problem $Hz\geq h$, $z\geqslant 0$, $z^T(Hz-h)=0$.

Obviously, if H^o is fixed then the multifunction $F(H^o,\cdot)$ is polyhedral and hence, by Proposition 3, locally upper Lipschitzian.

In the following theorem we develop a sufficient condition for the multifunction F to be upper Lipschitzian. The proof is based on a standard technique of handling perturbed equations.

Theorem 1: Let F be defined as in (11), and let $H^o\in R^{s\times r}$ and $h^o\in R^s$ be given. Suppose that Y is a compact subset of R^r. Then there are real numbers $L>0$ and $\beta>0$, both depending on Y and H^o, such that

$$F(H,h)\cap Y\subset F(H^o,h^o) + L\max\{\|H-H^o\|\;,\;\|h-h^o\|\}\,B_r \tag{12}$$

if $\|H-H^o\| < \beta$ and $\|h-h^o\| < \beta$.

Proof: Since dom F is closed, in the case $F(H^o,h^o)=\emptyset$ the theorem trivially holds.
Suppose $F(H^o,h^o)\neq\emptyset$. Since $F(H^o,\cdot)$ is polyhedral, Proposition 3 implies that

$$F(H^o,h)\subset F(H^o,h^o) + \delta\|h-h^o\|\ B_r \qquad (\forall h\in U_\varepsilon(h^o)) \tag{13}$$

for some positive real numbers δ and ε. Set

$$\beta := \varepsilon(l(Y)+1)^{-1}, \text{ where } l(Y) := \max\left\{\|z\| \ / \ z\in Y\right\}.$$

Choose any (H,h) satisfying $\|H-H^o\|<\beta$ and $\|h-h^o\|<\beta$, and let $z\in F(H,h)\cap Y$. Hence,

$$0\in H^oz - h + (H-H^o)z + G(z)\ , \quad z\in Z\cap Y.$$

Setting

$$h' := h-(H-H^o)z,$$

we then have $z\in F(H^o,h')$ and

$$\|h'-h^o\| = \|h-(H-H^o)z-h^o\| < \beta + \beta l(Y) = \varepsilon\ .$$

Property (13) then yields

$$\operatorname{dist}(z,F(H^o,h^o)) \le \delta\,\|h'-h^o\| \le \delta\,\|h-h^o+(H^o-H)z\|$$
$$\le \delta\,(\,\|h-h^o\| + \|H^o-H\|\ l(Y)).$$

With $L := \delta(1+l(Y))$, this completes the proof. //

Remark 3: The question arises whether, if it was assumed that $F(H^o,h^o)$ is nonempty and that some regularity condition holds, one could guarantee that F(H,h) is nonempty for each (H,h) near (H^o,h^o) ("persistence of solvability"). In the case of parametric quadratic programs, an answer will be given in §3. In [22, Theorem 2] Robinson has shown the following result which concerns a special class of linear generalized equations. Consider the multifunction

$$F(H,h) := \left\{z\in Z \ / \ 0\in Hz - h + N_Z(z)\right\}\ , \tag{14}$$

where $N_Z(z)$ denotes the normal cone of Z at z, Z is a nonempty polyhedral convex set, h varies over R^s, and H varies over $R^{s\times s}$.
Suppose that H^o is positive semidefinite, and let $h^o\in R^s$. If $F(H^o,h^o)$ is nonempty and bounded, then for any open bounded set Q containing $F(H^o,h^o)$, there is some real number $\beta>0$ such that

$$F(H,h)\cap Q\neq\emptyset \quad \text{if} \quad \max\left\{\|H-H^o\|\ ,\ \|h-h^o\|\right\} < \beta\ .$$

An example will demonstrate that Robinson's result is no longer valid, when the assumption of positive semidefiniteness fails.

Example 2 ([14,15]): Given the parametric quadratic program

$$\min\left\{xy - x^2 \ / \ x\ge t,\ y\le 1\right\}\ , \quad t\in R,$$

necessary conditions for optimality are

$$-2x+y-u \;=0,\ u \geq 0,\ x \geq t,\ u(x-t)=0$$
$$x \qquad +v=0,\ v \geq 0,\ y \leq 1,\ v(y-1)=0.$$

This system can be written in the form (14) by setting

$$H = \begin{pmatrix} -2 & 1 & -1 & 0 \\ 1 & 0 & 0 & 1 \\ 1 & 0 & 0 & 0 \\ 0 & -1 & 0 & 0 \end{pmatrix},\ h(t) = \begin{pmatrix} 0 \\ 0 \\ -t \\ 1 \end{pmatrix},\quad Z = R^2 \times R^2_+$$

(cf. [22, p.140]). It is easy to see that, with

$$G(t) \;:=\; F(H,h(t)),$$

where F is defined as in (14), we have

$$G(0) = \left\{(0,w,w,0) \;/\; 0 \leq w \leq 1\right\},\ \text{but } G(t) = \emptyset \text{ for } t>0.$$

Here G(0) is nonempty, polyhedral, convex and bounded, but the matrix H (which is fixed) is indefinite. This shows: Though we have right-hand side perturbations only, the persistence property does not hold in the indefinite case.

3. Consequences for arbitrarily perturbed quadratic programs

In this paragraph we are concerned with stability properties of Kuhn-Tucker sets, stationary solutions, local and global minimizing sets and infima of arbitrarily perturbed problems in quadratic optimization. Consider the parametric optimization problem

$$\min_x \;\left\{ f(x,t) \;/\; x \in M(t) \right\} \qquad (15)$$

with the parameter tuple $t=(C,p,A,b)$, where

$$f(x,t) := \tfrac{1}{2}\, x^T C x + p^T x \;,\quad M(t) := \left\{ x \in R^n \;/\; Ax \leq b \right\},$$

and C varies over all symmetric (n,n)-matrices, A varies over $R^{m \times n}$, and the parameters p and b are vectors in R^n and R^m, respectively. As before, the set of all these parameter tuples is denoted by T. For $t = (C,p,A,b)$ let

$$\| t \|_T \;:=\; \max\left\{ \|C\| \;,\; \|p\| \;,\; \|A\| \;,\; \|b\| \right\} \;.$$

Again we note that the analysis of this parametric program could be carried out similarly, when M is defined by a system of linear inequalities, linear equations, constrained variables, etc. .

First we introduce some notations. Given any nonempty set $Q \subset R^n$ and any $t \in T$, we set

$$M_Q(t) \;:=\; M(t) \cap \operatorname{cl} Q,$$
$$\varphi_Q(t) \;:=\; \inf\left\{ f(x,t) \;/\; x \in M_Q(t) \right\},$$
$$\psi_Q(t) \;:=\; \left\{ x \in M_Q(t) \;/\; f(x,t) = \varphi_Q(t) \right\}.$$

Further, let $\psi_{loc}(t)$ denote the set of all local minimizers for $f(\cdot,t)$ w.r. to $M(t)$.

Following Robinson [25] we shall say that a nonempty set $X \subset R^n$ is a strict local minimizing set (SLM set) for $f(\cdot,t)$ w.r. to $M(t)$, if there is an open set $Q \supset X$ such that $X = \psi_Q(t)$; here t is supposed to be fixed. Obviously, such a SLM set is a subset of $\psi_{loc}(t)$. Regarding the parametric problem (15) we observe that, in this case, a SLM set is always closed.

Typical examples of strict local minimizing sets are

(i) the set of global minimizers $\psi(t)$ if $\psi(t) \neq \emptyset$ (set $Q:=R^n$),

(ii) $X = \{x^o\}$ if x^o is a strict local minimizer for $f(\cdot,t)$ w.r. to $M(t)$ (i.e., $f(x^o,t) < f(x,t)$ for all $x \in \mathrm{cl}\, U \cap M(t)$, where U is some neighborhood of x^o); in this case, one may choose $Q:=U$.

Let KT(t) denote the set of Kuhn-Tucker solutions of the program (15) for fixed $t=(C,p,A,b)$:

$$KT(t) := \left\{ (x,u) \in R^n \times R^m \;/\; \begin{array}{l} Cx+A^Tu+p = 0 \\ Ax-b \leq 0, u \geq 0 \\ u^T(Ax-b) = 0 \end{array} \right\} . \tag{16}$$

The corresponding set of stationary points, denoted SP(t), is

$$SP(t) := \pi_n(KT(t)), \tag{17}$$

where $\pi_n(\cdot)$ is the canonical projection to R^n. The set of Lagrange multipliers associated with $x \in SP(t)$ is given by

$$LM(x,t) := \left\{ u \in R^m \;/\; (x,u) \in KT(t) \right\} .$$

It is well-known that, in quadratic programming, $x \in \psi_{loc}(t)$ implies that $LM(x,t)$ is nonempty, without assuming any constraint qualification. As usual, $KT(\cdot)$, $SP(\cdot)$ and $LM(\cdot,\cdot)$ are considered to be multifunctions.

In the next theorems, we give sufficient conditions (cf. also [14,15]) for the persistence of local minimizers of the program (15) under data perturbations, and for some Lipschitz behavior of the mappings $KT(\cdot)$, $SP(\cdot)$, ψ_{loc}, ψ and φ.

Theorem 2 (Stability of the Kuhn-Tucker solutions): Consider the parametric problem (15). Let $t^o = (C^o,p^o,A^o,b^o) \in T$ be a given data tuple. Suppose that $X \subset R^n$ is a bounded SLM set for $f(\cdot,t^o)$ w.r. to $M(t^o)$. Further, suppose that the Slater condition holds w.r. to $M(t^o)$, i.e., there is a point x^o satisfying $A^o x^o < b^o$.

Then $K:= KT(t^o) \cap (X \times R^m)$ is a nonempty compact set, and one has:

(i) For each open set D containing K there exists a constant $\varepsilon > 0$ such that

$$KT(t) \cap D \neq \emptyset \qquad (\forall t \in U_\varepsilon(t^o)).$$

(ii) For each open bounded set D containing K there exist constants $\delta > 0$ and $L > 0$ such that

$$KT(t) \cap D \subset KT(t^{o}) + L \, \|t-t^{o}\|_{T} \, B_{n+m} \qquad (\forall t \in U_{\delta}(t^{o})).$$

<u>Proof</u>: First we note that for each $x \in X$, the set $LM(x,t^{o})$ is nonempty in virtue of $X \subset \psi_{loc}(t^{o})$, hence $K \neq \emptyset$. The Slater condition entails that $LM(x,t^{o})$ for $x \in X$ is bounded. By Robinson [24, Th. 2.3], the multifunction $LM(\cdot,t^{o})$ is upper semicontinuous on X. This, together with the compactness of X, implies that

$$K = KT(t^{o}) \cap (X \times R^{m}) \subset X \times \bigcup_{x \in X} LM(x,t^{o}) \subset X \times Y,$$

where Y is some compact subset of R^{m}. Since K is obviously closed, K is compact.

To show (i), we first demonstrate that for any open set Q containing X one can find a real number $\beta > 0$ such that

$$\psi_{loc}(t) \cap Q \neq \emptyset \qquad (\forall t \in U_{\beta}(t^{o})). \tag{18}$$

Since X is a bounded SLM set, there exists an open bounded set $Q' \supset X$ satisfying $X = \psi_{Q'}(t^{o})$. However, this implies that also $X = \psi_{Q \cap Q'}(t^{o})$. Set $Q'' := Q' \cap Q$. Hence

$$X = \psi_{Q''}(t^{o}) \subset Q''.$$

Since the Slater condition holds, we have $M_{Q''}(t) \neq \emptyset$ for t near t^{o}, and, further, $M_{Q''}$ is lower semicontinuous at t^{o}. Moreover, since the sets $M_{Q''}(t)$, $t \in T$, are subsets of the compact set cl Q'', then the Weierstrass theorem and classical upper semicontinuity results (cf., e.g., [1,§4.2] or [2,Chapter 6]) provide that for some $\beta > 0$,

$$\emptyset \neq \psi_{Q''}(t) \subset Q'' \qquad (\forall t \in U_{\beta}(t^{o})). \tag{19}$$

Hence,

$$\emptyset \neq \psi_{Q''}(t) \subset \psi_{loc}(t) \cap Q'' \subset \psi_{loc}(t) \cap Q \qquad (\forall t \in U_{\beta}(t^{o})),$$

and so, (18) is shown.

Let now D be any open set containing K, and set $\bar{Q} := \pi_{n}(D)$. Since $LM(\cdot,\cdot)$ is upper semicontinuous on $X \times \{t^{o}\}$ ([24, Th. 2.3]) and since X is compact, there are constants $\eta > 0$ and $\delta > 0$ such that $Q := X + \text{int}\,(\delta B_{n}) \subset \bar{Q}$ and

$$\{x\} \times LM(x,t) \subset D \qquad (\forall x \in Q \quad \forall t \in U_{\eta}(t^{o})). \tag{20}$$

By (18), there is a constant $\beta > 0$ such that $\psi_{loc}(t) \cap Q \neq \emptyset$ for all $t \in U_{\beta}(t^{o})$. Choose $t \in U_{\varepsilon}(t^{o})$ and $z \in \psi_{loc}(t) \cap Q$ with $\varepsilon := \min\{\beta, \eta\}$ we then have

$$\emptyset \neq \{z\} \times LM(z,t) \subset KT(t),$$

which completes the proof of (i).

Assertion (ii) immediately follows from Theorem 1. //

Remark 4: Example 2 may serve to illustrate the essence of the assumption that X has to be a SLM set. In this example, we obviously have

$$\psi_{loc}(0) = \{(0,y) \in R^2 \;/\; 0 < y \leq 1\}, \text{ but } SP(0) = \{(0,y) \in R^2 / 0 \leq y \leq 1\}.$$

The sets SP(t) are empty for $t > 0$. It is not difficult to verify that there is no strict local minimizing set for $f(x,y,0) = xy - x^2$ w.r. to $M(0) = \{(x,y) \;/\; x \geq 0,\ y \leq 1\}$. Indeed, if we assume, on the contrary, that there exists some SLM set X, the continuity of $f(\cdot,\cdot,0)$ and the closedness of M(0) ensure that X is a nonempty closed subset of $\psi_{loc}(0)$, by definition of a SLM set. Hence, there are real numbers $0 < y', y'' \leq 1$ such that $X = \{(0,y) \in R^2 \;/\; y' \leq y \leq y''\}$. Let Q be an arbitrary open set containing X. Since $xy - x^2 = 0 = \varphi_Q(0)$ if $x=0$, and since $\{(0,y) \in R^2 \;/\; y \leq 1\} \subset M(0)$, it then holds $(0,y_Q) \in \psi_Q(0) \setminus X$ for some $y_Q < y'$, which contradicts the definition of a SLM set.

Remark 5: Proposition (18) may be shown also for much more general situations in nonlinear parametric optimization. We refer, e.g., to [25] and [14, Part I].

Corollary 2.1 ([22, Th. 2]): Consider the problem (15), and let $t^o = (C^o, p^o, A^o, b^o)$ T. Suppose that C^o is positive semidefinite and that $KT(t^o)$ is nonempty and bounded.
Then there exists a constant $L > 0$ such that for each open bounded set $D \supset KT(t^o)$ there is a real number $\delta > 0$ such that

$$\emptyset \neq KT(t) \cap D \subset KT(t^o) + L \,\|t-t^o\|_T \, B_{n+m} \qquad (\forall t \in U_\delta(t^o)).$$

Proof: Since the initial problem at $t=t^o$ is a convex one, a classical result from convex optimization theory says that $KT(t^o)$ is nonempty and bounded if and only if both the Slater condition holds w.r. to $M(t^o)$ and $\psi(t^o)$ is nonempty and bounded. Therefore, $\psi(t^o)$ is a bounded SLM set, and Theorem 2 applies to this situation. It is not difficult to conclude from Theorem 1 and Theorem 2 that L can be chosen independently of D, the details are omitted here. //

Remark 6: In the case that $KT(\cdot)$ which corresponds to problem (15) is a single-valued mapping defined on some convex subset T' of T, Hager[10] has shown that, under suitable assumptions, $KT(\cdot)$ is even Lipschitzian on T'. Hager's assumptions involve as well second-order sufficient optimality as regularity conditions which particularly guarantee the continuity of the mapping $KT(\cdot)$ on T'.
Similar results (but with respect to the optimal set mapping ψ associated with the parametric quadratic program (15)) have been obtained by Daniel [6] who has even supposed that the matrices C are positive definite for all C near C^o.
Moreover, both authors have also given explicit roles for computing the Lipschitz modulus in this case.

Theorem 3 (Stability of stationary solutions and local minimizers): Assume the hypotheses of Theorem 2. Then for each open set $Q \supset X$ there exists some $\beta > 0$ such that

$$\psi_{loc}(t) \cap Q \neq \emptyset \qquad (\forall t \in U_\beta(t^o)).$$

Further, there are an open bounded set $Q' \supset X$ and a constant $L > 0$ such that for each open set Q satisfying $X \subset Q \subset Q'$ there exists some $\varepsilon > 0$ with

$$\psi_{loc}(t) \cap Q \subset SP(t) \cap Q \subset X + L \;\|t-t^o\|_T \; B_n \qquad (\forall t \in U_\varepsilon (t^o)). \tag{21}$$

Proof: The first assertion has been shown in Theorem 2, cf. property (18). The left-hand inclusion of (21) is trivially fulfilled. It remains to prove the Lipschitz property. First we show that there is an open bounded set $Q'' \supset X$ such that

$$X = SP(t^o) \cap Q''. \tag{22}$$

The representations (16) and (17) tell us that $SP(t^o)$ is a union of finitely many polyhedral convex sets $X_1,\dots,X_N$ with $X_k \neq \emptyset$ for all k. Recalling the well-known fact that if $y \in SP(t^o)$ then the inequality $h^T(C^o y+p^o) \geq 0$ holds for each feasible direction h, we have

$$(z-y)^T(C^o y+p^o) \geq 0 \qquad (\forall y \in SP(t^o) \qquad \forall z \in M(t^o)).$$

Since the sets X_k $(k=1,\dots,N)$ are convex subsets of $SP(t^o)$, for any two points $x,z \in X_k$, one has $y := \frac{1}{2}x + \frac{1}{2}z \in SP(t^o)$, and so

$$(x-y)^T(C^o y+p^o) \geq 0 \qquad \text{and} \qquad (z-y)^T(C^o y+p^o) \geq 0.$$

We thus obtain for any $x,z \in X_k$ that

$$0=(x-z)^T(C^o(\tfrac{1}{2}x+\tfrac{1}{2}z)+p^o)=\tfrac{1}{2}x^TC^ox+p^{oT}x+\tfrac{1}{2}x^TC^oz-\tfrac{1}{2}z^TC^ox-\tfrac{1}{2}z^TC^oz-p^{oT}z \;,$$

hence

$$f(x,C^o,p^o) = f(z,C^o,p^o) \qquad (\forall x,z \in X_k \;\; \forall k \in \{1,\dots,N\}).$$

Taking into account that X is a SLM set, this implies that $X_k \subset X$ for each $k \in I(X)$, where $I(X) := \{k \in \{1,\dots,N\} / X \cap X_k \neq \emptyset\}$. Because of the compactness of X there is a real number $\beta' > 0$ such that

$$(X + \beta' B_n) \cap X_j = \emptyset \qquad (\forall j \in \{1,\dots,N\} \setminus I(X)), \tag{23}$$

and setting $Q'' := X + \text{int}\,(\beta' B_n)$, we thus have

$$X = \bigcup_{i \in I(X)} X_i = SP(t^o) \cap Q'',$$

hence (22) is verified.

Property (20) (cf. the proof of Theorem 2) says that there are constants $\eta > 0$ and $\nu > 0$ such that

$$\{x\} \times LM(x,t) \subset D := K + B_{n+m}$$

for all $x \in X + \text{int}\,(\nu B_n)$ and for all $t \in U_\eta(t^o)$, where K is defined as in Theorem 2: $K = KT(t^o) \cap (X \times R^m)$. Set

$$Q' := X + \text{int}\,(\beta B_n) \quad , \quad \beta := \min\{\beta'/2\ ,\ \nu\}. \tag{24}$$

Hence,

$$Q' \subset Q''$$

and

$$Q' \times \bigcup_{z \in Q'} IM(z,t) \subset D \qquad (\forall t \in U_\eta(t^o)). \tag{25}$$

Since D is an open bounded set containing K, we can apply Theorem 2, and we have

$$KT(t) \cap D \subset KT(t^o) + L\ \|t-t^o\|_T\ B_{n+m} \qquad (\forall t \in U_\delta(t^o)) \tag{26}$$

for some $L > 0$ and some $\delta > 0$.

Let now Q be any open subset of Q' with $Q \supset X$. Let β, η, δ and L be as given in (24), (25) and (26). Set

$$\varepsilon := \min\{\eta,\ \delta,\ \tfrac{\beta}{L}\}\ .$$

Choose any $t \in U_\varepsilon(t^o)$ and any $z \in SP(t) \cap Q$. Hence, by (25), there exists a multiplier vector $u \in IM(z,t)$ satisfying

$$(z,u) \in KT(t) \cap D.$$

Applying (26), one can find a pair $(y,v) \in KT(t^o)$ with

$$\|(z,u) - (y,v)\| \leq L\ \|t-t^o\|_T\ .$$

Thus, by definition of Q' and ε,

$$\begin{aligned} \text{dist}\,(y,X) \leq \text{dist}\,(z,X) + \|z-y\| &< \beta + \|(z,u) - (y,v)\| \\ &\leq \beta + L\ \|t-t^o\|_T \\ &< \beta + \beta \\ &\leq \beta', \end{aligned}$$

and so $y \in Q''$. Hence, by (22),

$$y \in X.$$

This implies that

$$\text{dist}(z,X) \leq \|z-y\| \leq \|(z,u)-(y,v)\| \leq L\ \|t-t^o\|_T\ .$$

Since $t \in U_\varepsilon(t^o)$ and $z \in SP(t) \cap Q$ were chosen arbitrarily, this completes the proof. //

Remark 7: The inclusion $\psi_{loc}(t) \cap Q \subset SP(t) \cap Q$ in the preceding theorem may be strict, we refer to an example in [23, p. 213].

Corollary 3.1 (Stability of φ and ψ): Consider the problem (15), and let $t^o = (C^o,p^o,A^o,b^o) \in T$. Suppose $\psi(t^o)$ is nonempty, and there are a real number $\delta > 0$ and a compact set $S \subset R^n$ such that $M(t) \subset S$ for all $t \in U_\delta(t^o)$. Furthermore, suppose that the Slater condition is satisfied w.r. to $M(t^o)$. Then

(i) $\psi(t) \neq \emptyset$ for all t belonging to some neighborhood U of t^o,

(ii) ψ is upper Lipschitzian at t^o,

(iii) φ is Lipschitzian on some neighborhood of t^o.

Proof: If we set $X := \psi(t^o)$ then all assumptions of Theorem 3 are satisfied: to see this, we have only to note that $\psi(t^o)$ is a bounded SLM set. Further, let $Q' \supset X$ and $L \in R$ be as in the statement of Theorem 3.
The Slater condition implies that for some $\nu > 0$, $M(t)$ is nonempty if $t \in U_\nu(t^o)$, and so, since $M(t) \subset S$ for $t \in U_\delta(t^o)$ (S is supposed to be compact, and, without loss of generality, let $\nu \leq \delta$),

$$\psi(t) \neq \emptyset \qquad (\forall t \in U_\nu(t^o)).$$

Further, a classical upper semicontinuity theorem (cf., e.g., [1, Th. 4.2.2]) applies to this situation, hence for some $\eta > 0$,

$$\psi(t) \subset Q' \qquad (\forall t \in U_\eta(t^o)).$$

By (21) in Theorem 3 it follows that

$$\psi(t) \subset \psi_{loc}(t) \cap Q' \subset X + L \; \|t-t^o\|_T \, B_n$$

for t near t^o. Thus, (i) and (ii) hold. To show (iii) we only note that the hypotheses of this Corollary (namely, S compact, $M(t) \subset S$ for t near t^o, Slater's condition) imply that M is Lipschitzian on some ε-neighborhood of t^o ($\varepsilon \leq \delta$), cf. Blatt [3] or Klatte/Kummer [15 , Th. 1]. Further, f is Lipschitzian on $S \times U_\varepsilon(t^o)$. Hence, by a known theorem which says that the Lipschitz continuity of the constraint set mapping and of the objective function imply the Lipschitz continuity of φ (cf., e.g., [4] , [15, Lemma 1] , [14 , Part I, Th.1]), it immediately follows that φ is Lipschitzian on $U_\varepsilon(t^o)$. //

In Corollary 3.1 it has turned out that the properties (5a), (5b) and (5c) can be "saved" for parametric (generally non-convex) quadratic optimization problems. However, one has to add the assumption of local boundedness of the constraint set mapping M, but the necessity of this assumption is a fact well-known from the persistence and upper semicontinuity analysis in parametric optimization, cf. [1, §4.2].

4. Consequences for linearly perturbed quadratic programs

Let A be a given (m,n)-matrix, and let C be some given (n,n)-matrix which is symmetric. We consider the parametric program

$$\min \left\{ f(x,p) \,/\, x \in M(b) \right\} , \quad (p,b) \in R^n \times R^m, \tag{27}$$

where p and b are regarded as parameters, M is a multifunction from R^m to R^n defined by

$$M(b) := \left\{ x \in R^n \,/\, Ax \leq b \right\} , \quad b \in R^m,$$

and $f\colon R^n \times R^n \longrightarrow R$ is given by

$$f(x,p) := \frac{1}{2} x^T C x + p^T x \quad , \quad (x,p) \in R^n \times R^n.$$

We again note that the following analysis may be similarly pursued, if M is an arbitrary polyhedral convex multifunction.
For $t=(p,b)$, let $\varphi(t)$, $\psi(t)$, $KT(t)$ and $SP(t)$ be defined as in (1), (16) and (17). Obviously, $KT(\cdot)$ and $SP(\cdot)$ are polyhedral multifunctions from $R^n \times R^m$ to $R^n \times R^m$ and R^n, respectively. As an immediate consequence of Proposition 3, it then follows

<u>Theorem 4 ([20,23])</u>: The multifunctions $KT(\cdot)$ and $SP(\cdot)$ (associated with the parametric program (27)) are locally upper Lipschitzian on $R^n \times R^m$ with some uniform modulus.

A simple example shows that the optimal set mapping ψ of (27) is not in general polyhedral, cf. [13]. Nevertheless, it can be concluded from Lemma 1 and Lemma 2 that ψ is locally upper Lipschitzian on $R^n \times R^m$.
Before proving this result, we shall carry out an auxiliary construction. Since $KT(\cdot)$ is a polyhedral multifunction from $R^n \times R^m$ to $R^n \times R^m$, there are nonempty polyhedral convex sets $S_1,\ldots,S_N \subset R^n \times R^m \times R^n \times R^m$ such that

$$\text{graph } KT(\cdot) = \bigcup_{i=1}^{N} S_i$$

where, as defined above,

$$\text{graph } KT(\cdot) = \left\{ (p,b,x,u) \;/\; \begin{matrix} Cx+A^Tu+p=0 & , \; u \geq 0 \\ Ax \leq b & , \; (Ax-b)^Tu=0 \end{matrix} \right\} .$$

Set in (8) and (9), $G:=KT(\cdot)$, $r:=n+m$, $s:=n+m$; then (8) and (9) define polyhedral convex multifunctions $G_1,\ldots,G_N$ satisfying

$$KT(p,b) = \bigcup_{i=1}^{N} G_i(p,b) \qquad (\forall (p,b)).$$

Since each of the sets S_i, $i=1,\ldots,N$, may be represented as a system of linear inequalities in p, b, x and u, the corresponding multifunction G_i may be considered as a solution set mapping of a system of linear inequalities parameterized in the right-hand side of the inequalities. Hence, for each $i \in \{1,\ldots,N\}$,

$$\min \left\{ p^Tx - b^Tu \;/\; (x,u) \in G_i(p,b) \right\} , \quad (p,b) \in R^n \times R^m, \qquad (28)_i$$

defines a parametric linear optimization problem with fixed technological matrix.
Let $\bar{\psi}_i$ and $\bar{\varphi}_i$ be the optimal set mapping and the infimum function of the problem $(28)_i$, respectively ($i=1,\ldots,N$). Thus, the properties (4a), (4b) and (4c) for $\bar{\psi}_i$ and $\bar{\varphi}_i$ hold; we refer to the discussion concerning the parametric problem (4) in §1. Further, let $\bar{\psi}$ and $\bar{\varphi}$ be the optimal set mapping and the infimum function of the

parametric program

$$\min \left\{ p^T x - b^T u \ / \ (x,u) \in KT(p,b) \right\} \ , \quad (p,b) \in R^n \times R^m. \tag{29}$$

From quadratic optimization theory (cf., e.g., [5,9]) we then know that the mappings ψ and φ which are associated with the quadratic program (27) satisfy

$$\varphi(p,b) = \frac{1}{2}\bar{\varphi}(p,b) \quad (\forall (p,b) \in \operatorname{dom}\psi) \quad \text{and} \quad \operatorname{dom}\psi \subset \operatorname{dom}\bar{\psi}. \tag{30}$$

Further, set for $(p,b) \in R^n \times R^m$,

$$I(p,b) := \left\{ i \in \{1,\dots,N\} \ / \ \emptyset \neq \bar{\psi}_i(p,b) \subset \bar{\psi}(p,b) \right\}.$$

It is easy to see that for $(p,b) \in \operatorname{dom}\bar{\psi}$, we then have that $\bar{\varphi}_i(p,b) > -\infty$ $(\forall i \in \{1,\dots,N\})$ and

$$\bar{\varphi}(p,b) = \min_{1 \leq i \leq N} \bar{\varphi}_i(p,b)$$

hold. Hence, for $(p,b) \in \operatorname{dom}\bar{\psi}$,

$$I(p,b) = \left\{ i \in \{1,\dots,N\} \ / \ \bar{\psi}_i(p,b) \neq \emptyset \ , \ \bar{\varphi}_i(p,b) \leq \bar{\varphi}(p,b) \right\}. \tag{31}$$

Putting all these informations together, we obviously obtain

$$\bar{\psi}(p,b) = \bigcup_{i \in I(p,b)} \bar{\psi}_i(p,b) \subset \bigcup_{i=1}^{N} \bar{\psi}_i(p,b) \qquad (\forall (p,b)), \tag{32a}$$

$$\bar{\varphi}(p,b) = \bar{\varphi}_i(p,b) \qquad (\forall (p,b) \in \operatorname{dom}\bar{\psi} \quad \forall i \in I(p,b)\ ), \tag{32b}$$

$$\bar{\psi}_i \text{ is locally upper Lipschitzian on } R^n \times R^m \quad (\forall i\ ), \tag{32c}$$

by property (4c),

$$\bar{\varphi}_i \text{ is continuous on } \operatorname{dom}\bar{\psi}_i \quad (\forall i), \tag{32d}$$

by property (4b), and

$$\operatorname{dom}\bar{\psi}_i \text{ is closed} \quad (\forall i), \tag{32e}$$

by property (4a). Finally, we recall a result on the infimum function φ of problem (27):

$$\varphi \text{ is upper semicontinuous on } \operatorname{dom}\psi, \tag{33}$$

since M is a lower semicontinuous multifunction, and f is continuous, cf. [1, Thms. 4.2.1 and 4.2.2].

<u>Theorem 5 ([13])</u>: The optimal set mapping ψ of the parametric quadratic optimization problem (27) is locally upper Lipschitzian on $R^n \times R^m$.

<u>Proof</u>: Since $\operatorname{dom}\psi$ is closed (cf. [7] or Lemma 3 below), it suffices to prove that the theorem holds for any (p^o,b^o) belonging to $\operatorname{dom}\psi$. Choose any $(p^o,b^o) \in \operatorname{dom}\psi$. We shall apply Lemma 2 to the multifunctions $\bar{\psi}, \bar{\psi}_1, \dots, \bar{\psi}_N$ which are considered being restricted to the

domain dom ψ. Set in Lemma 2 $t^o:=(p^o,b^o)$, r:=n+m, s:=n+m, T:= dom ψ, $G:=\bar{\psi}$, $G_i:=\bar{\psi}_i$ for i=1,...,N. Recall that, in order to apply Lemma 2, we have to verify whether

(i) the hypotheses of Lemma 1 are satisfied, and

(ii) $G(t^o)$ is the union of the sets $G_i(t^o)$, $i \in I(t^o)$, where $I(t^o)=I(p^o,b^o)$ is defined as above, and

(iii) all the functions G_i, $i \in I(t^o)$, are upper Lipschitzian at t^o.

Obviously, in our case, (ii) is satisfied in virtue of property (32a). Further, (iii) is true because of (32c). It remains to verify that for each $i \in \{1,\dots,N\}$ the set

$$\text{dom}\,\psi \cap \text{dom}\,(\bar{\psi} \cap \bar{\psi}_i) \text{ is closed,} \tag{34}$$

because the other assumptions of Lemma 1 are trivially fulfilled. Of course, supposing that (34) is shown, we immediately obtain the desired result from Lemma 2.

To show (34), we take any $i \in \{1,\dots,N\}$ and any sequence $\{t^k\}$ with $\{t^k := (p^k,b^k)\} \subset \text{dom}\,\psi \cap \text{dom}\,(\bar{\psi} \cap \bar{\psi}_i)$ converging to t^o. Then we may conclude from the closedness of dom ψ and from (32e) and (30):

$$t^o \in \text{dom}\,\psi \cap \text{dom}\,\bar{\psi}_i \subset \text{dom}\,\bar{\psi} \cap \text{dom}\,\bar{\psi}_i .$$

Further,

$$\begin{aligned}
\bar{\varphi}_i(t^o) &= \lim_k \bar{\varphi}_i(t^k) && \text{(by (32d) and } t^o, t^k \in \text{dom}\,\bar{\psi}_i), \\
&= \lim_k \bar{\varphi}(t^k) && \text{(by (32b) and } i \in I(t^k)), \\
&= \lim_k 2\varphi(t^k) && \text{(by (30) and } t^k \in \text{dom}\,\psi), \\
&\leq 2\varphi(t^o) && \text{(by (33))}, \\
&= \bar{\varphi}(t^o) && \text{(by (30) and } t^o \in \text{dom}\,\psi).
\end{aligned}$$

From Property (31) we then derive that $i \in I(p^o,b^o)$, thus

$$t^o \in \text{dom}\,(\bar{\psi} \cap \bar{\psi}_i).$$

Hence, the set $\text{dom}\,\psi \cap \text{dom}\,(\bar{\psi} \cap \bar{\psi}_i)$ is closed. This completes the proof. //

<u>Corollary 5.1</u>: Consider the parametric program (27). If ψ is single-valued on some convex subset Y of dom ψ, then ψ is Lipschitzian there.

<u>Proof</u>: Recall that the Lipschitz constants for the multifunctions $\bar{\psi}_1,\dots,\bar{\psi}_N$ are uniform, cf. property (4c). Since ψ is single-valued on $Y \subset \text{dom}\,\psi$, ψ is then locally Lipschitzian on Y with some uniform modulus, i.e., there exists some constant $L > 0$ such that for each $t^o \in Y$ there is a real number $\beta > 0$ with $\|\bar{x}(t) - \bar{x}(t^o)\| \leq L\,\|t - t^o\|$ if $t \in Y \cap U_\beta(t^o)$ (where $\psi = \{\bar{x}(\cdot)\}$). Using the convexity of the set Y, we thus may conclude, by a standard technique, that ψ is even Lipschitzian on Y. //

Remark 8: At this place, we refer to a recent publication concerning a related question: Mangasarian/Shiau [18] study the Lipschitz continuity of the solution of certain uniquely solvable linear complementarity problems (e.g. those with so-called P-matrices and Z-matrices) with right-hand side perturbations. The authors give an explicit role for determining the Lipschitz modulus, in these cases.

The following two results complete the analysis of global optimal solutions and global infima of the parametric program (27). For the detailed proofs we refer to [13].

Lemma 3 ([13]): The effective domain dom ψ of the optimal set mapping of (27) is a union of finitely many convex polyhedra.

In the proof of Lemma 3, Eaves' characterization of global optimal solutions for quadratic programs ([7]) is essentially used, beside an application of the Basic Decomposition Theorem in parametric linear optimization ([19,26]). The proof of the following theorem goes back to Theorem 5, Lemma 3 and a criterion for the Lipschitz continuity of a real-valued function over a polyhedral (but not necessarily convex) domain ([13]):

Theorem 6: The infimum function φ of (27) is Lipschitzian on bounded subsets of dom ψ.

As announced in §1, it has turned out that the properties (4b) and (4c) of a linearly perturbed linear program can be completely extended to linearly perturbed quadratic programs. Concerning the fact that dom ψ is generally not convex one may consult an example in [1 , §5.5].

Acknowledgement: The author is indebted to Bernd Kummer (Humboldt-Universität zu Berlin) and to Peter Kall (Universität Zürich, Institut für OR) for stimulating discussions on the subject of the paper. He also wishes to thank an anonymous referee for the careful reading of the manuscript and some valuable hints.

References

[1] Bank, B., J. Guddat, D. Klatte, B. Kummer, K. Tammer: Non-Linear Parametric Optimization. Akademie-Verlag, Berlin, 1982.

[2] Berge, C.: Topological Spaces. Macmillan, New York, 1963.

[3] Blatt, H.-P.: Lipschitz-Stabilität von Optimierungs- und Approximationsaufgaben. Numerical Methods of Approximation Theory, ISNM 52 (1980), 9-28.

[4] Cornet, B.: Sensitivity analysis in optimization. CORE Discussion paper No. 8322, Univ. Catholique de Louvain, 1983.

[5] Cottle, R.W.: The principal pivoting method of quadratic programming. In: G.B. Dantzig, A.F. Vernott (eds.): Mathematics of the Decision Sciences, Part I. American Mathematical Society, Providence, 1968, pp. 144-162.

[6] Daniel, J.W.: Stability of the solution of definite quadratic programs. Math. Programming 5 (1973), 41-53.

[7] Eaves, C.: On quadratic programming. Management Science 17 (1971) 698-711.

[8] Eremin, I.I., N.N. Astafiev: Introduction to the Theory of Linear and Convex Programming. Nauka, Moscow, 1976. (In Russian)

[9] Gianessi, F., E. Tomasin: Nonconvex quadratic programs, linear complementarity problems, and integer linear programs. In: P.L. Hammer, G. Zoutendijk (eds.): Mathematical Programming in Theory and Practice. North-Holland, Amsterdam-Oxford-New York, 1974, pp. 161-199.

[10] Hager, W.W.: Lipschitz contimuity for constrained processes. SIAM J. Control Optim. 17 (1979) 321-338.

[11] Hoffman, A.J.: On approximate solutions of systems of linear inequalities. J. Res. Nat. Bur. Standards 49 (1952), 263-265.

[12] Klatte, D.: Eine Bemerkung zur parametrischen quadratischen Optimierung. Seminarbericht Nr. 50, Teil 1, Sektion Mathematik, Humboldt-Universität Berlin, pp. 174-185, 1983.

[13] Klatte, D.: On the Lipschitz behavior of optimal solutions in parametric problems of quadratic optimization and linear complementarity. optimization (formerly a series of Math. Operationsforsch. Statistik) 16 (1985), 819-831.

[14] Klatte, D.: On the stability of local and global optimal solutions in parametric problems of nonlinear programming. Part I and II. Seminarbericht Nr. 75, Sektion Mathematik, Humboldt-Universität Berlin, November 1985.

[15] Klatte, D., B. Kummer: Stability properties of infima and optimal solutions of parametric optimization problems. In: V. Demyanov, D. Pallaschke (eds.): Nondifferentiable Optimization - Motivations and Applications. Springer-Verlag, Berlin-Heidelberg-New York, 1985, pp. 215-229.

[16] Kleinmann, P.: Quantitative Sensitivitätsanalyse bei parametrischen Optimierungsaufgaben. Seminarbericht Nr. 9, Sektion Mathematik, Humboldt-Universität Berlin, 1978.

[17] Mangasarian, O.L.: A condition number for linear inequalities and linear programs. Methods of Operations Research No. 43, 1981, 3-15.

[18] Mangasarian, O.L., T.-H. Shiau: Lipschitz contimuity of solutions of linear inequalities, programs and complementarity problems. Technical Report No. 599, Computer Science Department, University of Wisconsin-Madison, 1985.

[19] Nožička, F., J. Guddat, H. Hollatz, B. Bank: Theorie der linearen parametrischen Optimierung. Akademie-Verlag, Berlin, 1974.

[20] Robinson, S.M.: An implicit-function theorem for generalized variational inequalities. MRC Technical Summary Report No. 1672, Mathematics Research Center, University of Wisconsin-Madison, 1976.

[21] Robinson, S.M.: A characterization of stability in linear programming. Operations Research 25 (1977), 435-447.

[22] Robinson, S.M.: Generalized equations and their solutions, Part I: Basic Theory. Math. Programming Study 10 (1979), 128-141.

[23] Robinson, S.M.: Some continuity properties of polyhedral multifunctions. Math. Programming Study 14 (1981), 206-214.

[24] Robinson, S.M.: Generalized equations and their solutions, Part II: Applications to nonlinear programming. Math. Programming Study 19 (1982), 200-221.

[25] Robinson, S.M.: Local epi-continuity and local optimization. MRC Technical Summary Report No. 2798, University of Wisconsin-Madison, 1985.

[26] Walkup, D., R. Wets: Lifting projections of convex polyhedra. Pacific J. Math. 28 (1969), 465-475.

[27] Walkup, D., R. Wets: A Lipschitzian characterization of convex polyhedra. Proceed. Amer. Math. Society 23 (1969), 167-173.

[28] Wets, R.: On the continuity of the value of a linear program and of related polyhedral-valued multifunctions. Math. Programming Study 24 (1985), 14-29.

LINEARLY AND NONLINEARLY PERTURBED OPTIMIZATION PROBLEMS IN FINITE DIMENSION

Bernd Kummer [1)]

0. Introduction

The theory of parametric optimization as well as the classic analysis show that many questions concerning the stability of the solutions of a given (perturbed) optimization problem or of a system of nonlinear equations can be answered by studying the behavior of the solutions under linear (RHS) perturbations only. Moreover, the original problem may be replaced by some sufficiently close linearization.
The resulting statements in the classic analysis are well-known, in optimization this approach leads to duality, convergence of approximation methods (e.g. Wilson's method), S.M. Robinson's stability theory for smooth optimization problems, marginal theory and several other local and qualitative statements including optimality conditions.
By A.V. Fiacco (1983) the strong analytic way of dealing with stability in parametric optimization has been clearly demonstrated.

The present paper will show that the approach of studying RHS - perturbations of the original problem allows to investigate the stability of quite general optimization problems from the qualitative as well as the quantitative point of view.

Our feasible sets may be described in the usual way by equations and inequalities, may consist of all solutions of some complementarity problem or of all Kuhn - Tucker points of some (second) optimization problem. The approximation we will use don't need explicitly differentiability (but better than first order approximation) of the mappings describing the original problem and allows statements even if the perturbed problems are "nonsmooth" but near according to the C^0 - norm.
The basic underlying statements are the theorems 1.2 and 4.1. They state the fact that the set $G^{-}(0)$ of all solutions to a generalized equation $0 \in G(x)$, $x \in A$ is non-empty and meets some set $A' \subset A$ if a certain distance $\rho(G,F)$ between G and some second (well- behaved) multifunction F is sufficiently small. Of course, we must specify the class of multifunctions G and the kind of distance ρ to be considered.
The first one is the class of Kakutani-mappings (see section 1) which covers Clarke's subdifferential mapping $\partial f(.)$ of a locally Lipschitz function f and the corresponding normal cone mapping $N_M(.)$ of a closed

1) Humboldt Universität zu Berlin, Sekt. Math.,Berlin 1o86,DDR

set M as well as the (Minkovsky-) sum of both. The second one is defined as the supremum

$$\rho(G,F) = \sup_{x\in A} d_H(G(x), F(x))$$

of the Hausdorff- distance d_H of the images. Using the standard notations of the theory of multifunctions we have

$$\rho(G,F) = \sup_{x\in A} \inf\{\varepsilon>0 \,/\, G(x)\subset F(x)+\varepsilon B \,,\; F(x)\subset G(x)+\varepsilon B\}$$

$$= \sup_{x\in A} \sup\{\sup_{y\in F(x)} \operatorname{dist}(y, G(x)) \,,\; \sup_{y\in G(x)} \operatorname{dist}(y, F(x))\}$$

where B denotes the unit-ball of the underlying normed space.
The stability we regard is relatad to optimization problems of the kind

$$(f,F) \qquad \inf\{f(x) \,/\, x\in A,\;\; 0\in F(x)\}$$

where we suppose

$A \subset R^n$	closed, non-empty
$f\colon A \longrightarrow R$	locally Lipschitz
$F\colon A \longrightarrow R^m$	multifunction, Kakutani- mapping.

Let the inf - value, the set of (global optimal) solutions and the set of feasible points of the problem be denoted by φ, ψ and M, respectively. We consider them as functions (multifunctions) of f and F and form a δ- neighbourhood of (f,F) by all problems (g,G) satisfying the above assumptions as well as

$$(1)_\delta \qquad \begin{array}{l} \rho(g, f) := \sup_{x\in A} |g(x) - f(x)| \le \delta_1 \\ \rho(G, F) \le \delta_2 \end{array} \qquad \delta\in R^2_+$$

<u>Definition:</u> A fixed problem (f,F) is said to be stable if for each $\varepsilon>0$ there is some $\delta\in \operatorname{int} R^2_+$ such that

$$|\varphi(g,G) - \varphi(f,F)| \le \varepsilon$$

and

$$\psi(g,G) \subset \psi(f,F) + \varepsilon\cdot B$$

hold whenever (g,G) belongs to the δ- neighbourhood of (f,F).

With other words, stability means nothing but continuity of φ and metrically upper semicontinuity of ψ at the point (f,F).
After some preliminaries concerning mainly Kakutani- mappings we will state sufficient and (if card $\psi(f,F) = 1$) necessary stability conditions. They help to quantify the $(\varepsilon - \delta)$ relation and are based on three functions related to (f,F) - the so called stability functions.
In the sections 3 and 4 we will estimate and simplify the stability functions b and c for convex and (or) polyhedral problems (f,F). In section 5 we take the question of approximation into account. Some simple applications will conclude this paper.
The author thanks Mr. D. Klatte for many helpful discussions to this work.

1. Kakutani- mappings and some properties of multifunctions

Let A be a non-empty, closed subset of R^n and let $F: A \longrightarrow R^m$ be any multifunction. We will use the following notations.

$$\text{graph } F = \{(y,x) \;/\; x \in A ,\; y \in F(x)\}$$

$$F^{-}(y) = \{ x \in A \;/\; y \in F(x) \}$$

$$\text{dom } F = \{ x \in A \;/\; F(x) \neq \emptyset \}.$$

F is called

closed if graph F is a closed set in R^{m+n}

polyhedral if graph F is a polyhedral set (This is any union of a finite number of polyhedra)

convex-valued if all images $F(x)$ are convex sets.

We call a closed and convex-valued multifunction F Kakutani-mapping if, for each $r > 0$, there is some $s > 0$ such that

$$F(x) \cap s\,B \neq \emptyset \qquad \text{whenever } x \in A,\; \|x\| < r.$$

It is not difficult to see that a closed and convex-valued multifunction F is a Kakutani-mapping if for each $x^0 \in A$ there are $\varepsilon > 0$ and $s > 0$ such that

$$F(x) \cap s\,B \neq \emptyset \qquad \text{for all } x \in A,\; \|x - x^0\| < \varepsilon .$$

Important examples of such mappings are multifunctions F of the type

(i) $F(x) = F_1(x) + F_2(x)$

where F_1 and F_2 are closed and convex-valued, dom F_1 = A = dom F_2 , F_1 is metrically upper semicontinuous and the sets $F_2(x)$ are cones having the vertex 0_m or

(ii) graph F is closed and convex, dom $F = A$.

The terminology just introduced is, of course, motivated by the Kakutani- fixed point theorem: Each Kakutani-mapping doing from a non-empty, convex and compact subset of R^n into itself has at least one fixed point ($x^0 \in F(x^0)$). The following two propositions present a further characterization of these mappings.

Lemma 1.1. Let $F: A \longrightarrow R^m$ be a Kakutani-mapping, $f: R^m \longrightarrow R$ be a convex function and let $S(x) = \arg\min \{ f(y) \;/\; y \in F(x) \}$. If, for some $x^0 \in A$, the set $S(x^0)$ is non-empty and bounded then there are $\varepsilon > 0$ and $\delta > 0$ such that $\emptyset \neq S(x) \subset \delta \cdot B$ for all $x \in A$, $\|x - x^0\| < \varepsilon$.

Proof: We take any $r > \|x^0\|$ and find some $s > 0$ such that $F(x) \cap s \cdot B \neq \emptyset$ for all $x \in A \cap r \cdot B$.

Now, assume the proposition to be false. Then there are sequences $x_k \longrightarrow x^0$ and $y_k \in F(x_k)$ satisfying $\|y_k\| \longrightarrow \infty$, $f(y_k) \leq f(z_k)$ where z_k are arbitrarily fixed points in $F(x_k) \cap s \cdot B$.

Setting

$$u_k = (y_k - z_k) \cdot \|y_k - z_k\|^{-1}$$

we may without loss of generality (otherwise some appropriate subsequence could be considered) assume that

$$u_k \longrightarrow u \;, \qquad s_k \longrightarrow s \;.$$

Since F is closed and convex-valued the points $s + \lambda u$, $\lambda \geqslant 0$ belong to $F(x^o)$. Since f is convex and $f(y_k) \leqslant f(s_k)$ we obtain

$$f(s + \lambda u) \leqslant f(s) \qquad \forall \lambda \geqslant 0 \;.$$

Again by the convexity of f we deduce from this inequality

$$f(y + \lambda u) \leqslant f(y) \qquad \forall \lambda \geqslant 0 \quad \forall y \in S(x^o).$$

Thus, $S(x^o)$ cannot be bounded. This contradiction proves the proposition.//

<u>Theorem 1.2.</u> Let $F: A \longrightarrow R^m$ be a Kakutani-mapping and assume $0_m \notin F(x)$ for all $x \in A$. Then, there is a continuous function $z: A \longrightarrow R^m$ such that $\| z(x) \| = 1$ and $\inf_{y \in F(x)} \langle z(x), y \rangle > 0$ hold for all $x \in A$. (Continuous separation of 0 and F(x))

<u>Proof:</u> The sets $Z(x) := \{ z \in R^m \;/\; \inf_{y \in F(x)} \langle z \,, y \rangle > 0 \}$ are convex and non-empty. In view of E. Michael (1956) there is a continuous selection $z = z(x)$ of the mapping Z if it is lower semicontinuous. After dividing $z(x)$ by its norm we would in this case obtain the function desired. Therefore, assume Z to be not lower semicontinuous, i. e. there exist some $(z^o, x^o) \in$ graph Z, some $\delta > 0$ and some sequence $x_k \longrightarrow x^o$ such that

(1.1) $\quad (z^o + \delta \cdot B) \cap (Z(x_k)) = \emptyset \qquad$ for all k.

Since F is a Kakutani-mapping there are $s > 0$ and $\eta_k \in F(x_k)$ satisfying $\| \eta_k \| \leqslant s$ for all k. Let

$2\alpha_o = \inf_{y \in F(x^o)} \langle z^o, y \rangle$ and $0 < \varepsilon < \min\{\delta, \alpha_o/s\}$.

The condition (1.1) holds obviously with ε instead of δ, too. The bounded sequence η_k may assumed to be convergent $\eta_k \rightarrow \eta^o$. Because of (1.1) we observe

$$\sup_{z \in z^o + \varepsilon B} \inf_{y \in F(x_k)} \langle z \,, y \rangle \leqslant 0 \qquad \text{for all } k$$

and recalling the minimax - theorem (one set is compact)

$$\inf_{y \in F(x_k)} \sup_{z \in z^o + \varepsilon B} \langle z, y \rangle \leqslant 0 \qquad \text{for all } k.$$

Therefore, points $y_k \in F(x_k)$ exist satisfying

(1.2) $\quad \sup_{z \in z^o + \varepsilon B} \langle z \,, y_k \rangle \leqslant 1/k \quad$ and $\| y_k \| \longrightarrow \infty$.

For $u_k = (y_k - \eta_k) \cdot \| y_k - \eta_k \|^{-1}$ we may again assume the convergence $u_k \longrightarrow u^o$. Since F is closed and convex-valued the points $\eta^o + \lambda u^o$ $(\lambda \geqslant 0)$ belong to $F(x^o)$ whereafter $z^o \in Z(x^o)$ implies

(1.3) $\quad \langle z^o, u^o \rangle \geqslant 0$.

On the other hand, each $z \in z^o + \varepsilon B$ fulfils

$$\langle z \,, \eta^o \rangle \geqslant \langle z^o, \eta^o \rangle - \| z - z^o \| \cdot \| \eta^o \| \geqslant 2\alpha^o - \varepsilon \| \eta^o \| \geqslant$$
$$\geqslant 2\alpha^o - \varepsilon \cdot s \geqslant \alpha^o > 0 \;.$$

Recalling (1.2) this yields (for large k)

$\langle z , y_k - \eta_k \rangle \leq 1/k - \langle z , \eta_k \rangle \leq 0$

and

(1.4) $\langle z , u^o \rangle = \lim \langle z , u_k \rangle \leq 0$ for all $z \in z^o + \varepsilon B$.

Having in mind that $\| u^o \| = 1$ the inequalities (1.3) and (1.4) contradict each other. Thus, the mapping Z is lower semicontinuous. The theorem is true. //

It should be noted that a similar but less general version of the proposition above has been shown in B. Kummer (1984).
The nicest property of a multifunction - Lipschitz continuity - generally fails to hold even if graph F is convex or polyhedral. In the first case, F will not be upper semicontinuous (also metrically) without additional requirements like compactness or polyhedrality. In the second one, the lower semicontinuity is difficultly to ensure. However, the next two statements often suffice in order to describe the continuity of the mappings in question.

Lemma 1.3 Let $F: A \to R^m$ be a closed multifunction with convex graph F and let x^o belong to the interior int dom F.
Then, for sufficiently small $\varepsilon > 0$ and large $s > 0$, there is $L > 0$ such that the mapping F_s as $F_s(x) = F(x) \cap s \cdot B$ fulfils $d_H(F_s(x_1) , F_s(x_2)) \leq L \| x_1 - x_2 \|$ for all $x_1, x_2 \in x^o + \varepsilon B$.

Lemma 1.4 Let $F: A \to R^m$ be a polyhedral multifunction, let $u \in R^n$ and $x^o \in$ dom F. Then, there are ε_o, k_1 and k_2 (all positive) such that

$F(x^o + \varepsilon u + v) \subset F(x^o + \varepsilon u) + k_1 \cdot \| v \| \cdot B$

whenever

$0 < \varepsilon < \varepsilon_o$ and $\| v \| \leq k_2 \cdot \varepsilon$, $x^o + \varepsilon u + v \in A$.

The proofs of these statements can here be omitted. We only note that Lemma 1.4 can be proven without difficulties by using the results of S.M. Robinson (1981) or also D. Walkup and R. Wets (1969). As the Lemma 1.3 is concerned we refer (in the same sense) to S.M. Robinson (1976).

2. Stability Functions

We return to the question of stability of a fixed problem

(f, F) $\inf \{ f(x) \ / \ x \in A , \ 0 \in F(x) \}$

under the suppositions

$A \subset R^n$ closed, non-empty

$f: A \to R$ locally Lipschitz

$F: A \to R^m$ Kakutani-mapping

with respect to (1) δ- neighbourhoods defined in the introduction. As an additional requirement let the solution set $\Psi(f,F)$ which we will briefly denote by S be non-empty and let $x^o \in S$ be arbitrarily fixed.

Now, define the following functions:

$$\alpha(y) = \inf\{f(x) \mid x \in F^-(y)\}$$
$$\beta(\varepsilon, y) = \inf\{f(x) \mid x \in F^-(y),\ \operatorname{dist}(x, S) \geq \varepsilon\}$$
$$\gamma(G) = \operatorname{dist}(x^0, G^-(0))$$

where $y \in R^m$, $\varepsilon \geq 0$, $G: A \longrightarrow R^m$ Kakutani-mapping.

Further, put for $t \geq 0$

$$a(t) = \inf_{\|y\| \leq t} \alpha(y)$$
$$b(\varepsilon, t) = \inf_{\|y\| \leq t} \beta(\varepsilon, y)$$
$$c(t) = \sup_{\varrho(G,F) \leq t} \gamma(G) \quad .$$

We will call a, b, c <u>stability functions</u>.

Obviously, c depends of $x^0 \in S$, $F^-(y)$ is the feasible set related to the Kakutani-mapping $G_y(x) = -y + F(x)$ (fixed translation of F) and α is the inf - function of the linearly perturbed problem (f,F). We have

$$\varphi(f,F) = \alpha(0_m) = a(0)$$
$$a(t) \leq b(\varepsilon, t), \qquad G^-(0) = M(G) .$$

The cases

$$\gamma(G) = +\infty \qquad \text{and} \qquad c(t) = +\infty$$

are possible.

After these settings we are able to consider the problems (g,G) in any δ - neighbourhood of (f,F).

<u>Theorem 2.1</u> Let $\delta = (\delta_1, \delta_2) \in \operatorname{int} R^2_+$, let $c(\delta_2) < \infty$ and let f be Lipschitz with some modul L on $(x^0 + c(\delta_2)\cdot B) \cap A$. If (g,G) satisfies $(1)_\delta$ the following holds:

(2.1) $\quad \varphi(g,G) \geq a(\delta_2) - \delta_1$

(2.2) $\quad \varphi(g,G) \leq a(0) + \delta_1 + L \cdot c(\delta_2)$

and

(2.3) $\quad \psi(g,G) \subset S + \varepsilon \cdot B \qquad$ whenever

(2.4) $\quad 2\delta_1 + L\, c(\delta_2) < b(\varepsilon, \delta_2) - a(0)$.

If, additionally, the set S is compact then (2.4) implies

(2.3)′ $\quad \psi(g,G) \neq \emptyset$.

The <u>proof</u> is nearly a one-way street.

(2.1): For any $x \in G^-(0)$ there is some y such that $y \in F(x)$ and $\|y\| \leq \delta_2$. Hence: $g(x) \geq f(x) - \delta_1 \geq a(\delta_2) - \delta_1$. //

(2.2): Since $\operatorname{dist}(x^0, G^-(0)) \leq c(\delta_2)$ there is some z in $G^-(0) \cap (x^0 + c(\delta_2) B$. Therefore $g(z) \leq f(z) + \delta_1 \leq f(x^0) + L\cdot c(\delta_2) + \delta_1$ .//

(2.3): If $x \in G^-(0) \setminus (S + \varepsilon B)$ we obtain in view of (2.4) $g(x) \geq f(x) - \delta_1 \geq b(\varepsilon, \delta_2) - \delta_1 > a(0) + \delta_1 + L\cdot c(\delta_2)$. Because of (2.2) this yields $x \notin \psi(g,G)$. //

(2.3)': The estimations above show also that there exists
$s \in G^{-}(0) \cap (S + o(\delta_2) B)$ satisfying
$g(s) \leq \inf \{ g(x) / x \in G^{-}(0) \setminus (S + \varepsilon \cdot B)\}$.
This reason the compact set
$G^{-}(0) \cap (S + (\varepsilon + o(\delta_2))\cdot B)$
includes a solution of the problem (g,G). //

The reader easily sees that the proposition 2.1 makes use of the continuity but not of the Lipschitz property of the function g.
The purpose of the theorem is to show how the stability functions allow to estimate the variation of the inf - function and the solution set mapping with respect to general nonlinear perturbations of (f,F). Examples which show that these estimates are sharp can be found in the one - dimensional space. So it seems to be natural to study the functions b and c more detailed (sections 3 and 4) whereas the function a should be analysed anyway. Without dealing with this function one cannot hope to estimate even $\varphi(f,G)$.
As an immediate consequence of theorem 2.1 we obtain

<u>Corollary 2.2</u> The problem (f , F) is stable if the relations

(2.5) $\lim_{t \to 0} a(t) = a(0)$

(2.6) $\lim_{t \to 0} b(\varepsilon, t) > a(0)$ for all $\varepsilon > 0$

and

(2.7) $\lim_{t \to 0} c(t) = 0$ for some $x^0 \in S$

hold.

These stability conditions are "almost necessary".

<u>Theorem 2.3</u> Suppose (f, F) to be stable, $S = \Psi(f, F)$ to be non-empty and $x^0 \in S$. Then (2.5), (2.6) as well as

(2.8) $b(\varepsilon, 0) > a(0)$ for all $\varepsilon > 0$

are true.
If, moreover, $S = \{x^0\}$ then also (2.7) holds.

<u>Proof:</u> First of all we note again that c depends of x^0. Viewing corollary 2.2 the equation (2.7) don't need to hold for each x^0 in S if (f,F) is stable.

(2.5): That means nothing but lower semicontinuity of α at 0_m and is obviously satisfied. //

(2.6): We assume conversely $a(0) = \lim_{k \to \infty} \beta(\varepsilon, y_k)$ for some $\varepsilon > 0$ and some sequence $y_k \to 0$. With corresponding $x_k \in F^{-}(y_k)$ we then have
$\text{dist}(x_k, S) \geq \varepsilon$ and $f(x_k) \to a(0)$.
Because of (2.5) we obtain
$$f(x_k) - \alpha(y_k) \longrightarrow +0 .$$

Now, we put $G_k(x) = -y_k + F(x)$ and construct locally Lipschitz functions g_k such that

$|g_k(x) - f(x)| \leq 2\,(f(x_k) - \alpha(y_k))$ for all $x \in A$

and $x_k \in \Psi(g_k, G_k)$.

Since $\operatorname{dist}(x_k, S) \geq \varepsilon$ then (f,F) would not be stable.

Construction of g_k:

If $\alpha(y_k) = f(x_k)$ we simply put $g_k = f$.

If $\alpha(y_k) < f(x_k)$ we take $r_k > 0$ such that

$f(x) < f(x_k) + (f(x_k) - \alpha(y_k))$ for all $x \in x_k + r_k B$, $x \in A$

and put

$$\lambda_k(x) = r_k^{-1} \cdot \| x - x_k \|$$

and

$$g_k(x) = \begin{cases} \lambda_k(x)\, f(x) + (1 - \lambda_k(x)) \cdot \alpha(y_k) & \text{if } x \in x_k + r_k B \\ f(x) & \text{otherwise} \end{cases}$$

The desired properties of g_k are then easily to verify. //

(2.8): Setting $y_k = 0$ and $\alpha(y_k) = \alpha(0)$ one may repeat the proof to (2.6). //

(2.7): We assume, finally, $S = \{x^0\}$, $r_k := \rho(G_k, F) \longrightarrow 0$ and $\gamma(G_k) = \operatorname{dist}(x^0, G^-(0)) \geq 2\varepsilon > 0$.

Viewing the inclusion

$$G_k^-(0) \subset \{x \in A \,/\, \operatorname{dist}(x, S) \geq \varepsilon,\ r_k B \cap F(x) \neq \emptyset\}$$

we have

$$\varphi(f, G_k) \geq b(\varepsilon, r_k)$$

and, because of (2.6)

$$\underline{\lim}\ \varphi(f, G_k) \geq \underline{\lim}\ b(\varepsilon, r_k) > a(0).$$

This inequality says that the function $\varphi(f, .)$ is not continuous at F. //

At this point some remarks are appropriate.

1. The conditions (2.5), (2.6) and (2.8) are already satisfied if both $S \neq \emptyset$ and the problem (f,F) is stable alone with respect to variations of f and linear variations of F (i.e. $G(.) = -y + F(.)$).

This is seen by the proof of theorem 2.3.

2. If A is compact the conditions (2.5), (2.6) and (2.8) are fulfilled and $\lim_{t \to 0} b(\varepsilon, t) = b(\varepsilon, 0)$.

This corresponds with C. Berge's (1963) well-known stability results.

Moreover, if S is non-empty and compact or if f is uniformly continuous on A the equation (2.5) follows from (2.6).

The proof may here be omitted.

3. In order to quantify the stable behavior of φ and Ψ near (f,F) one has to find appropriate estimates of the kind

$a(t) - a(0) \geq \ldots$, $b(\varepsilon,t) - b(\varepsilon,0) \geq \ldots$, $b(\varepsilon,0) - a(0) \geq \ldots$ and $c(t) \leq \ldots$.

For technical reasons we will deal with the difference

$$b(2\varepsilon, t) - b(\varepsilon, 0).$$

4. All we have said up to now, except the part (2.3)' of proposition 2.1, remains valid under weaker assumptions:
   A closed subset of X; X, Y Banach spaces; $F: A \longrightarrow Y$ closed; $f: A \longrightarrow R$ locally Lipschitz.
   In this case only two difficulties arise.
   First, the inequality $\operatorname{dist}(y, F(x)) \leq t$ will not imply the existence of $z \in F(x)$ such that $\|z - y\| \leq t$. Therefore, the theorem 2.1 holds only in the limit, i.e. after replacing δ_2 by $\delta_2' > \delta_2$ and $c(\delta_2)$ by any $c' > c(\delta_2')$. Since the set $S + \varepsilon B$ is no longer compact the statement (2.3)' may fail to hold for compact S.
   Second, the function $c = c(t)$ can become a very badly behaved one if all closed multifunctions G are included in its definition.
   In particular, the theorem 4.1 is then false.

3. The Function $b = b(\varepsilon, t)$

This stability function differs from $a = a(t)$ only by the additional constraint $\operatorname{dist}(x, S) \geqslant \varepsilon$ where $S = \Psi(f, F)$. Of course, this constraint is generally a complicated nonconvex one disregarding the fact that most often the set S is unknown at all.
However, if one knows that each ε-neighbourhood $S + \varepsilon B$ can be approximated by a polyhedral set as good as necessary then $\beta(\varepsilon, y)$ can be approximated by a finite number of infima

$$\alpha_\mu(y) = \inf \{ f(x) \,/\, y \in F(x), \quad x \in A \cap Q_\mu \}$$

where Q_μ is always a convex polyhedron. The difficulties to deal with $\alpha(y)$ or $\alpha_\mu(y)$ from the viewpoint of continuity may be seen as the same. To be more precise we suppose

(3.1) There is a finite number m' of convex polyhedra P_ν (depending of ε) such that

$$A \cap (S + \varepsilon \cdot B) \subset A \cap (\bigcup P_\nu) \subset A \cap (S + (3/2) \cdot \varepsilon \cdot B).$$

Without loss of generality one may then assume

$$\operatorname{int} P_\nu \neq \emptyset, \quad P_\nu = \{ x \in R^n / \; l_{\nu j}(x) \geqslant 0 \quad j = 1, \dots, k \}$$

where $l_{\nu j}$ are affin - linear functions.
For each y we observe

(3.2) $$\beta(\varepsilon, y) \leq v := \inf \{ f(x) \,/\, y \in F(x), x \in A, x \notin \bigcup \operatorname{int} P_\nu \} \leq \leq \beta(2\varepsilon, y).$$

Since $x \in R^n \setminus \operatorname{int} P_\nu$ if and only if there is some $j = j(\nu)$ satifying $l_{\nu j}(x) \leq 0$ we obtain

$$v = \inf \quad \inf \{ f(x) \,/\, x \in A, y \in F(x), \quad l_{\nu j_\nu}(x) \leq 0 \quad \nu = 1, \dots, m' \}.$$

Here the first infimum must be taken over all m'- tuples

$(j_1, \ldots, j_{m'})$ with $j_\nu \in \{1,\ldots,k\}$ and the set Q_μ above denotes the convex polyhedron being described by the m′ constraints

$$l_{\nu j_\nu}(x) \leq 0 .$$

Since

$$\beta(\varepsilon, y) \leq \min_\mu \alpha_\mu(y) \leq \beta(2\varepsilon, y)$$

we see that the inequality

$$(3.3) \qquad b(2\varepsilon, t) - b(\varepsilon, 0) \geq \min_{a_\mu(0) < \infty} (a_\mu(t) - a_\mu(0))$$

holds with

$$a_\mu(t) := \inf_{\|y\| \leq t} \alpha_\mu(t) .$$

The requirement (3.1) is fulfilled for compact S. However, the best case is the one of polyhedral S. Then, the number m′ of polyhedra P_ν^ε in (3.1) will not increase for vanishing ε, and P_ν^ε as well as Q_μ^ε may be written as

$$P_\nu^\varepsilon = \{x \;/\; A^\nu x \leq b^\nu + \varepsilon d^\nu\}$$

$$Q_\mu^\varepsilon = \{x \;/\; D^\mu x \leq p^\mu + \varepsilon q^\mu\}$$

with some fixed matrices and vectors. This leads to

$$(3.4) \qquad \alpha_\mu(y) = \alpha_\mu^\varepsilon(y) = \inf\{f(x) \;/\; x \in A,\; y \in F(x),\; D^\mu x \leq p^\mu + \varepsilon q^\mu\}$$

so that perturbation of ε means linear perturbation of the inf-function in terms of ε, too.

<u>Theorem 3.1</u> Suppose the sets A, S and graph F to be polyhedral and f to be globally Lipschitz on A. Then there exist ε_0, k_1 and k_2 (all positive) such that

$$(3.5) \qquad a(t) \geq a(0) - k_1 \cdot t \qquad \text{and}$$

$$(3.6) \qquad b(2\varepsilon, t) \geq b(\varepsilon, 0) - k_1 \cdot t$$

whenever $0 < \varepsilon \leq \varepsilon_0$ and $0 < t \leq k_2$.

<u>Proof:</u> Note that we don′t need the Kakutani-property of F.
The multifunctions H_μ defined by

$$H_\mu(\varepsilon, y) = \{x \in A \;/\; y \in F(x) ,\; D^\mu x \leq p^\mu + \varepsilon q^\mu\}$$

are polyhedral. Because of lemma 1.4 there are $k_{1\mu}$, $k_{2\mu}$ and ε_μ such that

$$H_\mu(\varepsilon, y) \subset H_\mu(\varepsilon, 0) + k_{1\mu} \|y\| \cdot B$$

for $0 < \varepsilon \leq \varepsilon_\mu$ and $\|y\| \leq k_{2\mu} \cdot \varepsilon$.
By (3.4) and the Lipschitz continuity of f, say with the modul L, we obtain

$$\alpha_\mu^{2\varepsilon}(y) \geq \alpha_\mu^\varepsilon(y) \geq \alpha_\mu^\varepsilon(0) - L\, k_{1\mu} \|y\| .$$

Setting

$$k_1 = \max L\, k_{1\mu},\; k_2 = \min k_{2\mu} \quad \text{and} \quad \varepsilon_0 = \min \varepsilon_\mu$$

(3.6) follows from (3.3). The proof of (3.5) is clear after applying directly that

$$F^-(y) \subset F^-(0) + \|y\| \cdot k \cdot B \qquad \text{for all } y$$

with some k holds. //

It should be mentioned that the above theorem remains true if the set S is empty.

<u>Remark:</u> If the statement of theorem 3.1 and, additional, for some $c_0 > 0$ the estimation

(3.7) $c(t) \le c_0 \cdot t$ for sufficiently small $t > 0$

hold there are k, l, ε_0 such that both the difference $|\varphi(g,G) - \varphi(f,F)|$ can be estimated above by $k(\delta_1 + \delta_2)$ and the inclusion

$\Psi(g,G) \subset S + 2\varepsilon \cdot B$

is true for $k(\delta_1 + \delta_2) \le b(\varepsilon, 0) - a(0)$ and $l \cdot \delta_2 \le \varepsilon$, $\varepsilon \le \varepsilon_0$ whenever (g,G) satisfies $(1)_\delta$.

This immediatly follows from the theorems 2.1 and 3.1.

The function Δ as $\Delta(\varepsilon) = b(\varepsilon, 0) - a(0)$ characterizes then the modul of metric upper semicontinuity of Ψ at (f,F) which is - disregarding a Lipschitz function- of the type Δ^{-1}.

If, e.g., S is polyhedral and bounded, $F^-(0)$ is polyhedral, $f: R^n \to R$ is polynomial of degree p and if, finally, (3.7) holds we obtain: There is $r > 0$ such that

$\Delta(\varepsilon) \ge r \cdot \varepsilon^p$ for $0 < \varepsilon < \varepsilon(r)$

and Ψ is upper semicontinuous (metrically) with some modul of the type $r' \cdot \sqrt[p]{\varepsilon}$.

<u>Theorem 3.2:</u> Let the Kakutani-mapping F have a convex graph and let $f: R^n \to R$ be convex. Assume, further, that S is non-empty and bounded and $0_m \in \text{int dom } F^-$.

Then, it holds the same as in theorem 3.1.

<u>Proof:</u> The restricted mapping $F^-/_{rB}$ fulfils, for small $r > 0$, the hypotheses of lemma 1.3 and, together with f, the suppositions of lemma 1.1. Setting

$S(\varepsilon, y) = \arg\min \{f(x) \,/\, x \in F^-(y),\ \text{dist}(x, S) \ge \varepsilon\}$

we thus obtain, with some positive ε_1, s_1,

(3.8) $\emptyset \ne S(0, y) \subset s\,B$

whenever $\|y\| \le \varepsilon_1$, $s \ge s_1$.

Let s be sufficiently large such that $S + B \subset s \cdot B$ and assume L to be a Lipschitz modul of f on sB.

We consider any point $x \in F^-(y)$ satisfying

$\text{dist}(x, S) \ge 2\varepsilon$ where $\|y\| \le \varepsilon_1$, $\varepsilon < 1/2$.

Because of (3.8) one finds $z \in S(0, y)$.

The line segment $[x, z]$ then contains a point ξ with

$\text{dist}(\xi, S) \ge 2\varepsilon$, $\xi \in sB$.

Moreover, we observe

$f(\xi) \le f(x)$, $\xi \in F^-(y)$.

For $\varepsilon < 1/2$ and $\|y\| \le \varepsilon_1$ this implies :

Either the set

$$M(2\varepsilon, y) = \{x \in F^{-}(y) \,/\, \text{dist}(x, S) \geq 2\varepsilon\}$$

is empty or

$$S(2\varepsilon, y) \cap sB \neq \emptyset .$$

In the second case we can take any point η of the intersection and find - by lemma 1.3 - some $u \in F^{-}(0)$ such that

$$\| \eta - u \| \leq l(s) \cdot \| y \| .$$

Therefore, we conclude

$$\beta(2\varepsilon, y) = f(\eta) \geq f(u) - L\, l(s)\, \| y \| .$$

With $\varepsilon = 0$ this yields the Lipschitz estimation (3.5) for the function a. With $\varepsilon > 0$ we find, additionally,

$$\text{dist}(u, S) \geq \varepsilon \quad \text{whenever} \quad l(s) \cdot \| y \| < \varepsilon$$

and, by considering $\inf f(u)$,

$$\beta(2\varepsilon, y) \geq \beta(\varepsilon, 0) - L\, l(s) \| y \| .$$

Since, in the first case , this inequality holds trivially, also (3.6) is verified. //

As it will be shown in the next section the estimation (3.7) really holds under the assumptions of theorem 3.2. The nice consequences of this fact are summarized in the remark after theorem 3.1.

4. The Function o = o(t)

In order to estimate

$$o(t) = \sup_{\varrho(G,F) \leq t} \text{dist}(x^0, G^{-}(0))$$

where the multifunctions G and F map $A \subset R^n$ into R^m we suppose, additionally, that A is a non-empty, closed and **convex** set. Further, we make essentially use of the requirement that G varies within the set of Kakutani-mappings only.
The key of what follows is the next

Theorem 4.1: Let $x^0 \in R^n$, $r > 0$ and suppose that, for some $t > 0$, the mapping H as

$$H(y) = F^{-}(y) \cap (x^0 + r \cdot B) \quad , \quad \| y \| = t$$

possesses a multivalued selection $h = h(y) \subset H(y)$ such that

$$h: \quad t \cdot \text{bd}\, B \longrightarrow A$$

is a Kakutani-mapping.
Then the relation $G^{-}(0) \cap (x^0 + r.B) \neq \emptyset$ holds for each Kakutani-mapping $G : A \longrightarrow R^m$ satisfying

(4.1) $\quad F(x) \subset G(x) + t \cdot B \quad$ for all $x \in A \cap (x^0 + r.B)$.

Remarks: Obviously, we have then $o(t) \leq r$. Further, no additional assumtions concerning F are needed. If F is closed and if the sets H(y) are non-empty and convex for all y , $\| y \| = t$ one may simply put $h = H$. Finally, the ball $x^0 + r \cdot B$ may be replaced by any compact, convex subset of R^n as it will be seen by the

<u>Proof:</u> We put $A_o = A \cap (x^o + r \cdot B)$ and assume contrarily to the assertion

$$G^-(0) \cap A_o = \emptyset$$

for some Kakutani-mapping G satisfying (4.1). Applying theorem 1.2 to the restricted mapping $G/_{A_o}$ we obtain:
There is a continuous function $z \overset{o}{=} z(x)$ which maps A_o into R^m and fulfils

(4.2) $\inf_{y \in G(x)} \langle z(x), y \rangle > 0$ for all $x \in A_o$

and

$$\|z(x)\| = 1 \quad \text{for all } x \in A_o .$$

The multifunction

$$x \longrightarrow h(-t \cdot z(x))$$

maps A_o into itself and is a Kakutani-mapping. Consequently, it possesses some fixed point x' which then satisfies

$$x' \in h(-t \cdot z(x')) \subset F^-(-t \cdot z(x')), \qquad x' \in A_o$$

and

$$-t \cdot z(x') \in F(x') .$$

Because of (4.1) there is also some $u \in t \cdot B$ such that

$$-t \cdot z(x') + u \in G(x') .$$

Since

$$\langle z(x'), -t\, z(x') + u \rangle \leq 0$$

we get a contradiction to (4.2). //

The theorem above allows several conclusions among of them the next two should be noted.

<u>Corollary 4.2:</u> Under the hypotheses of theorem 3.2 and the assumption $x^o \in S$ there is $c_o > 0$ such that $c(t) \leq c_o \cdot t$ holds for sufficiently small positive t.

<u>Proof:</u> By lemma 1.3 the distance dist $(x^o, F^-(y))$ can be estimated as not greater than $k \|y\|$ (for small $\|y\|$). Therefore, theorem 4.1 may be applied setting $r = k\, t$ and $h = H$.//

In this context we refer to the fact that the mappings G included in the definition of $c(t)$ don't need to have a convex graph.

<u>Corollary 4.3:</u> If $A \subset R^n$ is a non-empty, convex and compact set and $G : A \rightarrow R^n$ is any Kakutani.mapping then there is some $x \in A$ such that $0 \in G(x) + N_A(x)$ where N_A is the normal cone mapping associated with A.

<u>Proof:</u> We take t such that $G_t(x) = G(x) \cap t \cdot B \neq \emptyset$ for all $x \in A$ and r such that $A \subset r \cdot B$. Now, theorem 4.1 may be applied to the mappings N_A (as F) and $G_t + N_A$ (as G) with $x^o = 0$ and $h = H$.//

5. Approximation of Multifunctions

Generally, the nonlinear optimization problem (f, F) fulfils neither the requirements of the theorems 3.1 or 3.2 nor these of theorem 4.1. In this case the common idea is to replace the mapping F by a "better" multifunction L which approximates F with higher than linear order near to an interesting subset of the domain (see J.P. Aubin and H.Frankowska (1984); V.Demyanov, C.Lemarechal and J.Zowe (1985); S.M.Robinson (1979, 1982, 1983) ; B.Kummer (1984)). Whereas the assumptions needed through the well-known literature concerned with this question are rather strong and the perturbations of the multifunction F are less general than here the main problem (at least for finite dimensional mappings) can be reduced to the next two propositions.

<u>Lemma 5.1:</u> (locally Lipschitz upper semicontinuity)

Let L, $F : A \longrightarrow R^m$ be closed multifunctions and suppose there are $k > 0$, $\varepsilon_0 > 0$ such that

(5.1) $L^-(y) \subset L^-(0) + k\|y\| \cdot B$ for all $y \in R^m$, $\|y\| \leq \varepsilon_0$

where $L^-(0) \neq \emptyset$. Further, let a real function $o = o(t)$ exist satisfying

$\lim_{t \to +0} o(t)\, t^{-1} = 0$ and

(5.2) $F(x) \subset L(x) + o(t) \cdot B$ whenever $\mathrm{dist}(x, L^-(0)) \leq t$.

Finally, let $F = L$ on $L^-(0)$ and let $k' > k$.

Then, there exists $\varepsilon' > 0$ such that, for each ε with $0 < \varepsilon \leq \varepsilon'$, the mapping F^-_ε as

$F^-_\varepsilon(y) := F^-(y) \cap (L^-(0) + \varepsilon \cdot B)$ fulfils $F^-_\varepsilon(0) = L^-(0)$ and

(5.3) $F^-_\varepsilon(y) \subset L^-(0) + k'\|y\| \cdot B$ for all y, $\|y\| \leq \varepsilon'$.

<u>Proof:</u> Since $k' > k$ and $\lim o(t)\, t^{-1} = 0$ there is some $\varepsilon' > 0$ such that $\varepsilon' < (1/2) \cdot \varepsilon_0$, $t - k\, o(t) \geq (k/k')t$ and $o(t) < t$ hold for each t satisfying $0 < t \leq \varepsilon'$.

Consider now

$x \in F^-_\varepsilon(y)$ where $\varepsilon \leq \varepsilon'$, $\|y\| \leq \varepsilon'$

and assume $t := \mathrm{dist}(x, L^-(0)) > 0$.

Because of (5.2) the inclusion $y + z \in L(x)$ holds with some z, $\|z\| \leq o(t)$. Since $\|y + z\| \leq \varepsilon' + o(t) \leq \varepsilon_0$ we may apply (5.1) to get

$t \leq k\|y + z\| \leq k\|y\| + k\, o(t) \leq k\|y\| + t - (k/k')\, t$

and $t \leq k'\|y\|$.

Setting $y = 0$ we obtain $F^-_\varepsilon(0) \subset L^-(0)$.

Together with $F = L$ on $L^-(0)$ this inclusion shows $F^-_\varepsilon(0) = L^-(0)$, and the lemma is verified. //

The reader certainly finds that this lemma including the proof is a translation of the usual implicit function theorem into the set-valued case. In fact, the common estimations are to write as inclusions only. As a simple consequence of the lemma one sees that the inclusion

(5.3)' $\quad G^{-}(0) \cap (L^{-}(0) + \varepsilon \cdot B) \subset L^{-}(0) + k' \cdot r \cdot B$

holds for each closed mapping $G: A \longrightarrow R^m$ which satisfies (with $\varepsilon \leq \varepsilon'$, $r \leq \varepsilon'$)

$$G(x) \subset F(x) + r \cdot B \qquad \text{for all } x \in A \cap (L^{-}(0) + \varepsilon B).$$

However, as the most important fact we note that (5.1) holds globally with some k for every polyhedral mapping L having a non-empty set of zeros $L^{-}(0)$. If L is of this type then also (5.3) is globally valid, i.e. it holds for some k and all $k' > k$, $y \in R^m$. This can be seen immediately by the proof.

Unfortunately, the above lemma does not guarantee $F^{-}_{\varepsilon}(y) \neq \emptyset$ let alone $G^{-}(0) \neq \emptyset$ for small distances $\varrho(G, F)$. This statement needs a second set of requirements to L.

<u>Theorem 5.2:</u> (solvability in neighbourhoods)

Let $L, F : A \longrightarrow R^m$ be closed, $A \subset R^n$ be closed and convex and $C \subset L^{-}(0)$ be a non-empty, convex and compact subset. Suppose there are $k > 0$ and $\varepsilon_0 > 0$ such that the mappings H_t as $H_t(y) = L^{-}(y) \cap (C + k \cdot t \cdot B)$ $\quad (\|y\| = t)$ have selections $h_t : t \cdot \text{bd } B \longrightarrow A$ being Kakutani-mappings for each t, $0 < t \leq \varepsilon_0$. Finally, let

(5.4) $\quad L(x) \subset F(x) + o(t) \cdot B \quad$ whenever $\text{dist}(x, C) \leq t$, $x \in A$,

where $o = o(t)$ is as in lemma 5.1.

Then, for each $k' > k$, there is some $\varepsilon' > 0$ such that if $0 < t \leq \varepsilon'$ the relation

$$G^{-}(0) \cap (C + k' t \cdot B) \neq \emptyset$$

holds whenever $G : A \longrightarrow R^m$ is a Kakutani-mapping satisfying

(5.5) $\quad F(x) \subset G(x) + t \cdot B \qquad$ for all $x \in A \cap (C + k' t \cdot B)$.

<u>Proof:</u> One finds $\varepsilon' > 0$ such that $(k'/k)\, t \leq \varepsilon_0$ and $t + o(k' t) \leq (k'/k)\, t$ are true for all t, $0 < t \leq \varepsilon'$. With such fixed t we put $s = (k'/k)\, t$ $\;(\leq \varepsilon_0)$ and obtain for each x in $A \cap (C + k\, s \cdot B)$ in view of (5.5) and (5.4):

$F(x) \subset G(x) + t \cdot B$; $\qquad L(x) \subset G(x) + t \cdot B + o(k't) \cdot B \subset$

$\qquad \subset G(x) + (k'/k)\, t \cdot B = G(x) + s \cdot B.$

The assertion now follows easily from theorem 4.1 after replacing there x^0, F, t, r by C, L, s and ks, respectively. //

Setting $C = \{x^0\}$ the theorem presents a possibility to estimate the stability function $o = o(t)$ by $k't$. The existence of a mapping L which fulfils the suppositions of the theorem may be seen as a regularity property of the multifunction F at the set C.

6. Some Applications

6.1. Nonconvex pertubations of a convex problem

Suppose

(P) $\min \{f_0(x) \,/\, f_i(x) \le 0 \quad (i = 1,\dots,m)\,, \quad D\,x = d\}$

to be a regular convex optimization problem, i. e.

$f_k : R^n \longrightarrow R$ are convex, $k = 0, 1, \dots, m$;

D is a $(p\,,\,n)$ matrix , rank $D = p$, $d \in R^p$;

There is $u \in R^n$ such that $D\,u = d$ and $f_i(u) < 0 \quad (i = 1,\dots,m)$;

The solution set Ψ of (P) is non-empty and bounded.

We fix any closed, convex subset $A \subset R^n$ including $\Psi + B$ and form with <u>continuous</u> functions $g_k\,,\, h_j : A \longrightarrow R \quad (k = 0, 1,\dots,m\,;\ j = 1,\dots,p)$ the problem

(P)′ $\min \{g_0(x) \,/\, g_i(x) \le 0 \quad (i = 1,\dots,m)\,,\ h_j(x) = 0 \quad (j = 1,\dots,p)\}$

Let φ and φ' be the corresponding inf - values and Ψ' be the set of solutions to (P)′. Further, consider the numbers

$$\delta = \sup_{x \in A} \max \{\| f(x) - g(x) \|\ ,\ \| h(x) - (D\,x - d) \| \}$$

where the related vector functions are regarded and

$$b(\varepsilon) = \inf \{f_0(x) \,/\, f_i(x) \le 0 \quad (i = 1,\dots,m)\,,\ D\,x = d\,,\ \operatorname{dist}(x\,,\,\Psi) \ge \varepsilon\}.$$

After setting

$$F(x) = \begin{pmatrix} f_1(x)+R_+ \\ \cdots\cdots \\ f_m(x)+R_+ \\ Dx - d \end{pmatrix} \subset R^{m+p}$$

we may apply theorem 3.2, corollary 4.2 and recall the remark to theorem 3.1. They yield the existence of $k > 0$ and $\varepsilon_0 > 0$ such that

(6.1) $\quad | \varphi' - \varphi | \le k\,\delta \qquad$ for all $\quad \delta \le \varepsilon_0$

and

(6.2) $\quad \emptyset \ne \Psi' \subset \Psi + 2\,\varepsilon \cdot B$

hold whenever $\varepsilon \le \varepsilon_0$ and $k\,\delta < \min \{\varepsilon,\ \Delta(\varepsilon)\}$.
Here $\Delta(\varepsilon)$ denotes the difference $b(\varepsilon) - \varphi$.

If, particularily, $f_0 = 0$ the set Ψ coincides with the feasible domain of (P) and $\Delta(\varepsilon)$ becomes $+\infty$. Then, (6.2) can be seen as an upper Lipschitz characterization of the feasible points of (P)′. Similar results concerning the set of Kuhn - Tucker points to (P) and (P)′ have been established by R. Schultz (1985) where, however (P)′ has to be a convex problem without equality-constraints.

Finally, it should be noted without a proof that, in the case of convex perturbed problems (P)′ only, the numbers k and ε_0 will not be better than in the general one.

6.2. Nondifferentiable perturbation of a differentiable problem

Let $A \subset R^n$ be non-empty, convex and compact, let $f: R^n \longrightarrow R$ be a C^1- function and let $g: A \rightarrow R$ be locally Lipschitz.
We form the functions

$$h_t(x) = t\, g(x) + (1 - t)\, f(x) \qquad 0 \leq t \leq 1$$

and study the set $X(t)$ of the stationar points to the problem

$$(P(t)) \qquad \min \{ h_t(x) \;/\; x \in A \} \quad ,$$

or, with other words, the set of solutions to

$$0 \in \partial h_t(x) + N_A(x)$$

where ∂ denotes the Clarke subdifferential and N_A is the usual normal cone mapping (see F.H.Clarke (1984)).
Because of

$$h_t(x) = t \cdot \partial g(x) + (1 - t)\, Df(x)$$

there is some $L > 0$ such that

$$r(t) := \sup_{x \in A} d_H(\partial h_t(x) , Df(x)) \leq L\, t.$$

Let us introduce the set $S(y)$ of all stationar points to the problem

$$(P_y) \qquad \min \{ f(x) - \langle y , x \rangle \;/\; x \in A \} \qquad y \in R^n$$

and assume $S \subset S(0)$ to be a non-empty, convex and compact subset.
Finally, suppose the existence of positive ε_0 and δ_0 such that the mapping H as

$$H(y) = S(y) \cap (S + \varepsilon_0 B) \qquad \| y \| \leq \delta_0$$

has non-empty, convex ranges $H(y)$ or has a Kakutani selection h.

Then, theorem 4.1 ensures the existence of some $t_0 > 0$ such that

$$\emptyset \neq X(t) \cap (S + \varepsilon_0 B) \qquad \text{for all } t, \; 0 < t \leq t_0 .$$

The same holds true in the case that a non-linear homotopy between g and f is considered. In the case discussed above one may put $t_0 = \delta_0 \cdot L^{-1}$. If, additionally, the inclusion

$$H(y) \subset S + k \| y \| \cdot B \qquad \text{for all } y , \; \| y \| \leq \delta_0$$

holds with some $k > 0$ one obtains by obvious estimations

$$(6.3) \qquad X(t) \cap (S + \varepsilon_0 B) \subset S + k\, L\, t \cdot B \qquad \text{for } 0 \leq t \leq t_0 .$$

6.3 Approximations

To verify the above assumptions concerning the problems (P_y) and the related mapping H can be very hard.
If, however $S = \{x^o\}$ is a singleton and if the Frechet derivative $D^2 f(x^o)$ of $Df(x)$ at x^o exists then the function

$$L(x) = Df(x^o) + D^2 f(x^o)\, (x - x^o)$$

approximates $Df(x)$ near x^o of better than linear order. Recalling the propositions (5.1) and (5.2) it is enough to analyse the problems (P_y) after replacing the objective f by its quadratic approximation

at x^o. Under the assumptions of section 6.2, modified in this way, we will obtain the same statements up to a small difference:
In (6.3) one has to replace k by any $k' > k$,and the actual t_o is now smaller than there.

6.4. Concluding remarks

The arguments in the sections 6.2 and 6.3 can be used also for studying the sets of (generalized) Kuhn - Tucker points of differentiable optimization problems which are perturbed in a nondifferentiable way. The condition imposed upon H in section 6.2 then can be seen as some generalized second order condition.
An application of the statements in the sections 4 and 5 to complementarity problems of the type

$$0 \in F(x) + N_A(\dot{x}) \quad , \quad x \in A \quad ; \quad A \subset R^n \quad \text{closed convex}$$

where F varies within the family of Kakutani-mappings is also possible.

References

Aubin, J.P.; Frankowska, H. (1984) On inverse function theorems for set-valued maps. IIASA Working Paper, WP - 84 - 68.

Berge, C. (1963) Topological Spaces . Mc Millan, New York.

Clarke, F.H. (1984) Optimization and Nonsmooth Analysis. John Wiley.

Demyanov, V; Lemarechal, C.; Zowe, J. (1985) A first order approximation of a set-valued mapping. preprint.

Fiacco, A.V. (1983) Introduction to Sensitivity and Stability Analysis in Nonlinear Programming. Academic Press, New York.

Klatte, D.; Kummer, B. (1984) Stability properties of infima and optimal solutions of parametric optimization problems. Proceed. of the IIASA Workshop on Nondiff. Optimiz., Sopron, Hungary 1984

Kummer, B. (1984) Generalized equations: solvability and regularity. Mathem. Programming Study 21 , North-Holland, 199- 212.

Michael, E. (1956) Continuous selections I . Ann. of Math., Vol. 63, 361 - 382.

Robinson, S.M. (1976) Regularity and stability for convex multivalued functions. Math. of Operations Research, Vol. 1, 130 - 143.

Robinson, S.M. (1979) Generalized equations and their solutions, Part I: Basic theory. Mathem. Programming Study 10 , North-Holland, 128 - 141.

Robinson, S.M. (1981) Some continuity properties of polyhedral multifunctions. Mathem. Programming Study 14, North-Holland, 206 - 214.

Robinson, S.M. (1982) Generalized equations and their solutions, Part II. Applications in nonlinear programming. Mathem. Programming Study 17, 200 - 221 .

Robinson, S.M. (1983) Persistence and continuity of local minimizers. (Dedicated to Professor Dr. Karl Nickel on his 60th birthday). Preprint, Univ. of Wisconsin-Madison.

Schultz, R. (1985) Abschätzungen für Kuhn - Tucker- Punkte gestörter konvexer Optimierungsaufgaben. Dissertation (A), Humboldt - Universität zu Berlin.

Walkup, D.; Wets, R. (1969) A Lipschitzian characterization of convex polyhedra. Proceed. Amer. Mathem. Soc., Vol 23, 167 -173.

AN ALGORITHM FOR SOLVING ONE-PARAMETRIC OPTIMIZATION PROBLEMS BASED ON AN ACTIVE-INDEX SET STRATEGY

Ralph Lehmann[1)]

Abstract

An algorithm for solving one-parametric nonlinear optimization problems is introduced. It is based on an active-constraints strategy, thus requiring only the solution of linear equations. A method for automatic stepsize control is included.

1. Introduction

In this paper we present an algorithm for solving a parametric optimization problem of the following structure:

$$(P(t)) \quad \min_{x} \{f(x,t) \mid g(x,t) \leq 0,\ h(x,t) = 0\},\ t \in T, \qquad (1.1)$$

with $T := [0,1]$,

$f: \mathbb{R}^n \times T \longrightarrow \mathbb{R},\ g: \mathbb{R}^n \times T \longrightarrow \mathbb{R}^m,\ h: \mathbb{R}^n \times T \longrightarrow \mathbb{R}^l.$

The interest in such nonlinear programming problems depending on a real parameter arises for several reasons:

a) Physical and chemical systems, especially states of equilibrium, can often be described advantageously by some variational principle (e.g. minimal potential energy, minimal surface, minimal surface, minimal enthalpy). Investigating the dependence of the state of the system on a certain parameter (e.g. length of some part of the system, enclosed volume, quantity of reacting substances) one obtains (one-) parametric optimization problems, which can be transformed into the form (1.1) (see Examples 5.3, 5.4). Such problems also occur in the field of engineering, e.g. in the construction of transmissions (/20/).

b) In particular, economic decision problems can be reflected by means of stochastic or multi-objective optimization. The latter two branches are closely related to parametric optimization (e.g. /27/, /9/, /1/, /19/). Thus the decision problems can

1) Humboldt-Universität zu Berlin, Sektion Mathematik, PF 1297, Berlin, DDR 1086

finally be solved with the aid of a sequence of problems of the structure (1.1).

c) Moreover, a problem like (1.1) arises if an imbedding method is applied for globalizing a locally convergent nonlinear programming algorithm (e.g. ROBINSON's or WILSON's method) (/6/, /7/, /8/, /10/, /17/, /21/).

We assume that the problems P(t) are solvable in the following sense:

(E1) There exists a continuous function $x : T \longrightarrow \mathbb{R}^n$ such that $x(t)$ is a local minimizer for P(t).

(E2) x(0) is explicitly known (or approximations for it can be determined with any accuracy).

Remark: We assume (E1) for l o c a l minimizers since in the case of general nonlinear programming problems without any additional information it is only possible to determine local solutions. Nevertheless, we want to emphasize that the physical or chemical quantities mentioned in a) attain only a l o c a l minimum. (Note that for leaving such a minimum and changing over to another, possibly global one, the quantity mentioned above should be temporarily increased.)

In general, we are interested in tracing the path $\{(x(t),t) \mid t \in T\}$ by means of numerical approximation. If we apply a locally (but fast) convergent algorithm (A) for solving the problems P(t), this can be done as follows: Having chosen a partition of T, $0 = t_o < t_1 < \ldots < t_N = 1$, we compute approximations x_k for $x(t_k)$, $k = 1,\ldots,N$, consecutively. For this, x_{k-1} is used as a starting point for the algorithm (A) determining the approximate solution x_k of $P(t_k)$ within a finite number (j_k) of steps, $k = 1,\ldots,N$.
A simple method for constructing a suitable partition of T and appropriate integers j_k (simultaneously to the computation of the sequence $\{x_k\}$) is introduced in Chapter 4 of the present paper.
While we trace the path $\{(x(t),t) \mid t \in T\}$, the index set I(t) of the active constraints changes only for certain values of the parameter t (so-called critical points) and remains unchanged between two such

points (under suitable assumptions on the continuity of f,g,h). (For an illustration see Example 5.4). That is why it seems to be useful to apply an active-constraints strategy. Such a strategy is presented in Chapter 3. It offers the following advantage: If WILSON's method or one of its modifications is employed as the locally convergent algorithm (A), then the numerical effort is reduced to the solution of systems of linear equations.

In order to obtain clearer proofs, we will confine ourselves to considering only WILSON's method in the roll of algorithm (A). However, it should be noticed that analogous results can be obtained for a larger class of optimization algorithms containing also ROBINSON's method and WILSON's method with a consistent approximation of the Hessian matrix of the Lagrangian function or with special updating techniques (e.g. PSB) (see /18/).

Assumptions and theoretical results needed are introduced in Chapter 2. Finally, first numerical results are presented in Chapter 5.

2. Assumptions

At first we introduce some notations used in this paper: First and second order derivatives with respect to x are denoted by ∇_x and ∇_{xx}, respectively. $\|.\|$ represents an arbitrary norm. If a and b are vectors, then (a,b) denotes $\binom{a}{b}$. Kuhn-Tucker triples are denoted by

$$z := (x,u,v) \text{ or } \tilde{z} := (\tilde{x},\tilde{u},\tilde{v}) \text{ or } z(t) = (x(t),u(t),v(t)). \qquad (2.1)$$

Furthermore we use

$L(x,u,v,t) := f(x,t) + u^T g(x,t) + v^T h(x,t)$

(Lagrangian function corresponding to P(t))

$M := \{1,\dots,m\}$

(Index set of inequality constraints)

$I(t) := \{i \in M \mid g_i(x(t),t) = 0\}$

(Inequality constraints which are active at x(t))

$I^+(t) := \{ i \in M \mid u_i(t) > 0 \}$

(Positive Lagrange-multipliers; u(t) as in (2.1))

$Lg_i(x,\tilde{x},t) := g_i(\tilde{x},t) + \nabla_x g_i(\tilde{x},t)\,(x-\tilde{x})$

(Linearization of the function g_i)

$J(z,\tilde{z},t) := \{ i \in M \mid Lg_i(x,\tilde{x},t) = 0 \}$

(Active inequality constraints for a linearized problem; $z,\tilde{z}$ as in (2.1))

$J^+(z) := \{ i \in M \mid u_i > 0 \}$

(Positive Lagrange-multipliers for a fixed Kuhn-Tucker point; z as in (2.1))

$U_r(y) := \{ \xi \; \| \xi - y \| < r \}$

(neighbourhood of radius r)

$[a,b] := \{ \lambda a + (1-\lambda) b \mid \lambda \in [0,1] \}$

(segment between a and b)

card $\mathfrak{M}$:= number of elements of the set $\mathfrak{M}$

The objective function of the subproblems generated by WILSON's method is denoted by

$$\varphi(x,\tilde{z},t) := \nabla_x f(\tilde{x},t)(x-\tilde{x}) + \frac{1}{2}(x-\tilde{x})^T \; \nabla_{xx} L(\tilde{z},t)(x-\tilde{x}) \tag{2.2}$$

For the description of WILSON's method and the active-constraints strategy we introduce the following optimization problems:

$$(P(\tilde{z},t)) \quad \min \left\{ \varphi(x,\tilde{z},t) \;\middle|\; \begin{array}{l} Lg_i(x,\tilde{x},t) \le 0, \; i = 1,\dots,m, \\ Lh_j(x,\tilde{x},t) = 0, \; j = 1,\dots,l \end{array} \right\} \tag{2.3}$$

$$(P^I(t)) \quad \min \left\{ f(x,t) \;\middle|\; \begin{array}{l} g_i(x,t) = 0, \; i \in I, \\ h_j(x,t) = 0, \; j = 1,\dots,l \end{array} \right\} \tag{2.4}$$

$$(P^I(\tilde{z},t)) \quad \min \left\{ \varphi(x,\tilde{z},t) \;\middle|\; \begin{array}{l} Lg_i(x,\tilde{x},t) = 0, \; i \in I, \\ Lh_j(x,\tilde{x},t) = 0, \; j = 1,\dots,l \end{array} \right\} \tag{2.5}$$

Thus WILSON's method, applied for determining a Kuhn-Tucker point of P(t), can be formulated briefly:

(A): Start with z^0.

Having z^i, let z^{i+1} be a Kuhn-Tucker point of $P(z^i,t)$. (If there is more than one such point, choose z^{i+1} to be closest in norm to z^i.)

Knowing the structure of (A) we shall not only trace the path $\{(x(t),t) \mid t \in T\}$ as described in Chapter 1, but the path of Kuhn-Tucker points $\{(z(t),t) \mid t \in T\}$. In order to avoid double indices we prefer the following formulation:

Given a suitable partition of T

$$0 = t_o \leq t_1 \leq \ldots \leq t_N = 1 \tag{2.6}$$

and the point $z_o = z(0)$, compute z_k, $k = 1,\ldots,N$, as a Kuhn-Tucker point of $P(z_{k-1},t_k)$. (If there is more than one such point, choose z_k to be closest in norm to z_{k-1}.)

This conception implies only one step of (A) per level t_k. However, condition (2.6) does not exclude equal levels. Consequently, this approach is sufficiently general to cover the situation described in Chapter 1.

Now we introduce the assumptions posed for the optimization problems P(t):

(E1), (E2) - cf. Chapter 1.

(V1) There exists a neighbourhood U of $\{(x(t),t) \mid t \in T\} \subset R^n \times T$ such that for all $(x,t) \in U$ the functions f, g_i, and h_j ($i=1,\ldots,m$, $j=1,\ldots,l$) are twice continuously differentiable with respect to x.

(V2) The gradients of the active constraints

$$\{\nabla_x g_i(x(t),t)\}_{i \in I(t)} \cup \{\nabla_x h_j(x(t),t)\}_{j=1,\ldots,l}$$

are linearly independent for each $t \in T$.

(V1), (V2) imply that there exist functions $u : T \longrightarrow \mathbb{R}^m$, $v : T \longrightarrow \mathbb{R}^l$, which are uniquely defined such that $z(t) := (x(t),u(t),v(t))$ (which has already been mentioned above without proof of existence) satisfies the first-order Kuhn-Tucker conditions:

$$\nabla_x L(x,u,v,t) = 0$$

$$g_i(x,t) \leq 0,\ u_i \geq 0,\ u_i g_i(x,t) = 0,\ i = 1,\ldots,m,$$

$$h_j(x,t) = 0, \qquad j = 1,\ldots,l.$$

Additionally we need the following strong second-order sufficient condition:

(V3) Let $t \in T$. For each non-zero y with

$$\nabla_x g_i(x(t),t)\, y = 0 \quad \forall i \in I^+(t)$$

and $\nabla_x h_j(x(t),t)\, y = 0 \quad j = 1,\dots,l$

we assume that

$$y^T \; \nabla_{xx} L(z(t),t)\, y > 0.$$

Saying "(V1), (V2), (V3) hold for $P(t_*)$" we mean that in (V2), (V3) "for each t" should be read "for $t = t_*$" and $\{x(t),t) \mid t \in I$ in (V1) should be replaced by $(x(t_*,t_*)$.

(V1), (V2), (V3) represent usual assumptions for convergence proofs for WILSON's method. A discussion of them can be found, for instance, in /7/, /18/; further theoretical results about the existence of a continuous solution curve, i.e., topics related to (E1), are presented in /12/, /13/, /14/, /16/.

3. An active- constraints strategy

Active-constraints strategies for nonlinear programming algorithms have already been investigated in /15/, (for hybrid algorithms), in /26/ (for algorithms with damping), and in /6/, /7/, /8/ (for imbedding methods). The advantage of such strategies consists in the fact that only equality constraints occur in the subproblems generated by the algorithm. In the case of quadratic subproblems (i.e., for WILSON's method or one of its modifications) the corresponding Kuhn-Tucker conditions form a system of linear equations. By appropriately controlling some parameter one can ensure that these systems of linear equations are regular, i.e., uniquely solvable, (under suitable assumptions). Moreover, in general the subproblems generated by an active-constraints algorithm have less constraints than the original problem, i.e., the dimension of the corresponding Kuhn-Tucker system is reduced.

In Chapter 1 we have already mentioned an important aspect of an active-constraints strategy for solving one-parameter-dependent optimization problems: While we trace the path $\{x(t),t) \mid t \in T\}$ (or $\{(z(t),t) \mid t \in T\}$, respectively), the index set $I(t)$ of the active constraints changes only for certain values of the parameter t (critical points) and remains unchanged on whole (open) intervals (under suitable assumptions). Now suppose that we have already solved (approximately) a problem $P(t_{k-1})$, for some $t_{k-1} \in T$, and have determined a set I_{k-1} of active (or "almost active") constraints of $P(t_{k-1})$. Going on to the solution of $P(t_k)$, for some $t_k > t_{k-1}$, we shall test whether I_{k-1} represents the "almost") active constraints of $P(t_k)$ or not. If not, then I_{k-1} has to be modified in an appropriate way. Since the algorithm for solving $P(t_k)$ (WILSON's method) generates a sequence of subproblems, there are two possible "levels" for testing and modifying the index set of active constraints:

a) We first determine the set $I_k = I(t_k)$ of active constraints of $P(t_k)$ and then apply WILSON's method for solving $P^{I_k}(t_k)$ (see /6/, /7/, /8/). This procedure has the advantage of working with the exact set $I(t_k)$, but it has the disadvantage that the numerical effort for the determination of this set is rather high for parameters t_k near critical points. Furthermore, additional assumptions are necessary for obtaining theoretical results.

b) We apply WILSON's method to the "whole" problem $P(t_k)$, but solve the occuring subproblems with the aid of an active-constraints strategy (related to ZANGWILL's "manifold-suboptimization procedure" /29/, cf. also FLETCHER's quadratic programming algorithm /4/). The application of such an active-constraints strategy to the subproblems is possible since the subproblems are linearly constrained.

Now we are going to explain procedure b) in more detail: ZANGWILL's "manifold-suboptimization procedure" applied to a subproblem $P(\tilde{z},t)$ starts with a feasible point x^o and the corresponding set I^1 of active constraints (active with respect to $P(\tilde{z},t)$). For $\nu = 1,2,3,\ldots,$

the problems $P^{I^\nu}(\tilde{z},t)$ (cf. (2.5)) are solved. Having obtained a Kuhn-Tucker point $\hat{z}^\nu = (\hat{x}^\nu, u^\nu, v^\nu)$ for $P^{I^\nu}(\tilde{z},t)$, we test whether $\hat{x}^\nu$ is feasible for $P(\tilde{z},t)$ or not. If it is not, then we determine the feasible point x^ν on $[x^{\nu-1}, \hat{x}^\nu]$ lying nearest to $\hat{x}^\nu$. This is not difficult since all constraints of $P(\tilde{z},t)$ are linear. At x^ν at least one more constraint is active as compared to $\hat{x}^\nu$. After enlarging I^ν by such an index (thus obtaining $I^{\nu+1}$) we go on to solve $P^{I^{\nu+1}}(\tilde{z},t)$. If, however, $\hat{x}^\nu$ was feasible for $P(\tilde{z},t)$, we set $x^\nu := \hat{x}^\nu$ and check the sign of the corresponding Lagrange multipliers. If they are all non-negative, we have found a Kuhn-Tucker point for $P(\tilde{z},t)$. Otherwise, we delete an index corresponding to a negative Lagrange multiplier from I^ν (thus obtaining $I^{\nu+1}$) and go on to solve $P^{I^{\nu+1}}(\tilde{z},t)$.

Now the question about a feasible starting point x^o remains to be answered: If the algorithm just described is applied to solving $P(z_{k-1},t_k)$ within an algorithm tracing $\{(z(t),t) \mid t \in T\}$, we test whether the active-constraints set $I^o := I_{k-1}$ obtained for $P(z_{k-2},t_{k-1})$ is also suitable for $P(z_{k-1},t_k)$, i.e., whether the solution x^o of $P^{I^o}(z_{k-1},t_k)$ is feasible for $P(z_{k-1},t_k)$. If it is not, then we enlarge I^o by a constraint not satisfied by x^o (not changing the upper index 0 of I^o!) and solve $P^{I^o}(z_{k-1},t_k)$ again. This procedure must lead to a feasible point (at latest for $I^o = M$!) or an unsolvable subproblem $P^{I^o}(z_{k-1},t_k)$. However, it can be proved that for a suitable partition of T, $0 = t_o \leq t_1 \leq \ldots \leq t_N = 1$, unsolvable subproblems do not occur.

Based on the procedure outlined we obtain the following algorithm:

<u>Algorithm 3.1:</u>

Suppose that a partition $0 = t_o \leq t_1 \leq \ldots \leq t_N = 1$ of T is given.

<u>S0:</u> $z_o := z(0)$, determine I_o with $I^+(0) \subseteq I_o \subseteq I(0)$.

For $k := 1,\ldots,N$ carry out the following steps: (For simplifying the notation we shall write "$\sim$" instead of the index k-1 and omit the index k.)

<u>S1:</u> $I^o := \tilde{I}$, $\nu := 0$

<u>S2:</u> Determine a Kuhn-Tucker point

$z^{\nu} = (\hat{x}^{\nu}, u^{\nu}, v^{\nu})$ of $P^{I^{\nu}}(\tilde{z}, t)$. Let $(u^{\nu})_i := 0$ for $i \in M \setminus I^{\nu}$.
Does an index $i \in M \setminus I^{\nu}$ exist with $Lg_i(\hat{x}^{\nu}, \tilde{x}, t) > 0$?

S2.1: Yes:

If $\nu = 0$, then $I^0 := I^0 \cup \{i\}$ *), go to S2.

If $\nu > 0$, then:

$$\alpha := \min_j \left\{ - \frac{Lg_j(x^{\nu-1}, \tilde{x}, t)}{\nabla_x g_j(\tilde{x}, t)(\hat{x}^{\nu} - x^{\nu-1})} \;\middle|\; Lg_j(\hat{x}^{\nu}, \tilde{x}, t) > 0 \text{ and } j \in M \setminus I^{\nu} \right\} \quad (3.1)$$

(The minimum shall be attained for $j = i$.)

$x^{\nu} := x^{\nu-1} + \alpha(\hat{x}^{\nu} - x^{\nu-1})$

$I^{\nu+1} := I^{\nu} \cup \{i\}$ *)

$\nu := \nu + 1$, go to S2.

S2.2: No: Continue.

S3: Does an index $i \in I^{\nu}$ exist with $(u^{\nu})_i < 0$?

S3.1: Yes: $x^{\nu} := \hat{x}^{\nu}$

$I^{\nu+1} := I^{\nu} \setminus \{i\}$ *)

$\nu := \nu + 1$, go to S2.

S3.2: No: $I := I^{\nu}$

$z := z^{\nu}$ ($P(\tilde{z}, t)$ has been solved.)

Remark 3.2:

It is possible that in the operations marked by *) (in steps 2.1 and 3.1 of Algorithm 3.1) the index i is not uniquely defined. If several such indices exist, one may choose an arbitrary one of them.

Remark 3.3:

If step S2.1 is reached with $\nu = 0$, then a feasible point for $P(z_{k-1}, t_k)$ has not been obtained yet, and the search for such a point is continued by enlarging I^0 (not changing $\nu = 0$). If step S2.1 is reached with $\nu > 0$, then $x^{\nu-1}$ has already been feasible for $P(z_{k-1}, t_k)$ (strong proof by mathematical induction) and x^{ν} is constructed in S2.1 in such a way that it is feasible for $P(z_{k-1}, t_k)$, too. As a consequence we obtain $\alpha \in [0, 1)$ (cf. (3.1) taking into consideration that

$\nabla_x g_j(\tilde{x},t)(\hat{x}^\nu - x^{\nu-1}) = Lg_j(\hat{x}^\nu,\tilde{x},t) - Lg_j(x^{\nu-1},\tilde{x},t)$, and $Lg_j(x^{\nu-1},\tilde{x},t) \le 0$ for all $j \in M$ because of the feasibility of $x^{\nu-1}$).

The main result of this chapter states that, for a sufficiently "fine" partition of T, Algorithm 3.1 is well-defined and provides an approximation of the path $\{(z(t),t) \mid t \in T\}$ with a prescribed accuracy:

Theorem 3.4:

Assume (E1), (E2), (V1), (V2), (V3) for the optimization problems P(t). $\rho > 0$ shall be arbitrarily fixed. Then there exists a constant $\Delta > 0$ such that for each partition of T, $0 = t_o \le t_1 \le \ldots \le t_N = 1$, with $t_k - t_{k-1} \le \Delta$, $k = 1,\ldots,N$, we have: All subproblems occuring in Algorithm 3.1 are uniquely solvable, and the algorithm terminates after a finite number of steps, having generated a sequence $\{z_k\}_{k=1,\ldots,N}$ with

$$\|z_k - z(t_k)\| < \rho \quad , \; k = 1,\ldots,N.$$

In the proof of Theorem 3.4 we will first examine the behaviour of Algorithm 3.1 "locally", i.e., for values t_k in a neighbourhood of a fixed parameter t_*. In preparation, we note that some properties of $P(t_*)$, including regularity, can also be proved for "neighbouring" problems $P(\tilde{z},t)$:

Lemma 3.5:

Assume (V1), (V2), (V3) and let $t_* \in T$ be arbitrarily fixed. Then there exist constants $R > 0$, $\sigma > 0$, and $\delta > 0$ such that for all $\tilde{z} \in U_R(z(t_*))$, $z \in U_\sigma(z(t_*))$, $t \in U_\delta(t_*)$ the following holds:

(i) The vectors

$$\{\nabla_x g_i(\tilde{x},t)\}_{i \in I(t_*)} \cup \{\nabla_x h_j(\tilde{x},t)\}_{j=1,\ldots,l}$$

are linearly independent.

(ii) For each non-zero $y \in \mathbb{R}^n$ with

$$\nabla_x g_i(\tilde{x},t)y = 0 \qquad \forall i \in I^+(t_*) \tag{3.2}$$

and $$\nabla_x h_j(\tilde{x},t)y = 0 \qquad j = 1,\dots,l \tag{3.3}$$

one has

$$y^T \nabla_{xx} L(\tilde{z},t)\; y > 0. \tag{3.4}$$

(iii) $\forall i \in M:\; u_i(t_*) > 0 \longrightarrow u_i > 0$ (3.5)

(iv) $\forall i \in M:\; g_i(x(t_*),t_*) < 0 \longrightarrow Lg_i(x,\tilde{x},t) < 0,$ (3.6)

i.e. $\forall i \in M:\; Lg_i(x,\tilde{x},t) = 0 \longrightarrow g_i(x(t_*),t_*) = 0$ (3.7)

Proof:

(i) Because of (V2) the matrix

$$\begin{pmatrix} (\nabla_x g_i(x(t_*),t_*))_{i\in I(t_*)} \\ (\nabla_x h_j(x(t_*),t_*))_{j=1,\dots,l} \end{pmatrix}$$

contains a regular quadratic submatrix of rank $\mathrm{card}(I(t_*)) + l$. Applying the pertubation lemma (cf. /25/ 2.1.13) to this matrix and regarding (V1) we obtain the desired result.

(ii) Obviously, it is sufficient to prove (3.4) for $y \in R^n$ with $\|y\| = 1$. Let us consider the auxiliary function $\hat{\varphi}$ defined by

$$\hat{\varphi}(\tilde{z},t) := \min_{I} \left\{ y^T \nabla_{xx} L(\tilde{z},t)y \;\middle|\; y \text{ satisfies (3.2), (3.3), and } \|y\| = 1 \right\}.$$

As $\{y \in R^n \mid \|y\| = 1\}$ is compact, (V3) implies:

$$\hat{\varphi}(z(t_*),t_*) > 0.$$

Moreover, $\hat{\varphi}$ is lower semicontinuous by /1/ (Theorem 4.2.2 and 3.1.2), which provides the desired result.

(iii) It is sufficient to choose

$$\sigma < \min_{i\in M} \{u_i(t_*) \mid u_i(t_*) > 0\}.$$

(iv) (3.6) can be proved because of the continuity of $g_i(.,.)$ and $\nabla_x g_i(.,.)$ assumed by (V1). We obtain (3.7) as an immediate consequence if we regard that $x(t_*)$ is feasible for $P(t_*)$ by definition, i.e., $g_i(x(t_*),t_*) \le 0,\; \forall i \in M$.

In preparation of Lemma 3.7 we first formulate the following lemma, which is essentially based on results by S.M. ROBINSON:

Lemma 3.6:

Assume (V1), (V2), (V3) for an optimization problem $P(t_*), t_* \in T$ fixed. $\alpha > 0$ shall be arbitrarily fixed.
Then there exist constants $R > 0$, $\sigma > 0$, and $\delta > 0$ such that for all $\tilde{z} \in U_R(z(t_*))$, $t \in U_\delta(t_*)$, the problem $P(\tilde{z}, t)$ has a unique Kuhn-Tucker point $\hat{z}(\tilde{z}, t)$ in $U_\sigma(z(t_*))$, and the following relation is fulfilled under the additional assumption (E1):

$$\|\hat{z}(\tilde{z}, t) - z(t)\| \leq \alpha \|\tilde{z} - z(t)\|,$$

i.e., particularly, $\hat{z}(\tilde{z}, t)$ depends continuously on $\tilde{z}, t$.

Proof:

The proof consists in an application of /24/ Theorems 4.1 and 2.1, cf. /17/ Proof of Theorem 3.2 (especially equation (3.21)) and the subsequent Remark there.

Now we provide a tool for analyzing the solvability of the subproblems occuring in Algorithm 3.1:

Lemma 3.7:

Assume (E1), (V1), (V2), (V3) and let $t_* \in T$ be arbitrarily fixed. Then there exist constants $R > 0$, $\sigma > 0$, and $\delta > 0$ which are not greater than the corresponding constants in Lemma 3.5 such that for all $\tilde{z} \in U_R(z(t_*))$, $t \in U_\delta(t_*)$ the following holds:

(i) Each of the problems $P(\tilde{z}, t)$ and $P^I(\tilde{z}, t)$, with $I^+(t_*) \subseteq I \subseteq I(t_*)$, has a unique Kuhn-Tucker point $\hat{z}(\tilde{z}, t)$ or $\hat{z}^I(\tilde{z}, t)$, respectively, in $U_\sigma(z(t_*))$.

(ii) The problems $P^I(\tilde{z}, t)$ in (i) are (globally!) uniquely solvable.

(iii) $\|\hat{z}(\tilde{z}, t) - z(t)\| \leq \frac{1}{4} \|\tilde{z} - z(t)\|$

Proof:

For $P(t_*)$ the assumptions (V1), (V2), (V3) have been supposed. By writing down the corresponding Kuhn-Tucker conditions one easily proves that, for $I^+(t_*) \subseteq I \subseteq I(t_*)$, $z(t_*)$ is a Kuhn-Tucker point of $P^I(t_*)$ and (V1), (V2), (V3) are fulfilled also for $P^I(t_*)$. Thus it follows from Lemma 3.6 that, for each I with $I^+(t_*) \subseteq I \subseteq I(t_*)$, there are constants $R^I > 0$, $\sigma^I > 0$, $\delta^I > 0$ such that, for all $\tilde{z} \in U_{R^I}(z(t_*))$, $t \in U_{\delta^I}(t_*)$, the problem $P^I(\tilde{z},t)$ has a unique Kuhn-Tucker point $\hat{z}^I(\tilde{z},t)$ in $U_{\sigma^I}(z(t_*))$. Analogous constants R^0, σ^0, δ^0 exist for $P(z,t)$; the corresponding Kuhn-Tucker point is denoted by $\hat{z}(\tilde{z},t)$.

Now let

$$\sigma := \min \left\{\sigma_{3.5}, \sigma^0; \sigma^I \,\middle|\, I^+(t_*) \subseteq I \subseteq I(t_*)\right\} .$$

(The index 3.5 denotes the corresponding constant in Lemma 3.5.)

Because of the continuity of $\hat{z}^I(.,.)$ (according to Lemma 3.6) we can restrict $R^I > 0$ and $\delta^I > 0$ so that the Kuhn-Tucker point $\hat{z}^I(\tilde{z},t)$ lies in $U_\sigma(z(t_*))$ for each $\tilde{z} \in U_{R^I}(z(t_*))$, $t \in U_{\delta^I}(t_*)$. As this Kuhn-Tucker point was unique in $U_{\sigma^I}(z(t_*))$, it is also unique in $U_\sigma(z(t_*)) \subseteq U_{\sigma^I}(z(t_*))$. The constants R^0 and δ^0 can be restricted analogously.

Setting $R := \min\left\{R_{3.5}, R^0; R^I \,\middle|\, I^+(t_*) \subseteq I \subseteq I(t_*)\right\}$

and $\delta := \min \left\{\delta_{3.5}, \delta^0; \delta^I \,\middle|\, I^+(t_*) \subseteq I \subseteq I(t_*)\right\}$

we obtain assertion (i).

For $\tilde{z} \in U_R(z(t_*))$, $t \in U_\delta(t_*)$, $I^+(t_*) \subseteq I \subseteq I(t_*)$, the problem $P^I(\tilde{z},t)$ is strongly convex by Lemma 3.5 (ii). This provides (ii).

(iii) is a direct consequence of Lemma 3.6.

Now we are going to demonstrate that the termination of Algorithm 3.1 must occur after a finite number of iterations. For this, the corresponding proofs in /29/, /4/ cannot be applied directly since the assumption of "no-degeneracy" (i.e., $x^\nu \neq x^{\nu-1}$, $\nu = 1,2,\ldots$) supposed there is not necessarily fulfilled in the case of Algorithm 3.1. First we prove:

<u>Lemma 3.8:</u>

Assume (V1), (V2), (V3) and let $t_* \in T$ be arbitrarily fixed. The constants $R > 0$ and $\delta > 0$ shall be chosen according to Lemma 3.7. Suppose that the sequence $\{x^\nu\}_{\nu=1,2,\ldots}$ has been generated by S1 - S3 of Algorithm 3.1 with $\tilde{z} \in U_R(z(t_*))$, $t \in U_\delta(t_*)$ so that

$$I^+(t_*) \subseteq I^\nu \subseteq I(t_*), \quad \nu = 1,2,\dots.$$

Then there is no $\nu_0 \in \mathbb{N}$ such that

$$x^\nu = x^{\nu-1} \text{ for all } \nu \geq \nu_0.$$

Proof:

Let us assume that the opposite is true, i.e., there is a $\nu_0 \in \mathbb{N}$ with

$$x^\nu = \bar{x}, \quad \forall \nu \geq \nu_0.$$

An infinite sequence $\{x^\nu\}$ can be generated only if both, enlarging I^ν and deleting indices from I^ν, occur infinitely often. As a consequence we obtain that the algorithm must run through the sequence of steps S2.1 ⟶ S2 ⟶ S2.2 ⟶ S3 infinitely often. Particularly, for some $\nu = \varkappa > \nu_0$, the algorithm enters this sequence of steps such that in S2.1 the set I^ν, $\nu = \varkappa$, is enlarged by an index k which has already been deleted from I^ν (in S3.1) for some $\nu = \mu < \varkappa$, $\mu > \nu_0$.

This means:

$$\nabla_x \varphi(\hat{x}^\nu, \tilde{z}, t) + \sum_{i \in I^\nu} u_i^\nu \nabla_x g_i(\tilde{x}, t) + (v^\nu)^T \nabla_x h(\tilde{x}, t) = 0^T, \quad \nu = \mu, \varkappa, \varkappa+1, \tag{3.8}$$

$$Lg_i(\hat{x}^\nu, \tilde{x}, t) = 0 \qquad \forall i \in I^\nu, \ \nu = \mu, \varkappa, \varkappa+1 \tag{3.9}$$

$$Lh_j(\hat{x}^\nu, \tilde{x}, t) = 0, \qquad j = 1, \dots, l, \ \nu = \mu, \varkappa, \varkappa+1 \tag{3.10}$$

$$\hat{x}^\nu = x^\nu = \bar{x}, \qquad \nu = \mu, \varkappa+1, \tag{3.11}$$

since S3 has been reached,

$$u_k^\mu < 0, \tag{3.12}$$

since the index k has been deleted from I^ν in S3.1 for $\nu = \mu$,

$$Lg_k(\hat{x}^\varkappa, \tilde{x}, t) > 0 \tag{3.13}$$

$$\text{and} \quad I^{\varkappa+1} = I^\varkappa \cup \{k\}, \tag{3.14}$$

since the index k becomes an element of I^ν for $\nu = \varkappa$.

Lemma 3.5 (i) supplies the linear independence of the vectors

$\{\nabla_x g_i(\tilde{x},t)\}_{i\in I(t_*)} \cup \{\nabla_x h_j(\tilde{x},t)\}_{j=1,\ldots,l}$.

Taking into account that $I^\nu \subseteq I(t_*)$, $\nu = \mu, \varkappa+1$, this yields together with (3.8) and (3.11):

$$u_i^\mu = u_i^{\varkappa+1} \quad \forall i \in I(t_*),$$

especially, $u_k^{\varkappa+1} = u_k^\mu < 0$ (3.15)

Subtracting (3.9) for $\nu=\varkappa+1$ from (3.9) for $\nu=\varkappa$ we obtain, taking into account $\hat{x}^{\varkappa+1} = \bar{x}$:

$$\nabla_x g_i(\tilde{x},t)\,(\hat{x}^\varkappa - \bar{x}) = 0 \quad \forall i \in I^\varkappa (\subseteq I^{\varkappa+1}). \tag{3.16}$$

Analogously, from (3.10),

$$\nabla_x h_j(\tilde{x},t)\,(\hat{x}^\varkappa - \bar{x}) = 0, \; j = 1,\ldots,l, \tag{3.17}$$

and from (3.13), (3.14) and (3.9):

$$\nabla_x g_k(\tilde{x},t)(\hat{x}^\varkappa - \bar{x}) > 0. \tag{3.18}$$

Substraction of (3.8) for $\nu=\varkappa+1$ from (3.8) for $\nu=\varkappa$ followed by multiplication by $(\hat{x}^\varkappa - \bar{x})$ yields, under consideration of (3.11), (3.16) and (3.17):

$$(\nabla_x \varphi(\hat{x}^\varkappa,\tilde{z},t) - \nabla_x \varphi(\bar{x},\tilde{z},t))\,(\hat{x}^\varkappa - \bar{x})$$
$$- u_k^{\varkappa+1}\,\nabla_x g_k(\tilde{x},t)\,(\hat{x}^\varkappa - \bar{x}) = 0.$$

Now it follows from (3.15) and (3.18):

$$(\nabla_x \varphi(\hat{x}^\varkappa,\tilde{z},t) - \nabla_x \varphi(\bar{x},\tilde{z},t))\,(\hat{x}^\varkappa - \bar{x}) < 0$$

And hence, from the definition of φ in (2.2):

$$(\hat{x}^\varkappa - \bar{x})^T \nabla_{xx} L(\tilde{z},t)\,(\hat{x}^\varkappa - \bar{x}) < 0.$$

But this contradicts Lemma 3.5 (ii) together with (3.16), (3.17) and $I^+(t_*) \subseteq I^\varkappa$!

<u>Lemma 3.9:</u>

Assume the suppositions of Lemma 3.8. Then it holds for $\nu \geq 1$: If $x^\nu \neq x^{\nu-1}$, then $\varphi(x^\nu,\tilde{z},t) < \varphi(x^{\nu-1},\tilde{z},t)$.

<u>Proof:</u>

It follows from the construction of the sets I^ν that $x^{\nu-1}$ is feasible for $P^{I^\nu}(\tilde{z},t)$, $\nu=1,2,\ldots$. Moreover, $\hat{x}^\nu$ is feasible for $P^{I^\nu}(\tilde{z},t)$ because of its definition in S2. Subtracting the corresponding equations we obtain

$$\nabla_x g_i(\tilde{x},t)\,(x^{\nu-1}-\hat{x}^\nu)=0 \quad \forall i\in I^\nu , \tag{3.19}$$

$$\nabla_x h_j(\tilde{x},t)\,(x^{\nu-1}-\hat{x}^\nu)=0,\ j=1,\ldots,l. \tag{3.20}$$

As x^ν is a convex combination of $x^{\nu-1}$ and $\hat{x}^\nu$ (cf. Remark 3.3) and we have assumed $x^\nu \neq x^{\nu-1}$, it follows $x^{\nu-1}\neq \hat{x}^\nu$. Thus Lemma 3.5 (ii) together with (3.19), (3.20), and the assumption $I^+(t_*)\subseteq I^\nu$ yields that the function $\varphi(.,\tilde{z},t)$ is strongly convex on $[\hat{x}^\nu,x^{\nu-1}]$. Since $(\hat{x}^\nu,u^\nu,v^\nu)$ is a Kuhn-Tucker point for $P^{I^\nu}(\tilde{z},t)$ and $x^{\nu-1}$ is feasible for this problem, we obtain

$$\varphi(x^{\nu-1},\tilde{z},t)>\varphi(\hat{x}^\nu,\tilde{z},t). \tag{3.21}$$

Further,

$\varphi(x^\nu,\tilde{z},t)\le\alpha\,\varphi(\hat{x}^\nu,\tilde{z},t)+(1-\alpha)\,\varphi(x^{\nu-1},\tilde{z},t)$

since $\varphi(.,\tilde{z},t)$ is convex on $[\hat{x}^\nu,x^{\nu-1}]$
and $\alpha\in[0,1)$

$<\varphi(x^{\nu-1},\tilde{z},t)$

because of (3.21) together with $\alpha\neq 0$
(because $x^\nu\neq x^{\nu-1}$)

<u>Lemma 3.10:</u>

Assume the suppositions of Lemma 3.8. Then there is a $\nu\ge 0$ for which step S3.2 of Algorithm 3.1 is reached, i.e., the cycle $\rightarrow$S2 $\longrightarrow$ S3 (returning to S2) is left.

<u>Proof:</u>

Let us assume that the opposite is true, i.e., an infinite sequence $\{x^\nu\}$ is generated by steps S2, S3 of Algorithm 3.1.
It follows from Lemma 3.8 and Lemma 3.9 that

$$\varphi(x^\nu,\tilde{z},t)\le\varphi(x^{\nu-1},\tilde{z},t),\quad \nu=1,2,\ldots,$$

where the relation "$<$" holds for infinitely many numbers ν . Thus it is possible to determine a subsequence $\{\nu_i\}_{i=1,2,\ldots}$ of those ν

for which step S3 is reached in such a way that

$$\varphi(x^{\nu_{i+1}},\tilde{z},t) < \varphi(x^{\nu_i},\tilde{z},t), \quad i=1,2,\ldots \tag{3.22}$$

As step S3 is reached, it holds

$$x^{\nu_i} = \hat{x}^{\nu_i}, \quad i = 1,2,\ldots \tag{3.23}$$

Obviously, among the index sets I^{ν_i}, $i=1,2,\ldots$, there are equal sets, i.e.,

$I^{\nu_j} = I^{\nu_k}$ for some j,k with $j \neq k$, which provides $\hat{x}^{\nu_j} = \hat{x}^{\nu_k}$. This contradicts (3.22), (3.23).

Lemma 3.11:

Assume (E1), (V1), (V2), (V3) and let $t_* \in T$ be arbitrarily fixed. Then there are constants $\bar{R} = \bar{R}(t_*)$ and $\bar{\delta} = \bar{\delta}(t_*)$ such that:

If $\tilde{z} \in U_{\bar{R}}(z(t))$, $\tilde{\tilde{z}} \in U_{\bar{R}}(z(\tilde{t}))$, $t,\tilde{t} \in U_{\bar{\delta}}(t_*)$, and $\tilde{I}$ is an index set with

$$J^+(\tilde{z}) \subseteq \tilde{I} \subseteq J(\tilde{z},\tilde{\tilde{z}},\tilde{t}), \tag{3.24}$$

then all subproblems $P^{I^\nu}(\tilde{z},t)$ generated by steps S1 - S3 of Algorithm 3.1 are uniquely solvable, and, after a finite number of iterations, step S3.2 is reached with a vector z and an index set I, which satisfy the following relations:

$$\|z-z(t)\| \leq \frac{1}{4}\,\|\tilde{z}-z(t)\| \tag{3.25}$$

$$J^+(z) \subseteq I \subseteq J(z,\tilde{z},t) \tag{3.26}$$

Remark: Afterwards $\tilde{\tilde{z}}$ will represent z_{k-2}, but at this stage an interpretation of $\tilde{\tilde{z}}$ is not necessary because of the logical structure of Lemma 3.11.

Proof of Lemma 3.11:

Let the constants $R>0, \sigma>0$, and $\delta>0$ be chosen according to Lemma 3.7. Suppose $\tilde{z},\tilde{\tilde{z}} \in U_R(z(t_*))$, $t,\tilde{t} \in U_\delta(t_*)$ and assume that $\tilde{I}$ satisfies (3.24).

α) First we prove that under these assumptions all subproblems $P^{I^\nu}(\tilde{z},t)$ generated by S1 - S3 of Algorithm 3.1 are uniquely solvable. Because of Lemma 3.7 it is sufficient to verify that

$$I^+(t_*) \subseteq I^\nu = I(t_*), \quad \nu = 0,1,2,\ldots\ . \tag{3.27}$$

We prove this by mathematical induction:
Lemma 3.5 (iii), (iv) (relation (3.7)) yields:

$$I^+(t_*) \subseteq J^+(\tilde{z}) \tag{3.28}$$

$$J(\tilde{z},\tilde{\tilde{z}},\tilde{t}) \subseteq I(t_*) \tag{3.29}$$

Because of $I^0 := \tilde{I}$ (in step S1) and (3.24) it follows that

$$I^+(t_*) \subseteq I^0 \subseteq I(t_*)$$

holds for the set I^0 defined in S1.

Now let us suppose that (3.27) is valid for some $\nu \geq 0$. We shall prove (3.27) for $\nu+1$:
The problem $P^{I^\nu}(\tilde{z},t)$ has a unique Kuhn-Tucker point z^ν, which lies in $U_\sigma(z(t_*))$, according to Lemma 3.7 (i), (ii).
First case: $I^{\nu+1}$ is determined in S2.1. Then $Lg_i(x^\nu,\tilde{x},t) > 0$ for the index i joining I^ν. This implies $i \in I(t_*)$ (because of Lemma 3.5 (iv) (relation (3.7)) and together with (3.27):

$$I^+(t_*) \subseteq I^{\nu+1} \subseteq I(t_*) \tag{3.30}$$

We emphasize that analogous considerations have to be applied also in the case that I^0 is altered in S2.1 (while $\nu = 0$ remains constant); they provide that the relation

$$I^+(t_*) \subseteq I^0 \subseteq I(t_*)$$

remains valid for the altered set I^0.

Second case: $I^{\nu+1}$ is determined in S3.1. Then $u_i^\nu < 0$ for the index i, which is deleted from I^ν. This implies $i \notin I^+(t_*)$ (because of Lemma 3.5 (iii)), and together with (3.27) we obtain (3.30) again.

ß) According to Lemma 3.10, after a finite number of iterations, step 3.2 is reached. The vector $z := z^\nu$ obtained there satisfies the following relations:

$$Lg_i(x,\tilde{x},t) = 0 \qquad \forall i \in I^\nu \tag{3.31}$$

$$Lg_i(x,\tilde{x},t) \leq 0 \qquad \forall i \in M \setminus I^\nu \tag{3.32}$$

$$u_i \geq 0 \qquad \forall i \in I^\nu \tag{3.33}$$

$$u_i = 0 \qquad \forall i \in M \setminus I^\nu \tag{3.34}$$

As z^ν is a Kuhn-Tucker point of $P^{I^\nu}(\tilde{z},t)$ it follows from (3.31) - (3.34) that $z = z^\nu$ is a Kuhn-Tucker point of $P(\tilde{z},t)$. Thus (3.25)

is an immediate consequence of Lemma 3.7 (iii).
For $I := I^{\nu}$ (according to S3.2) we obtain:

$I \subseteq J(z,\tilde{z},t)$ because of (3.31),

$J^{+}(z) \subseteq M \setminus (M \setminus I) = I$ because of (3.34).

This supplies (3.26).

γ) We can finish the proof defining $\bar{R} := \frac{R}{2}$ and $\bar{\delta} \in (0, \delta)$ such that $z(t) \in U_{\bar{R}}(z(t_*))$ for all $t \in U_{\bar{\delta}}(t_*)$, since this provides

$$\tilde{z} \in U_R(z(t_*)) \text{ for } \tilde{z} \in U_{\bar{R}}(z(t)),\ t \in U_{\bar{\delta}}(t_*).$$

On the basis of Lemma 3.11 we can prove the main result of this chapter:

<u>Proof of Theorem 3.4:</u>

Let $\bar{R}(t_*)$ and $\bar{\delta}(t_*)$ be the constants mentioned in Lemma 3.11. The neighbourhoods $\{U_{\bar{\delta}(t_*)/2}\}_{t_* \in T}$ form a system of open sets covering the compact set T. Hence, there exists a finite number of covering sets $U_{\bar{\delta}(t_1^*)/2}, \ldots, U_{\bar{\delta}(t_s^*)/2}$.

We define

$$r := \min\{2\rho,\ \bar{R}(t_1^*), \ldots, \bar{R}(t_s^*)\}. \tag{3.35}$$

$z(.)$ is continuous on T (cf. /17/ Lemma 3.3); hence there is a constant $\delta > 0$ such that

$$\|z(\tilde{t}) - z(t)\| < \frac{r}{2} \text{ for all } \tilde{t}, t \in T \text{ with } |\tilde{t}-t| < \delta. \tag{3.36}$$

We define

$$\Delta := \min\left\{\delta, \frac{\bar{\delta}(t_1^*)}{2}, \ldots, \frac{\bar{\delta}(t_s^*)}{2}\right\} \tag{3.37}$$

and assume $0 = t_o \leq t_1 \leq \ldots \leq t_N = 1$ to be a partition of T with

$$t_k - t_{k-1} \leq \Delta,\ k = 1, \ldots, N.$$

Suppose that the sequence $\{z_k\}_{k=0,1,\ldots,N}$ has been generated by Algorithm 3.1. Additionally, we set $z_{-1} := z(0)$. In S0 of Algorithm 3.1 we defined: $z_o := z(0)$, $I^{+}(0) \subseteq I_o \subseteq I(0)$.

Thus we obtain

$$J^{+}(z_o) = I^{+}(0) \subseteq I_o \subseteq I(0) = J(z_o, z_{-1}, t_o).$$

Moreover, we have

$$\|z_{-1}-z(t_o)\| = 0 < r$$

and, because of (3.36),

$$\|z_o-z(t_1)\| < \frac{r}{2} < r.$$

In order to apply mathematical induction we suppose for some $k \geq 1$:

$$J^+(z_{k-1}) \subseteq I_{k-1} \subseteq J(z_{k-1}, z_{k-2}, t_{k-1}) \quad (3.38)$$

and $\|z_{k-2} - z(t_{k-1})\| < r$ (3.39)

and $\|z_{k-1} - z(t_k)\| < r.$ (3.40)

For t_k there exists a t_i^* with $t_k \in U_{\bar{\delta}(t_i^*)/2}(t_i^*)$. (3.41)

From $| t_k - t_{k-1} | \leq \Delta \leq \frac{\bar{\delta}(t_i^*)}{2}$ it follows that

$$t_{k-1} \in U_{\bar{\delta}(t_i^*)}(t_i^*). \quad (3.42)$$

(3.35) provides $r \leq \bar{R}(t_i^*)$. (3.43)

Because of (3.38) - (3.43), Lemma 3.11 is applicable (with $t_* \hat{=} t_i^*$, $\tilde{z} \hat{=} z_{k-1}$, $\tilde{\tilde{z}} \hat{=} z_{k-2}$, $t \hat{=} t_k$, $\tilde{t} \hat{=} t_{k-1}$). It states that all subproblems occurring in S1 - S3 are uniquely solvable and, moreover,

$$J^+(z_k) \subseteq I_k \subseteq J(z_k, z_{k-1}, t_k) \quad (3.44)$$

and $\| z_k-z(t_k)\| \leq \frac{1}{4} \|z_{k-1} - z(t_k)\| \quad (< \frac{r}{2})$. (3.45)

According to $| t_{k+1} - t_k | \leq \Delta \leq \delta$, we obtain from (3.45) together with (3.36):

$$\|z_k - z(t_{k+1})\| < r. \quad (3.46)$$

The relations (3.44), (3.40), and (3.46) allow us to proceed with mathematical induction until $k = N$.

The inequality $\|z_k - z(t_k)\| < \rho$, $k=1,\ldots,N$, follows from (3.45) together with (3.35).

4. Stepsize Control

In Theorem 3.4 the existence of a suitable partition of T, $0 = t_o \leq t_1 \leq \ldots \leq t_N = 1$, was proved. Now we want to introduce an algorithm for determining such a partition. Corresponding algorithms for the case of nonlinear equations instead of optimization problems can be found, for instance, in /2/, /3/, /5/, /11/, /25/. They cannot be applied directly because of special effects occurring for parameter-dependend optimization problems, especially the change of active constraints and the non-differentiability of z(.) at critical points. The procedure presented here is based on suggestions of C. RICHTER (private communication, cf. /22/). Though the application in a more general framework is possible, we introduce the algorithm in a form adapted to Algorithm 3.1.

The basic ideas are the following:

We set $t_{k+1} > t_k$ if the vector computed last, z_k, can be accepted as an approximate solution of $P(t_k)$, i.e., we would like to test whether

$$\|z_k - z(t_k)\| \leq \varepsilon$$

holds for some previously fixed $\varepsilon > 0$.

Since $z(t_k)$ is unknown, we have to modify this criterion. Instead of $z_k - z(t_k)$ we shall apply the difference of the two successive iterates just computed, i.e., we check the relation

$$\|z_k - z_{k-1}\| \leq \varepsilon \,. \tag{4.1}$$

If this is true, the corresponding value of k is stored: $\bar{k} := k$. Otherwise we test whether for fixed t, i.e., $t_{k+1} := t_k$, the norm difference of two successive iterates can be reduced, more precisely:

$$\|z_{k+1} - z_k\| \leq \frac{1}{2} \|z_k - z_{k-1}\| \tag{4.2}$$

(That means we test whether the sequence $\{z_k\}$ "seems to converge". The factor $\frac{1}{2}$ could be replaced by any $\beta \in (0,1)$, cf. /18/). As we shall formulate the algorithm so that the vector computed last is z_k, (4.2) will be replaced by

$$\|z_k - z_{k-1}\| \le \frac{1}{2} \|z_{k-1} - z_{k-2}\| \tag{4.3}$$

preceded by the test if t_k was equal to t_{k-1}.
If (4.3) holds for $k = k', k'+1, k'+2, \ldots$, then (4.1) will be fulfilled for some $k \in \mathbb{N}$. But if (4.3) is violated for some $k \in \mathbb{N}$, then we shall reject t_k and test whether the vector accepted last, $z_{\bar{k}}$, seems to be an appropriate starting point for solving $P(t_{\bar{k}} + \tau)$, $\tau = \frac{1}{2}(t_k - t_{\bar{k}})$. As a criterion for this we apply (4.3) again. If, after bisecting $t_k - t_{\bar{k}}$ several times, we obtain $t_k - t_{\bar{k}} < \tau_{min}$ (with a previously fixed bound τ_{min}), then we shall suppose that $z_{\bar{k}}$ has not been a good approximation for $z(t_{\bar{k}})$. In this case we repeat the whole imbedding procedure from the beginning with a higher accuracy ($\varepsilon := \varepsilon/2$, $\tau_{min} := \tau_{min}/2$). Thus we obtain the following algorithm:

<u>Algorithm 4.1:</u>

Suppose that values $\varepsilon > 0$ and $0 < \tau_{min} \le \tau_{max} \le 1$ are given.

<u>S0:</u> $z_o := z(0)$, $t_o := 0$, $k := 0$.
Determine I_o with $I^+(0) \subseteq I_o \subseteq I(0)$ (for the application of Algorithm 3.1).

<u>S1:</u> Choose $\tau_k \in [\tau_{min}, \tau_{max}]$.
If $1 - t_k \le \tau_{max}$, then $\tau_k := 1 - t_k$.

<u>S2:</u> $t_{k+1} := t_k + \tau_k$, $k := k+1$.

<u>S3:</u> Determine the Kuhn-Tucker point z_k of $P(z_{k-1}, t_k)$ (by Algorithm 3.1, steps S1 - S3). If unsolvable subproblems occur, then go to S6.

<u>S4:</u> If $\|z_k - z_{k-1}\| \le \varepsilon$ then:
If $t_k = 1$, then stop.
If $t_k < 1$, then $\bar{k} := k$, go to S1.

<u>S5:</u> If $t_{k-1} < t_k$, then $\tau_k := 0$, go to S2.
If $\|z_k - z_{k-1}\| \le \frac{1}{2} \|z_{k-1} - z_{k-2}\|$,
then $\tau_k := 0$, go to S2.

<u>S6:</u> $\tau_k := \tau_{\bar{k}}/2$.

If $\tau_{\bar{k}} \geq \tau_{min}$, then $k := \bar{k}$, go to S2.

S7: $\varepsilon := \varepsilon/2$, $\tau_{min} := \tau_{min}/2$, $\tau_{max} := \tau_{max}/2$, go to S0.

The following theorem states that Algorithm 4.1 is well-defined:

Theorem 4.2:

Assume (E1), (E2), (V1), (V2), (V3) for the optimization problems P(t).
$\varepsilon > 0$ and $0 < \tau_{min} \leq \tau_{max} = 1$ shall be arbitrarily fixed.
Then Algorithm 4.1 terminates after a finite number of iterations with

$$t_k = 1,$$
$$\|z_k - z_{k-1}\| \leq \varepsilon, \text{ for some } k \in N.$$

Proof:

α) Let Δ be the constant defined in (3.37) in the proof of Theorem 3.4; for this, the value $\rho > 0$ (influencing Δ by (3.35), (3.36)) can be chosen arbitrarily. First we prove: If $\tau_{max} \leq \Delta$, then Algorithm 4.1 terminates in the manner desired:
Because of $\tau_{max} \leq \Delta$, we have for the sequence $\{t_k\}$ generated by Algorithm 4.1 (it might be finite or not):

$$t_k - t_{k-1} \leq \Delta \quad , \; k = 1,2,\ldots$$

Thus, the proof of Theorem 3.4 yields (cf. (3.45)):

$$\|z_k - z(t_k)\| \leq \frac{1}{4}\|z_{k-1} - z(t_k)\|, \; k=1,2,\ldots \tag{4.4}$$

It follows:

$$\begin{aligned} \|z_k - z_{k-1}\| &\leq \|z_k - z(t_k)\| + \|z_{k-1} - z(t_k)\| \\ &\leq \frac{5}{4} \|z_{k-1} - z(t_k)\| \end{aligned} \tag{4.5}$$

and

$$\begin{aligned} \|z_{k-1} - z(t_k)\| &\leq \|z_{k-1} - z_k\| + \|z_k - z(t_k)\| \\ &\leq \|z_{k-1} - z_k\| + \frac{1}{4}\|z_{k-1} - z(t_k)\|, \end{aligned}$$

which yields

$$\|z_{k-1} - z(t_k)\| \le \frac{4}{3} \|z_k - z_{k-1}\|. \tag{4.6}$$

In the case of $t_k = t_{k-1}$ we obtain from (4.5), (4.4), and (4.6):

$$\begin{aligned}\|z_k - z_{k-1}\| &\le \frac{5}{4} \|z_{k-1} - z(t_k)\| \\ &\le \frac{5}{4} \cdot \frac{1}{4} \|z_{k-2} - z(t_{k-1})\| \\ &\le \frac{5}{4} \cdot \frac{1}{4} \cdot \frac{4}{3} \|z_{k-1} - z_{k-2}\| \\ &\le \frac{1}{2} \|z_{k-1} - z_{k-2}\|\end{aligned}$$

This means: For $t_{k-1} = t_k$ as well as for $t_{k-1} < t_k$, in S5 the algorithm decides for going to S2. Thus S6 and S7 are not reached, and therefore ε, τ_{min}, and τ_{max} remain unchanged. Moreover, after passing through the loop S2 ⟶ S3 ⟶ S4 ⟶ S5 ⟶ S2 several times (thus $\|z_k - z_{k-1}\| \le \frac{1}{2} \|z_{k-1} - z_{k-2}\|$ for $k = k', k'+1, k'+2, \dots$ with some k') it holds that $\|z_k - z_{k-1}\| \le \varepsilon$ for some k.
Consequently S1 is reached and t_k is increased at least by τ_{min}. Since τ_{min} remains unchanged, $t_k = 1$ must be obtained once. The arguments just applied yield that then , after a finite number of iterations, the algorithm terminates in S4 with $\|z_{k'} - z_{k'-1}\| \le \varepsilon$ for some k'.

β) The algorithm has the following structure:

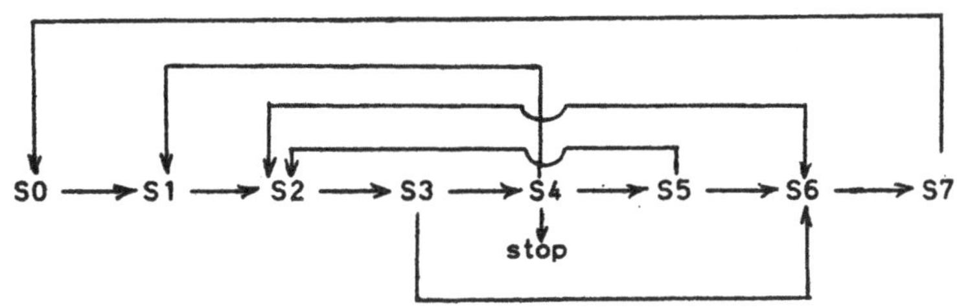

If S7 is not reached, then it holds $\tau_k \ge \tau_{min}$ = const. This implies that the steps S1 and S6 can be passed through only finitely many times. The only remaining loop after deleting S1 and S6 from the algorithm is S2 ⟶ S3 ⟶ S4 ⟶ S5 ⟶ S2, which must be left after a finite number of iteration according to the arguments applied in α).

Thus the following situation arises: The algorithm either terminates after a finite number of iterations (in S4) or, after several iterations, always reaches S7. In the latter case $\tau_{max} \leq \Delta$ is once obtained, by which the assertion follows from α).

Remark 4.3:

α) The norm-difference of Kuhn-Tucker points in S4, S5 could be replaced by the norm-difference of primal variables, i.e., $\|x_k - x_{k-1}\|$ or $\|x_{k-1} - x_{k-2}\|$, respectively (cf. /18/). This would be desirable if one was only interested in information about the primal variables or if a quasi-NEWTON variant of WILSON's method was applied.

β) The factors $\frac{1}{2}$ in steps S5, S6, S7 can be replaced by any numbers in the interval (0,1).

γ) As one can easily see from the proof of Theorem 4.2, the Algorithm 4.1 would work correctly also without the reduction of ε in S7. However, in general, this would not be efficient since ε controls the accuracy of the approximations z_k for $z(t_k)$: If ε was chosen too large and kept constant after this, then the (numerically) necessary accuracy would be achieved by multiple reduction of τ_{max}, which would produce uniformly small steps over the whole parameter interval T.

δ) We cannot give up the reduction of τ_{max} in S7 since it yields the possibility of leaving a solution branch $\{(z^*(t),t) \mid t \in T_1 \subset T\}$ which does not reach t = 1. For illustration see Fig. 4.1.

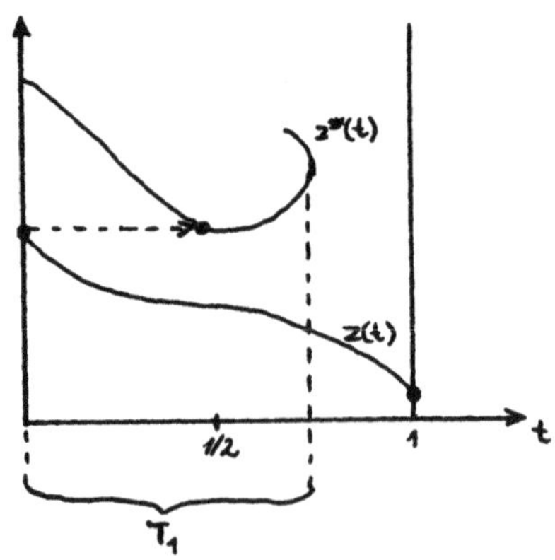

Fig. 4.1:
z(t) and $z^*(t)$ denote Kuhn-Tucker points of P(t)

We assume $\tau_o = \tau_{max} = \frac{1}{2}$. Independently of ε and $\tau_{min} \leq \frac{1}{2}$, the algorithm accepts $z(0) = z^*(\frac{1}{2})$ as a starting point for solving $P(\frac{1}{2})$ and obtains $z_1 = z^*(\frac{1}{2})$. After that, even a reduction of ε and τ_{min} cannot prevent that the curve $\{(z^*(t),t) | (t \in T_1\}$ is tracked, which makes a successful termination at t = 1 impossible.

ε) If Algorithm 4.1. is applied in a continuation method for globalizing a locally convergent algorithm (i.e., we are interested only in a solution of P(1)), then two parameters for controlling the accuracy can be used: ε_1 for $t_k \in [0,1)$ and ε_2 for $t_k = 1$, where ε_1 is controlled as ε in Algorithm 4.1.

φ) In S1 one can choose $\tau_k \in [\tau_{min}, \tau_{max}]$ arbitrarily. For instance, it is possible to assume that a suitable value of τ_k is of the same magnitude as τ_{k-1}. Since, in general, it would not be good to prevent an increase of the step size, one could define $\tau_k := \min\{2\cdot\tau_{k-1}, \tau_{max}\}$ (cf. /3/).

Another strategy could consist in starting with a large interval $[\tau_{min}, \tau_{max}]$ and determining $\tau_k \in [\tau_{min}, \tau_{max}]$ by a parameter-control procedure known for continuation methods for nonlinear equations (cf. /11/, /5/, /2/). In this case Algorithm 4.1 would only "supervize" the work of the other algorithm.

η) The proof of Theorem 4.2 provides an estimate of the accuracy of Algorithm 4.1: From (4.4), (4.6) it follows: If $\tau_{max} \leq \Delta$, then

$$\|z_k - z(t_k)\| \leq \frac{1}{3} \|z_k - z_{k-1}\|. \tag{4.7}$$

Of course, (4.7) is predominantly of theoretical value since, in general, Δ is not known explicitly.

5. Numerical results

Basing on Algorithms 3.1 and 4.1, a computer program for solving a one-parametric optimization problem was implemented (second derivatives occurring in WILSON's method were approximated by finite differences). As we have already mentioned in Chapter 3, the numerical effort has been reduced to solving only systems of l i n e a r equations.

Now we present the results of numerical tests performed with problems from /7/ and physical examples:

Example 5.1: (/7/ Example 1)

$$(P(t))\ \min \left\{ x_1^2 + (4t-2)x_2 + \frac{1}{2} x_2^2 \mid \begin{array}{l} g_1(x,t) \equiv x_1^2 + x_2^2 - 1 \leq 0 \\ g_2(x,t) \equiv x_1 - x_2 - t - \frac{1}{4} \leq 0 \end{array} \right\}$$

Exact solution for $t = 1$: $x^* = (0,-1)$.

This example demonstrates an interesting behaviour of the set of active constraints:

$$I(t) = \begin{cases} \{1\}, & t \in [0, 0.25], \\ \emptyset & t \in (0.25, 0.75), \\ \{1,2\}, & t = 0.75, \\ \{1\}, & t \in (0.75, 1]. \end{cases}$$

Table 5.1 presents the results of the numerical algorithm. We choose $\varepsilon = 10^{-6}$, $\tau_{min} = 10^{-6}$, $\tau_{max} = 0.2$ for Algorithm 4.1. ACT.CON. denotes the set of "active constraints" determined by the algorithm, i.e., I_k in Algorithm 3.1; LIN.EQU. is the number of systems of linear equations solved (it is 1 for $k = 0$ because the Lagrange multipliers u(0) were computed by the algorithm).

k	t_k	x_1	x_2	ACT.CON.	LIN.EQU.
0	0.0	0.0	1.00000000	1	1
1	0.2	0.0	1.00000000	1	1
2	0.4	0.0	0.50000010	-	2
3	0.4	0.0	0.40000010	-	1
4	0.4	0.0	0.40000010	-	1
5	0.6	0.0	-0.39999986	-	1
6	0.6	0.0	-0.39999986	-	1
7	0.8	-0.04999995	-1.09999990	2	3
8	0.8	0.00441176	-1.00588234	1	3
9	0.8	0.00000000	-1.00002708	1	1
10	0.8	-0.00000000	-1.00000000	1	1
11	0.8	0.00000000	-1.00000000	1	1
12	1.0	0.00000000	-1.00000000	1	1

Table 5.1

Notice that for solving one quadratic subproblem only o n e system of linear equations has to be solved, in general; solely for a changing set of active constraints the number of systems required goes up to 2 - 3.

Example 5.2: (/7/ Example 4)

$$(P(t)) \quad \min\left\{\frac{1}{2}\sum_{i=1}^{5} x_i^2 \,\middle|\, h_1(x,t) \equiv x_1 + 4t - 4 = 0,\right.$$

$$g_i(x,t) \equiv -(x_1-2)^2 - \sum_{\substack{j=2\\ j\neq i+1}}^{5} x_j^2 - (x_{i+1}-0.2)^2 + 1 \leq 0, \quad i = 1,2,3,$$

$$\left. g_4(x,t) \equiv -(x_1-2-0.4\sqrt{6})^2 + \sum_{j=2}^{5} x_j \leq 0\right\}.$$

Exact solution for $t = 1$: $x^* = (0,0,0,0,0)$

Numerical solution for $t = 1$: $x_i = 0.00000000$, $i = 1,\ldots,5$.

Numerical effort: Solution of 81 systems of linear equations.

In this example the set of active (inequality) constraints is given by

$$I(t) = \begin{cases} \emptyset, & t \in [0,t^*), \\ \{1,2,3,4\}, & t = t^*, \\ \{1,2,3\}, & t \in (t^*,t^{**}], \\ \emptyset & t \in (t^{**},1], \end{cases}$$

with $t^* := 0.2550511$, $t^{**} := 0.7449489$.

For determining the active constraints $\{1,2,3\}$ for $t > t^*$ the algorithm solves 2 systems of linear equations; the constraint 4, which is active only at t^*, is ignored. The determination of $I(t) = \emptyset$ for $t > t^{**}$ requires the solution of 4 systems of linear equations; however, this happens four times, because a step size reduction is necessary in the neighbourhood of the critical point t^{**}, which results in a multiple crossing over this point (cf. Algorithm 4.1).

Note that no problems arise from the change of several active constraints at one critical point, i.e., in a case in which assumptions of investigations like /16/, for instance, are not satisfied.

Example 5.3:

A quadratic column, which can reflect light, rotates around its vertical axis. A source S of light and a receiver R are situated as shown in Fig. 5.2 (at the same vertical coordinate). Find the course of a light beam from S to R which is reflected on its way in dependence on the position of the column.

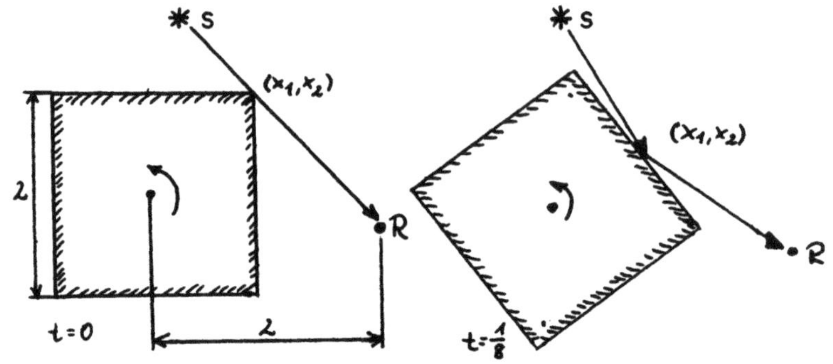

Fig. 5.2: View from above

This example can be formulated as a parametric optimization problem by means of FERMAT's principle. The latter states that a light beam "chooses" such a way between two points S and R that the time for passing it is extremal, i.e., minimal in our example. An advantage of such a formulation of the problem consists in the fact that the face of the column which reflects the light beam is automatically determined by the algorithm (active-constraints strategy!).

Mathematical formulation:

$$(P(t)) \min \left\{ \sqrt{(x_1-2)^2 + x_2^2} + \sqrt{x_1^2 + (x_2 - 2)^2} \;\middle|\; x_1 \cdot \cos(2t\pi + \tfrac{i}{2}\pi) + x_2 \cdot \sin(2t\pi + \tfrac{i}{2}\pi) \leq 1, \quad i = 0,1,2,3 \right\}$$

In order to solve this problem for 64 values of t ($\tau_{max} = \frac{1}{64}$), the algorithm required the solution of 187 systems of linear equations.

Example 5.4:

A chain consisting of 6 links (of length 1 cm) is hung up between two points which are situated in a distance of 4 cm from each other at a height h over a (horizontal) table (cf. Fig. 5.3). Determine the shape of the chain for h = 1, 1.1, 1.2, ..., 3 cm.

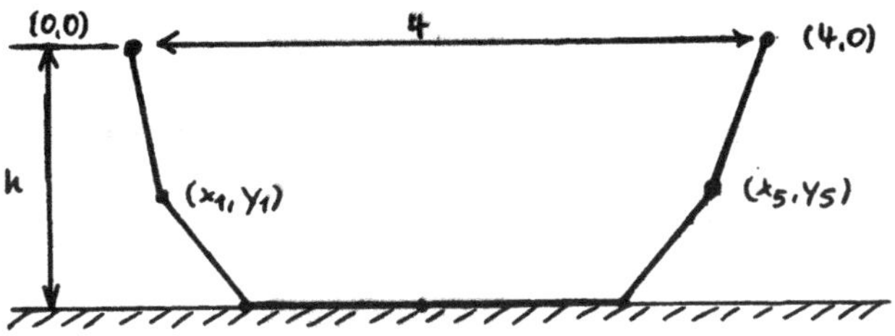

Fig. 5.3

For the formulation of this example as an optimization problem we take advantage of the fact that the chain aims at a state with minimal potential energy.

Mathematical formulation:

$$(P(t)) \quad \min \{ y_1 + y_2 + y_3 + y_4 + y_5 \mid$$
$$x_1^2 + y_1^2 = 1$$
$$(x_{i+1} - x_i)^2 + (y_{i+1} - y_i)^2 = 1, \quad i = 1,2,3,4,$$
$$(4 - x_5)^2 + y_5^2 = 1,$$
$$y_i \geq -1 - 2t \; (= -h), \quad i = 2,3,4 \}$$

In order to solve this problem for 20 values of t ($\tau_{max} = \frac{1}{20}$) the algorithm required the solution of 58 systems of linear equations, i.e., again about 3 per nonlinear optimization problem.
Note that this example illustrates the concept of active constraints very well: A constraint is active iff the corresponding hinge of the chain touches the surface of the table.

The test results presented show that the proposed active-constraints strategy works surely and with low numerical effort. The method introduced for controlling the parameter t also operates reliablely, but it is likely to become still more efficient by further sophistications. Additional test results can be found in /18/.

Acknowledgement

This work was developed during my research study at the Humboldt-University in Berlin, which included a one-year stay at the Technical University of Dresden.
I want to thank Prof. J. GUDDAT, Berlin, and Prof. C. RICHTER, formerly Dresden, for valuable discussions about the subject presented.

References:

/1/ Bank, B.; Guddat, J.; Klatte, D.; Kummer, B.; Tammer, K.: Non-Linear Parametric Optimization. - Berlin: Akademie-Verlag, 1982

/2/ Den Heijer, C.; Rheinboldt, W.C.: On steplength algorithms for a class of Continuation methods. - SIAM J. on Numerical Analysis 18(1981)5, 925-948

/3/ Eder, G.; Zulehner, W.: Ein Algorithmus zur Berechnung von Lösungen nichtlinearer Gleichungssysteme. - Univ. Linz, Mathem. Inst., Institutsbericht, 1980

/4/ Fletcher, R.: A general quadratic programming algorithm. - J. Inst. Math. Appl. 7(1971), 1-20

/5/ Georg, L.: Zur numerischen Realisierung von Kontinuitätsmethoden mit Prädiktor-Korrektor - oder simplizialen Verfahren. - Univ. Bonn, Sonderforschungsbereich 72, Approximation und Optimierung, 1982, Preprint Nr. 526

/6/ Gfrerer, H.; Guddat, J.; Wacker, Hj., Zulehner, W.: Path-following methods for Kuhn-Tucker Curves by an active index set strategy. In Bagchi, A.; Jongen, H.T. (eds.): System and Optimization, Proceedings of the Twente Workshop, Lecture Notes in Control and Information Sciences 66, Springer Verlang Berlin-Heidelberg-New York-Tokyo 1985, 111-132.

/7/ Gfrerer, H.; Guddat, J.; Wacker, Hj.: A globally convergent algorithm based on imbedding and parametric optimization. - Univ. Linz, Mathem. Inst., Institutsbericht Nr. 209; Computing 30(1983)3, 225-252

/8/ Guddat, J.; Wacker, Hj.; Zulehner, W.: On imbedding and parametric optimization - a concept of a globally convergent algorithm for nonlinear optimization problems. - Univ. Linz, Mathem. Inst., Institutsbericht Nr. 200, 1981, Math. Programming Study (1984)21, 79-96

/9/ Guddat, J.; Guerra, F.; Tammer, K.; Wendler, K.: Multiobjective and Stochastic Optimization Based on Parametric Optimization. - Berlin: Akademie-Verlag, 1985

/10/ Hackl, J.: Solution of optimization problems with nonlinear restrictions via continuation methods. In: Continuation Methods / Ed. Wacker, Hj. - New York: Academic Press, 1977, 95-127

/11/ Hackl, J.; Wacker, Hj,; Zulehner, W.: An efficient stepsize control for continuation methods - BIT 20(1980)4, 475-485

/12/ Jongen, H.T.; Jonker, P.; Twilt, F.: On deformation in optimization. Methods of Operations Research 37(1980), 171-184.

/13/ Jongen, H.T.; Jonker, P.; Twilt, F.: On one-parameter families of optimization problems: Equality constraints. JOTA 48, 141-161(1986).

/14/ Jongen, H.T.; Jonker, P.; Twilt, F.: Critical sets in parametric optimization. Math. Programming 35, 1-21 (1986)

/15/ Kleinmichel, H.; Richter, C.; Schönefeld, K.: On the global and superlinear convergence of a discretized version of Wilson's method. Computing 29(1982)4, 289-307

/16/ Kojima, M.; Hirabayashi, R.: Continuous deformations of nonlinear programs. - Tokyo Inst. of Technology, Dept. of Information Science, Techn. Rep. 101, 1981

/17/ Lehmann, R.: On the numerical feasibility of continuation methods for nonlinear programming problems. - Math. Operationsforsch. u. Statistik, Ser. Optim. 15(1984)4, 517-530

/18/ Lehmann, R.: Zu Einbettungsverfahren in der nichtlinearen Optimierung. - 1984. - Humboldt-Universität zu Berlin, Sekt. Math., Diss. A

/19/ Lommatzsch, K. (ed.): Anwendungen der linearen parametrischen Optimierung. - Berlin: Akademie-Verlag, 1979

/20/ Modler, K.-H.; Richter. C.: Einbettungsverfahren der nichtlinearen Optimierung und ihre Anwendung zur Lösung parameterabhängiger Aufgaben in der Mechanik. - 27. Intern. Wiss. Koll., TH Ilmenau, 1982

/21/ Richter, C.: Über die numerische Behandlung von Einbettungsverfahren in der nichtlinearen Optimierung. TU Dresden, Preprint, 1980

/22/ Richter, C.: Ein implementierbares Einbettungsverfahren der nichtlinearen Optimierung. Math. Operationsforsch. u. Statist., Ser. Optimization 15(1984)4, 545-553

/23/ Robinson, S.M.: A quadratically convergent algorithm for general nonlinear programming problems. - Math. Programming 3(1972), 145-156

/24/ Robinson, S.M.: Strongly regular generalized equations. Math. of Operations Research 5(1980), 43-62

/25/ Schwetlick, H.: Numerische Lösung nichtlinearer Gleichungen. - Berlin, Deutscher Verlag der Wissenschaften, 1979

/26/ Spelucci, P.: Han's method without solving QP. - In: Optimization and Optimal Control. - Proc. of a Conf. held at Oberwolfach, March 16-22, 1980/eds. Auslender, A.; Oettli, W., Stoer, J. - Springer-Verlag, 1981, 123-141 (Lecture Notes in Control and Information Sciences; 30)

/27/ Tammer, K.: Beiträge zur Theorie der parametrischen Optimierung, zu den mathematischen Grundlagen ihrer Anwendung und zu Lösungsverfahren. - 1979. - Humboldt-Universität zu Berlin, Sekt. Math., Diss. B

/28/ Wilson, R.B.: A simplicial method for concave programming. - 1963. Cambridge, Harvard Univ., Ph. D. Diss.

/29/ Zangwill, W.J.: Nonlinear Programming: A Unified Approach. - New Jersey: Prentice-Hall, Inc., Englewood Cliffs, 1969.

SMOOTH HOMOTOPIES FOR MATHEMATICAL PROGRAMMING

Pavol Meravý[1)]

Abstract. In the most papers where homotopy approaches for solving mathematical programming problems are described one imbeds the original problem into a one-parametric family of mathematical programming problems and follows the curve of Kuhn-Tucker points of this one-parametric family of problems. This contribution is based on an alternative homotopy approach, which consists in solving necessary optimality conditions using a homotopy approach. More precisely, the methods of one-parametric transformation functions (barrier and penalty) are modified and obtained homotopy methods for finding Kuhn-Tucker points are discussed.

1. Introduction

In this contribution we give a short characteristic of two main approaches in the application of homotopy methods to solve nonlinear mathematical programming problems and describe our approach in detail.

Homotopy methods for solving equations are closely related to the famous fixed point theorem by Brouwer 1912 [4]. First, this approach was considered as a theoretical and not a constructive tool for finding fixed points (zeros) of maps. Several decades passed untill some constructive elements of this approach were used to develop homotopy algorithms, e.g. [8, 9, 16, 20, 21, 27, 29] . Since then the research in homotopy methods became very intensive (e.g. see surveys [3, 10, 11, 22, 30, 31]). They have been widely used in recent years on a variety of problems: solving systems of nonlinear equations, complementarity problems, boundary value problems, mathematical programming problems, finding branches at a bifurcation point and finding all isolated solutions of systems of polynomial equations.

Before we go to the application of homotopy methods in nonlinear programming (Sec.2) we briefly characterise the main idea of a homotopy method (HM):

Suppose, we have to find a solution of the (nonlinear) system of equations

$$f(x) = 0 \qquad (1),$$

where $f\colon R^n \longrightarrow R^n$ is a continuous map. HM consists in the imbedding of the original system (1) into a one-parametric family of systems

1) Institute of Applied Mathematics and Computing, Comenius University, Mlynská dolina - mat. pavilon, 842 15 Bratislava, Czechoslovakia

$$h(x,t) = 0 \qquad 0 \leq t \leq 1 \qquad (2)$$

(where $h: R^n \times I \longrightarrow R^n$ we call the homotopy map and I denotes the interval $[0,1]$ of the homotopy variable t) such that

$$h(x,1) \equiv f(x) \qquad (3)$$

(i.e. the "terminal" system in the family (2) for $t=1$ is the original one) and the solution x_0 of the "initial" system of (2) for $t=0$ is known

$$h(x_0,0) = 0 \qquad (4).$$

Having a homotopy h satisfying (3), (4) the solution $\bar{x}$ of (1) is found by following the homotopy path $(x(s),t(s))$ for $s \in J=[0,\bar{s}]$ (where $h(x(s),t(s))=0$ for each $s \in J$, i.e. the homotopy path lies in the zero-set $h^{-1}(0)$) from the known initial point $(x(0),t(0))=(x_0,0)$ to the terminal point in the hyperplane $t=1$ $(x(\bar{s}),t(\bar{s}))=(\bar{x},1)$.

If such terminal point exists on the path then we shall obtain in this way a solution $\bar{x}$ of (1). However, there are various reasons why a homotopy path may not reach the hyperplane $t=1$ (turning back to the hyperplane $t=0$ or diverging to infinity at some $\bar{t} \in I$ (i.e. $\|x(s)\| \longrightarrow \infty$, $t(s) \longrightarrow \bar{t}$ for $s \longrightarrow \bar{s}$)). Therefore, the main problem when constructing a homotopy method is to guarantee the intersection of a homotopy path (beginning at the known solution x_0) with the hyperplane $t=1$.

Moreover, from the algorithmical point of view it is desirable to construct a homotopy map h in such a way, that the homotopy path to be followed would have certain good properties which will enable us to follow it efficiently. In practice this means that the homotopy path is a simple nonbifurcating curve, piecewise linear or (piecewise) continuously differentiable.

From the above summary of the main idea it is clear, that the standard notion of convergence of a (usually iterative) numerical method should be adapted for homotopy methods. The key-property to be proved for the convergence of HM is that the homotopy path starting at the known point in the hyperplane $t=0$ has nonempty intersection with the terminal hyperplane $t=1$. Or in slightly modified formulation (as used e.g. in Sec.3) it may suffice to prove

$$\emptyset \neq \partial(h^{-1}(0) \cap \{(x,t) \mid t<1\}) \subset (f^{-1}(0) \times \{1\}) \qquad (5),$$

where $\partial A = \text{cl}\, A \setminus A$ denotes the set of all limit points (outside A) of A .

We note, that unlike other (usually iterative) numerical methods, the notion of the "speed" of convergence is not clear for homotopy methods. (None of the usually used approaches to measure the speed of convergence, see e.g. [28], seems to be suitable for HM.) The efficiency of the path-following in HM depends not only on the lenght but also on the qualitative properties of the homotopy path (existence of turning points, bifurcation points, monotonicity in t). These properties influence the complexity and robustness of the path-following algorithm used and hence its speed (efficiency).

HM are applied in concrete computations mostly due to their usually better global convergence properties (convergence under bad initial approximation). The main disadvantage of HM is their low efficiency in comparison with other standard, usually iterative numerical methods, which, however, are only effective in a neighbourhood of a solution. But there are also problems for which no other useful numerical method than HM could be constructed (e.g. finding all isolated solutions of a system of polynomial equations [5]).

2. Two homotopy approaches in mathematical programming

Let us suppose the nonlinear programming problem (NP-problem)

$$\text{Min} \left\{ f(x) \;\middle|\; \begin{array}{ll} g_i(x) = 0 & i=1,\dots,r \\ g_j(x) \geq 0 & j=r+1,\dots,m \end{array} \right\} \qquad \text{(NP)},$$

where functions f, g_i $(i=1,\dots,m)$ are in general nonlinear and three times continuously differentiable on R^n. (In the sequel C^k will mean "k-times continuously differentiable".)

Under certain regularity assumptions (see e.g. [15]) there are wellknown Kuhn-Tucker necessary optimality conditions (KT-conditions) for a solution x of the NP-problem: there exist Lagrangian multipliers u_i $(i=1,\dots,m)$ such that

$$\nabla f(x) - \sum_{i=1}^{r} u_i \nabla g_i(x) - \sum_{j=r+1}^{m} u_j \nabla g_j(x) = 0 \qquad (6)$$

$$g_i(x) = 0 \qquad i=1,\dots,r \qquad (7)$$

$$\left.\begin{array}{rl} g_j(x) \geq 0 & (8) \\ u_j\, g_j(x) = 0 & (9) \\ u_j \geq 0 & (10). \end{array}\right\} \quad j=r+1,\dots,m$$

One way to apply the homotopy approach to NP-problems is to reduce the NP-problem to a suitable associated system of equations. Then the associated system can be solved using a homotopy approach as explained above. Due to the correspondence between the original problem and the associated system we can then make some statements about the solution of the original problem.

The KT-conditions (6-10) can not be used in general as an associated system to (NP), because it includes in the case $r<m$ also inequalities. There are several options for an associated system of equations (ASE) based on KT-conditions, e.g.:

ASE 1 - transform the original NP-problem into a NP-problem with equality constraints only (e.g. replacing $g_j(x) \geq 0$ by $g_j(x) - y_j^2 = 0$). The KT-conditions to the transformed problem do not include any inequalities.

ASE 2 - reduction of KT-conditions (which have continuously differentiable left-hand sides with respect to x, u) to a piecewise continuously differentiable system (Kojima [19])

$$\nabla f(x) - \sum_{i=1}^{r} u_i \nabla g_i(x) - \sum_{j=r+1}^{m} y_j^+ \nabla g_j(x) = 0 \qquad (11)$$

$$g_i(x) = 0 \qquad i=1,\dots,r \qquad (12)$$

$$y_j^- + g_j(x) = 0 \qquad j=r+1,\dots,m \qquad (13)$$

(i.e. the inequality-multipliers are given by $u_j = y_j^+$), where $y_j^+ = \max(0,y_j)$ and $y_j^- = \min(0,y_j)$. This approach is characterised by the loss of continuity o the differential of the functions defining the associated system (11-13).

ASE 3 - reduction of KT-conditions due to Mangasarian [24]

$$\nabla f(x) - \sum_{i=1}^{r} u_i \nabla g_i(x) - \sum_{j=r+1}^{m} \phi(y_j - \mathcal{L} g_j(x))^+ \nabla g_j(x) = 0 \quad (14)$$

$$g_i(x) = 0 \qquad i=1,\dots,r \qquad (15)$$

$$\theta(|y_j - \mathcal{L} g_j(x)|) - \theta(\mathcal{L} g_j(x)) - \theta(y_j) = 0 \qquad j=r+1,\dots,m \qquad (16)$$

(i.e. the inequality-multipliers are given by $u_j = \phi(y_j)$), where $\mathcal{L} > 0$, the functions ϕ, θ : R $\longrightarrow$ R are strictly increasing, satisfy $\phi(0) = \theta(0) = 0$ and ϕ is surjective (i.e. ϕ maps R onto R).

Possible choices of ϕ and θ are:

a) $2\mathcal{L}\,\theta(z) = \phi(z) = z$ (this choice leads to the Rockafellar's augmented Lagrangian which is not everywhere continuously differentiable),

b) another choice from [24] leading to C^1 left-hand sides of (14-16)

$$\theta(z) = \frac{1}{\mathcal{L}}\phi(z) = \begin{cases} \dfrac{z}{\mathcal{L}} + \dfrac{1}{2\mathcal{L}^2} & \text{if} \quad z \leq -\dfrac{1}{\mathcal{L}} \\ \dfrac{1}{2} z|z| & \text{if} \quad |z| < \dfrac{1}{\mathcal{L}} \\ \dfrac{z}{\mathcal{L}} - \dfrac{1}{2\mathcal{L}^2} & \text{if} \quad z \geq \dfrac{1}{\mathcal{L}} \end{cases}$$

By the appropriate choice of the functions ϕ, θ the associated system (14-16) can be made C^k for any $k \geq 1$ (e.g. $\theta(z) = \phi(z) = z|z|^k$).

ASE 4 - for the case r=0 (i.e. only inequality constraints are in (NP)) Fukushima [12] gave an approximate reduction of KT-conditions

$$\nabla_x F(x,u) = 0$$
$$\nabla_u F(x,u) = 0 \qquad (17)$$

where

$$F(x,u) = f(x) + \sum_{j=1}^{m} \psi(u_j, -g_j(x))$$

$$\psi(a,b) = \begin{cases} ab + \frac{1}{2}cb^2 & \text{if} \quad a+cb \geq \varepsilon \\ \frac{(a-\varepsilon)^2}{12\varepsilon c} + \frac{(a+\varepsilon)^2 b}{4\varepsilon} + \frac{c(a+\varepsilon)b^2}{4\varepsilon} + \frac{c^2b^3}{12\varepsilon} & \text{if} \quad |a+cb| < \varepsilon \\ -\frac{3a^2+\varepsilon^2}{6c} & \text{if} \quad a+cb \leq -\varepsilon \end{cases}$$

$c, \varepsilon > 0$ are constants in (17). Here approximate means that (17) is equivalent to KT-conditions only if for each $j=1,\dots,m$ it holds

either $u_j+cg_j(x) \geq \varepsilon$ or $u_j+cg_j(x) \leq -\varepsilon$.

If for each KT-point the strong complementarity condition holds then for each $c > 0$ fixed a sufficiently small ε exists such that the above condition is satisfied. This approach was motivated by the modification of the Rockafellar´s augmented Lagrangian in order to obtain its global continuous differentiability.

We may now summarise this approach as follows: First Reduce the original NP-problem to an associated system of equations ASE and then imbed it into a one-parametric family of equations ASE(t) for $t \in I$ such that ASE(0) has known solution and ASE(1) is the original associated system ASE (see [2,25]).

The other possibility to apply the homotopy approach to solve a NP-problem consists in Imbedding it into a one-parametric family NP(t) of NP-problems for $t \in I$ and then to follow the homotopy path of solutions (KT-points) of the family NP(t) (see [19, 13, 6]). The family NP(t) is of course constructed in such a way that NP(0) has a known solution and NP(1) is the original NP-problem.

The first approach we shall call the R-approach and the second one the I-approach.

The I-approach may also lead to a path-following procedure of a one-parametric family of equations, when the family NP(t) is reduced to an one-parametric associated system of equations ASE(t) which determines the path of solutions (KT-points) of NP(t) to be numerically followed.

We illustrate the two approaches leading to one-parametric family of equations in the following simple examples.

I-approach: Imbedding of the NP-problem into a family

$$\text{Min} \left\{ \bar{f}(x,t) \;\middle|\; \begin{array}{ll} \bar{g}_i(x,t) = 0 & i=1,\dots,r \\ \bar{g}_j(x,t) \geq 0 & j=r+1,\dots,m \end{array} \right\} \qquad \text{NP(t)}$$

where

$$\bar{f}(x,t) = (1-t)\|x-a\|^2 + t\, f(x)$$

$$\bar{g}_i(x,t) = (1-t)(b_i^T x - c_i) + t\, g_i(x) \qquad i=1,\dots,m$$

$t \in I$, $a, b_i \in R^n$, $c_i \in R$, $b_i^T a = c_i$ $(i=1,\dots,r)$, $b_j^T a > c_j$ $(j=r+1, \dots, m)$, b_i $(i=1,\dots,r)$ are linearly independent. The only solution of NP(0) gives a KT-point $x=a$ with all zero-valued multipliers. Using e.g. the ASE 2 reduction we obtain

$$\left.\begin{aligned} &\nabla_x \bar{f}(x,t) - \sum_{i=1}^{r} u_i \nabla_x \bar{g}_i(x,t) - \sum_{j=r+1}^{m} y_j^+ \nabla_x \bar{g}_j(x,t) = 0 \\ &\bar{g}_i(x,t) = 0 \qquad i=1,\dots,r \\ &y_j^- + \bar{g}_j(x,t) = 0 \qquad j=r+1,\dots,m \end{aligned}\right\} \text{ASE}(t)$$

(the known solution of ASE(0) is $(x,u_i,y_j) = (a,0,-b_j^T a + c_j)$).

R-approach: Let us assume a NP-problem reduced to an associated system ASE 2. Using a simple homotopy imbedding we obtain

$$(1-t)(x-a) + t\left[\nabla f(x) - \sum_{i=1}^{r} u_i \nabla g_i(x) - \sum_{j=r+1}^{m} y_j^+ \nabla g_j(x)\right] = 0$$

$$(1-t)(u_i - b_i) + t\, g_i(x) = 0 \qquad (i=1,\dots,r)$$

$$(1-t)(y_j - c_j) + t\left[y_j^+ + g_j(x)\right] = 0 \qquad (j=r+1,\dots,m)$$

where $a \in R^n$, $b \in R^r$, $c \in R^{m-r}$ (the known solution for $t=0$ is $(x,u_i,y_j) = (a,b_i,c_j)$).

As shown in the above examples it is not difficult to construct a parametric system of equations defining the homotopy h such that conditions (3), (4) are satisfied (even with a unique solution for $t=0$). It is also usually not difficult to prove that the homotopy path is a (piecewise) differentiable manifold for almost all choices of the homotopy parameters (like a, b, c in the examples above); see e.g. [3,2,7]. The key-problem still remains the convergence proof of HM (i.e. to guarantee the intersection of the homotopy path with the hyperplane $t=1$). This reduces usually to the proof of boundness of the homotopy path or to the proof of boundness of the feasible sets for the family NP(t).

3. Homotopy methods based on barrier and penalty functions

In this section we describe main ideas from [25]. In the case of the barrier homotopy we use here a weaker assumption - the Mangasarian-Fromovitz constraint qualification: we say that (MF) holds at a feasible solution x of the NP-problem iff following two conditions are satisfied:

$$\{\nabla g_i(x) \mid i=1,\dots,r\} \text{ is a linear independent set of vectors (gradients of equality-constraints)} \tag{MF1}$$

there exists a solution $z \in R^n$ of the system

$$\left.\begin{array}{ll} \nabla g_i^T(x)\, z = 0 & i=1,\dots,r \\ \nabla g_j^T(x)\, z > 0 & j \in J(x) \end{array}\right\} \tag{MF2}$$

where $J(x) = \{ j \mid g_j(x) = 0 \}$ is the active inequality set at x

We note that (MF) implies that the set of Lagrangian multipliers at a KT-point is bounded.

3.1. Barrier homotopy map

In this subsection we assume the NP-problem with inequality constraints only (i.e. $r=0$) and $\emptyset \neq K^i = \{ x \in R^n \mid g_j(x) > 0 \quad (j=1,\dots,m)\}$.

The barrier (transformation) method consists in computing a sequence of (unconstrained) minima x^k of the barrier function

$$B(x,w^k) = f(x) + w^k \sum_{j=1}^{m} \beta(g_j(x)) \tag{18}$$

for $w^k \to 0$. B is defined for $x \in K^i$, $w^k > 0$, where the function $\beta : R_+ \to R$ $(R_+ = \{ c \in R \mid c > 0 \})$ satisfies the following barrier slope-property

$$\lim_{c \downarrow 0} \beta'(c) = -\infty \quad .$$

Moreover, we suppose that β is C^3 and

$$\beta'(c) < 0 \qquad \text{for all} \quad c > 0 \tag{19}.$$

We shall now combine the barrier and homotopy method. Minima x^k of (18) satisfy the stationarity equation

$$\nabla f(x) + w^k \sum_{j=1}^{m} \beta'(g_j(x)) \nabla g_j(x) = 0 \tag{20}.$$

We imbed (20) into

$$(1-t)(x-a) + t\left[\nabla f(x) + w^k \sum_{j=1}^{m} \beta'(g_j(x)) \nabla g_j(x) \right] = 0 \quad .$$

Using the substitution $w^k = 1-t$ we decrease the barrier parameter w^k to zero at the same time as the homotopy variable t tends to 1 . So we obtain the <u>barrier homotopy (map)</u> $h: K^i \times I \to R^n$

$$h(x,t) = (1-t)(x-a) + t\,\nabla f(x) + t(1-t) \sum_{j=1}^{m} \beta'(g_j(x)) \nabla g_j(x) \tag{21},$$

where $x \in K^i$, $t \in I$ and the homotopy parameter a is from K^i .

To overcome the problems with convergence of the corresponding HM we assume also that K^1 is bounded (i.e. the closure ol K^1 is a compact subset of the set of feasible solutions).

It is well known [15] that barrier functions can be obtained from the KT-conditions as follows: Each triple of conditions (8,9,10) for j fixed is replaced by one perturbed condition

$$u_j \frac{1}{-\beta'(g_j(x))} = w \tag{22}$$

for small $w > 0$. Then expressing u_j from (22) and inserting it into (6) one obtains

$$\nabla_x B(x,w) = 0 \quad .$$

The same can be done with imbedded equation (6)

$$(1-t)(x-a) + t\left[\nabla f(x) - \sum_{j=1}^{m} u_j \nabla g_j(x)\right] = 0 \tag{6'}$$

instead of (6). In this way we obtain the barrier homotopy which can be hence considered as the result of imbedding of the barrierly perturbed KT-conditions.

We may also write the barrier homotopy in another way:

$$h(x,t) = \nabla_x T_t(x,w) \quad ,$$

where $w=t(1-t)$ and

$$T_t(x,w) = \tfrac{1}{2}(1-t)\|x-a\|^2 + t\, f(x) + w \sum_{j=1}^{m} \beta(g_j(x)) \quad .$$

This allows another interpretation of $h(x,t)=0$ for the barrier homotopy map h : it is the stationarity equation of a barrier function for a one-parametric family of NP-problems ($t \in I$)

$$\mathrm{Min}\left\{ F(x,t) \;\middle|\; g_j(x) \geq 0 \quad (j=1,\dots,m) \right\} ,$$

where

$$F(x,t) = \tfrac{1}{2}(1-t)\|x-a\|^2 + t\, f(x) \quad .$$

The next three theorems give the convergence result for the barrier homotopy method.

<u>Theorem 1.</u> Let us assume the NP-problem (r=0) with $K^1 \neq \emptyset$ bounded, let h_a be the barrier homotopy with $a \in K^1$ fixed. Let for each feasible solution x it hold :

there exists $z \in R^n$ such that $g_j^T(x)\, z > 0$ for $j \in J(x)$ (23).

Then each limit point $(\bar{x},\bar{t}) \in \partial(h_a^{-1}(0) \cap K^1 \times (0,1))$ of the barrier homotopy (21) satisfies:

either $\bar{t}=1$ and $\bar{x}$ is a KT-point of the original NP-problem,

or $\bar{t}=0$ and $\bar{x}$ is a KT-point of the following problem

$$\mathrm{Min}\left\{ \tfrac{1}{2}\|x-a\|^2 \;\middle|\; g_j(x) \geq 0 \quad (j=1,\dots,m) \right\} \tag{24}$$

Proof. Let $(x^k,t^k)\in K^i\times(0,1)$, $h_a(x^k,t^k)=0$ and $(x^k,t^k)\xrightarrow[k\to\infty]{}(\bar{x},\bar{t})$. First let us prove by contradiction that $\bar{t}\notin(0,1)$.

If $\bar{t}\in(0,1)$ then the limit point from $\partial(h_a^{-1}(0)\cap K^i\times(0,1))$ must have $J(\bar{x})\neq\emptyset$. For $j\in J(x)$ the limit

$$\lim_{k\to\infty}\beta'(g_j(x))$$

is infinite. Dividing (21) by $\|v^k\|$ (where $v_j^k=\beta'(g_j(x))$) and going to the limit for $k\to\infty$ we obtain, that there exist finite nonpositive numbers $\bar{v}_j$ (not all zero) such that

$$\sum_{j\in J(\bar{x})}\bar{v}_j\,\nabla g_j(\bar{x})=0 \quad .$$

By the scalar multiplication of the above equation with z from (23) we obtain a contradiction, which proves $\bar{t}=0$ or $\bar{t}=1$.

Let $\bar{t}=1$. Going to the limit in (21) we obtain

$$\nabla f(\bar{x})+\sum_{j=1}^{m}\nabla g_j(\bar{x})\lim_{k\to\infty}\left[(1-t^k)\,\beta'(g_j(x^k))\right]=0 \quad .$$

By an analogous technique, as used above in this proof, we can prove that the limits in the previous equation are finite and by (19) nonpositive. Denoting this limits by $-\bar{u}_j$ we have that $\bar{x}$ is a KT-point of the original NP-problem with multipliers $\bar{u}_j$ (obviously $\bar{x}$ is feasible since $\bar{x}\in\mathrm{cl}\,K^i$ and the complementarity condition is satisfied as for $j\in J(\bar{x})$ the limits $-\bar{u}_j$ are zero).

The case $\bar{t}=0$ can be proved in an analogous way as the case $\bar{t}=1$, since from (21) one obtains

$$(\bar{x}-a)+\sum_{j=1}^{m}\nabla g_j(\bar{x})\lim_{k\to\infty}\left[(1-t^k)\,\beta'(g_j(\bar{x}))\right]=0 \quad . \qquad \text{Q.E.D.}$$

Theorem 1 gives properties which are satisfied at each limit point of the zero-set $h_a^{-1}(0)$ on the boundary of $K^i\times(0,1)$ and motivates the construction of the barrier homotopy method.

Theorem 2. For almost all choices of $a\in K^i$ (i.e. for all $a\in A\subset K^i$ where the Lebesque measure of the set $K^i\setminus A$ is zero) the zero-set $h_a^{-1}(0)$ of the barrier homotopy (21) is a differentiable submanifold in $K^i\times(0,1)$ of dimension 1 .

Proof. Using some basic results from differential topology (regular values of maps, parametric Sard's theorem, see e.g. [7,17,1]) it can be seen that it is sufficient to prove that $0\in R^n$ is a regular value of the barrier homotopy map h (considered as a map with variables x, t, a) on the open set $K^i\times(0,1)\times K^i$. (The C^2-differentiability of h is needed in the parametric Sard's theorem.) This regularity of the zero value of h is implied by the regularity of the Jacobi matrix Dh because the submatrix

$$D_a\,h(x,t,a)=-(1-t)\,E$$

(D_a denotes the differential with respect to a and E is identity matrix) is regular on $K^i \times (0,1) \times K^i$. Q.E.D.

This type of statement and the idea of its proof was first used in connection with HM in [7]. The statement of Theorem 2 can be explained as follows: a random choice of the value of the homotopy parameter $a \in K^i$ will with probability one lead to $a \in A$ and for this choice the zero-set $h_a^{-1}(0)$ consists in $K^i \times (0,1)$ of connected components being either homeomorphic to a circle or to an open interval. This property of the zero-set excludes the bifurcation of the homotopy path.

Theorem 3. Let K^i be a convex set (g_j need not be convex) and $a \in A$ (A as in Th.2). Then the barrier homotopy h defines a homotopy method converging to a KT-point of NP-problem ($r=0$). More precisely, there is a unique differentiable homotopy path beginning at $(a,0)$ leading to a point $(\bar{x},1)$, where $\bar{x}$ is a KT-point of the original NP-problem.

Proof. As $D_x\, h_a(a,0)=E$ is regular, it follows from the implicit function theorem that the homotopy path beginning at $(a,0)$ does not turn back to this point. As $a \in A$ we conclude from Th.2 that the homotopy path (as a subset of $K^i \times (0,1)$) is homeomorphic to an open interval and hence has to have another limit point. As K^i is convex, it can be easily proved, that the NP-problem (24) has an unique KT-point $x=a$ (for each $a \in K^i$). This implies (using Th.1) that the other limit point must lie in the hyperplane $t=1$, i.e. it is a KT-point of the original NP-problem. Q.E.D.

3.2. Penalty homotopy map

In this subsection we assume the equality-constrained NP-problem (i.e. $r=m$).

The penalty method consists in computing a sequence of unconstrained minima x^k of the penalty function

$$P(x,w^k) = f(x) + \frac{1}{w^k}\sum_{i=1}^{m} \alpha(g_i(x))$$

for $w^k \longrightarrow 0$. P is considered to be defined for all x and $w>0$, where the function $\alpha : R \longrightarrow R$ is C^3 and satisfies

$$\alpha'(c) \neq 0 \quad \text{for all} \quad c \neq 0 \tag{25}$$

$$\alpha'(0) = 0 \tag{26}$$

Analogous as in the barrier case we imbed the stationarity equation $\nabla_x P(x,w^k)=0$ and substitute $w^k=1-t$ to force the penalty parameter w to tend to zero as t tends to one. We obtain

$$(1-t)(x-a) + t\nabla f(x) + \frac{t}{1-t}\sum_{i=1}^{m} \mathcal{L}'(g_i(x))\nabla g_i(x) = 0 \qquad (27)$$

After multiplying the left-hand side by $1-t$ we obtain the <u>penalty homotopy (map)</u>

$$h(x,t) = (1-t)^2(x-a) + t(1-t)\nabla f(x) + t\sum_{i=1}^{m} \mathcal{L}'(g_i(x))\nabla g_i(x) \qquad (28)$$

where $x,\ a \in R^n$, $t \in I$.

The solution set of (27) for $t<1$ and the zero-set $h^{-1}(0) \cap \{(x,t) \mid t<1\}$ are identical so the corresponding homotopy methods (one based on the left-hand side of (27) and the other based on the penalty homotopy map) produce the same homotopy paths. Moreover, (28) is more convenient as it is defined and bounded at $t=1$.

The penalty homotopy (more precisely the equation (27)) allows the same interpretations as in the barrier case:

- it can be considered to be the imbedding of the KT-conditions (from (6) we obtain (6') and for u_i is substituted from
$$(1-t)\,u_i + \mathcal{L}'(g_i(x)) = 0 \quad (i=1,\dots,m),$$
which can be considered as a perturbation of (7) for $1-t\ (>0)$ small),
- it can be interpreted as a stationarity equation of a penalty function ($w=(1-t)/t$) applied to a one-parametric family of NP-problems
$$\mathrm{Min}\ \{F(x,t) \mid g_i(x) = 0 \quad (i=1,\dots,m)\},$$
where F is the same as in Sec.3.1.

The next theorem says under which assumptions a homotopy path of the penalty homotopy (28) leads to a KT-point of a NP-problem.

<u>Theorem 4</u>. Let h_a be the penalty homotopy (28) with $a \in R^n$ fixed. Let the NP-problem ($r=m$) to be solved satisfy: at <u>every</u> $x \in R^n$ the set of vectors $\{\nabla g_i(x) \mid i=1,\dots,m\}$ is linearly independent. Then each limit point $(\bar{x},1)$ of the zero-set $h^{-1}(0) \cap R^n \times \{t<1\}$ is a KT-point of the original NP-problem.

<u>Proof</u>. Let $(x^k,t^k) \in R^n \times (0,1)$, $h_a(x^k,t^k)=0$ such that $(x^k,t^k) \xrightarrow[k\to\infty]{} (\bar{x},1)$. From $h_a(x^k,t^k)=0$ for $k\to\infty$ we obtain

$$\sum_{i=1}^{m} \mathcal{L}'(g_i(\bar{x}))\nabla g_i(\bar{x}) = 0 ,$$

which due to the assumption of Th.4 implies that $\mathcal{L}'(g_i(\bar{x}))=0$, i.e. $\bar{x}$ is feasible. Going to the limit for $k\to\infty$ in (27) we obtain

$$\nabla f(\bar{x}) + \sum_{i=1}^{m} \nabla g_i(\bar{x}) \lim_{k\to\infty} \frac{\mathcal{L}'(g_i(x^k))}{1-t^k} = 0 \qquad (29)$$

(we suppose a subsequence (x^k,t^k) was chosen in such a way that the limits in (29) exist - may be finite or $\pm\infty$). Dividing (29) by $\|v^k\|$ (where $v_i^k = \mathcal{L}'(g_i(x^k))/(1-t^k)$) we can in an analogous way as in Th.1

prove that the limits in (29) are finite (due to the assumption on linear independence of gradients of the constraints). Denoting the limits by $-\bar{u}_i$ we have that $\bar{x}$ is a KT-point of the original NP-problem with multipliers u_i . Q.E.D.

Theorem 4 is not a convergence theorem for the penalty homotopy method because it does not assert the existence of such a limit point on the homotopy path beginning at $(a,0)$. We can only say that if the homotopy path of the penalty homotopy meets the hyperplane $t=1$ then it leads to a KT-point.

Analogously to Th.2 it can be proved, that the zero-set $h_a^{-1}(0)$ of the penalty homotopy is for almost all $a \in R^n$ a submanifold of $R^n \times (0,1)$ of dimension 1 .

4. Conclusions

Barrier and penalty homotopy derived in Sec.3 were shown to have common features with both the R-approach (the perturbed KT-conditions play here the role of an associated system of equations) as well as the I-approach (the imbedding is of special form - objective function is imbedded and constraints remain the same).

The barrier homotopy method differs from other homotopy methods in the literature by the following:

- its x-region of definition is not the whole space R^n ,
- the concept of imbedding is not characterised by (3) because $h(x,1)=0$ is not equivalent to the original problem (i.e. (perturbed) KT-conditions of the NP-problem ($r=0$)).

One can formulate another modification of Th.3 which does not include any convexity assumption (see [26]).

The homotopy methods derived here inherit an unpleasant property of the transformation methods, namely, the bad conditioning of $D_x h_a$ for points (x,t) with t approaching 1 . We note, that in general it may be also a non-trivial problem to choose the parameter vector a into the barrier homotopy map, because it has to satisfy $a \in K^1$. As the experiences show, the randomness of the choice of a makes no trouble - we have never been forced to change the first feasible choice of a during numerical experiments.

For a special NP-problem

$$\text{Min} \left\{ f(x) \;\middle|\; c_k \leq x_k \leq d_k \quad (k=1,\ldots,n) \right\}$$

(where $c_k < d_k$) and a special choice of β ($\beta(c)=-\ln c$) we have modified the barrier homotopy to

$$h_k(x,t)=\left[(1-t)(x_k-a_k) + t\frac{\partial f}{\partial x_k}(x)\right](x_k-c_k)(d_k-x_k) + t(1-t)(2x_k-c_k-d_k)$$

for all components $k=1,\ldots,n$. This modified barrier homotopy map has no more the above mentioned unpleasant property of bad conditioning of

$D_x h_a$ for t near 1 (under strong second order sufficient conditions, see [26]). For the modified barrier homotopy we can not in general find a map T such that $\nabla_x T(x,t) \equiv h(x,t)$;c.f. Sec.3.1.

The linear independence assumption in Th.4 is rather strong. In practice it means that the equality constraints are linear.

One possible way of generalising our results would be to find barrier homotopy corresponding to a one-parametric family of NP-problems, where also the feasible set would vary depending on the parameter t. However the recent results of H.Th.Jongen et al. [14,18] show, that considering a one-parametric family of feasible sets defined by inequalities one can not generically exclude the occurence of points at which (MF) are not satisfied. At these points the topological structure of the feasible set can change and these points may be the limit points of the homotopy path. This may cause failing of the barrier homotopy method. Henceforth, this generalised approach can be used (if at all) only for special NP-problems.

Literature

[1] Abraham, R., Robbin J.: Transversal mappings and flows. Benjamin Inc., New York - Amsterdam 1967.

[2] Alexander, J.C., Li T.-Y., Yorke J.A.: Piecewise smooth homotopies. In: Eaves B.C., Gould F.J., Peitken H.O., Todd M.J. (eds.): Homotopy methods and global convergence. Plenum P.C. 1983.

[3] Allgower,E., Georg K.: Simplicial and continuation methods for approximating fixed points and solutions to systems of equations. SIAM Review, 22 (1980), 1, 28-85.

[4] Brouwer, L.E.J.: Ueber Abbildung von Mannigfaltigkeiten. Math.Ann., 71 (1912), pp.97-115.

[5] Brunovský, P., Meravý P.: Solving systems of polynomial equations by bounded and real homotopy. Numer.Math., 43 (1984), pp. 397-418.

[6] Dontchev, A.L., Jongen H.Th.: On the regularity of the Kuhn-Tucker curve. Report 460, Tech.Hogeschool Twente, Enschede, Netherlands 1984.

[7] Chow, S.N., Mallet-Paret J., Yorke J.A.: Finding zeros of maps: Homotopy methods that are constructive with probability one. Math. of Comput., 32 (1978), 143, 887-899.

[8] Eaves, B.C.: Homotopies for computation of fixed points. Ibid., 3 (1972), 1, 1-22.

[9] Eaves, B.C., Saigal R.: Homotopies for computation of fixed points on unbounded regions. Math.Progr., 3 (1972), 2, 225-237.

[10] Eaves, B.C.: A short course in solving equations with PL-homotopies. SIAM-AMS Proc., 9 (1976), pp. 73-143.

[11] Eaves, B.C., Scarf H.: The solution of systems of piecewise linear equations. Math.Oper.Res., 1 (1976), 1, 1-27.

[12] Fukushima M.: Solving inequality constrained optimization problems by differential homotopy continuation methods. Preprint, Dept. Appl.Math.Phys. Fac.Eng., Kyoto Univ. 1982.

[13] Gfrerer, H., Guddat J., Wacker Hj.: A globally convergent algorithm based on imbedding and parametric optimisation. Computing, 30 (1983), pp. 225-252.

[14] Guddat, J., Jongen H.Th., Rueckmann J.: On stability and stationary points in nonlinear optimization. Memorandum Nr.526, Tech. Hogeschool Twente, Enschede, Netherlands 1985.

[15] Hamala, M.: Nelineárne programovanie. Alfa, Bratislava 1976.

[16] Hansen, T.: On the approximation of a competitiveequilibrium. Ph.D. thesis, Yale University 1968.

[17] Hirsch, M.V.: Differential topology. Russian transl., Mir, Moskva 1979.
[18] Jongen, H.Th., Jonker P., Twilt F.: On one-parametric-families of sets defined by (in)equality constraints. Nieuw Archief voor Wiskunde (3), XXX (1982), pp. 307-322.
[19] Kojima, M.: Strongly stable stationary solutions in nonlinear programs. In: Robinson S.M. (ed.): Analysis and Computation of fixed points. Academic Press 1980, pp. 93-138.
[20] Kuhn, H.W.: Approximate search for fixed points. Computing Methods in Optimization Problems 2, Academic Press, New York 1969.
[21] Kuhn H.W., Mac Kinnon J.G.: Sandwich method for finding fixed points. J.Optim.Theory and Appl., 17 (1975), pp. 189-204.
[22] Luethi, H.J.: Komplementaritaets- und Fixpunktalgorithmen in der mathematischen Programierung. Springer Lec.Notes in Economics and Math.Systems, 129 (1976).
[23] Mangasarian, O.L., Fromovitz S.: The Fritz-John necessary optimality conditions in the presence of equality and inequality constraints. J.Math.Anal. and Appl., 17 (1967), pp. 37-47.
[24] Mangasarian, O.L.: Unconstrained methods in nonlinear programming. In: SIAM-AMS Proc., 9 (1976), pp. 169-184.
[25] Meravý, P.: A note about some relations between the methods of parametric transformation functions and the smooth homotopy methods for solving constrained optimisation problems. In: Seminarbericht Nr. 71, Humboldt-Universitaet zu Berlin, Sektion Mathematik 1985, pp. 50-60.
[26] Meravý, P.: Homotopy methods for solving systems of equations and mathematical programming problems. Thesis, Comenius University Bratislava 1984 (in slovak).
[27] Merrill, O.H.: Applications and extensions of an algorithm that computes fixed points of a certain upper semicontinuous point to set mapping. Ph.D. thesis, Dept. of Industrial Engineering, Univ. of Michigan 1972.
[28] Ortega, J.M., Rheinboldt W.C.: Iterative solution of nonlinear equations in several variables. Academic Press, New York 1970.
[29] Scarf, H.: The approximation of fixed points of a continuous mapping. SIAM J.Appl.Math., 15 (1967), 5, 1328-1343.
[30] Todd, M.J.: The computation of fixed points and applications. Springer Lec Notes in Economics and Math.Systems, 124 (1976); russian transl., Nauka, Moskva 1983.
[31] Wacker, H.: A summary of developments on imbedding methods. In: Wacker H. (ed.): Continuation methods. Academic Press, New York 1978.

Strong Unicity in Nonlinear Parametric Optimization

Günther Nürnberger [1]

Abstract

The following question of stability is investigated: which nonlinear semi-infinite optimization problems have a strongly unique local solution and are stable under small perturbations of the restriction function such that strong unicity is preserved? A sufficient condition for this property is derived. Under certain assumptions this condition is also necessary. Applications to global best approximation by splines with free knots and rational functions are given.

Introduction

A fundamental question in optimization and approximation is: when does a given problem has a unique solution? In numerical considerations the notion of strong unicity plays an important role. Moreover, since in the computation of solutions the parameters which determinate a problem are only known up to some error we are led to consider strong unicity in connection with stability.

In this context we give conditions for nonlinear semi-infinite optimization problems which have a strongly unique local solution and for which strong unicity is preserved under small changes of the restriction function.

The results are applied to best approximation of continuous functions. In the case of Kolmogorov-sets of the second kind a characterization of the above property is obtained. Further applications include stability results for global rational approximations and spline approximations with free knots.

1 Semi-Infinite Optimization

Let U be an open subset of $\mathbb{R}^N$ and T be a compact metric space. Moreover, let fixed mappings $p : U \to \mathbb{R}$ and $B : T \times U \to \mathbb{R}$ be given. We assume that the mappings

$$x \to \text{grad } p(x) = \left(\frac{\partial p(x)}{\partial x_1}, \ldots, \frac{\partial p(x)}{\partial x_N}\right) \quad (x \in U)$$

and

$$(t, x) \to \text{grad } B(t, x) = \left(\frac{\partial B(t, x)}{\partial x_1}, \ldots, \frac{\partial B(t, x)}{\partial x_N}\right) \quad ((t, x) \in T \times U)$$

are continuous.

By $C(T)$ we denote the set of all continuous functions $b : T \to \mathbb{R}$, endowed with the supremum norm $\|b\|_\infty = \sup\{|b(t)| : t \in T\}$.

For a given function $b \in C(T)$ we consider the following nonlinear semi-infinite optimization problem, denoted by MP(b):

$$\text{Minimize } p(x), \text{ subject to } x \in U \text{ and } B(t, x) \leq b(t) \text{ for all } t \in T.$$

Given $b \in C(T)$, we define the set of **feasible points** by

$$Z(b) = \{x \in U : B(t, x) \leq b(t) \text{ for all } t \in T\},$$

---

[1]Mathematisches Institut, Universität Erlangen-Nürnberg, 8520 Erlangen, W.-Germany

the **minimum value** by

$$E(b) = \inf\{p(x) : x \in Z(b)\}$$

and the set of **solutions** by

$$P(b) = \{x \in Z(b) : p(x) = E(b)\}.$$

A vector $x \in Z(b)$ is called **local solution** of MP(b), if there exists an $\varepsilon > 0$ such that for all $y \in Z(b)$ with $\|y - x\|_2 \leq \varepsilon$, $p(y) \geq p(x)$. Furthermore, a vector $x \in Z(b)$ is called **strongly unique local solution** of MP(b), if there exist $\varepsilon > 0$ and $K > 0$ such that for all $y \in Z(b)$ with $\|y - x\|_2 \leq \varepsilon$, $p(y) - p(x) \geq K\|y - x\|_2$. (Here $\|\cdot\|_2$ denotes the Euclidean norm of $\mathbb{R}^N$.)

In these cases we say that $x \in Z(b)$ is a local (respectively strongly unique local) solution of MP(b) on $B_\varepsilon(x)$. (Here $B_\varepsilon(x)$ denotes the closed ball with center x and radius ε .) Moreover, given $\varepsilon > 0$ and $x \in \mathbb{R}^N$, we set

$$\mathrm{SUL}_\varepsilon(x) = \{b \in C(T) : \mathrm{MP}(b) \text{ has a strongly unique local solution on } B_\varepsilon(x)\}.$$

For $x \in Z(b)$ the set of **active points** is defined by

$$N(b, x) = \{t \in T : B(t, x) = b(t)\}.$$

We say that MP(b) satiesfies **condition (CQ)** in $x \in Z(b)$, if there exists a vector $y \in \mathbb{R}^N$ such that

$$< \mathrm{grad}\, B(t, x), y > \quad < \quad 0 \text{ for all } t \in N(b, x).$$

For $x \in Z(b)$ a subset $\{t_1, \ldots, t_q\}$ of $N(b, x)$ is called **critical** (with respect to x), if $q \leq N$ and

$$-\mathrm{grad}\, p(x) \in \mathrm{cone}(\{\mathrm{grad}\, B(t_i, x) : i = 1, \ldots, q\}).$$

(Here *cone* denotes the convex cone generated by a set.)

We first note that unique and strongly unique solutions are not the same in general. Moreover, if we speak of a unique solution in practical computations, we actually mean a strongly unique solution (see Example 2.1 in [11]).

In the following we will give conditions for those functions $b \in \mathrm{SUL}_\varepsilon(x)$ which are stable under small perturbations such that strong unicity is preserved, i.e. for functions from int $\mathrm{SUL}_\varepsilon(x)$. (Here *int* denotes the interior of a set.)

For doing this we need the following results. The first theorem is the famous Kuhn-Tucker-Theorem for local solutions.

Theorem 1.1 *Let MP(b) satisfy condition (CQ) in some point $x \in Z(b)$. We consider the following statements.*

(i) x is a local solution of MP(b).

(ii) There exists a critical subset of $N(b, x)$ **(Kuhn-Tucker-condition)**.

Then (i) $\Rightarrow$ (ii).

The following characterization of strongly unique local solutions is well known (see Hettich & Zencke [7]), except the equivalence of (i) and (ii).

Theorem 1.2 *Let MP(b) satisfy condition (CQ) in some point $x \in Z(b)$. The following statements are equivalent.*

(i) x is a strongly unique local solution of MP(b).

(ii) There exist $\varepsilon > 0$ and $K > 0$ such that for all $y \in U$ with $\|y - x\|_2 < \varepsilon$,

$$\max\{p(y) - p(x), \max_{t \in N(b,x)} (B(t, y) - b(t))\} \quad \geq \quad K\|y - x\|_2.$$

(iii) $-\mathrm{grad}\, p(x) \in \mathrm{int\, cone}(\{\mathrm{grad}\, B(t, x) : t \in N(b, x)\})$.

(iv) There does not exist a point $z \in \mathbf{R}^N, z \neq 0$, such that

$$< grad\, p(x), z > \quad \leq \quad 0$$

and

$$< grad\, B(t,x), z > \quad \leq \quad 0 \text{ for all } t \in N(b,x).$$

Proof. The equivalence (i) $\Leftrightarrow$ (iii) $\Leftrightarrow$ (iv) is given in Hettich & Zencke [7] on p.49 and p. 63. The implication (ii) $\Rightarrow$ (i) is obvious. Thus, it remains to prove that (iv) $\Rightarrow$ (ii). Suppose that (ii) fails, i.e. there exists a sequence (y_n) in $\mathbf{R}^N$ such that $y_n \to x$ and for all n,

$$p(y_n) - p(x) \quad < \quad \frac{1}{n}\|y_n - x\|_2 \tag{1.1}$$

and

$$B(t,y_n) - b(t) \quad < \quad \frac{1}{n}\|y_n - x\|_2 \text{ for all } t \in N(b,x). \tag{1.2}$$

We set for all n, $\varepsilon_n = \|y_n - x\|_2$ and $z_n = (y_n - x)/\|y_n - x\|_2$. Then we have $y_n = x + \varepsilon_n z_n$ for all n. Since $\|z_n\|_2 = 1$ for all n, we may assume that the sequence (z_n) converges to some $z \in \mathbf{R}^N$. It follows from (1.1) that

$$p(y_n) - p(x) \quad \leq \quad o(\varepsilon_n) \text{ for } n \to \infty.$$

Then by the mean value theorem

$$\varepsilon_n < z_n, \text{grad } p(x) > \quad = \quad p(y_n) - p(x) + o(\varepsilon_n) \quad \leq \quad o(\varepsilon_n) \text{ for } n \to \infty$$

and thus

$$< z, \text{grad } p(x) > \quad \leq \quad 0 \tag{1.3}$$

Moreover, it follows from (1.2) that for all $t \in N(b,x)$,

$$B(t,y_n) - b(t) \quad \leq \quad o(\varepsilon_n) \text{ for } n \to \infty.$$

Again by the mean value theorem, for all $t \in N(b,x)$,

$$\varepsilon_n < z_n, \text{grad } B(t,x) > \quad = \quad B(t,y_n) - B(t,x) + o(\varepsilon_n)$$

$$= B(t,y_n) - b(t) + o(\varepsilon_n) \quad \leq \quad o(\varepsilon_n) \text{ for } n \to \infty,$$

and thus

$$< z, \text{grad } B(t,x) > \quad \leq \quad 0 \tag{1.4}$$

Finally, (1.3) and (1.4) show that (iv) fails which proves Theorem 1.2.

The following result is a consequence of Theorem 1.2 (see Hettich & Zencke [7]).

Corollary 1.3 *Let MP(b) satisfy condition (CQ) in some point $x \in Z(b)$. We consider the following statements.*

(i) x is a strongly unique local solution of MP(b).

(ii) There exist a critical subset $\{t_1, \ldots, t_N\}$ of $N(b,x)$ such that each N vectors from

$$\{grad\, p(x), grad\, B(t_1,x), \ldots, grad\, B(t_N,x)\}$$

are linearly independent.

Then (ii) $\Rightarrow$ (i). If the set $N(b,x)$ consists of exactly N points, then (i) $\Rightarrow$ (ii). Moreover, if (i) holds, then the set $N(b,x)$ contains at least N points.

(The implication (i) $\Rightarrow$ (ii) does not hold in general.)

We now consider the following question of stability. Let a function $b \in C(T)$ be given such that MP(b) has a strongly unique local solution x. If $\varepsilon > 0$ is sufficiently small, does there exist a $\delta_\varepsilon > 0$ such that for all $\tilde{b} \in C(T)$ with $\|b - \tilde{b}\|_\infty < \delta_\varepsilon$, MP($\tilde{b}$) has a strongly unique local solution $\tilde{x} \in B_\varepsilon(x)$ on $B_{\tilde{\varepsilon}}(\tilde{x})$?

The relevant case is that the real numbers $\tilde{\varepsilon} > 0$ are bounded away from zero, otherwise in practice $\tilde{x}$ would not be considered as a strongly unique local solution of MP($\tilde{b}$). Thus, for the results in this paper (except Lemma 1.6, Lemma 1.7, Corollary 2.4 and Corollary 2.5) we make the following

General assumption. Let MP(b) have a strongly unique local solution x. We only consider those $b \in C(T)$ such that, if $\varepsilon > 0$ is sufficiently small and MP($\tilde{b}$) has a strongly unique local solution $\tilde{x} \in B_\varepsilon(x)$, then $\tilde{x}$ is a strongly unique local solution of MP($\tilde{b}$) on $B_\varepsilon(x)$.

The reader interested in statements without this assumption should study the proof of Theorem 1.4.

We now give the following stability result.

Theorem 1.4 *Let MP(b) satisfy condition (CQ) in $x \in Z(b)$, where x is a local solution of MP(b). We consider the following statements.*

(i) $b \in int\ SUL_\varepsilon(x)$ for sufficiently small $\varepsilon > 0$.

(ii) There exists a critical subset $\{t_1, \ldots, t_N\}$ of $N(b, x)$ consisting of N distinct points and for every such set $\{t_1, \ldots, t_N\}$ each N vectors from

$$\{grad\ p(x), grad\ B(t_1, x), \ldots, grad\ B(t_N, x)\}$$

are linearly independent.

Then (ii) ⇒ (i). Moreover (i) ⇒ (ii), if the Kuhn-Tucker-condition characterizes local solutions for all $b \in C(T)$.

In the case of linear semi-infinite optimization problem Theorem 1.4 was proved in [11].

Before proving Theorem 1.4 we give the following result which is an immediate consequence of Corrollary 1.3 and Theorem 1.4.

Corollary 1.5 *Let MP(b) satisfy condition (CQ) in $x \in Z(b)$, where x is a local solution of MP(b), and let $N(b, x)$ consist of exactly N points. The following statements are equivalent.*

(i) $b \in int\ SUL_\varepsilon(x)$ for sufficient small $\varepsilon > 0$.

(ii) x is a strongly unique local solution of MP(b).

For the proof of Theorem 1.4 we need two lemmas.

Lemma 1.6 *Let MP(b) satisfy condition (CQ) in some point $x \in Z(b)$. Then there exists a real number $\varepsilon > 0$ such that for all $\tilde{b} \in C(T)$ with $\|b - \tilde{b}\|_\infty < \varepsilon$ and all $\tilde{x} \in Z(\tilde{b})$ with $\|x - \tilde{x}\|_2 < \varepsilon$, MP($\tilde{b}$) satisfies condition (CQ) in $\tilde{x}$.*

Proof. Since MP(b) satisfies condition (CQ) in $x \in Z(b)$, there exists a vector $y \in \mathbb{R}^N$ such that for all $t \in N(b, x)$,

$$< \text{grad } B(t, x), y > \quad < \quad 0.$$

Then by compactness arguments there exists a neighborhood V of $N(b, x)$ and a constant $C > 0$ such that for all $t \in V$,

$$< \text{grad } B(t, x), y > \quad < \quad -C.$$

Moreover, by continuity arguments there exists a real number $\varepsilon > 0$ such that for all $\tilde{b} \in C(T)$ with $\|b - \tilde{b}\|_\infty < \varepsilon$ and all $\tilde{x} \in Z(\tilde{b})$ with $\|x - \tilde{x}\|_2 < \varepsilon$, $N(\tilde{b}, \tilde{x}) \subset V$ and thus, for all $t \in N(\tilde{b}, \tilde{x})$,

$$< \text{grad } B(t, x), y > \quad < \quad -\frac{C}{2}.$$

Thus proves Lemma 1.6.

Lemma 1.7 *Let MP(b) satisfy condition (CQ) in some point $x \in Z(b)$ and let (b_n) be a sequence in $G(T)$ converging to b. Then there exists a sequence (x_n) such that $(x_n) \in Z(b_n)$ for all n and $x_n \to x$.*

Proof. Since MP(b) satisfies condition (CQ) in $x \in Z(b)$, there exists a vector $y \in \mathbb{R}^N$ such that for all $t \in N(b,x)$,

$$< \text{grad } B(t,x), y > \quad < \quad 0$$

We set for all n, $x_n = x + \epsilon_n y$, where $\epsilon_n > 0$ for all n and $\epsilon_n \to 0$. It suffices to show that $x_n \in Z(b_n)$ for sufficiently large n. By compactness arguments there exists a neighborhood V of $N(b,x)$ and a constant $C_1 > 0$ such that for all $t \in V$,

$$< \text{grad } B(t,x), y > \quad < \quad -C_1.$$

Moreover, by the mean value theorem, for each $t \in V$ there exists a real number $\epsilon_n(t) \in [0, \epsilon_n]$ such that

$$\begin{aligned} B(t,x_n) &= B(t, x + \epsilon_n y) = B(t,x) + \epsilon_n < \text{grad } B(t, x + \epsilon_n(t) y), y > \\ &\leq b(t) - \epsilon_n \tfrac{C_1}{2} \quad < \quad b_n(t) \text{ for all } t \in V, \end{aligned}$$

if n is sufficiently large. Furthermore, there exists a real number $C_2 > 0$ such that for all $t \in T \setminus V$,

$$B(t,x) - b(t) \quad \leq \quad -C_2.$$

Then by continuity arguments for all $t \in T \setminus V$,

$$B(t,x_n) - b_n(t) \leq -\frac{C_2}{2},$$

if n is sufficiently large. This proves Lemma 1.7.

Proof of 1.4. (ii) $\Rightarrow$(i). Suppose that (ii) holds, but (i) fails. It follows from Corollary 1.3 that $x \in Z(b)$ is a strongly unique local solution of MP(b), i.e. there exist $\epsilon > 0$ and $K > 0$ such that for all $y \in Z(b) \cap B_\epsilon(x)$, $p(y) - p(x) \geq K\|y - x\|_2$. Since (i) fails, there exists a sequence (b_n) in $C(T)$ converging to b such that for all n, MP(b_n) does not have a strongly unique local solution on $B_\epsilon(x)$. Since U is open, we may assume that $B_\epsilon(x) \subset U$. By Lemma 1.7 there exists a sequence (z_n) such that $z_n \in Z(b_n)$ for all n and $z_n \to x$. Therefore, we may assume that $Z(b_n) \cap B_\epsilon(x) \neq \emptyset$ for all n. Since $Z(b_n) \cap B_\epsilon(x)$ is compact, there exists a local solution $x_n \in Z(b_n)$ of MP(b_n) on $B_\epsilon(x)$ for all n. We will show that $x_n \to x$. Suppose that this is not true.Since $(x_n) \subset B_\epsilon(x)$, there exists a subsequence of (x_n), which we again denote by (x_n), such that $x_n \to \tilde{x}$ and $\tilde{x} \neq x$. Since $x_n \in Z(b_n)$ for all n and $x_n \to \tilde{x}$, it is easy to verify that $\tilde{x} \in Z(b) \cap B_\epsilon(x)$. Now, let $y \in Z(b) \cap B_\epsilon(x)$ be given. By Lemma 1.7 there exists a sequence (y_n) such that $y_n \in Z(b_n)$ for all n and $y_n \to y$. Since $x_n \in Z(b_n)$ is a local solution of MP(b_n) on $B_\epsilon(x)$, $y_n \in Z(b_n)$ and $y_n \to y \in B_\epsilon(x)$, it followes from continuity arguments that there exists a sequence (ϵ_n) with $\epsilon_n > 0$ for all n and $\epsilon_n \to 0$ such that $p(x_n) \leq p(y_n) + \epsilon_n$ for all n. By taking limits we get $p(\tilde{x}) \leq p(y)$. This shows that $\tilde{x} \in Z(b)$ is a local solution of MP(b) on $B_\epsilon(x)$. On the other hand, since $x \in Z(b)$ is the only local solution of MP(b) on $B_\epsilon(x)$ and $x \neq \tilde{x}$, we get a contradiction.

Now let n be given. Since $x_n \to x$, by Lemma 1.7 we may assume that MP(b_n) satisfies condition (CQ) in x_n . Moreover, since x_n is a local solution of MP(b_n), it follows from Theorem 1.1 that there exists a critical subset $M_n = \{t_{1,n}, \ldots, t_{q_n,n}\}$ of $N(b_n, x_n)$. By going to a subsequence we may assume that $q_n = q \in \{1, \ldots, N\}$ for all n and $t_{i,n} \to t_i \in T$, $i = 1, \ldots, q$. We will show that $M = \{t_1, \ldots, t_q\}$ is a critical subset of $N(b,x)$. Since M_n is a subset of $N(b_n, x_n)$ for all n, by taking limits it follows that M is a subset of $N(b,x)$. Now, let an integer n and $y \in \mathbb{R}^N$ be given. Since M_n is critical with respect to x_n, it follows from Farkas' lemma (see Hettich & Zencke [7, p.55]) that

$$< \text{grad } p(x_n), y > \quad \geq \quad 0$$

or there exists an integer $j_n \in \{1, \ldots, q\}$ with

$$< \text{grad } B(t_{j_n}, x_n), y > \quad > \quad 0.$$

By taking limits, it follows that

$$< \text{grad } p(x), y > \quad \geq \quad 0$$

or there exists an integer $j \in \{1, \ldots, q\}$ with

$$< \text{grad } B(t_j, x), y > \quad \geq \quad 0.$$

It is well known (see Hettich & Zencke [7, p.47]) that, since MP(b) satisfies condition (CQ) in x, by Farkas' lemma M is critical with respect to x. By omitting some points of M, if necessary, we may assume that the points of M are distinct.

Case 1. $q \in \{1, \ldots, N-1\}$

By (ii) we may choose points $t_{q+1}, \ldots, t_N$ in $N(b, x)$ such that the points of $t_1, \ldots, t_N$ are distinct. Moreover, it follows from (ii) that the vectors

$$\text{grad } p(x), \text{grad } B(t_1, x), \ldots, \text{grad } B(t_{N-1}, x)$$

are linearly independent. Thus, there exists a vector $y \in \mathbb{R}^N$ such that

$$< \text{grad } p(x), y > \quad = \quad -1$$

and

$$< \text{grad } B(t_i, x), y > \quad = \quad -1, \qquad i = 1, \ldots, N-1.$$

This implies that M is not critical with respect to x, a contradiction.

Case 2. $q = N$

Let n be given. Since $b_n \notin \text{SUL}_\varepsilon(x)$, it follows from Corollary 1.3 that there exist N vectors from

$$\text{grad } p(x_n), \text{grad } B(t_{1,n}, x_n), \ldots, \text{grad } B(t_{N,n}, x_n)$$

which are linearly dependent. By taking limits it follows that there exist N vectors from

$$\text{grad } p(x), \text{grad } B(t_1, x), \ldots, \text{grad } B(t_N, x)$$

which are linearly dependent. Since $M = \{t_1, \ldots, t_N\}$ is a critical subset of $N(b, x)$ we get a contradiction to (ii).

This proves (ii) $\Rightarrow$ (i).

We now show that (i) $\Rightarrow$ (ii), if the Kuhn-Tucker condition is sufficient for local solutions Suppose that (i) holds, i.e. $b \in int\ \text{SUL}_\varepsilon(x)$ for sufficiently small $\varepsilon > 0$. Therefore, it follows from Theorem 1.1 and Corollary 1.3 that there exists a critical subset of $N(b, x)$ consisting of N distinct points. We now assume that (ii) fails, i.e. there exists such a set $\{t_1, \ldots, t_N\}$ such that N vectors from

$$\{\text{grad } p(x), \text{grad } B(t_1, x), \ldots, \text{grad } B(t_N, x)\} \tag{1.5}$$

are linearly dependent. We will show that there exists a sequence (b_n) in $C(T)$ such that $b_n \to b$ and for all n, $x \in Z(b_n)$ and $N(b_n, x) = \{t_1, \ldots, t_N\}$. For each $i \in \{1, \ldots, N\}$ we choose an open neighborhood V_i of t_i such that the neighborhoods are disjoint. Then by Urysohn's lemma for each n there exists a function $h_n \in C(T)$ such that

$$h_n(t) = 0 \qquad \text{for all } t \in \{t_1, \ldots, t_N\},$$
$$h_n(t) = \tfrac{1}{n} \qquad \text{for all } t \in T \setminus \textstyle\bigcup_{i=1}^N V_i$$

and

$$0 < h_n(t) < \tfrac{1}{n} \qquad \text{for all } t \in \textstyle\bigcup_{i=1}^N V_i \setminus \{t_1, \ldots, t_N\}.$$

For all n we set $b_n = b + h_n$. Then (b_n) has the desired property. By Lemma 1.6 we may assume that MP(b_n) satisfies condition (CQ) in x for all n. Since $N(b_n, x) = \{t_1, \ldots, t_N\}$ is a critical set, by assumption x is a local solution of MP(b_n) for all n. But, it follows from (1.5) and Corollary 1.3 that x is not a strongly unique local solution of MP(b_n) for all n. This contradicts (i) and proves Theorem 1.4.

2 Best Approximation

In this section we consider applications of the previous results to best approximation of functions in the supremum norm.

Let V be an open set of $\mathbb{R}^{N-1}$ and S be a compact metric space. Moreover, let a fixed mapping $A : V \to C(S)$ be given such that the mapping

$$(s,x) \to \text{grad } A(x)(s) = (\frac{\partial A(x)(s)}{\partial x_1}, \ldots, \frac{\partial A(x)(s)}{\partial x_{N-1}}) \qquad ((s,x) \in S \times V)$$

is continuous. For $f \in C(S)$ a function $g_f \in G = A(V)$ is called (global) **best approximation** of f, if $\|f - g_f\|_\infty = \min \{\|f - g\|_\infty : g \in G\}$. This approximation problem can be written as a special semi-infinite optimization problem:

> Minimize $p(x_1, \ldots, x_{N-1}, x_N) = x_N$, subject to $(x_1, \ldots, x_{N-1}) \in V$ and $\eta A(x_1, \ldots, x_{N-1})(s) - x_N \leq \eta f(s)$ for all $(\eta, s) \in \{-1, 1\} \times S$.

For details see e.g. Brosowski [4] and Hettich & Zencke [7]. Therefore, the results in section 1 can be applied to this situation except Theorem 1.4 and Corollary 1.5, since the restriction functions in the approximation problem are of special type.

Let a function $f \in C(S)$ and a local best approximation $g_f = A(x_f) \in G$ of f be given. The **tangent space** of G at g_f is defined by

$$TG(g_f) = \text{span } \{\frac{\partial A(x_f)}{\partial x_1}, \ldots, \frac{\partial A(x_f)}{\partial x_{N-1}}\}$$

(Meinardus [9]).

The set of extreme points of $f - g_f$ is defined by

$$E(f - g_f) = \{s \in S : |f(s) - g_f(s)| = \|f - g_f\|_\infty\}.$$

A subset $\{s_1, \ldots, s_q\}$ of $E(f - g_f)$ is called **critical**, if $q \leq N$ and

$$\min_{i=1,\ldots,q} (f(s_i) - g_f(s_i))\, h(s_i) \quad \leq \quad 0 \text{ for all } h \in TG(g_f).$$

Moreover, a subset $\{s_1, \ldots, s_{N-1}\}$ of S is called **poised** w.r.t. $TG(g_f)$, if the determinant

$$\det(\frac{\partial A(x_f)(s_i)}{\partial x_j})_{i,j=1}^{N-1} \neq 0.$$

A function $g_f = A(x_f) \in G$ is called **local best approximation (w.r.t. the parameter)** of $f \in C(S)$, if there exists an $\epsilon > 0$ such that for all $g = A(x) \in G$ with $\|x - x_f\|_2 \leq \epsilon$,

$$\|f - A(x)\|_\infty \quad \geq \quad \|f - A(x_f)\|_\infty.$$

A function $g_f = A(x_f) \in G$ is called **strongly unique local best approximation (w.r.t. the parameter)** of $f \in C(S)$, if there exist $\epsilon > 0$ and $K_f > 0$ such that for all $g = A(x) \in G$ with $\|x - x_f\|_2 \leq \epsilon$,

$$\|f - A(x)\|_\infty \quad \geq \quad \|f - A(x_f)\|_\infty + K_f\|x - x_f\|_2.$$

Moreover, given $\epsilon > 0$ and $g_f = A(x_f) \in G$, we set

$$\text{SUL}_\epsilon(G, g_f) = \{\tilde{f} \in C(S) : \tilde{f} \text{ has a strongly unique local best approximation from G on } B_\epsilon(x_f)\}.$$

It follows from Theorem 1.1 that if $g_f \in G$ is a local best approximation of $f \in C(S)$, then there exists a critical subset of $E(f - g_f)$. The set G is called **Kolmogorov set of 2. kind**, if for all $f \in C(S)$ the above condition implies that $g_f \in G$ is a best approximation of f. Standard examples of such sets are the set of rational functions and the set of exponential functions (see Meinardus [9]).

Analogously as in section 1 we make the following

General assumption. Let $f \in C(S)$ have a local best approximation $g_f = A(x_f) \in G$. We only consider those $f \in C(S)$ such that, if $\varepsilon > 0$ is sufficiently small and $\tilde{f} \in C(S)$ has a strongly unique local best approximation $g_{\tilde{f}} = A(x_{\tilde{f}}) \in G$ with $x_{\tilde{f}} \in B_\varepsilon(x_f)$, then $g_{\tilde{f}}$ is a strongly unique local best approximation of $\tilde{f}$ on $B_\varepsilon(x_f)$.

We first prove an analog of Theorem 1.4 for the approximation problem. Since the arguments are similar we only give a sketch of proof.

Theorem 2.1 *Let $g_f \in G$ be a local best approximation of $f \in C(S)$. We consider the following statements.*

(i) $f \in int\ SUL_\varepsilon(G, g_f)$ for sufficiently small $\varepsilon > 0$.

(ii) There exists a critical subset $\{s_1, \ldots, s_N\}$ of $E(f - g_f)$ consisting of N distinct points and for every such set

$$\{s_1, \ldots, s_{i-1}, s_{i+1}, \ldots, s_N\}, \qquad i = 1, \ldots, N,$$

is poised w.r.t. $TG(g_f)$.

Then (ii) $\Rightarrow$ (i). Moreover, (i) $\Rightarrow$ (ii), if G is a Kolmogorov set of 2. kind.

Sketch of proof. (ii) $\Rightarrow$ (i). Suppose that (ii) holds, but (i) fails. It follows from Corollary 1.3 that $g_f = A(x_f) \in G$is a strongly unique local best approximation of f, i.e. there exist $\varepsilon > 0$ and $K_f > 0$ such that for all $g = A(x) \in G$ with $\|x - x_f\|_2 \leq \varepsilon$,

$$\|f - A(x)\|_\infty \ \geq \ \|f - A(x_f)\|_\infty \ + \ K_f \ \|x - x_f\|_2.$$

Since (i) fails, there exists a sequence (f_n) in $C(S)$ converging to f such that for all n, f_n does not have a strongly unique local best approximation from G on $B_\varepsilon(x_f)$. Analogously as in the proof of Theorem 1.4 for each n there exists a local best approximation $g_n = A(x_n) \in G$ of f_n on $B_\varepsilon(x_f)$ such that $x_n \to x_f$. This implies that $g_n \to g_f$. Moreover, by Theorem 1.1 for each n there exists a critical subset $M_n = \{s_{1,n}, \ldots, s_{q_n,n}\}$ of $E(f_n - g_n)$, i.e.

$$\min_{i=1,\ldots,q_n} (f_n(s_{i,n}) - g_n(s_{i,n}))\, h(s_{i,n}) \ \leq \ 0 \qquad \text{for all } h \in TG(g_n).$$

By going to a subsequence we may assume that $q_n = q \in \{1, \ldots, N\}$ for all n and $s_{i,n} \to s_i$, $i = 1, \ldots, q$. By taking limits it follows that

$$\min_{i=1,\ldots,q} (f(s_i) - g_f(s_i))\, h(s_i) \ \leq \ 0 \qquad \text{for all } h \in TG(g_f).$$

This shows that $M = \{s_1, \ldots, s_q\}$ is a critical subset of $E(f - g_f)$.

Case 1. $q \in \{1, \ldots, N-1\}$

By (ii) we may choose points $s_{q+1}, \ldots, s_N$ in $E(f - g_f)$ such that the points $s_1, \ldots, s_N$ are distinct. Moreover, it follows from (ii) that $\{s_1, \ldots, s_{N-1}\}$ is poised w.r.t. $TG(g_f)$. This implies that there exists a function $h \in TG(g_f)$ such that

$$h(s_i) = \text{sgn}\ (f(s_i) - g_f(s_i)), \qquad i = 1, \ldots, N-1$$

which contradicts the fact that M is a critical subset of $E(f - g_f)$.

Case 2. $q = N$

By going to a subsequence of (g_n) the following holds. Since $g_n \in G$ is not a strongly unique local best approximation of f_n, by Corollary 1.3 there exists an integer $j \in \{1, \ldots, N\}$ such that for all n,

$$\{s_{1,n}, \ldots, s_{j-1,n}, s_{j+1,n}, \ldots, s_{N,n}\}$$

is not poised w.r.t. $TG(g_n)$. By taking limits it follows that

$$\{s_1, \ldots, s_{j-1}, s_{j+1}, \ldots, s_N\}$$

is not poised w.r.t. $TG(g_f)$. Since M is a critical subset of $E(f - g_f)$ this contradicts (ii). This proves (ii) $\Rightarrow$ (i).

We now show that (i) $\not\Rightarrow$ (ii), if G is a Kolmogorov set of 2.kind. Suppose that (i) holds, i.e. $f \in \operatorname{int} \mathrm{SUL}_\varepsilon(G, g_f)$ for sufficiently small $\varepsilon > 0$. Therefore, it follows from Theorem 1.1 and Corollary 1.3 that there exists a critical subset of $E(f - g_f)$ consisting of N distinct points. We now assume that (ii) fails, i.e. there exists such a set $\{s_1, \ldots, s_N\}$ and an integer $j \in \{1, \ldots, N\}$ such that

$$\{s_1, \ldots, s_{j-1}, s_{j+1}, \ldots, s_N\}$$

is not poised w.r.t. $TG(g_f)$. We will show that there exists a sequence (f_n) in $C(S)$ such that $f_n \to f$ and for all n,

$$E(f_n - g_f) = \{s_1, \ldots, s_N\}.$$

Suppose for the moment that this is true. Since the set $\{s_1, \ldots, s_N\}$ is critical, by assumption $g_f \in G$ is a best approximation of f_n. On the other hand, since

$$\{s_1, \ldots, s_{j-1}, s_{j+1}, \ldots, s_N\}$$

is not poised w.r.t. $TG(g_f)$, it follows from Corollary 1.3 that $g_f \in G$ is not a strongly unique local best approximation of f_n. This contradicts (i).

Thus it remains to construct the sequence (f_n). For each $i \in \{1, \ldots, N\}$ we choose an open neighborhood V_i of s_i such that the neighborhoods are disjoint. Then by Urysohn's lemma for each n there exists a function $z_n \in C(S)$ such that

$$\begin{aligned} z_n(s) &= 1 && \text{for all } s \in \{s_1, \ldots, s_N\}, \\ z_n(s) &= 1 - \tfrac{1}{n} && \text{for all } s \in S \setminus \textstyle\bigcup_{i=1}^N V_i \\ \text{and} \\ 1 - \tfrac{1}{n} &< z_n(s) < 1 && \text{for all } s \in \textstyle\bigcup_{i=1}^N V_i \setminus \{s_1, \ldots, s_N\}. \end{aligned}$$

For all n, we set $f_n = z_n \cdot h + g_f$. Then the sequence (f_n) has the desired property. This proves Theorem 2.1.

The next result follows from Theorem 2.1 and Corollary 1.3.

Corollary 2.2 *Let $g_f \in G$ be a local best approximation of $f \in C(S)$ such that $f - g_f$ has exacty N extreme points. The following statements are equivalent.*

(i) $f \in int\ SUL_\varepsilon(G, g_f)$ for sufficiently small $\varepsilon > 0$.

(ii) g_f is a strongly unique local best approximation of f from G.

We now prove the following result.

Corollary 2.3 *Let $g_f \in G$ be a local best approximation of $f \in C[a, b]$ such that $TG(g_f)$ is an $(N - 1)$-dimensional Haar space. The following statements are equivalent.*

(i) $f \in int\ SUL_\varepsilon(G, g_f)$ for sufficiently small $\varepsilon > 0$.

(ii) g_f is a strongly unique local best approximation of f from G.

(iii) $f - g_f$ has at least N alternating extreme points.

Proof. Let $TG(g_f)$ be an $(N - 1)$-dimensional Haar space.Then it follows from Satz 4.2.2 in Hettich & Zencke [9] that (ii) is equivalent to (iii). Since $TG(g_f)$ is an $(N - 1)$-dimensional Haar space, $\{s_1, \ldots, s_N\}$ is a critical subset of $E(f - g_f)$ if and only if $s_1, \ldots, s_N$ are alternating extreme points of $f - g_f$, and therefore (ii) implies condition (ii) in Theorem 2.1. This shows that (iii) $\Rightarrow$ (i) and proves Corollary 2.3.

The set of **rational functions** on $[-1,1]$ is defined by

$$R_{m,n} = \{\frac{p}{q} : p \in \Pi_m,\ q \in \Pi_n,\ q(s) = 1 + \sum_{i=1}^{n} a_i s^i \neq 0 \text{ for all } s \in [-1,1]\},$$

where Π_m denotes the space of polynomials of degree at most m. It is well known that for $G = R_{m,n}$ all tangent spaces $TG(g_f)$ are Haar spaces of dimension less or equal to $N-1 = m+n+1$ (see Meinardus [9]). A function $f \in C[-1,1]$ is called **normal**, if its best approximation g_f from $R_{m,n}$ has full degree after dividing by common factors. In this case $TG(g_f)$ has dimension $N-1$.

In the following we deduce stability results for global rational approximations and spline approximations with free knots.

A function $g_f \in G$ is called **strongly unique best approximation** of $f \in C(S)$, if there exists a constant $C_f > 0$ such that for all $g \in G$,

$$\|f-g\|_\infty \quad \geq \quad \|f-g_f\|_\infty + C_f\|g-g_f\|_\infty.$$

We set

$$SU(G) = \{f \in C(S) : f \text{ has a strongly unique best approximation from G}\}.$$

The next result on rational approximation is known (compare Cheney[5], Cheney & Loeb [6], Rice [1] and Bartke [2]).

Corollary 2.4 *A function $f \in C[-1,1]$ is normal if and only if $f \in int\ SU(R_{m,n})$.*

Proof. It is well known that f is normal if and only if $TG(g_f)$ is an $N-1$-dimensional Haar space and $f-g_f$ has at least N alternating extreme points, where $g_f \in R_{m,n}$ is the best approximation of f. Moreover, f is normal if and only if f has a strongly unique best approximation from $R_{m,n}$ (see Cheney [5]). Therefore, the result follows from Corollary 2.3. This proves Corollary 2.4.

We now consider best approximation by splines with free knots. By $S_{m,k}$ we denote the set of **splines** of degree m with k **free knots**. Moreover, by $\tilde{S}_{m,k}$ we denote the set of splines of degree m with k free simple knots. (For the definitions see Schumaker[17]).

Corollary 2.5 *Let $f \in C[a,b]$, $m \geq 2$, $g_f \in S_{m,k} \setminus S_{m,k-1}$ have k simple knots and g_f be a best approximation of f from $S_{m,k}$ such that $f-g_f$ has $N = m+2k+2$ alternating extreme points. The following statements are equivalent.*

(i) $f \in$ int $SU(S_{m,k})$

(ii) g_f is a strongly unique best approximation of f from $S_{m,k}$.

In particular, if $f^{(m+1)} > 0$ on (a,b), then $f \in int\ SU(S_{m,k})$.

Proof. It was shown in [14] that for $g_f \in \tilde{S}_{m,k}$, g_f is a strongly unique local best approximation of f from $\tilde{S}_{m,k}$ if and only if g_f is a strongly unique best approximation of f from $S_{m,k}$. Therefore, (i) $\Leftrightarrow$ (ii) follows from Corollary 2.2. Moreover, it follows from Braess [3,p.364] and Schaback [16,p.389] that if $f^{(m+1)} > 0$ on (a,b) , then f has a strongly unique best approximation $g_f \in S_{m,k}$ and $f-g_f$ has exactly $m+2k+2$ alternating extreme points. This proves Corollary 2.5.

References

[1] B.Bank, J.Guddat, D.Klatte, B.Kummer, K.Tammer, Nonlinear parametric optimization, Akademie Verlag, Berlin 1982.

[2] B.Bartke, Beste Approximation mit varisolventen Familien, Archiv Math. 32 (1979), 79-85.

[3] D.Braess, Chebyshev approximation by spline functions with free knots, Numer. Math. 17 (1971), 357-366.

[4] B.Brosowski, Parametric semi-infinite optimization, Peter Lang, Frankfurt 1982.

[5] E.W.Cheney, Introduction to approximation theory, McGraw Hill, New York 1966.

[6] E.W.Cheney and H.L.Loeb, Generalized rational approximation, SIAM J. Numer. Anal. 1 (1964), 11-25.

[7] R.Hettich and H.Zenke, Numerische Methoden der Approximation und semi-infiniten Optimierung, Teubner, Stuttgart 1982.

[8] W.Krabs, Optimierung und Approximation, Teubner, Stuttgart 1975.

[9] G.Meinardus, Approximation of functions: theory and numerical Methods, Springer, Berlin 1967.

[10] G.Nürnberger, Strong unicity of best approximations: a numerical aspect, Numer. Funct. Anal. Optim. 6 (1983), 399-421.

[11] G.Nürnberger, Unicity in semi-infinite optimization and approximation, in Parametric optimization and approximation, B.Brosowski and F.Deutsch eds., ISNM 72, Birkhäuser, Basel 1985, 231-247

[12] G.Nürnberger, Global unicity in semi-infinite optimization, Numer. Funct. Anal. Optim. 8 (1985), 173-191.

[13] G.Nürnberger, Best approximation by spline functions: theory and numerical methods, in Delay equations, approximation and application, G.Meinardus and G.Nürnberger eds., ISNM 74, Birkhäuser, Basel 1985, 180-212.

[14] G.Nürnberger, Strongly unique spline approximations with free knots, preprint.

[15] J.R.Rice, The approximation of functions II, Addison & Wesley, Reading, Massachusetts 1969.

[16] R.Schaback, On alternation numbers in nonlinear Chebyshev approximation, J.Approx.Theory 23 (1978), 379-391.

[17] L.L.Schumaker, Spline functions: basic theory, Wiley-Interscience, New York 1981.

OBTAINING CONVERGENCE RATES FOR APPROXIMATIONS IN STOCHASTIC PROGRAMMING *

Werner Römisch[1)] and Anton Wakolbinger[2)]

Abstract:

Stochastic programming problems are viewed as parametric programs with respect to the underlying probability distribution. General results about continuity properties of expectation functionals and of certain set-valued mappings (defined by probabilistic constraints) with respect to the weak topology on the set of probability distributions are proved. They are applied to obtain quantitative distribution sensitivity results for the optimal values of recourse and chance-constrained problems, respectively, of stochastic (linear) programming. These results lead to convergence rates for approximations of such problems (e.g. via certain discrete probability measures or via empirical measures).

1. Introduction

In the (theoretical and numerical) treatment of stochastic programming problems, their stability with respect to a perturbation of the distribution of the underlying random variables plays an essential role.

If, e.g., this distribution (or some of its parameters) is not known, the stability w.r. to the distribution is important when estimates for the distributions resp. its parameters are used. Questions of this kind were the starting point in [10], [11], [18], [20], [27], [28] and [30]. There, "stability" means continuity of optimal values, optimal solutions and (or) Kuhn-Tucker points with respect to the distribution resp. its parameters. (For a survey on stability results in parametric optimization, see, e.g., [2].)

Another motivation for the interest in such stability results comes from the application of numerical methods for the solution of stochastic programming problems. One of the main lines in the development of numerical methods consists in approximating the (known) probability distribution by "simpler" (e.g. discrete) distributions. ([5], [14], [16], [17], [20], [22], [26], see also the survey [31])

The present paper gives a contribution to the quantitative distribution sensitivity analysis of stochastic programming problems, which, though suitable for both of the directions mentioned above, stresses, as an application, the investigation of convergence rates of approximative methods. We explain our aim by the following rather general class of stochastic programming problems:

---

* Research supported by the Scientific Exchange Treaty between Austria and the GDR.

1) Sektion Mathematik
Humboldt-Universität
DDR-1086 Berlin

2) Institut für Mathematik
Johannes-Kepler-Universität
A-4040 Linz, Austria

$$\min\{ \int_Z f(z,x)\mu(dz) \,|\, x \in \mathbb{R}^m,\ \mu(\{z \in Z \,|\, x \in X(z)\}) \geq \alpha\} \tag{1.1}$$

where Z is a Borel subset of $\mathbb{R}^s$, f a mapping from $Z \times \mathbb{R}^m$ into $\mathbb{R}$, X is a set-valued mapping from Z into $\mathbb{R}^m$, $\alpha \in [0,1]$ and μ is a probability distribution on Z.

Note that a number of stochastic programs with recourse or with chance (i.e. probabilistic) constraints fit into this class (see [15], [31] and Sections 3,4).

We consider (1.1) as a parametric optimization problem with parameter μ varying in $\mathcal{P}(Z)$, the set of all Borel probability measures on Z, equipped with the topology of weak convergence ([4], [3, Ch. 1], [9]). It is well known that (even if Z is a seperable metric space) this topology is metrizable, e.g. by the Prokhorov metric ρ and the bounded Lipschitz (or Dudley) metric β defined by ([7], [9], [32])

$$\begin{aligned} \rho(\mu,\nu) &:= \inf\{\varepsilon > 0 \,|\, \mu(B) \leq \nu(B^\varepsilon) + \varepsilon, \text{ for all Borel sets } B \subseteq Z\} \\ \beta(\mu,\nu) &:= \sup\{ |\int g(z)\mu(dz) - \int g(z)\nu(dz)| \,/\, g \in BL(Z,d), \|g\|_{BL} \leq 1\} \end{aligned} \tag{1.2}$$

where $B^\varepsilon := \{z \in Z \,|\, \inf\{d(z,\tilde{z}) \,|\, \tilde{z} \in B\} < \varepsilon\}$ $\quad (B \subseteq Z,\ \varepsilon > 0)$,

$$BL(Z,d) := \{g : Z \to \mathbb{R} \,|\, \|g\|_{BL} := $$

$$:= \sup_{z \in Z} |g(z)| + \sup_{z \neq \tilde{z}} \frac{|g(z) - g(\tilde{z})|}{d(z,\tilde{z})} < \infty\},$$

d denotes the metric on Z.

In this paper, we will impose only mild regularity assumptions on f and μ, but will restrict ourselves to the stability of the optimal value $\varphi(\mu)$ of (1.1) with respect to μ.

More precisely, we will prove results on the Hölder- and local Lipschitz continuity of $\varphi : \mathcal{P}(Z) \to \mathbb{R}$ with respect to the metrics β and ρ. (Let us recall here that the metrics β and ρ are equivalent in view of $\frac{1}{2}\beta(\mu,\nu) \leq \rho(\mu,\nu) \leq (\frac{3}{2}\beta(\mu,\nu))^{1/2}$; $\mu,\nu \in \mathcal{P}(Z)$ ([7]))

From [1, Theorem 7, p. 53] resp. [19] it is rather obvious that Hölder- resp. local Lipschitz properties of the mappings

$$(x,\mu) \mapsto \int_Z f(z,x)\mu(dz)$$

$$\mu \mapsto C_\alpha(\mu) = \{x \in \mathbb{R}^m \,|\, \mu(\{z \in Z \,|\, x \in X(z)\}) \geq \alpha\}$$

are essential for the desired continuity properties of $\varphi : \mathcal{P}(Z) \to \mathbb{R}$. This will be discussed in more detail in Sections 2 and 4. In Section 2, we obtain a result about the Hölder continuity (with respect to the metric β) of the mapping

$$\mu \mapsto \int_Z g(z)\mu(dz)$$

on certain subsets of $\mathcal{P}(Z)$ (defined by moment conditions which are related to a quantitative local Lipschitz property ([25], [12]) imposed on g).

In Section 3, we apply this result to stochastic linear programming problems with complete fixed recourse, obtaining a convergence rate

$$|\varphi(\mu) - \varphi(\mu_n)| = O(\beta(\mu,\mu_n)^{1-1/p})$$

(where p is linked with moment conditions on μ and μ_n) if (μ_n) converges weakly towards μ. Especially, we investigate in Section 3 discrete approximations of μ by means of conditional expectations and the approximation of μ by empirical measures.

In Section 4, we prove a result on the local Lipschitz continuity of the mapping

$$\mu \mapsto C_\alpha(\mu)$$

with respect to the Hausdorff-distance and the Prokhorov-metric ρ and apply it to a stochastic linear programming problem with chance constraints (Corollary 4.7).

2. Continuity properties of expectation functionals of probability measures with respect to the weak topology

Let (Z,d) be a separable metric space and $\mathcal{P}(Z)$ be the set of all Borel probability measures on Z equipped with the topology of weak convergence (which will be denoted by $\xrightarrow{w}$). Let g be a mapping from Z into IR such that

$$|g(z)-g(\tilde{z})| \leq L(\max\{d(z,0),d(\tilde{z},0)\})d(z,\tilde{z}) \quad (z,\tilde{z} \in Z) \tag{2.1}$$

where $L: \mathbb{R}_+ \to \mathbb{R}_+$ is continuous and monotonically increasing, and $0 \in Z$ is some distinguished element.
For the following we put

$$L_1(t) := L(t).t \qquad (t \in \mathbb{R}_+),$$

$$M_p(\mu) := (\int_Z L_1(d(z,0))^p \, \mu(dz))^{1/p} \qquad (\mu \in \mathcal{P}(Z),\ 1 \leq p < \infty).$$

Note that $\int_Z |g(z)|\mu(dz)$ is finite if $M_1(\mu) < \infty$.

We are aiming at "quantitative continuity results" of the type

$$\left| \int_Z g(z)\mu(dz) - \int_Z g(z)\nu(dz) \right| \leq \psi(\mu,\nu) \quad (\mu,\nu \in \mathcal{P}(Z)) \tag{2.2}$$

If $g: Z \to \mathbb{R}$ is a bounded Lipschitz function, then (2.2) holds with
$\psi(\mu,\nu) := \|g\|_{BL}\, \beta(\mu,\nu)$.
The main result of this section is an estimate of the type (2.2), where ψ depends on g, on the generalized moments $M_p(\mu)$ and $M_p(\nu)$, and on $\beta(\mu,\nu)$:

2.1 Theorem: For all $p \in]1,\infty[$, there holds (with notations and general assumptions as above)

$$\left| \int_Z g(z)\mu(dz) - \int_Z g(z)\nu(dz) \right| \le C(1 + M_p(\mu) + M_p(\nu))\, \beta(\mu,\nu)^{1-1/p} \tag{2.3}$$

for all $\mu,\nu \in \mathcal{P}(Z)$ such that $M_p(\mu) + M_p(\nu) < \infty$ (where $C := \max\{4, 2L(1) + 10\,|g(0)|\}$).

Proof: Let $\mu,\nu \in \mathcal{P}(Z)$ be such that $M_p(\mu) + M_p(\nu) < \infty$. For arbitrary $r > 0$, g is Lipschitz continuous on $K_r := \{z \in Z \mid d(z,0) \le r\}$ with Lipschitz constant $L(r)$, and there holds $\sup_{z \in K_r} |g(z)| \le |g(0)| + L_1(r)$.

Any bounded Lipschitz function from a subset of Z into IR has a bounded Lipschitz extension to the whole of Z with the same BL-norm ([9, Corollary 7.4]).

Let g_r be such an extension of $g|_{K_r}$, i.e.

$$\|g_r\|_{BL} = \|g|_{K_r}\|_{BL} \le |g(0)| + L_1(r) + L(r).$$

Hence we obtain

$$\begin{aligned}
&\left|\int_Z g\, d(\mu-\nu)\right| \le \\
&\le \left|\int_Z (g-g_r)d\mu\right| + \left|\int_Z g_r d(\mu-\nu)\right| + \left|\int_Z (g-g_r)d\nu\right| \le \\
&\le \int_{Z\setminus K_r} (|g| + |g_r|)\, d(\mu+\nu) + \|g_r\|_{BL} \cdot \beta(\mu,\nu) \le \qquad (2.4) \\
&\le \int_{Z\setminus K_r} (L_1(d(z,0)) + L_1(r) + 2|g(0)|)(\mu(dz)+\nu(dz)) + (|g(0)| + L_1(r) + L(r))\beta(\mu,\nu) = \\
&= \int_{Z\setminus K_r} L_1(d(z,0))(\mu(dz)+\nu(dz)) + (L_1(r) + 2|g(0)|)(\mu+\nu)(Z\setminus K_r) + \\
&\qquad + (|g(0)| + L_1(r) + L(r))\beta(\mu,\nu)
\end{aligned}$$

From Chebyshev's inequality we get

$$\mu(Z\setminus K_r) \le L_1(r)^{-p} \int_Z L_1(d(z,0))^p \mu(dz) = \left(\frac{M_p(\mu)}{L_1(r)}\right)^p \tag{2.5}$$

this estimate together with Hölder's inequality yields

$$\begin{aligned}
\int_{Z\setminus Kr} L_1(d(z,0))\mu(dz) &\le \left(\int_{Z\setminus K_r} L_1(d(z,0))^p \mu(dz)\right)^{1/p} \mu(Z\setminus K_r)^{1-1/p} \le \\
&\le M_p(\mu)^p L_1(r)^{1-p} \qquad (2.6)
\end{aligned}$$

Using (2.5), (2.6) and the same estimates for ν instead of μ, we continue (2.4) by

$$\begin{aligned} &\left|\int_Z g\,d(\mu-\nu)\right| \le \\ &\le (M_p(\mu)^p + M_p(\nu)^p)L_1(r)^{1-p} + \\ &+ (M_p(\mu)^p + M_p(\nu)^p)(L_1(r) + 2|g(0)|)L_1(r)^{-p} + \\ &+ (L_1(r) + L(r) + |g(0)|)\beta(\mu,\nu) \le \\ &\le (M_p(\mu) + M_p(\nu)).2A_p(L_1(r)^{1-p} + |g(0)|L_1(r)^{-p}) + \\ &+ (2L_1(r) + L(1) + |g(0)|)\beta(\mu,\nu) \end{aligned} \tag{2.7}$$

where we put $A_p := \max\{M_p(\mu)^{p-1}, M_p(\nu)^{p-1}\}$.

Now we choose $r > 0$ such that

$$L_1(r) = \max\{M_p(\mu), M_p(\nu)\}.\beta(\mu,\nu)^{-1/p} = A_p^{1/(p-1)}\,\beta(\mu,\nu)^{-1/p}$$

(note that $L_1(0) = 0$, $\lim_{t\to+\infty} L_1(t) = +\infty$ and L_1 is continuous, hence L_1 is surjective).

We then have $A_pL_1(r)^{1-p} = \beta(\mu,\nu)^{1-1/p}$ and
$A_pL_1(r)^{-p} = A_p^{-1/(1-p)}\,\beta(\mu,\nu) = (\max\{M_p(\mu),M_p(\nu)\})^{-1}\,\beta(\mu,\nu)$
and continue (2.7) by

$$\begin{aligned} &\left|\int_Z g\,d(\mu-\nu)\right| \le \\ &\le 4(M_p(\mu)+M_p(\nu))\beta(\mu,\nu)^{1-1/p} + (L(1)+5|g(0)|)\beta(\mu,\nu) \le \\ &\le \max\{4,2L(1)+10|g(0)|\}(1+M_p(\mu)+M_p(\nu))\beta(\mu,\nu)^{1-1/p} \end{aligned}$$

(note that $\beta(\mu,\nu)^{1/p} \le 2^{1/p} \le 2$) □

<u>2.2 Remark:</u> In view of $\beta(\mu,\nu) \le 2\rho(\mu,\nu)$ ($\mu,\nu \in \mathcal{P}(Z)$), Theorem 2.1 improves the estimate in [24, Theorem 1] considerably.
Theorem 2.1 states that the mapping $\mu \mapsto \int_Z g\,d\mu$ is, for each $p > 1$,
Hölder continuous on the subset $\{\mu \in \mathcal{P}(Z) | M_p(\mu) < \infty\}$ of $\mathcal{P}(Z)$ with exponent $1-1/p$.

The next proposition (which, in some sense, is a converse to Theorem 2.1), states that this exponent is "optimal"

<u>2.3 Proposition:</u> Let $Z = \mathbb{R}$, $g(z) := z^k$ ($k \in \mathbb{N}$ fixed), $L(t) := k\,t^{k-1}$ Let further $\mu \in \mathcal{P}(Z)$

and $p \geq 1$ be such that $M_p(\mu) < \infty$. Then there exists a sequence (μ_n) in $\mathcal{P}(Z)$ converging weakly towards μ such that $(M_p(\mu_n))$ is bounded and

$$\left|\int_Z g(z)\mu(dz) - \int_Z g(z)\mu_n(dz)\right| \geq C \,.\, \beta(\mu,\mu_n)^{1-1/p} \tag{2.8}$$

for all $n \in \mathbb{N}$ and some constant $C > 0$.

Proof: Put $\mu_n := (1-\frac{1}{n})\mu + \frac{1}{n}\delta_{n^{1/(kp)}}$ $(n \in \mathbb{N})$.

Then we observe that $M_p(\mu_n) = ((1-\frac{1}{n})\int_Z k^p\,|z|^{kp}\,\mu(dz) + \frac{1}{n}\,k^p n)^{1/p}$

is bounded, and that

$$\left|\int_Z gd(\mu-\mu_n)\right| = \left|\frac{1}{n}\int_Z z^k\mu(dz) - \frac{1}{n}\,n^{1/p}\right| \geq C_1 n^{1/p-1} \tag{2.9}$$

for a suitable $C_1 > 0$.

On the other hand there holds $\beta(\mu,\mu_n) \leq 2\rho(\mu,\mu_n) \leq \frac{2}{n}$, which implies that

$$\beta(\mu,\mu_n)^{1-1/p} \leq C_2 n^{1/p-1} \tag{2.10}$$

for a suitable $C_2 > 0$.
Combining (2.9) and (2.10) we arrive at (2.8). □

2.4 Remark: In a similar way as in Proposition 2.3 it can be shown that for general Z, unbounded $g: Z \to \mathbb{R}$ and $\mu \in \mathcal{P}(Z)$ such that $\int |g(z)|^p\mu(dz) < \infty$, there exists a sequence (μ_n) in $\mathcal{P}(Z)$ converging weakly towards μ such that (2.8) holds.

3. Convergence rates for approximations in stochastic linear programming with complete fixed recourse

We consider the following class of problems, which has been studied in detail in [29] and [15]:

$$\min\{c^T x + \int_Z Q(z,x)\mu(dz)\,|\,x \in X_0\} \tag{3.1}$$

where $X_0 \subseteq \mathbb{R}^m$ is a convex, compact polyhedron, and $c \in \mathbb{R}^m$,

$$Q(z,x) := \eta(a,b-Ax), \quad \forall z = (a,b,A) \in Z,\ \forall x \in X_0, \tag{3.2}$$

$$Z := \{(a,b,A)\,|\,b \in \mathbb{R}^r,\ A \in \mathcal{L}(\mathbb{R}^m,\mathbb{R}^r),\ a \in \mathbb{R}^d,\ \{u \in \mathbb{R}^r\,|\,W^T u \leq a\} \neq \emptyset\}, \tag{3.3}$$

$$\eta(a,v) := \inf\{a^T y \mid y \in \mathbb{R}^d,\ Wy = v,\ y \geq 0\}, \quad (3.4)$$
$$\forall v \in \mathbb{R}^r$$

$W \in \mathcal{L}(\mathbb{R}^d, \mathbb{R}^r)$ is the "fixed recourse" matrix, and $\mu \in \mathcal{P}(Z)$.

We generally assume that for all $v \in \mathbb{R}^r$ the set $\{y \in \mathbb{R}^d \mid Wy = v,\ y \geq 0\}$ is non-void ("complete fixed recourse"). This implies that $\eta(a,v) < +\infty$ for all $a \in \mathbb{R}^d$ and $v \in \mathbb{R}^r$. Besides, due to the duality theorem of linear optimization, for all $a \in \mathbb{R}^d$ such that $\{u \in \mathbb{R}^r \mid W^T u \leq a\} \neq \emptyset$ the dual problem

$$\max\{v^T u \mid u \in \mathbb{R}^r,\ W^T u \leq a\}$$

has a solution u_* with $\eta(a,v) = v^T u^*$. Hence $\eta(a,v)$ is finite for all $(a,v) \in D$ where

$$D := \{(a,v) \in \mathbb{R}^d \times \mathbb{R}^r \mid \{u \in \mathbb{R}^r \mid W^T u \leq a\} \neq \emptyset\}. \quad (3.5)$$

In the sequel we will use the following result from linear parametric optimization:

3.1 Lemma: [21, Theorem 8.8, p. 219]
η (as defined in (3.4)) is continuous on D (defined in (3.5)). Moreover, there exists a finite number of convex cones K_j $(j = 1,\ldots,N)$ such that $\bigcup_j K_j = D$, and there exist matrices $C_j \in \mathcal{L}(\mathbb{R}^d, \mathbb{R}^r)$ $(j = 1,\ldots,N)$ such that

$$\eta(a,v) = (C_j a)^T v \quad \text{for all } (a,v) \in K_j. \quad (3.6)$$

In addition to Lemma 3.1, we need a quantitative continuity result on η:

3.2 Lemma: η obeys the following Lipschitz condition on D:

$$|\eta(a_1,v_1) - \eta(a_2,v_2)| \leq MR(\|a_1 - a_2\| + \|v_1 - v_2\|),$$

if $\max\{\|a_1\| + \|v_1\|, \|a_2\| + \|v_2\|\} \leq R$,

where $M := \max_{j=1,\ldots,N} \|C_j\|$ and C_j is as in Lemma 3.1.

Proof: Obviously there holds in view of (3.6) for all $(a_1,v_1), (a_2,v_2) \in K_j$ $(j = 1,\ldots,N)$:

$$|\eta(a_1,v_1) - \eta(a_2,v_2)| \leq \|C_j\|(\|a_1\| \cdot \|v_1 - v_2\| + \|v_2\| \cdot \|a_1 - a_2\|) \leq$$

$$\leq M . \max\{\|a_1\| + \|v_1\|, \|a_2\| + \|v_2\|\}\ (\|a_1 - a_2\| + \|v_1 - v_2\|).$$

Now put, for $R > 0$,

$$D_R := D \cap \{(a,v) \in \mathbb{R}^d \times \mathbb{R}^r \mid\ \|a\| + \|v\| \leq R\}$$

Then η is continuous on D_R (Lemma 3.1), and, for all $a,v \in D_R$ there holds

$$\eta(a,v) \in \bigcup_{j=1}^{N} \{(C_j a)^T v\},$$

i.e. $\eta(a,v)$ takes one of the values of N mappings, all of which are Lipschitz continuous with constant MR. Using [13, Theorem 2.1], one shows now, analogously to [6, Theorem 2.2], that η itself is Lipschitz continuous on D_R with constant MR. □

We now consider the mapping $f: Z \times X_0 \longrightarrow \mathbb{R}$,

$$f(z,x) := c^T x + Q(z,x)$$
$$= c^T x + \eta(a,b-Ax) \quad \text{for all } z = (a,b,A) \in Z,\ x \in X_0,$$

where Z is the subset of $\mathbb{R}^d \times \mathbb{R}^r \times \mathcal{L}(\mathbb{R}^m, \mathbb{R}^r)$ defined in (3.3) (equipped with the norm $\|z\| := \|a\| + \|b\| + \|A\|$).
f is real valued, as $(a,b-Ax) \in D$ for all $x \in X_0$.

For $\mu \in \mathcal{P}(Z)$, we denote by

$$\varphi(\mu) := \inf\{ \int_Z f(z,\mu)\mu(dz) | x \in X_0\}$$

the optimal value of problem (3.1).

The following result deals with the quantitative continuity properties of φ with respect to the bounded Lipschitz metric on $\mathcal{P}(Z)$:

<u>3.3 Theorem:</u> Let (μ_n) be a sequence which converges weakly towards $\mu \in \mathcal{P}(Z)$, and assume that $\int_Z \|z\|^{2p}\mu(dz) < \infty$ and $\int_Z \|z\|^{2p}\mu_n(dz) < \infty$ ($n \in \mathbb{N}$). Then there holds

$$|\varphi(\mu) - \varphi(\mu_n)| = O((1 + (\int_Z \|z\|^{2p}\mu_n(dz))^{1/p}\ \beta(\mu,\mu_n)^{1-1/p}))$$

<u>Proof:</u> We are going to apply Theorem 2.1; hence we need an estimate of the type (2.1):
For arbitrary $x \in X_0$ and $z = (a,b,A)$, $\tilde{z} = (\tilde{a},\tilde{b},\tilde{A}) \in Z$ there holds according to Lemma 3.2:

$$|f(z,x) - f(\tilde{z},x)| = |\eta(a,b-Ax) - \eta(\tilde{a},\tilde{b}-\tilde{A}x)| \le$$
$$\le M \max\{\|a\| + \|b-Ax\|,\ \|\tilde{a}\| + \|\tilde{b}-\tilde{A}x\|\}(\|a-\tilde{a}\| + \|b-\tilde{b}+(A-\tilde{A})x\|) \le$$
$$\le M.K^2 \max\{\|z\|,\ \|\tilde{z}\|\}\ \|z-\tilde{z}\|$$

where $K := \max\{1, \max\{\|x\| | x \in X_0\}\}$.

Hence, for each $x \in X_0$, $g := f(.,x): Z \to \mathbb{R}$ fulfills the Lipschitz condition (2.1) with $L(t) := L_0 t$, $t \in \mathbb{R}_+$, and $L_0 := MK^2$, which yields, for each $x \in X_0$, the estimate.

$$|f(z,x)| \le L_0\|z\|^2 + |f(0,x)| \le L_0\|z\|^2 + K\|c\|$$

From this we get, due to our assumption, that

$$\int_Z |f(z,x)|\mu(dz) < \infty, \int_Z |f(z,x)|\mu_n(dz) < \infty \quad (n \in \mathbb{N}, x \in X_0)$$

and $\varphi(\mu) > -\infty$, $\varphi(\mu_n) > -\infty$.

Now take an arbitrary $x \in X_0$. Putting $g := f(.,x)$, we conclude from Theorem 2.1 that there exists a constant $C > 0$ (independent of x) such that

$$\left|\int_Z f(z,x)\mu(dz) - \int_Z f(z,x)\mu_n(dz)\right| \leq$$
$$\leq C(1 + M_p(\mu_n))\beta(\mu,\mu_n)^{1-1/p} \quad (n \in \mathbb{N}).$$

From this estimate our assertion results in view of

$$M_p(\mu_n) = L_0(\int_Z \|z\|^{2p} \mu_n(dz))^{1/p} \quad (n \in \mathbb{N}) \quad \text{and}$$

$$|\varphi(\mu) - \varphi(\mu_n)| \leq \sup_{x \in X_0} \left|\int_Z f(z,x)\mu(dz) - \int_Z f(z,x)\mu_n(dz)\right| \qquad \square$$

Theorem 3.3 improves the result of [24, Theorem 5] and can be used to investigate speed of convergence of approximations of different type for the problem (3.1). We demonstrate this first for the case of discrete approximations of μ by conditional expectations. This type of approximation has gained particular attention in the literature ([5], [16], [17], [31]).

<u>3.4 Corollary:</u> Let $\mu \in \mathcal{P}(Z)$ be such that $\int_Z \|z\|^{2p} \mu(dz)$ is finite for some $p > 1$. Let, for every $n \in \mathbb{N}$, $S_n := \{Z_{nk} | k = 1,\dots,m(n)\}$ be a partition of Z into Borel sets such that S_{n+1} is a refinement of S_n $(n \in \mathbb{N})$ and that $\bigcup_{n \in \mathbb{N}} S_n$ generates the σ-algebra of Borel subsets of Z.

Let $\mu_n := \sum_{k=1}^{m(n)} \mu(Z_{nk})\delta_{e_{nk}}$, where

$$e_{nk} := \mu(Z_{nk})^{-1} \int_{Z_{nk}} z\mu(dz) \quad \text{for } k = 1,\dots,m(n) \text{ and } n \in \mathbb{N}.$$

Then $|\varphi(\mu) - \varphi(\mu_n)| = O(\beta(\mu,\mu_n)^{1-1/p})$.

<u>Proof:</u> By martingale convergence we have $\mu_n \xrightarrow{w} \mu$, and from Jensen's inequality we get

$$\int_Z \|z\|^{2p}\mu_n(dz) \leq \int_Z \|z\|^{2p}\mu(dz) \quad (n \in \mathbb{N})$$

Now Theorem 3.3 yields the assertion. $\square$

As a further application we mention the approximation of $\mu \in \mathcal{P}(Z)$ by empirical measures. Let $(z_i)_{i \in \mathbb{N}}$ be a sequence of independent Z-valued random variables with distribution μ, and consider the empirical measures

$$\mu_n(\omega) := \frac{1}{n} \sum_{i=1}^{n} \delta_{z_i(\omega)}, \; n \in \mathbb{N}, \; \omega \in \Omega.$$

<u>3.5 Corollary:</u>

a) If $\mu \in \mathcal{P}(Z)$ is such that $\int_Z \|z\|^{2p} \mu(dz)$ is finite for some $p > 1$, we have

$$E[|\varphi(\mu) - \varphi(\mu_n(\omega))|] = O(E[\beta(\mu, \mu_n(\omega))]^{1-1/p})$$

b) If $\int_Z \|z\|^{\alpha} d\mu$ is finite for all $\alpha > 1$, and if the support of μ is a subset of a k-dimensional space, then we have

$$E[|\varphi(\mu) - \varphi(\mu_n(\omega))|] = O(n^{-1/\Delta}) \quad \text{for every } \Delta > \max(2,k).$$

<u>Proof:</u> From the strong law of large numbers there follows (cf. [7,9]) that $\mu_n(\omega) \xrightarrow{w} \mu$ almost surely.

Besides, there holds

$$\int_Z \|z\|^{2p} \mu_n(\omega)(dz) = \frac{1}{n} \sum_{i=1}^{n} \|z_i(\omega)\|^{2p} < \infty \qquad (\omega \in \Omega, \; n \in \mathbb{N}).$$

Hence Theorem 3.3 is applicable, and there exists a constant $C > 0$ such that for all $\omega \in \Omega$ there holds

$$|\varphi(\mu) - \varphi(\mu_n(\omega))| \le C(1 + (\int_Z \|z\|^{2p} \mu_n(\omega)(dz)))^{1/p}) \beta(\mu, \mu_n(\omega))^{1-1/p} =$$

$$= C(1 + (\frac{1}{n} \sum_{i=1}^{n} \|z_i(\omega)\|^{2p})^{1/p}) \beta(\mu, \mu_n(\omega))^{1/q} \qquad (\frac{1}{p} + \frac{1}{q} = 1)$$

Applying Hölder's inequality, we get the estimate

$$E[|\varphi(\mu) - \varphi(\mu_n(\omega))|] \le C(1 + E[(\frac{1}{n} \sum_{i=1}^{n} \|z_i(\omega)\|^{2p})]^{1/p}) E[\beta(\mu, \mu_n(\omega))]^{1/q}$$

$$= C(1 + (\int_Z \|z\|^{2p} \mu(dz))^{1/p}) E[\beta(\mu, \mu_n(\omega))]^{1-1/p}$$

This proves assertion a)

b) From part a) there follows for any $p > 1$.

$$E[|\varphi(\mu) - \varphi(\mu_n(\omega))|] = O(E[\beta(\mu, \mu_n(\omega))]^{1-1/p})$$

From [8, Prop. 3.4] we obtain that

$$E[\beta(\mu, \mu_n(\omega))] = O(n^{-1/\Delta}) \quad \text{for all } \Delta > \max\{2,k\}.$$

Hence the assertion results. □

The just stated results on the mean-convergence of the optimal values may, on one hand, be interpreted as a theoretical basis for methods of Monte-Carlo-type for the numerical treatment of recourse problems in stochastic linear programming. On the other hand they provide asymptotic properties of the optimal values, if estimates for the (unknown) distribution μ, based on empirical distributions, are used. For such a situation, similar asymptotic properties of the optimal values (and also of optimal solutions under stronger regularity assumptions on (3.1) - (3.4) and μ) are derived in [10], [11], [27], [30] and in [18], where a rate-of-convergence-estimate for the optimal values is obtained w.r. to convergence in probability ([18, Theorem 2], but for the case that the support of μ is bounded (since the integrand $f(z,x)$ is not bounded on $Z \times X_0$ as required in [18, p. 350]).

4. On distribution sensitivity of the optimal value of chance-constrained programming problems

We are going to investigate (quantitative) continuity results of the following set-valued mapping (induced by a probabilitstic constraint) from $\mathcal{P}(Z)$ into $\mathbb{R}^m$

$$\mu \mapsto C_\alpha(\mu) := \{x \in \mathbb{R}^m | \mu(X^{-1}(\{x\})) \geq \alpha\} \tag{4.1}$$

with respect to the weak topology in $\mathcal{P}(Z)$ and the Hausdorff distance (on the class of all nonempty subsets of $\mathbb{R}^m$), respectively.

Thereby, $Z := \mathbb{R}^s$ (equipped with the norm $\| \|_\infty$), X is a set-valued mapping from Z into $\mathbb{R}^m$ (where we denote, as usual, $X^{-1}(B) := \{z \in Z | X(z) \cap B \neq \emptyset\}$ for $B \subseteq \mathbb{R}^m$) and $\alpha \in]0,1[$.

We are aiming at a result on the (local) Lipschitz-property of the mapping (4.1), where (based on the following lemma) the Prokhorov- metric (1.2) offers itself as a metric for the weak topology in $\mathcal{P}(Z)$.

<u>4.1 Lemma:</u> Let $\mathcal{B}$ be a class of Borel subsets of Z and $\mu \in \mathcal{P}(Z)$ Assume that

$$\left.\begin{array}{l}\text{there exists a constant } M > 0 \text{ such that} \\ \sup\limits_{B \in \mathcal{B}} \mu((\partial B)^\varepsilon) \leq M\varepsilon \qquad \text{for all } \varepsilon > 0\end{array}\right\} \tag{4.2}$$

(where ∂B is the topological boundary of B).

Then we have for all $\nu \in \mathcal{P}(Z)$:

$$\alpha_{\mathcal{B}}(\mu,\nu) := \sup_{B \in \mathcal{B}} |\mu(B) - \nu(B)| \leq (M+1)\rho(\mu,\nu) \tag{4.3}$$

<u>Proof:</u> For $\varepsilon > 0$ and $B \subseteq Z$, put $B^{-\varepsilon} := B \setminus (Z \setminus B)^\varepsilon$ and note that $(\partial B)^\varepsilon = B^\varepsilon \setminus B^{-\varepsilon}$ ([3], p.14). Now take an arbitrary $\varepsilon > \rho(\mu,\nu)$ and $B \in \mathcal{B}$, and observe that

$$\mu(B^{-\varepsilon}) \leq \nu((B^{-\varepsilon})^\varepsilon) + \varepsilon \leq \nu(B) + \varepsilon \qquad \text{and}$$

$$\nu(B) \le \mu(B^{\varepsilon}) + \varepsilon$$

Hence follows

$$\mu(B) - \nu(B) \le \mu(B) - \mu(B^{-\varepsilon}) + \varepsilon \le \mu(B^{\varepsilon} \setminus B^{-\varepsilon}) + \varepsilon \le (M+1)\varepsilon$$

and

$$\nu(B) - \mu(B) \le \mu(B^{\varepsilon}) - \mu(B) + \varepsilon \le \mu(B^{\varepsilon} \setminus B^{-\varepsilon}) + \varepsilon \le (M+1)\varepsilon$$

which implies (4.3). □

4.2 Remark: Condition (4.2) implies that $\mathcal{B}$ is a μ-uniformity class ([3], p.14). We mention two important examples of classes $\mathcal{B}$ that fulfill (4.2) under certain assumptions on μ:

(i) $\mathcal{B} \subseteq \{]-\infty,z] := \bigtimes_{i=1}^{s}]-\infty,z_i] \,|\, z = (z_1,\dots,z_s) \in \mathbb{R}^s\}$.

If the distribution function F_μ belonging to μ is Lipschitz continuous (i.e. if all marginal distributions of μ have bounded densities), then (4.2) is fulfilled for $\mathcal{B}$ and μ. Indeed, for $B =]-\infty,z]$ ($z \in \mathbb{R}^s$) there holds:

$$\mu((\partial B)^{\varepsilon}) \le F_\mu(z+\bar{\varepsilon}) - F_\mu(z-\bar{\varepsilon}) \le \tag{4.4}$$
$$\le L_\mu \|(z+\bar{\varepsilon}) - (z-\bar{\varepsilon})\|_\infty = 2L_\mu \|\bar{\varepsilon}\|_\infty = 2L_\mu \varepsilon$$

(where $\bar{\varepsilon} = (\varepsilon,\dots,\varepsilon) \in \mathbb{R}^s$, and L_μ is a Lipschitz constant of F_μ).

(ii) $\mathcal{B} \subseteq \{B \subseteq \mathbb{R}^s \,|\, B \text{ is convex and Borel}\}$.

For this case, [3, Theorem 3.1] contains conditions on the density of $\mu \in \mathcal{P}(\mathbb{R}^s)$ that imply (4.2).

Now we state a first continuity result on the mapping (4.1):

4.3 Lemma: Let $\mu \in \mathcal{P}(Z)$, and $\mathcal{B} := \{X^{-1}(\{x\}) \,|\, x \in \mathbb{R}^m\}$ be a class of Borel subsets of Z. Assume that

$$\left.\begin{array}{l} \text{there are constants } L > 0 \text{ and } \delta_0 > 0 \text{ such that} \\ C_{\alpha'}(\mu) \subseteq C_{\alpha'+\delta}(\mu)^{L\delta} \quad \text{for all } \alpha' \in [\alpha-\delta_0,\alpha],\ \delta \in]0,\delta_0]. \end{array}\right\} \tag{4.5}$$

Then there holds

$$\left.\begin{array}{l} D(C_\alpha(\mu), C_\alpha(\nu)) \le L\alpha_{\mathcal{B}}(\mu,\nu) \\ \text{for all } \nu \in \mathcal{P}(Z) \text{ with } \alpha_{\mathcal{B}}(\mu,\nu) \le \delta_0. \end{array}\right\} \tag{4.6}$$

Proof: If $\alpha_{\mathcal{B}}(\mu,\nu) = 0$, we have $C_\alpha(\mu) = C_\alpha(\nu)$.

Hence we assume $0 < \alpha_{\mathcal{B}}(\mu,\nu) =: \delta \le \delta_0$.

For all $x \in C_{\alpha+\delta}(\mu)$, there holds
$\nu(X^{-1}(\{x\})) \ge \mu(X^{-1}(\{x\})) - \delta \ge \alpha + \delta - \delta = \alpha$,
thus we get the inclusion $C_{\alpha+\delta}(\mu) \subseteq C_\alpha(\nu)$
and analogously $C_\alpha(\nu) \subseteq C_{\alpha-\delta}(\mu)$.
Using (4.5), we obtain $C_\alpha(\nu) \subseteq C_{\alpha-\delta}(\mu) \subseteq C_\alpha(\mu)^{L\delta}$
and $C_\alpha(\mu) \subseteq C_{\alpha+\delta}(\mu)^{L\delta} \subseteq C_\alpha(\nu)^{L\delta}$,
which immediately yields (4.6). □

Combining Lemmata 4.1 and 4.3, we arrive directly at

<u>4.4 Proposition:</u> Let $\mu \in \mathcal{P}(Z)$, and $\mathcal{B} := \{X^{-1}(\{x\}) | x \in \mathbb{R}^m\}$ be a class of Borel sets of Z. Assume the validity of (4.2) and (4.5). Then there holds

$$D(C_\alpha(\mu), C_\alpha(\nu)) \le L(M+1)\rho(\mu,\nu)$$

for all $\nu \in \mathcal{P}(Z)$ with $\rho(\mu,\nu)$ sufficiently small.

<u>4.5 Remark:</u> Condition (4.2) in Proposition 4.4 was discussed in Remark 4.2. The essential new condition we impose is the local Lipschitzian property (4.5) of the mapping

$$\alpha' \mapsto C_{\alpha'}(\mu) = \{x \in \mathbb{R}^m \,|\, \mu(X^{-1}(\{x\})) \ge \alpha'\}$$

at $\alpha \in]0,1[$. [23] contains general results which seem to be useful in this particular case, too (e.g. [23, Corollary 3.5] states conditions on $t(x) := \mu(X^{-1}(\{x\}))$, $x \in \mathbb{R}^m$ (even for nonsmooth t) that imply (4.5).

But, for the particular case we shall discuss now, we prefer to give a self-contained proof.

For the rest of this section, we consider the case

$$X(z) := \{x \in \mathbb{R}^m \,|\, Ax \ge z\} \qquad (z \in Z = \mathbb{R}^s)$$

where $A \in \mathcal{L}(\mathbb{R}^m, \mathbb{R}^s)$, and check the applicability of Proposition 4.4 on the set-valued mapping

$$\left.\begin{aligned} \mu \mapsto C_\alpha(\mu) &:= \{x \in \mathbb{R}^m \,|\, \mu(X^{-1}(\{\alpha\})) \ge \alpha\} \\ &= \{x \in \mathbb{R}^m \,|\, \mu(\{z \in \mathbb{R}^s \,|\, Ax \ge z\}) \ge \alpha\} \\ &= \{x \in \mathbb{R}^m \,|\, F_\mu(Ax) \ge \alpha\} \end{aligned}\right\} \qquad (4.7)$$

where $0 < \alpha < 1$, and F_μ is the distribution function of $\mu \in \mathcal{P}(Z)$.

4.6 Theorem: Let $\alpha \in]0,1[$ and $A \in \mathcal{L}(\mathbb{R}^m, \mathbb{R}^s)$ have rank $s \le m$. Assume that the distribution function F_μ of $\mu \in \mathcal{P}(\mathbb{R}^s)$ is Lipschitzian and satisfies the following "(local) inverse Lipschitz condition at α":

$$\left.\begin{array}{l}\text{There exist positive constants } \gamma,\ \delta_1 \text{ and } \delta_2 \text{ such that} \\ F_\mu(z+\bar{\varepsilon}) \ge F_\mu(z) + \gamma\varepsilon \\ \text{for all } \varepsilon \in [0,\delta_2],\ \bar{\varepsilon} := (\varepsilon,\dots,\varepsilon),\ z \in F_\mu^{-1}([\alpha-\delta_1,\alpha+\delta_1])\end{array}\right\} \quad (4.8)$$

Then there is a constant $L_0 > 0$ such that

$$D(C_\alpha(\mu), C_\alpha(\nu)) \le L_0\rho(\mu,\nu)$$

for all $\nu \in \mathcal{P}(\mathbb{R}^s)$ with sufficiently small $\rho(\mu,\nu)$.

Proof: As $\mathcal{B} = \{X^{-1}(\{x\}) | x \in \mathbb{R}^m\} \subseteq \{]-\infty,z] | z \in \mathbb{R}^s\}$, and as F_μ is assumed to be Lipschitzian, condition (4.2) is fulfilled in view of Remark 4.2 (i). We now show that also condition (4.5) is satisfied:
Due to the assumption that A has rank $s \le m$, there exists an $x_0 \in \mathbb{R}^m$ such that $Ax_0 = (1,\dots,1) \in \mathbb{R}^s$
We now put $\delta_0 := \min\{\delta_1, \delta_2\gamma\}$ and $L := \frac{\|x_0\|}{\gamma} + 1$, and show that

$$C_{\alpha'}(\mu) \subseteq C_{\alpha'+\delta}(\mu)^{L\delta} \quad \text{for all } \alpha' \in [\alpha-\delta_0,\alpha],\ \delta \in [0,\delta_0].$$

Choose such an α' and δ, take an arbitrary $x \in C_{\alpha'}(\mu)$ and put $y := x + \frac{\delta}{\gamma} x_0$.

Then there holds $d(x,y) \le \frac{\|x_0\|}{\gamma}\delta < L\delta$ and

$$F_\mu(Ay) = F_\mu(Ax + \frac{\delta}{\gamma} Ax_0) = F_\mu(Ax + \frac{\delta}{\gamma}(1,\dots,1))$$

$$\ge F_\mu(Ax) + \frac{\delta}{\gamma}\gamma \ge \alpha' + \delta.$$

This means that $y \in C_{\alpha'+\delta}(\mu)$, and consequently, that $x \in C_{\alpha'+\delta}(\mu)^{L\delta}$ Hence Proposition 4.4 is applicable, yielding the assertion. □

4.7 Corollary: Assume that the conditions of Theorem 4.6 are satisfied for $\alpha \in]0,1[$, $A \in \mathcal{L}(\mathbb{R}^m, \mathbb{R}^s)$ and $\mu \in \mathcal{P}(\mathbb{R}^s)$. Let $c \in \mathbb{R}^m$ and $\varphi(\mu) := \inf\{c^T x | x \in \mathbb{R}^m,\ F_\mu(Ax) \ge \alpha\}$ (and likewise $\varphi(\nu)$ for $\nu \in \mathcal{P}(\mathbb{R}^s)$). Assume that $\varphi(\mu)$ and $\varphi(\nu)$ are finite, and that $\rho(\mu,\nu)$ is sufficiently small. Then there is a constant $L_1 > 0$ (not depending on ν) such that

$$|\varphi(\mu) - \varphi(\nu)| \le L_1\rho(\mu,\nu)$$

Proof: For all $\varepsilon > 0$, there exists an $x \in C_\alpha(\mu)$ such that $c^T x \le \varphi(\mu) + \varepsilon$. Due to Theorem 4.6 there exists a $y \in C_\alpha(\nu)$ such that $d(x,y) \le L_0\rho(\mu,\nu) + \varepsilon$. This implies

$$\begin{aligned}\varphi(\nu) - \varphi(\mu) &\le c^T y - c^T x + \varepsilon \le |c^T(x-y)| + \varepsilon \\ &\le \|c\| d(x,y) + \varepsilon \\ &\le \|c\| L_0 \rho(\mu,\nu) + \varepsilon(\|c\| + 1)\end{aligned}$$

Interchanging the roles of μ und ν, we arrive at our assertion. □

4.8 Remark: The proof of Theorem 4.6 (together with Lemma 4.3) shows that the mapping $\mu \mapsto C_\alpha(\mu)$ is (locally) Lipschitz continuous even w.r. to the Levy-metric

$$\lambda(\mu,\nu) := \inf\{\varepsilon > 0 \,|\, \mu(B) \le \nu(B^\varepsilon) + \varepsilon,\ \nu(B) \le \mu(B^\varepsilon) + \varepsilon, \text{ for every } B =]-\infty, z],\ z \in \mathbb{R}^s\}$$

(which also metrizes the weak topology on $\mathcal{P}(\mathbb{R}^s)$).

Thereby, the essential condition is the (local) inverse Lipschitz property (4.8) for F_μ in α. For this property we finally give a sufficent condition, which shows that, e.g. the multivariate normal distribution μ obeys all conditions of Theorem 4.6 (see also Remark 4.2 (i)).

4.9 Lemma: Assume that the distribution function F_μ of $\mu \in \mathcal{P}(\mathbb{R}^s)$ has a continuous, strictly positive density f_μ. Then condition (4.8) of Theorem 4.6 is satisfied for all $\alpha \in]0,1[$.

Proof: We choose $\delta_1 > 0$ such that $0 < \alpha - \delta_1 < \alpha + \delta_1 < 1$ and put $M := F_\mu^{-1}([\alpha - \delta_1, \alpha + \delta_1]) \subseteq \mathbb{R}^s$.

As $\lim\limits_{z \to (-\infty,\dots,-\infty)} F_\mu(z) = 0$, there exist a constant $C > 0$ such that

$$\Big|\min_{i=1,\dots,s} z_i\Big| \le C \quad (z = (z_1,\dots,z_s) \in M).$$

We now put $m := \min\limits_{z \in [-2C,2C]^s} f_\mu(z)$ and $\gamma := \delta_2 := mC^{s-1}$.

Then there holds for arbitrary $z \in M$ and $\varepsilon \in [0,\delta_2]$ (we are assuming without loss of generality that $z_1 = \min\limits_{i=1,\dots,s} z_i$):

$$\begin{aligned}F_\mu(z+\bar\varepsilon) - F_\mu(z) &\ge \int_{z_1}^{z_1+\varepsilon} \int_{-\infty}^{z_2} \cdots \int_{-\infty}^{z_s} f_\mu(t_1,\dots,t_s)\,dt_1 \dots dt_s \ge \\ &\ge \int_{z_1}^{z_1+\varepsilon} \int_{-2C}^{-C} \cdots \int_{-2C}^{-C} f_\mu(t_1,\dots,t_s)\,dt_1 \dots dt_s \ge m \cdot \varepsilon C^{s-1} = \gamma\varepsilon\end{aligned}$$

where $\bar\varepsilon := (\varepsilon,\dots,\varepsilon)$. Consequently, F_μ satisfies condition (4.8). □

References:

[1] J.-P. Aubin and A. Cellina, Differential Inclusions, Springer-Verlag, Berlin 1984.

[2] B. Bank, J. Guddat, B. Kummer, D. Klatte and K. Tammer, Nonlinear parametric optimization, Akademie-Verlag, Berlin 1982.

[3] R.N. Bhattacharya and R. Ranga Rao, Normal Approximation and Asymptotic Expansions, Wiley, New York 1976.

[4] P. Billingsley, Convergence of Probability Measures, Wiley, New York 1968.

[5] J. Birge and R. Wets, Designing approximation schemes for stochastic optimization problems, in particular for stochastic programs with recourse, IIASA Working Paper WP-83-111 (1983), Laxenburg, Austria and Mathematical Programming Study (to appear).

[6] A. Donchev and H.Th. Jongen, On the regularity of the Kuhn-Tucker curve, SIAM J. Control and Optimization 24 (1986), 169-176.

[7] R.M. Dudley, Distances of probability measures and random variables, Ann. Math. Statist. 39 (1968), 1563-1572.

[8] R.M. Dudley, The speed of mean Glivenko-Cantelli convergence, Ann. Math. Stat. 40 (1969), 40-50.

[9] R.M. Dudley, Probabilities and Metrics, Lecture Notes Series No. 45, Aarhus Universitet 1976.

[10] J. Dupačová, Stability in stochastic programming with recourse, Acta Univ. Carol. - Math. et Phys. 24 (1983), 23-34.

[11] J. Dupačová, Stability in stochastic programming with recourse-estimated parameters, Math. Programming 28 (1984), 72-83.

[12] H.W. Engl and A. Wakolbinger, Continuity properties of the extension of a locally Lipschitz continuous map to the space of probability measures, Monatsh. Math. 100 (1985), 85-103.

[13] W.W. Hager, Lipschitz continuity for constrained processes, SIAM J. Control and Optimiz. 17 (1979), 321-338.

[14] P. Kall, Approximations to stochastic programs with complete fixed recourse, Numer. Math. 22 (1974), 333-339.

[15] P. Kall, Stochastic Linear Programming, Springer-Verlag, Berlin 1976.

[16] P. Kall and D. Stoyan, Solving stochastic programming problems with recourse including error bounds, Math. Operationsforsch. Statist., Ser. Optimization, 13 (1982), 431-447.

[17] P. Kall, K. Frauendorfer and A. Ruszczynski, Approximation techniques in stochastic programming, Manuskript, Institut f. Operations Research, Universität Zürich, 1984.

[18] V. Kaňková, An approximative solution of a stochastic optimization problem, Trans. Eighth Prague Conf. on Information Theory, Statist. Decision Funct., Random Processes, Academia, Prague, Vol. A, 349-353.

[19] D. Klatte, Beiträge zur Stabilitätsanalyse nichtlinearer Optimierungsprobleme, Dissertation (B), Humboldt-Universität Berlin, Sektion Mathematik, 1984.

[20] K. Marti, Approximationen stochastischer Optimierungsprobleme, Verlag Anton Hain, Königstein 1979.

[21] F. Nožička, J. Guddat, H. Hollatz and B. Bank, Theorie der linearen parametrischen Optimierung, Akademie-Verlag, Berlin 1974.

[22] P. Olsen, Discretizations of multistage stochastic programming problems, Mathematical Programming Study 6 (1976), 111-124.

[23] R.T. Rockafeller, Lipschitzian properties of multifunctions, Nonlinear Analysis 9 (1985), 867-885.

[24] W. Römisch, On convergence rates of approximations in stochastic programming, Proc. 17. Jahrestagung "Mathematische Optimierung" (ed. K. Lommatzsch), Humboldt-Universität Berlin, Sektion Mathematik, Seminarbericht (to appear).

[25] W. Römisch and A. Wakolbinger, On Lipschitz dependence in systems with differentiated inputs, Math. Ann. 272 (1985), 237-248.

[26] G. Salinetti, Approximations for chance-constrained programming problems, Stochastics 10 (1983), 157-179.

[27] F. Solis and R. Wets, A statistical view of stochastic programming, manuscript, University of Kentucky, Lexington 1981.

[28] J. Wang, Distribution sensitivity analysis for stochastic programs with complete recourse, Math. Programming 31 (1985), 286-297.

[29] R. Wets, Stochastic programs with fixed recourse: the equivalent deterministic program, SIAM Review 16 (1974), 309-339.

[30] R. Wets, A statistical approach to the solution of stochastic programs with (convex) simple recourse, Preprint, University of Kentucky, Lexington, 1980.

[31] R. Wets, Stochastic programming: solution techniques and approximation schemes, in: Mathematical Programming: The State- of-the-Art 1982, Springer-Verlag, Berlin, 1983, 566-603.

[32] V.M. Zolotarev, Probability metrics, Theory Prob. Appl. 28 (1983), 278-302.

STOCHASTIC OPTIMIZATION AND STOCHASTIC PROCESSES: THE EPIGRAPHICAL APPROACH.

G.Salinetti(*)

Summary. Epi-convergence is the "weakest" convergence notion to attack convergence of infima. At the same time the topology of epi-convergence allows to develop a self-contained theory for convergence of normal integrands, a key ingredient in stochastic programming. This seems to delineate the appropriate framework for approximation schemes in stochastic programming which overcomes the bounds of the weak convergence of stochastic processes and, in this direction, it candidates also as an extended setting for convergence of stochastic processes.

Key words: epi-convergence, epi-topology, weak convergence, probability measures, stochastic processes, normal integrands.

1. Stochastic Optimization and Stochastic Processes.

For a family of functions $\{f;\ f_n, n=1,2,\dots\}$ on a topological space (X,τ) valued in the extended reals $\bar{R}$, epi-convergence, denoted $f_n \xrightarrow{epi} f$, means (topological) convergence of their epigraphs in the product space $X\times R$ (cfr. sec.2).

Its flexibility (the notion depends only on the topology τ of the domain) and the fact that it is the "weakest" notion to obtain convergence of the infima, fully justifies its relevant role in optimization theory (approximation schemes); see [1] for a comprehensive reference.

In stochastic optimization this role is further potentiated by the existence, guaranteed in many relevant and interesting cases, of a topology of epi- convergence; this allows to deal with normal integrands (a key ingredient of stochastic optimization problems) as Borel functions with a consequent self-contained theory for convergence of normal integrands: a natural "minimal" theoretical framework for approximation schemes in stochastic optimization.

Many stochastic optimization problems can be cast in the following (abstract)form:

$$(1.1)\qquad \mathrm{Min}_{x\in X} J(x); \quad J(x)= \int_\Omega u(f(x,\omega))\mu(d\omega)$$

where (Ω, A, μ) is a given probability space, $f: X\times\Omega \to \bar{R}$ is a normal integrand (n.i.) i.e. a function jointly depending on x and ω, lowersemicontinuous (l.sc.) in x for

(*) Dipartimento di Statistica Probabilità e Statistiche Applicate, Università di Roma "La Sapienza", Piazzale A.Moro 5, Rome, Italy.

every ω , and such that its epigraphical representation $\omega \to \text{epi}f(\omega,.)=\{(x,\alpha):f(x,\omega)\leq\alpha\}$ is a closed valued measurable multifunction.

Models which fit (1.1) are stochastic programs with recourse, certain classes of Markov decision processes, stochastic control problems in discrete time, statistical estimation problems and so on.

It is quite known that, unless we consider very special cases, a fruitful possible way to attack problem (1.1) is via approximations replacing the original problem by a sequence of approximating problems

$$\text{(1.2)} \qquad \text{Min}_{x\in X}J_n(x); \quad J_n(x)= \int_\Omega u(f_n(x,\omega))\mu(d\omega)$$

and try to obtain

$$\text{(1.3)} \qquad \text{Min}_{x\in X}J(x)=\lim \text{Min}_{x\in X}J_n(x).$$

On the base of the above, relation (1.3) immediately points out that the minimal setting for obtaining convergence of the infima (1.3) is the epi-convergence

$$\text{(1.4)} \qquad J_n \xrightarrow{epi} J.$$

Observe that the integrands in (1.1) and (1.2) are stochastic processes, i.e. for every $x\in X$, $u(x,.)$ is a random variable, however the classical theory of stochastic processes and in particular the available theory for convergence of stochastic processes does not provide the minimal requirements for the epi-convergence. This depends on the fact that, from the convergence point of view, the processes which can be actually handled are basically the ones with continuous or càd-làg (continues à droite-limites à gauche) realizations on intervals of the real line, a too narrow framework to contain the situation sketched above.

The situation is even more complicated when one realizes that the stochastic processes f and f_n are the results of functionals of other stochastic processes. Typically in the problem (1.1) f is obtained from a "recourse problem" and has the form

$$\text{(1.5)} \qquad f(x,\omega) = \text{Inf}_{y\in Y}g(x,y,\omega).$$

It is clear then that any step in the direction of (1.4) requires an adequate notion of convergence for $\{f_n\}$ to f and the development of a theory for it.

It is worthwhile to observe that a function of the type (1.5) is of interest in its own: in the so called "distribution problem" in stochastic programming the objective is to find the distribution function of the random variable

$$\text{(1.6)} \qquad Z(\omega) = \text{Inf}_{y\in Y}f(y,\omega)$$

where f is a normal integrand. Practically the only possible approach is to obtain $Z(.)$ as limit in distribution of the sequence of random variables

$$\text{(1.7)} \qquad Z_n(\omega) = \text{Inf}_{y\in Y}f_n(y,\omega).$$

Again the basic question boils down to design the minimal convergence of $\{f_n\}$ to f for obtaining $Z_n \to Z$ in distribution.

More generally, also in connection with (1.1), the distribution function of a functional of the normal integrand is often of interest and this takes us back to the convergence questions sketched above and not covered by the classical theory of stochastic processes.

Fortunately, in all these situations, the same notion of epi-convergence, but more exactly here the attached topology, allows to deal with the random elements f and f_n as Borel measurable functions and provides a self-contained theory for convergence.

It is relevant to observe that all the results so obtained for the convergence of random semicontinuous functions also incapsulate the known results of the classical theory of stochastic processes.

The notion of epi-convergence and the attached epi-topology are briefly reviewed in Section 2. In Section 3 the construction of a convergence theory for random semicontinuous functions is delineated according to [2]. In Section 4 we deal with convergence in distribution of random semicontinuous functions and finally in Section 5 we examine some implications for convergence in distribution of functionals.

2. Topological convergence of sets, epi-convergence and epi-topology.

A basic reference for this section is [1].

2.1 DEFINITION- A sequence $\{F_n, n=1,\dots\}$ of subsets of the topological space (E,τ) is said to topologically converge (sometime also said to converge in the Kuratowski sense) if

$$\limsup_{n\to\infty} F_n = \liminf_{n\to\infty} F_n \qquad (2.2)$$

where:

$$\limsup_{n\to\infty} F_n = \bigcap_{n\geq 1} \mathrm{cl} \bigcup_{m\geq n} F_m; \quad \liminf_{n\to\infty} F_n = \bigcap_{\{n_m\}} \mathrm{cl} \bigcup_{m} F_{n_m},$$

"cl" stays for the closure in the topology τ of E and in "liminf" the intersection is over all the subsequences $\{n_m\}$ of $\{n\}$ [3].

It is clear from the definition that we always have $\liminf F_n \subset \limsup F_n$ and the limit, when existing, is always a closed subset of E.

Let $f: X\times\Omega\to\bar{R}$ be a function on the topological space (X,τ) with values in the extended reals $\bar{R}$. The epigraph of f is the subset of $E=X\times R$ defined by

$$\mathrm{epi}f = \{(x,\alpha)\in E:\ f(x)\leq\alpha\}$$

and it is closed in the product topology of $E=X\times R$ if and only if f is lower semicontinuous (l.sc.) on (X,τ).

On the base of (2.2) we have:

2.3 DEFINITION- A sequence of functions $\{f_n, n=1,\dots\}$, $f_n: X\to\bar{R}$ is said to epi-converge if the sequence $\{\mathrm{epi}\, f_n\}$ of the epigraphs topologically converge.

The limit function, called epi-limit, when existing, is a l.sc. function on (X,τ).

The product space E=X×R with the topology η product of the topology τ of X and the standard topology of R inherits the properties of τ. Depending on them, different and more usable characterizations of epi-convergence can be given. In particular when (X,τ) is a first countable space we have [1,Proposition 1.14]:

$$(2.4)\qquad f_n \xrightarrow{epi} f \iff \begin{cases} \text{for any } x\in X, \text{ any } x_m\to x, \text{any } \{n_m\}\subset\{n\} \text{ we have } f(x)\le\liminf f_{n_m}(x_m) \\ \text{for any } x\in X \text{ there exists } x_n\to x \text{ such that } \limsup f_n(x_n)\le f(x). \end{cases}$$

When (X,τ) is locally compact then so is (E,η). In this case, denoting $\mathcal{K}(E)$ and $\mathcal{G}(E)$ respectively the compact and open subsets of E, the epi-convergence expressed as epigraphs convergence can be characterized as follows

$$(2.5)\qquad \mathrm{epi}f_n\to\mathrm{epi}f \iff \begin{cases} \text{for any } K\in\mathcal{K}(E) \text{ such that } \mathrm{epi}f\cap K=\emptyset \text{ we have} \\ \mathrm{epi}f_n\cap K=\emptyset \text{ for all n sufficiently large} \\ \\ \text{for any } G\in\mathcal{G}(E) \text{ such that } \mathrm{epi}f\cap G\neq\emptyset \text{ we have} \\ \mathrm{epi}f_n\cap G\neq\emptyset \text{ for all n sufficiently large.} \end{cases}$$

This characterization with the fact that the epi-limit is always a l.sc. function substantiates the question about the existence of a topology on the space of l.sc. functions where convergence is epi-convergence. The answer is positive as stated in the next proposition.

Let S^X be the space of the l.sc. functions on (X,τ). For any subset D of E denote:

$$\mathcal{E}^D=\{f\in S^X:\mathrm{epi}f\cap D=\emptyset\};\qquad \mathcal{E}_D=\{f\in S:\mathrm{epi}f\cap D\neq\emptyset\}.$$

In the topology of S^X generated by the subbase

$$(2.6)\qquad \{\mathcal{E}^K,\ K\in\mathcal{K}(E)\} \text{ and } \{\mathcal{E}_G,\ G\in\mathcal{G}(E)\}$$

convergence is epi-convergence if (X,τ) (and consequently (E,η)) is locally compact, as it can be derived from (2.5). We will refer to such topology as epi-topology and denote it "epi" ([1],[3]). We have:

2.7 PROPOSITION- <u>If</u> (X,τ) <u>is Hausdorff and locally compact then the topological space</u> (S^X,epi) <u>is Hausdorff and compact.</u>
<u>If, in addition,</u> (X,τ) <u>is second countable then also</u> (S^X,epi) <u>is second countable, then metrizable.</u>

2.8 REMARK- It has to be observed that the definition of epi-convergence only depends on the topology of X. Thus we can refer to the epi-convergence with respect to the topology τ in comparison with the epi-convergence with respect to another topology, say σ still on X, an interesting aspect in variational problems. In the particular case where σ is the discrete topology the epi-convergence is pointwise convergence.

The minimal setting for equivalence between epi-convergence and pointwise convergence is equisemicontinuity. This notion introduced in [11] and further developed in [3] in a more general case can be expressed as follows [3, definition 2.17].

2.9 DEFINITION- A sequence of l.sc. functions $\{f_n, n=1,...\}$ on the topological space (X,τ) is said to be <u>equi-lowersemicontinuous</u> (equi-l.sc.) at $x\in X$ if there exists

$\delta_x>0$ such that to every δ, $0<\delta<\delta_x$, there corresponds a neighbourhood $N_\delta(x)$ of x in the topology τ of X such that for all n

(2.10) $$\mathrm{Inf}_{y\in N_\delta(x)} f_n(y) \geq \mathrm{Min}\ [f_n(x)-\delta,\ \delta^{-1}]$$

The sequence is said equi-l.sc. if (2.10) is satisfied for every $x\in X$.

The notion of equi-lowersemicontinuity as the minimal setting for equivalence between epi-convergence and pointwise convergence is stated in [3,Theorem 2.17]. Of that we emphasize the local character of equi-l.sc.; in particular of interest here is the following:

2.11 PROPOSITION -Let $\{f;\ f_n,\ n=1,\ldots\}$ be a family of l.sc. functions and let $x\in X$. If $f_n \xrightarrow{epi} f$ then $f_n(x)\to f(x)$ if and only if $\{f;\ f_n, n=1,\ldots\}$ is equi-l.sc. at x.

3. Normal integrands as random semicontinuous functions.

A basic reference of this section is [2].

The integrand in (1.1) as it usually emerges in stochastic programming problems is normal; more precisely, given the probability space (Ω, A, μ) and the topological space (X,τ) we have:

3.1 DEFINITION-A normal integrand(n.i.) is a function $f:X\times\Omega \to \bar{R}$ such that
- for every ω the map $x \to f(x,\omega)$ is l.sc. on (X,τ), i.e. $f(.,\omega)\in S^X$;
- the epigraphical representation $\omega \to \mathrm{epi}\ f(.,\omega)$ is a measurable multifunction of Ω into $X\times R$, i.e. for every closed subset F of $X\times R$ we have:

(3.2) $$\{\omega\in\Omega:\ \mathrm{epi} f(.,\omega)\cap F\neq\ \} \in A.$$

If (X,τ) is locally compact the measurability condition (3.2) can be equivalently expressed by

(3.3) $$\{\omega\in\Omega:\ \mathrm{epi} f(.,\Omega)\cap K\neq\ \}\in A \qquad \text{for every } K\in K(E).$$

For the following we will assume that the topological space (X,τ) is locally compact, Hausdorff and second countable, a situation enough general to cover interesting cases in stochastic programming.

The theory of normal integrands was introduced and developed by Rockafellar ([4, 5]) to study variational problems involving constraints; it also provides the natural framework to analyze the convergence of stochastic optimization problems as we will see here (see also [6]).

Looking at the n.i. f as a map of Ω with values in S^X, the space of the l.sc. functions on X, with the epi-topology, on the base of (3.3) related to (2.6) and relying on the fact that the Borel field of the epi-topology on S^X is generated by the class $\{E^K, K\in K(X\times R)\}$ or equivalently by $\{E_G, G\in G(X\times R)\}$, we have:

3.4 PROPOSITION [2,p.26-27]- A normal integrand $f:X\times\Omega\to\bar{R}$ is a map of Ω into S^X, measurable with respect to the σ-algebras A of Ω and B_{epi}, the Borel field of

(S^X, epi), thus inducing a probability measure P on $\mathcal{B}_{epi}$.

As a consequence of this proposition, the standard tools of probability and measure theory provide the basic framework to deal with convergence of normal integrands. The major facts are the possibility to express the measurability with respect to the Borel field $\mathcal{B}_{epi}$, allows to exploit at the fullest the "well behaving" topological structure of S^X (as for the random variables); on the other hand the fact that the epi-topology is metrizable allows to approach all the questions about convergence of the probability measures induced by a sequence of normal integrands in the known context of convergence of probability measures in a metric space [7].

We concentrate here on almost sure convergence and convergence in distribution and refer to [2]. Convergence in probability can be easily derived from [8].

3.5 DEFINITION- A sequence of normal integrands $\{f_n: X\times\Omega\to\bar{R}, n=1,\ldots\}$ is said to almost surely (a.s.) converge to the normal integrand $f: X\times\Omega\to\bar{R}$, if for every $\omega\in\Omega$, except possibly for a null subset of Ω, we have $f_n(.,\omega) \xrightarrow{epi} f(.,\omega)$.

By (2.4) we can derive:

3.6 PROPOSITION [2, Proposition 3.8]- Suppose that $\{f_n: X\times\Omega \to \bar{R}, n=1,\ldots\}$ is a sequence of normal integrands. They converge a.s. to the normal integrand $f: X\times\Omega \to \bar{R}$ if and only if for every $\omega\in\Omega$, except possibly on a set of null measure, we have that for all $x\in X$

(3.7) for any subsequence $\{f_{n_k}, k=1,\ldots\}$ and any sequence $\{x_k, k=1,\ldots\}$ converging to x we have

$$f(x,\omega) \le \liminf f_{n_k}(x_k,\omega),$$

and

(3.8) there exists a sequence $\{x_n, n=1,\ldots\}$ converging to x such that

$$\limsup f_n(x_n,\omega) \le f(x,\omega).$$

3.9 REMARK- In connection with the convergence question posed in the form (1.4), Proposition 3.6 gives a first answer simply through the Fatou's Theorem. In effect when

$$J(x) = \int_\Omega f(x,\omega)\mu(d\omega) \quad \text{and} \quad J_n(x) = \int_\Omega f_n(x,\omega)\mu(d\omega),\ n=1,2,\ldots$$

and the sequence of n.i.'s $\{f_n\}$ almost surely converges to the n.i. f then, if $\int_\Omega f(,\omega)\mu(d\omega)$ is finite, by (3.7), (3.8) and the Fatou's Lemma 12 we obtain (1.4).

One can ask for (1.4) as a consequence of weaker convergence of $\{f_n\}$ to f for example convergence in distribution, as well as for convergence of functionals of f_n and f as "u" in (1.2) or "inf" in (1.5), (1.6) and (1.7). All this requires the introduction of convergence in distribution for normal integrands and their functionals.

4. <u>Convergence in distribution of normal integrands</u>.

A major fact emerging from looking at a n.i. as a random l.sc. function with values in S^X, is the metrizability of the topological space (S^X,epi) which incapsulates all the questions about convergence of probability measures in the known framework of weak convergence in a metric space and all the machinery developed there became applicable.

In addition to all that, the properties of the epi-topology and the fact that $\mathcal{B}_{epi}$ is the σ-algebra generated by $\{E_K, K \in \mathcal{K}(X\times R)\}$ allow to restrict the study of the convergence of probability measures to the convergence of their restrictions on $\{E_K, K \in \mathcal{K}(X\times R)\}$, basically with the same scheme for random variables, with relevant implications from the applications point of view.

For a n.i. $f: X\times\Omega \to \overline{R}$ let P denote the probability measure induced on $\mathcal{B}_{epi}$ by the random l.sc. function $\omega \to f(.,\omega): \Omega \to S^X$.

The restriction of $P: \mathcal{B}_{epi} \to [0,1]$ to $\{E_K, K \in \mathcal{K}(X\times R)\}$, defines a function, say D, on $\mathcal{K}(X\times R)$, through the correspondence

$$D(K) = P(E_K), \qquad K \in \mathcal{K}. \tag{4.1}$$

The function $D: \mathcal{K} \to [0,1]$ has the following properties:

(i) $D(\emptyset) = 0$

(ii) $D(K_1) \leq D(K_2)$ whenever $K_1 \subset K_2$

(iii) $D(K_n) \downarrow D(K)$ whenever $K_n \downarrow K$

(iv) the differences Δ_n, n=1,2,... defined recoursively by

$\Delta_o(K_o)) = 1 - D(K_o)$

$\Delta_1(K_o; K_1) = \Delta_o(K_o) - \Delta_o(K_o \cup K_1)$

....

$\Delta_n(K_o; K_1,\ldots,K_n) = \Delta_{n-1}(K_o; K_1,\ldots,K_{n-1}) - \Delta_{n-1}(K_o \cup K_n; K_1,\ldots,K_{n-1}), \quad n \geq 2$

take on their values in [0,1].

4.2 REMARK- The properties (i)-(iv) can be regarded as the natural extension of the properties of the distribution functions on finite dimensional euclidean spaces: (ii) can be viewed as monotonicity, (iii) as "right-continuity" and (iv) is the extension in infinite dimension of the standard non negativity of the differences.

The name <u>distribution function</u>(d.f.) given to any function $D: \mathcal{K} \to [0,1]$ with the properties (i)-(iv) is naturally justified by Remark 4.2, but more than that it finds a complete justification in view of the next theorem.

4.3 THEOREM ([2,Theorem 1.3(Choquet)]) - <u>Every probability measure</u> P on $\mathcal{B}_{epi}$ <u>determines a distribution function</u> D <u>on</u> $\mathcal{K}$ <u>through the correspondences</u> (4.1) <u>and</u>

$$\Delta_n(K_o; K_1,\ldots,K_n) = P(E^{K_o} \cap E_{K_1} \cap \ldots \cap E_{K_n}). \tag{4.4}$$

<u>Conversely, every distribution function</u> D <u>on</u> $\mathcal{K}$ <u>determines a probability measure</u> P <u>on</u> $\mathcal{B}_{epi}$ <u>through</u> (4.1) <u>or alternatively</u> (4.4).

Consider now a family of n.i.'s $\{f;\ f_n, n=1,2,\ldots\}$ and let $\{P;\ P_n, n=1,\ldots\}$ be the corresponding family of probability measures on $\mathcal{B}_{epi}$ and $\{D; D_n, n=1,2,\dots\}$ be the corresponding family of distribution functions.

According with the classical theory (see for example [7]), weak convergence of the probability measures $\{P_n\}$ to P, denoted $P_n \xrightarrow{w} P$, by definition means:

(4.5) $\int_{S^X} g\, dP_n \to \int_{S^X} g\, dP$, for every g bounded and continuous on S^X

or, equivalently, anyone of the following

(4.6) $\limsup P_n(F) \leq P(F)$ for any F closed subset of (S^X, epi)

(4.7) $P(G) \leq \liminf P_n(G)$ for any G open subset of (S^X, epi)

(4.8) $P(B) = \lim P_n(B)$ for any $B \in \mathcal{B}_{epi}$ with $P(\partial B)=0$

where ∂B stays for the boundary of B in (S^X, epi).

The one-to-one correspondence between probability measures on $\mathcal{B}_{epi}$ and distribution functions on $(X \times R)$ together with the (topological) properties of the epi-topology, do actually allow to rephrase the weak convergence of probability measures in terms of pointwise convergence of the distribution functions on the "continuity set" of the limit distribution function, as convergence in distribution for random variables. The only thing to clarify is the notion of "continuity set" for the limit distribution function.

Continuing with the analogy with random variables, once observed that any distribution function is "right continuous" in the sense of (iii) one is naturally lead to the following definition.

4.9 DEFINITION- The distribution function $D: \mathcal{K} \to [0,1]$ is distribution continuous at K if for any sequence $K_n \uparrow K$ (i.e. $K = \mathrm{cl} \cup K_n$) we have

$$D(K_n) \uparrow D(K);$$

the distribution continuity set of D is denoted $\mathcal{C}_D$.

For the sequences $\{P; P_n, n=1,2,\dots\}$ and corresponding $\{D; D_n, n=1,2,\dots\}$ we always have

$$D_n(K) \to D(K), \text{ for every } K \in \mathcal{C}_D \Rightarrow P_n \xrightarrow{w} P.$$

The viceversa does not necessarily hold over $\mathcal{C}_D$ but over "regular" elements in $\mathcal{C}_D$ which, however, are "convergence determining". An example is given by the subclass of $\mathcal{C}_D$ of the compact K such that $K = \mathrm{cl\ int}\ K$. But in the case where X is finite dimensional euclidean space we can actually limit ourselves to pointwise convergence on $\mathcal{C}_D^{ub} = \mathcal{C}_D \cap \mathcal{K}^{ub}$ where $\mathcal{K}^{ub} = \{K \in \mathcal{K}(X \times R)$: K finite union of closed balls with positive radii$\}$.

We have then

4.10 PROPOSITION - For the family of n.i.'s $\{f; f_n: X \times \Omega \to \bar{R}, n=1,\dots\}$ with probability measures $\{P; P_n, n=1,\dots\}$ on $\mathcal{B}_{epi}$ and $\{D; D_n, n=1,\dots\}$ distribution functions on $\mathcal{K}(X \times R)$ we have

$P_n \overset{w}{\to} P$ <u>if and only if</u> $D_n(K) \to D(K)$, <u>for every</u> $K \in C_D^{ub}$.

5. Functionals of normal integrands.

In the one-to-one correspondence between P on $\mathcal{B}_{epi}$ and D on $K(X \times R)$ there is an intermediate step extremely relevant in the applications involving functionals of normal integrands.

In the space (S^X, epi), let $\mathcal{G}(S)$ denote the space of all open subsets of (S^X, epi) and equip $\mathcal{G}(S)$ with the topology, say $\mathcal{I}$, where for any family $\{G, G_n, n=1,2,\ldots\}$ in $\mathcal{G}(S)$ convergence $G_n \to G$ means convergence of the complements $G_n^c \to G^c$ in the sense of topological convergence of closed sets as in the definition 2.1.

It can be proved that for every measure P on $\mathcal{B}_{epi}$ its restriction, say P^*, on $\mathcal{G}(S)$ is l.sc. on $(\mathcal{G}(S), \mathcal{I})$. Moreover for any family $\{P; P_n, n=1,2,\ldots\}$ of probability measures on $\mathcal{B}_{epi}$ with restrictions $\{P^*; P_n^*, n=1,\ldots\}$ we have

(5.1) $$P_n \overset{w}{\to} P \quad \text{if and only if} \quad P_n^* \overset{epi}{\to} P^*$$

[9, Proposition 3.1, Theorem 4.1].

In general we do not have pointwise convergence $P_n^*(G) \to P^*(G)$ on $\mathcal{G}(S)$ as consequence of $P_n \overset{w}{\to} P$ but only $P_n(G) \to P(G)$ on every $G \in \mathcal{G}(S)$ with $P(\partial G)=0$, by (4.8). However, by Proposition 2.11 the following result is immediate:

5.2 PROPOSITION - <u>Suppose that</u> $P_n \overset{w}{\to} P$ <u>and let</u> $G \in \mathcal{G}(S)$. Then $P_n(G) \to P(G)$ <u>if and only if the family of the restrictions</u> $\{P_n^*; P^*, n=1,\ldots\}$ <u>on</u> $(\mathcal{G}(S), \mathcal{I})$ <u>is equi-l.sc. at</u> G.

Now we examine some implications of this proposition for convergence in distribution of functionals.

Suppose that J is a functional on S^X and assume that J is upper semicontinuous (u.sc.) on (S^X, epi) (the same argument goes true if J is l.sc.); then in particular $J: S^X \to R$ is Borel measurable. The probability measure induced by J on $\mathcal{B}(R)$ is uniquely determined by the distribution function $\Phi: R \to [0,1]$

(5.3) $$\Phi(z) = P(\{f \in S^X: J(f) < z\}), \qquad z \in R.$$

As sketched in Section 1, in an approximation scheme $\{f_n, n=1,\ldots\}$ with probability measures $\{P_n, n=1,\ldots\}$ and

$(5.3)_n$ $$\Phi_n(z) = P_n(\{f \in S^X: J(f) < z\}), \qquad z \in R$$

the major convergence question is about the minimal conditions to obtain $\Phi_n \overset{distr}{\to} \Phi$, i.e.

(5.4) $$\Phi_n(z) \to \Phi(z) \quad \text{for every } z \text{ where } \Phi \text{ is continuous}$$

when $P_n \overset{w}{\to} P$.

The answer is given by Proposition 5.2. The set $\{f \in S^X: J(f) < z\}$ is an open subset of (S^X,epi) then $P_n \overset{w}{\to} P$ only implies $\Phi(z) \le \liminf \Phi_n(z)$ for every z. By Proposition 5.2 the convergence $\Phi(z)=P(\{f \in S^X: J(f)<x\})=\lim P_n(\{f \in S^X: J(f)<z\})=\lim \Phi_n(z)$ is guaranteed if and only if the family of probability measures $\{P; P_n, n=1,\ldots\}$ is equi-l.sc. at $\{f \in S^X: J(f)<z\}$.

We can conclude:

5.5 PROPOSITION - Let $\{f; f_n: X \times \Omega \to \bar{R}, n=1,2,\ldots\}$ be a sequence of n.i.'s with corresponding probability measures $\{P; P_n, n=1,2,\ldots\}$ on $\mathcal{B}_{epi}(S^X)$; let $J: S^X \to R$ be an u.sc. functional on (S^X,epi) and $\{\Phi; \Phi_n, n=1,\ldots\}$ as in (5.3) and $(5.3)_n$.

Suppose that $P_n \overset{w}{\to} P$ then $\Phi_n \overset{distr}{\to} \Phi$ if and only if for every z where Φ is continuous, the family of probability measures $\{P; P_n, n=1,\ldots\}$ is equi-l.sc. at $\{f \in S^X: J(f)<z\}$.

5.6 REMARK - Of course the same type of argument holds if J is l.sc. on (S^X,epi) because in this case $\{f \in S^X: J(f)<z\}$ is closed and then its complement is open.

It is clear that concrete applications will require a certain amount of machinery to test the equi-l.sc. or equi-u.sc.: that will depend on the nature of the problem or of the class of problems at hand. An example is worked out in [10] for the case $J(f)=\mathrm{Inf}_{x \in X} f(x)$.

REFERENCES

[1] H.Attouch- Variational Convergence for Functions and Operators, Pitman Advanced Publishing Program, Boston, 1984.

[2] G.Salinetti, R.J-B Wets - On the Convergence in Distribution of Closed-valued Multifunctions (random sets), Normal Integrands, Stochastic Processes and Stochastic Infima. To appear on Mathematics for Operation Research.

[3] S.Dolecki, G.Salinetti and R.J-B Wets - Convergence of Functions: Equi-semicontinuity. Trans.Amer.Math.Soc., Volume 276, No 1, March 1983, 409-429.

[4] R.T.Rockafellar - Integrals which are Convex Functionals, I& II. Pacific J.Math. 24(1968),525-539 and 39(1971), 439-469.

[5] R.T.Rockafellar - Integral Functionals, Normal Integrands and Measurable Selections. In Non Linear Operators and Calculus of Variations, Lecture Notes in Mathematics 543, Springer Verlag, Berlin 1977.

[6] R.T.Rockafellar , R.J-B Wets - Variational Systems, an Introduction. In Multifunctions and Integrands, Lecture Notes in Mathematics 1091, Springer Verlag, Berlin 1984.

[7] P.Billingsley - Convergence of Probability Measures. J.Wile y& Sons, New York, 1968.

[8] G.Salinetti, W.Vervaat, R.J-B Wets - On Convergence in Probability of Closed-valued Measurable Multifunctions. To appear on Mathematics for Operation Research.

[9] G.Salinetti, R.J-B- Wets - On the Hypo-convergence of Probability Measures. Proceedings of 3.rd Course in Optimization Theory and Related Fields, Erice (Italy) 1984.

[10] G.Salinetti - Convergence of Stochastic Infima: Equi- semicontinuity. Proceedings of International Conference on Stochastic Optimization, Kiev (USSR) 1984.

[11] G.Salinetti, R.J-B Wets - On the Relation between two Types of Convergence for Convex Functions. J.Math.Anal.Appl. 60, 1977, 211-226.

[12] R.Ash - Real Analysis and Probability. Academic Press, New York 1972.

ON DUALITY AND STABILITY OF PARAMETRIZED OPTIMIZATION PROBLEMS AND RELATED TOPICS

Ivan Singer[1)]

Abstract. We present a survey of some of our contributions to the theories of duality and stability of parametrized optimization problems, conjugations of functionals, and generalized convexity. We place the survey in a historical framework, giving an introduction to some of the ideas in these topics and showing also some related developments.

1. In his monograph of 1976, at the end of the introduction to the chapter on duality, Avriel ([3], p.106) wrote: "Little can be said about duality in general nonconvex programming, for this subject is at about the same stage as convex duality was in the early 1950 s. A few works on the subject exist, such as..., but results are not very satisfactory...". Also, in the introduction of their paper of 1980 on duality, Jefferson and Scott ([36],p.519) wrote: "An essential concept in the analysis of mathematical programs is the idea of duality. This has furnished new approaches and interpretational insights and has provided the basis for many powerful algorithms. To date, most of this theory has concentrated on convex mathematical programs, whereas the vast spectrum of nonconvex programs has remained relatively untouched...". On the other hand, in the Abstract of his paper of 1977, Balder [4] wrote: "By an effective extension of the conjugate function concept a general framework for duality-stability relations in nonconvex optimization problems can be studied...".

The aim of the present paper is to present a survey of some of our contributions to the above mentioned topics, i.e., to duality and stability theory of parametrized optimization problems, and to the theory of conjugations of functionals. We shall also mention some of our related results on generalized convexity. We shall place this survey in a historical framework, giving an introduction to some of the ideas in these topics, and showing also some related developments (naturally, from a subjective point of view).

In order to keep the paper within a reasonable size, we shall simplify some parts of the presentation (where it would have been too technical to give more details). Also, we shall omit deliberately many related subjects, e.g. Lagrangian functionals (for their general definition, see [82], definition 1.1), minimax theory, ε-subdifferentials, characterizations of solutions, local optima, etc., and some particular cases, e.g., semi-infinite optimization, differentiable optimization, discrete optimization, etc. We shall concentrate especially on explaining the development of various concepts, methods, and their connections, rather than on stating or proving theorems. Instead of aiming at completeness of bibliography, we shall only give some samples of references. We hope that the present paper will offer a picture of some recent developments and will stimulate further research.

2. Let F be a locally convex space, $G(\neq\emptyset)$ a subset of F and $h:F\to\overline{R}=[-\infty,+\infty]$, and let us consider the (global, scalar) "primal" infimization problem

$$(P)\qquad \alpha=\inf h(G)[=\inf_{y\in G} h(y)], \tag{2.1}$$

1) Department of Mathematics, INCREST, Bd.Păcii 220, 79622 Bucharest and Institute of Mathematics,Str.Academiei 14, 70109 Bucharest, Romania.

embedded into a family of parametrized (or "perturbed") infimization problems

$$(P_x) \qquad f(x)=\inf_{y\in F} p(y, x) \qquad (x\in X), \qquad (2.2)$$

where the parameter set X is a locally convex space and where the parametrization functional $p:F\times X\to\bar{R}$ satisfies

$$p(y, 0)=\begin{cases} h(y) & \text{if } y\in G \\ +\infty & \text{if } y\in F\setminus G; \end{cases} \qquad (2.3)$$

by (2.1)-(2.3), we have

$$\alpha=\inf_{y\in F} p(y, 0)=f(0). \qquad (2.4)$$

The functional $f:X\to\bar{R}$ of (2.2), called the optimal value (or, the"primal", or the "marginal") functional, is of great importance, since it shows how the optimal value of problem (P_x) varies when the parameter x varies. Also, the properties of the functionals p, f and $h+\chi_G$ are intimately related, where χ_G denotes the indicator functional of the set G (i.e., $\chi_G(y)=0$ for $y\in G$ and $=+\infty$ for $y\in F\setminus G$), and where $-\infty+\infty=+\infty$.

There also exist some other parametrization schemes. For example, instead of perturbing the objective functional h of (P), by (2.2), one can perturb the constraint set G of (P), embedding (P) into the family of problems

$$(P_x) \qquad f(x)=\inf_{y\in\Gamma(x)} h(y) \qquad (x\in X), \qquad (2.5)$$

where X is a locally convex space and $\Gamma:X\to 2^F$ (where 2^F denotes the collection of all subsets of F) is a multifunction, satisfying

$$\Gamma(0)=G. \qquad (2.6)$$

However, this scheme turns out to be a particular case of (2.2), by taking the "natural perturbation functional" ([67], [68])

$$p(y, x)=\begin{cases} h(y) & \text{if } y\in\Gamma(x) \\ +\infty & \text{if } y\in F\setminus\Gamma(x). \end{cases} \qquad (2.7)$$

Furthermore, one can perturb both h and G, embedding (P) into the family of problems

$$(P_x) \qquad f(x)=\inf_{y\in\Gamma(x)} p(y, x) \qquad (x\in X), \qquad (2.8)$$

but this turns out to be a particular case of (2.2), by taking

$$p'(y, x)=\begin{cases} p(y, x) & \text{if } y\in\Gamma(x) \\ +\infty & \text{if } y\in F\setminus\Gamma(x); \end{cases} \qquad (2.9)$$

for some further relations between these schemes, see [68] and [77]. In the sequel we shall consider only the embedding scheme (2.2)-(2.4). For a more general "semi-embedding" scheme, see [75], §6.

3. The Lagrangian dual problem to (P) relative to the parametrization (X, p), in the sense of Rockafellar [57] (see also [28], [55], [56] and [37]) is, by definition, the supremization problem

$$(Q) \qquad \beta=\sup \lambda(X^*), \qquad (3.1)$$

where X* is the conjugate space of X, endowed with the weak* topology, and

$$\lambda(w)=\inf_{(y,x)\in F\times X} \{p(y,x)-w(x)\} \qquad (w\in X^*). \qquad (3.2)$$

Actually, in the formulation of [57], there is $+w(x)$ instead of $-w(x)$ in λ of (3.2), which yields the same value for β of (3.1), but we shall find it more convenient to use (3.2) above (following e.g. [19]). This definition of a dual problem to (P) has turned out to be very useful, since it has yielded as particular cases (i.e., for suitable choices of (X, p)), the "usual" dual problems to various concrete problems (P) (defined, initially, in a direct way, without perturbations), as well as some "new" dual problems to them (see e.g. [57]). For an example of recovering the "usual" dual problems with this method, let us consider the convex programming problem

$$(P) \qquad \alpha = \inf_{\substack{y \in R^n \\ u(y) \leq 0}} h(y), \tag{3.3}$$

where $u: R^n \to R^m$ and $h: R^n \to \bar{R}$ are convex and $\leq$ is understood in the sense of the natural partial order of R^m; this is the particular case $F = R^n$, $G = \{y \in R^n \mid u(y) \leq 0\}$, of (2.1). Then, taking $X = R^m$ and

$$p(y, x) = \begin{cases} h(y) & \text{if } y \in R^n, x \in R^m, u(y) \leq x \\ +\infty & \text{if } y \in R^n, x \in R^m, u(y) \nleq x, \end{cases} \tag{3.4}$$

one can compute (see e.g. [57], p.23 or [19], p.64) that λ of (3.2) becomes the "usual" dual objective functional

$$\lambda(w) = \begin{cases} \inf\limits_{y \in F} \{h(y) + w(u(y))\} & \text{if } w \in (R^m)^*, \ w \geq 0 \\ -\infty & \text{if } w \in (R^m)^*, \ w \ngeq 0, \end{cases} \tag{3.5}$$

where $w \geq 0$ means that $w(x) \geq 0$ for all $x \in X = R^m$, $x \geq 0$. Note also that f of (2.2) becomes now

$$f(x) = \inf_{\substack{y \in R^n \\ u(y) \leq x}} h(y) \qquad (x \in X = R^m). \tag{3.6}$$

4. A major observation has been that λ and β of (3.2), (3.1), can be expressed with the aid of (Fenchel) conjugation. We recall that the <u>conjugate</u> of a functional $\varphi: X \to \bar{R}$ is the functional $\varphi^*: X^* \to \bar{R}$ defined by

$$\varphi^*(w) = \sup_{x \in X} \{w(x) - \varphi(x)\} \qquad (w \in X^*); \tag{4.1}$$

the <u>second conjugate</u> of φ is, by definition, $\varphi^{**} = (\varphi^*)^*: X \to \bar{R}$, i.e.,

$$\varphi^{**}(x) = \sup_{w \in X^*} \{w(x) - \varphi^*(w)\} \qquad (x \in X). \tag{4.2}$$

It is easy to compute (see e.g. [57], [19]) that

$$\lambda(w) = -p^*(0, w) \qquad (w \in X^*), \tag{4.3}$$

where we use the canonical identification $F^* \times X^* \cong (F \times X)^*$, given by

$$(v, w)(y, x) = v(y) + w(x) \qquad ((v, w) \in F^* \times X^*, (y, x) \in F \times X); \tag{4.4}$$

also, by (3.2), (2.2) and (4.1), we have

$$\lambda = -f^*, \tag{4.5}$$

whence, by (3.1), (4.1) and (4.2), we obtain

$$\beta = \sup_{w \in X^*} \inf_{x \in X} \{f(x) - w(x)\} = \sup(-f^*)(X^*) = f^{**}(0). \tag{4.6}$$

The importance of the above observations lies in the fact that they permit to use the well-developed machinery of Fenchel conjugation to study the dual problems (3.1), (3.2).

For example, by (4.5), λ is always concave and w*-upper semi-continuous (this also follows from (3.2)). Furthermore, since for any $\varphi:X\to\bar{R}$ there holds $\varphi\geq\varphi^{**}$, from (2.4) and (4.6) we obtain

$$\alpha\geq\beta \tag{4.7}$$

(this follows also directly from (2.4), (3.1) and (3.2)). Also, for any $\varphi:X\to\bar{R}$ we have (see [23] and formula (11.13) below)

$$\varphi^{**}=\varphi_{\overline{co}}\ , \tag{4.8}$$

the closed convex hull of φ (i.e., the greatest closed convex functional $\leq\varphi$; we recall that a functional $\psi:X\to\bar{R}$ is said to be "closed", if either ψ is lower semi-continuous, nowhere having the value $-\infty$, or ψ is the constant functional $-\infty$). Hence, by (4.6), we obtain

$$\beta=f_{\overline{co}}(0). \tag{4.9}$$

Formulae (4.5) and (4.9) are called in [57], p.19, "the central theorem about dual problems". Indeed, let us recall that the most useful cases are when weak duality holds (i.e., $\alpha=\beta$), or, when strong duality holds (that is, $\alpha=\beta$ and there exists a "solution" of the dual problem (Q) of (3.1), (3.2), i.e., a functional $w_o\in X^*$ such that $\lambda(w_o)=$ $=\sup\lambda(X^*)=\beta$; in other words, we have $\alpha=\beta$, with sup replaced by max in (3.1)), since then often the "solutions" of the primal problem (P) (i.e., the elements $g_o\in G$ such that $h(g_o)=$ $=\inf h(G)$) can be found via the dual problem (Q) (see e.g. [57], pp.4-5). Now, from (2.4) and (4.9) we see that weak duality $\alpha=\beta$ is equivalent to the "stability" relation $f(0)=$ $=f_{\overline{co}}(0)$; hence, in particular, when p of (2.2), (2.3) is convex and $\alpha=f(0)$ is finite, we have $\alpha=\beta$ if and only if f is lower semi-continuous at 0 (see e.g. [19], p.50, proposition 2.1). Also, it is known (see e.g. [57], theorem 16 or [19], p.50, proposition 2.2) that strong duality holds if and only if the subdifferential $\partial f(0)$ of f at 0 is non-empty, and then $\partial f(0)$ coincides with the set of all solutions w_o of the dual problem (Q). In the sequel, for brevity, we shall concentrate only on relations between weak duality $\alpha=\beta$ and stability, and we shall not mention the corresponding results for strong duality, involving various concepts of subdifferentials (e.g.,"quasi-subdifferential"[34],[93], "pseudo-subdifferential" [69], "semi-subddiferential" [72], [75], etc.) of f at 0.

The equalities (2.4), (4.9) are also useful when (weak or strong) duality does not hold, since they lead to formulae for the evaluation of the "duality gap"

$$\gamma=\alpha-\beta; \tag{4.10}$$

for results of this type, see e.g. [2].

5. As shown e.g. by the above mentioned characterization of weak duality $\alpha=\beta$, the dual problems (3.1), (3.2) are useful especially in the convex case. However, the assumption of convexity is too restrictive; for example, in mathematical economics, one often deals with infimization of quasi-convex functionals. Therefore, it has been necessary to develop more general concepts of dual problems.

For problem (3.3), with $u:R^n\to R^m$ convex and $h:R^n\to R$ quasi-convex, Luenberger [41] has considered the dual problem (3.1), with the dual objective functional $\lambda:(R^m)^*\to$ $\to\bar{R}$ defined by

$$\lambda(w)=\begin{cases}\inf\limits_{\substack{y\in R^n\\ w(u(y))\le 0}} h(y) & \text{if } w\in(R^m)^*,\ w\ge 0\\ -\infty & \text{if } w\in(R^m)^*,\ w\not\ge 0,\end{cases} \tag{5.1}$$

and has used it to construct a theory which parallels the results on the dual problem (3.1), (3.5) for convex h. The main difference between (3.5) and (5.1) is that in (3.5) the "penalty term" $w(u(y))$ is added to the objective functional h, in order to "compensate" that the constraint set of (3.5) is F (instead of $G=\{y\in R^n | u(y)\le 0\}$), while in (5.1), $w(u(y))$ is used to form new constraint sets

$$\{y\in R^n | w(u(y))\le 0\} \qquad (w\in(R^m)^*,\ w\ge 0), \tag{5.2}$$

for the unchanged objective functional h. Also [41], λ of (5.1) is quasi-concave and upper semi-continuous. The constraint sets (5.2) have been introduced initially by Glover [29] for 0-1 integer programming and have been called (see e.g. [29], [30], [33]) surrogate constraint sets; also, problem (3.1), (5.1) is called a surrogate dual problem to (P) of (3.3). Note that, in the above case, the sets (5.2) are convex and they contain the initial constraint set $\{y\in R^n | u(y)\le 0\}$. More general surrogate dual problems, involving surrogate constraint sets and penalizations in the objective functional, have been introduced in [32].

Greenberg and Pierskalla [33] have observed the following obvious connection between the Lagrangian dual problem (3.1), (3.5) and the surrogate dual problem (3.1), (5.1), to (P) of (3.3): we have

$$\begin{aligned}\alpha\ge\beta_{surr}&=\sup_{\substack{w\in(R^m)^*\\ w\ge 0}}\ \inf_{\substack{y\in R^n\\ w(u(y))\le 0}} h(y)\ge\\ \ge\beta_{Lagr}&=\sup_{\substack{w\in(R^m)^*\\ w\ge 0}}\ \inf_{y\in R^n}\{h(y)+w(u(y))\};\end{aligned} \tag{5.3}$$

hence, for the corresponding duality gaps, we have

$$0\le\gamma_{surr}=\alpha-\beta_{surr}\le\gamma_{Lagr}=\alpha-\beta_{Lagr}, \tag{5.4}$$

and thus, in particular, the equality $\alpha=\beta_{surr}$ holds for a larger class of problems (3.3) than the equality $\alpha=\beta_{Lagr}$. For evaluations of the "surrogate duality gap" γ_{surr} of (5.4), see e.g. [15].

Let us note that "surrogate duality" is useful even for some convex problems for which we have $\alpha=\beta_{Lagr}$ (whence also $\alpha=\beta_{surr}$). Indeed, for example, if F is a normed linear space, G a convex subset of F, $y_o\in F\setminus\overline{G}$ (where $\overline{G}$ is the closure of G), and

$$h(y)=\|y_o-y\| \qquad (y\in F) \tag{5.5}$$

(which is a finite, convex, continuous functional), then (P) of (2.1) becomes

$$(P)\qquad \alpha=\inf_{y\in G}\|y_o-y\|=\operatorname{dist}(y_o, G), \tag{5.6}$$

for which it is known that

$$\alpha=\max_{D\in\mathcal{D}_{G,y_o}}\operatorname{dist}(y_o, D)=\max_{D\in\mathcal{D}_{G,y_o}}\inf h(D), \tag{5.7}$$

where $\mathcal{D}_{G,y_o}$ denotes the collection of all support half-spaces $D=\{y\in F \mid w(y)\geq \inf w(G)\}\supset G$ $(0\neq w\in F^*)$, with $y_o\notin D$. Formula (5.7), used in approximation theory, may be regarded as a formula of "surrogate duality" $\alpha=\beta_{surr}$, with surrogate constraint sets $D\in\mathcal{D}_{G,y_o}$ and dual variables w (see e.g.[77] and the references therein).

Surrogate dual problems are also convenient for computations; for some recent results in this direction, see e.g. [18], [61].

6. Due to the importance of Fenchel conjugation for the study of Lagrangian duality, Greenberg and Pierskalla [34] have introduced a new concept of "quasi-conjugation", as a tool for the study of surrogate duality.

For any locally convex space X, and any $\varphi:X\to\overline{R}$ and $\nu\in R$, <u>the first</u> <u>ν-quasi-conjugate of</u> φ is the functional $\varphi_\nu^\gamma:X^*\to\overline{R}$ defined [34] by

$$\varphi_\nu^\gamma(w)=\nu-\inf_{\substack{x\in X\\ w(x)\geq\nu}}\varphi(x) \qquad (w\in X^*), \qquad (6.1)$$

<u>the second</u> <u>ν-quasi-conjugate of</u> φ is the functional $(\varphi_\nu^\gamma)_\nu^\gamma:X\to\overline{R}$ defined [34] by

$$(\varphi_\nu^\gamma)_\nu^\gamma(x)=\nu-\inf_{\substack{w\in X^*\\ w(x)\geq\nu}}\varphi_\nu^\gamma(w) \qquad (x\in X), \qquad (6.2)$$

and <u>the normalized second quasi-conjugate of</u> φ is defined [34] by

$$\varphi^{\gamma\gamma}=\sup_{\nu\in R}\ (\varphi_\nu^\gamma)_\nu^\gamma:X\to\overline{R}. \qquad (6.3)$$

Greenberg and Pierskalla [34] have shown, among other results, that for any $\varphi:X\to\overline{R}$ we have

$$\varphi^*=\sup_{\nu\in R}\varphi_\nu^\gamma\ , \qquad \varphi\geq\varphi^{\gamma\gamma}\geq\varphi^{**}, \qquad (6.4)$$

where φ^*, φ^{**} are the Fenchel conjugates (4.1), (4.2), and that

$$\beta_{surr}=f^{\gamma\gamma}(0), \qquad (6.5)$$

where β_{surr} and f are those of (5.3) and (3.6) respectively.

Let us also mention that a more symmetric approach to "quasi-convex conjugation", and applications to duality in quasi-convex optimization, have been given by Passy and Prisman ([51], [52]); for a different approach to conjugation and quasi-convex duality, see also Flachs [27]. In the sequel we shall consider only the quasi-conjugates (6.1)-(6.3).

7. The construction of a duality theory for quasi-convex infimization has been continued by Crouzeix ([9]-[11]), who has replaced problem (P) of (3.3) by the general primal problem (P) of (2.1), embedded into a parametrized family (2.2) (instead of (3.6)), with the aid of an arbitrary $p:F\times X\to\overline{R}$ satisfying (2.3), and has defined the <u>surrogate dual problem</u> relative to the parametrization (X, p), as the supremization problem (Q) of (3.1), with

$$\lambda(w)=\inf_{\substack{(y,x)\in F\times X\\ w(x)\geq 0}} p(y,\ x) \qquad (w\in X^*); \qquad (7.1)$$

the main difference between the λ's of (3.2) and (7.1) is similar to the one between (3.5) and (5.1), mentioned after (5.1). Taking, in particular, problem (P) of (3.3), embedded into the parametrized family (3.6), it is easy to compute [9] that λ of (7.1) reduces to (5.1); on the other hand, for the best approximation problem (P) of (5.6), in a normed linear space F, taking X=F and $p:F\times F\to\overline{R}$ defined by

$$p(y, x)=\begin{cases}\|y_o-y\| & \text{if } y\in G+x\\ +\infty & \text{if } y\notin G+x,\end{cases} \tag{7.2}$$

one can compute (see e.g. [74], theorem 3.4, applied to h of (5.5) above), that β of (3.1), with λ of (7.1), becomes the right hand side of (5.7), with max replaced by sup.

Furthermore, Crouzeix [9] has shown that λ and β of (7.1), (3.1) can be expressed with the aid of the Greenberg-Pierskalla quasi-conjugates (similarly to the case of λ and β of (3.2), (3.1) and Fenchel conjugates), namely,

$$\lambda(w)=-p^{\gamma}_{(y_o,0)}(0, w)=-f^{\gamma}_{o}(w) \qquad (w\in X^*), \tag{7.3}$$

$$\beta=\sup_{w\in X^*}\ \inf_{\substack{x\in X\\ w(x)\geq 0}} f(x)=\sup(-f^{\gamma}_{o})(X^*)=f^{\gamma\gamma}(0), \tag{7.4}$$

where $y_o\in F$ is arbitrary. Hence, λ of (7.1) is always quasi-concave and w*-upper semi-continuous, and the inequalities (5.4) remain valid in this general case (by (6.4)).

8. The next natural question has been whether weak duality $\alpha=\beta$, with β of (7.4), is equivalent to a "stability" property of problem (P) relative to the parametrization (X, p), i.e., whether $\alpha=\beta$ holds if and only if f(0) coincides with the value of a "hull" of f at 0 (following Moreau [47], p.149, by a hull of f is meant the functional - if it exists - which, among a given set of functionals, is the greatest minorant of f). More generally, it is natural to ask

Question A. For any $\varphi:X\to\overline{R}$, is $\varphi^{\gamma\gamma}$ some hull of φ (i.e., does $\varphi^{\gamma\gamma}$ satisfy a relation similar to (4.8), with $\varphi_{\overline{co}}$ replaced by a suitable hull of φ)?

In this direction, Crouzeix [9] has shown that, for any $\varphi:X\to\overline{R}$ we have

$$\varphi_{\overline{q}}\leq\varphi^{\gamma\gamma}\leq\varphi_{q}(\leq\varphi), \tag{8.1}$$

where $\varphi_{\overline{q}}$ is the lower semi-continuous quasi-convex hull, and φ_q is the quasi-convex hull, of φ; hence, in particular, if $\varphi_{\overline{q}}=\varphi$ (i.e., if φ is quasi-convex and lower semi-continuous), then $\varphi_{\overline{q}}=\varphi^{\gamma\gamma}=\varphi_q=\varphi$. In the general case, rather than answering question A, formula (8.1) has suggested, in view of applications to duality-stability relations, the following further questions:

Question B. Is there a concept of "conjugation" such that, for any $\varphi:X\to\overline{R}$, $\varphi_{\overline{q}}$ coincides with the "normalized second conjugate" of φ?

Question C. Same as B, with $\varphi_{\overline{q}}$ replaced by φ_q. Any conjugation with this property is called [45] "exact quasi-convex conjugation".

An affirmative answer to question A has been given, independently, by Martínez-Legaz ([42], [43]) and Passy and Prisman [51], who have shown that, for any $\varphi:X\to\overline{R}$, we have

$$\varphi^{\gamma\gamma}=\varphi_{eq} \tag{8.2}$$

where φ_{eq} is the "evenly quasi-convex hull" of φ; we recall that a function $\psi:X\to\overline{R}$ is said to be evenly quasi-convex [51], if all level sets

$$S_c(\psi)=\{x\in X\,|\,\psi(x)\leq c\} \qquad (c\in R) \tag{8.3}$$

of ψ are evenly convex sets, in the sense of Fenchel [26] (i.e., intersections of open half-spaces). From (8.2) it follows that for β of (7.4) we have

$$\beta=f_{eq}(0), \tag{8.4}$$

(which corresponds to (4.9)), and hence weak duality $\alpha=\beta$ is equivalent to the stability relation $f(0)=f_{eq}(0)$ (where f is the optimal value functional (2.2)).

In connection with question B, let us mention that Crouzeix [9] has introduced, for any $\varphi:X\to\bar{R}$ and $\nu\in R$, "another ν-quasi-conjugate" $\varphi_\nu^C:X^*\to\bar{R}$, and has shown that

$$\varphi^*=\sup_{\nu\in R}\varphi_\nu^C \ , \quad \varphi_q=\inf_{\nu\in R}(\varphi_\nu^C)^* \ , \tag{8.5}$$

where $\varphi^*,(\varphi_\nu^C)^*$ denote the Fenchel conjugates (4.1) of φ, φ_ν^C respectively; however, (8.5) expresses φ_q only as a "mixed" second conjugate, since at involves two different types of conjugates. Furthermore, El Qortobi [20] and Attéia and El Qortobi [1] have introduced, for any $\varphi:X\to\bar{R}$, the projective conjugate $\varphi^{\#}:X^*\to\bar{R}$ and the "projective biconjugate" $\varphi^{\#\#}:X\to\bar{R}$, of φ, and have shown that

$$\varphi_q=\varphi^{\#\#}; \tag{8.6}$$

however, they have observed ([20],[1]) that, in general, $\varphi^{\#\#}\neq(\varphi^{\#})^{\#}$. Also, Martínez-Legaz ([42], [43]) has introduced, for any $\varphi:X\to\bar{R}$, the "H-conjugate" f^O of f, with respect to a family H of functions $h:R\to\bar{R}$, as a certain mapping $\varphi^O:X^*\to H$, and the "H-conjugate" ψ^O of any $\psi:X^*\to H$, as a functional $\psi^O:X\to\bar{R}$ (whence $\varphi^{OO}=(\varphi^O)^O:X\to\bar{R}$), and has shown that if H is the family of all non-decreasing functions $h:R\to\bar{R}$, which are continuous from the left, then

$$\varphi^O(w)(\nu)=\varphi_\nu^C(w) \qquad (w\in X^*), \quad \varphi_q=\varphi^{OO}, \tag{8.7}$$

with φ_ν^C of (8.5); however, here the values of the H-conjugate φ^O are not in $\bar{R}$, but in the family H of functions $h:R\to\bar{R}$.

A natural solution to question B has been given in [72], defining, for any locally convex space X and any $\varphi:X\to\bar{R}$ and $\nu\in R$, the first ν-semi-conjugate $\varphi_\nu^\Theta:X\to\bar{R}$ of φ, by

$$\varphi_\nu^\Theta(w)=\nu-1-\inf_{\substack{x\in X\\ w(x)>\nu-1}}\varphi(x) \qquad (w\in X^*), \tag{8.8}$$

the second ν-semi-conjugate $(\varphi_\nu^\Theta)_\nu^\Theta:X\to\bar{R}$ of φ, by

$$(\varphi_\nu^\Theta)_\nu^\Theta(x)=\nu-1-\inf_{\substack{w\in X^*\\ w(x)>\nu-1}}\varphi_\nu^\Theta(w) \qquad (x\in X), \tag{8.9}$$

and the normalized second semi-conjugate $\varphi^{\Theta\Theta}:X\to\bar{R}$ of φ, by

$$\varphi^{\Theta\Theta}=\sup_{\nu\in R}(\varphi_\nu^\Theta)_\nu^\Theta; \tag{8.10}$$

indeed, in [72] it has been shown that, for any $\varphi:X\to\bar{R}$, we have

$$\varphi_q=\varphi^{\Theta\Theta}. \tag{8.11}$$

Note that definition (8.8) is similar to definition (6.1), with the difference that, instead of the closed half-spaces $\{x\in X|w(x)\geq\nu\}$ of (6.1), the open half-spaces $\{x\in X|w(x)>\nu-1\}$ are used in (8.8), and, instead of the added term ν in (6.1), the term $\nu-1$ is added in (8.8) (in order to ensure $\varphi^*\geq\sup_{\nu\in R}\varphi_\nu^\Theta\geq\varphi^*-1$, corresponding to (6.4)). Let us mention that, instead of closed or open half-spaces, one can also use (closed) hyperplanes, to define a concept of conjugation. Indeed, in [69], for any locally convex space X and any $\varphi:X\to\overline{R}$ and $\nu\in R$, <u>the first</u> ν-<u>pseudo-conjugate</u> $\varphi_\nu^\pi:X^*\to\overline{R}$ <u>of</u> φ has been defined by

$$\varphi_\nu^\pi(w)=\nu\text{-}\inf_{\substack{x\in X\\ w(x)=\nu}} \varphi(x) \qquad (w\in X^*), \tag{8.12}$$

and $(\varphi_\nu^\pi)_\nu^\pi$, $\varphi^{\pi\pi}$ have been defined correspondingly (i.e., replacing $\geq\nu$ by $=\nu$ and γ by π, in (6.2) and (6.3)). Then, as has been observed in [69], we have $\varphi^*\geq\varphi_\nu^\gamma\geq\varphi_\nu^\pi$, $\varphi^*=\sup\limits_{\nu\in R}\varphi_\nu^\gamma=\sup\limits_{\nu\in R}\varphi_\nu^\pi$ and $\varphi\geq\varphi^{\gamma\gamma}\geq\varphi^{\pi\pi}\geq\varphi^{**}$. For further results on pseudo-conjugates and semi-conjugates, see [69], [70] and [72], [75] respectively.

In connection with question C, Martínez-Legaz [45] has introduced, for any $\varphi:R^n\to\overline{R}$, the "H-conjugate" φ^∇ of φ, with respect to the family H of all lexicographically non-decreasing functions $h:R^n\to\overline{R}$ (i.e., $h\in H$ if for all $x,y\in R^n$ such that x is lexicographically less than y, we have $h(x)\leq h(y)$), as a certain mapping $\varphi^\nabla:\mathcal{L}(R^n)\to H$ (where $\mathcal{L}(R^n)$ denotes the set of all linear endomorphisms of R^n), and the "H-conjugate" ψ^∇ of any $\psi:\mathcal{L}(R^n)\to H$, as a functional $\psi^\nabla:R^n\to\overline{R}$ (whence $\varphi^{\nabla\nabla}=(\varphi^\nabla)^\nabla:R^n\to\overline{R}$), and has shown that for any $\varphi:R^n\to\overline{R}$ we have

$$\varphi_q=\varphi^{\nabla\nabla}. \tag{8.13}$$

9. Using the above results, there have been introduced some new dual problems to (P), relative to the parametrization (X, p), and the questions on equivalence of duality and stability for them (corresponding to the one raised, for β of (7.4), before question A above) have been answered.

Namely, one can define (see [69], [11], [72]) the <u>pseudo-dual</u> and <u>semi-dual problems to</u> (P) of (2.1), relative to the parametrization (X, p) of (2.2), (2.3), as the supremization problem (Q) of (3.1), with

$$\lambda(w)=\inf_{\substack{(y,x)\in F\times X\\ w(x)=0}} p(y,x) \qquad (w\in X^*), \tag{9.1}$$

$$\lambda(w)=\inf_{\substack{(y,x)\in F\times X\\ w(x)>-1}} p(y,x) \qquad (w\in X^*), \tag{9.2}$$

respectively. Then, one can show ([69], [11] [72]) that, for the β's of (3.1) corresponding to (9.1), (9.2), we have, respectively,

$$\alpha\geq\beta=\sup_{w\in X^*}\inf_{\substack{x\in X\\ w(x)=0}} f(x)=f^{\pi\pi}(0), \tag{9.3}$$

$$\alpha\geq\beta=\sup_{w\in X^*}\inf_{\substack{x\in X\\ w(x)>-1}} f(x)=f^{\theta\theta}(0)=f_{\overline{q}}(0), \tag{9.4}$$

with f of (2.2); hence, by (2.4) and (9.4), <u>weak duality</u> $\alpha=\beta$ <u>of</u> (3.1), (9.2) <u>is equivalent to the stability relation</u> $f(0)=f_{\overline{q}}(0)$.

Similarly, for $F=R^n$, one can define [45] a "dual" problem to (P) of (2.1), relative to the parametrization $(X=R^m,p)$ of (2.2), (2.3), as the supremization problem (Q) of (3.1), with X* replaced by $\mathcal{L}(R^m)$ and with

$$\lambda(w)=\inf_{\substack{(y,x)\in R^n\times R^m \\ w(x)\,\geq_L\, 0}} p(y,x) \qquad (w\in\mathcal{L}(R^m)), \quad (9.5)$$

where $\geq_L$ means "equal or lexicographically less than". Then, one can show, using (8.13) and [79], formula (2.30) (or [45], proposition 2.4 and corollary 2.5) that, for β of (3.1), with X^* replaced by $\mathcal{L}(R^m)$ and with λ of (9.5), we have

$$\alpha\geq\beta=\sup_{w\in\mathcal{L}(R^m)} \inf_{\substack{x\in R^m \\ w(x)\,\geq_L\, 0}} f(x)=f^{\nabla\nabla}(0)=f_q(0), \qquad (9.6)$$

where f is the functional (2.2). Hence, by (2.4) and (9.6), <u>weak duality</u> $\alpha=\beta$ <u>is equivalent to the stability relation</u> $f(0)=f_q(0)$.

10. The similarity of the definitions of quasi-conjugates, pseudo-conjugates and semi-conjugates and of the results on them (including applications to duality and stability in optimization) suggests to try to unify them. To this end, it is useful to observe that the terms ν and $\nu-1$ have been added in (6.1), (8.12), (8.8), (6.2), etc. only in order to compare "nicely" φ_ν^μ and $\varphi^{\mu\mu}$ $(\mu=\gamma,\pi,\theta)$ with the Fenchel conjugates φ^*, φ^{**} (see e.g. (6.4)) and that their omission does not alter the normalized second conjugates $\varphi^{\mu\mu}$ $(\mu=\gamma,\pi,\theta)$, nor the dual optimization problems defined with the aid of φ_ν^μ, but permits to write the conjugates φ_ν^μ in the unified form

$$\varphi_\nu^\mu(w)=-\inf\varphi(\Delta_{\nu,w}^\mu) \qquad (w\in X^*,\ \mu=\gamma,\pi,\theta), \quad (10.1)$$

where

$$\Delta_{\nu,w}^\gamma=\{x'\in X\,|\,w(x')\geq\nu\} \qquad (w\in X^*), \quad (10.2)$$

$$\Delta_{\nu,w}^\pi=\{x'\in X\,|\,w(x')=\nu\} \qquad (w\in X^*), \quad (10.3)$$

$$\Delta_{\nu,w}^\theta=\{x'\in X\,|\,w(x')>\nu-1\} \qquad (w\in X^*). \quad (10.4)$$

Now, given any family of sets $\Delta_{\nu,w}\subseteq X$ $(\nu\in R,\ w\in X^*)$, for any $\varphi:X\to\bar{R}$ and $\nu\in R$ one can define [73] the ν-Δ-<u>conjugate functional</u> $\varphi_\nu^\Delta:X^*\to\bar{R}$, by

$$\varphi_\nu^\Delta(w)=-\inf\varphi(\Delta_{\nu,w}) \qquad (w\in X^*); \quad (10.5)$$

in [73], these have been also called "surrogate conjugate functionals". Furthermore, it is convenient to consider the sets $\Delta_{\nu,w}$ as images of a multifunction $\Delta:(\nu,w)\to\Delta_{\nu,w}$ from $R\times X^*$ into 2^X and to assume that Δ induces a multifunction $\Delta:(\nu,x)\to\Delta_{\nu,x}$ from $R\times X$ into 2^{X^*}, defined "in the same way" (for the precise definition of a "universally defined multifunction" Δ, see [73], §1); this is satisfied for $\Delta_{\nu,w}=\Delta_{\nu,w}^\mu$ of (10.2)-(10.4), and it permits to define [73] "the second ν-Δ-conjugate" $(\varphi_\nu^\Delta)_\nu^\Delta$ and "the normalized second Δ-conjugate" $\varphi^{\Delta\Delta}$.

For $\Delta:R\times X^*\to 2^X$ as above, one can define [73] a "surrogate dual" problem to (P) of (2.1), relative to the parametrization (X, p) of (2.2), (2.3), as the supremization problem (Q) of (3.1), with

$$\lambda(w)=\inf_{\substack{(y,x)\in F\times X \\ x\in\Delta_{0,w}}} p(y,x) \qquad (w\in X^*); \quad (10.6)$$

in particular, for $\Delta_{\nu,w}=\Delta_{\nu,w}^\mu$ $(\mu=\gamma,\pi,\theta)$ above, (10.6) reduces to (7.1), (9.1) and (9.2), respectively. One can show [73] that, for β of (3.1) corresponding to λ of (10.6), we have

$$\alpha\geq\beta=\sup_{w\in X^*}\inf f(\Delta_{o,w})=f^{\Delta\Delta}(0), \tag{10.7}$$

with f of (2.2); hence, by (2.4) and (10.7), weak duality $\alpha=\beta$ of (3.1), (10.7) is equivalent to $f(0)=f^{\Delta\Delta}(0)$. The latter equality may be regarded as a stability relation, since $f^{\Delta\Delta}$ coincides [73] with the "Δ-quasi-convex hull" of f. We recall [73] that a subset M of X is said to be Δ-convex, if for each $x\notin M$ there exists $w\in X^*$ such that $M\cap\Delta_{w(x),w}=\emptyset$, $x\in\Delta_{w(x),w}$ (i.e., M can be separated from each $x\notin M$ by a set of the form $X\setminus\Delta_{w(x),w}$, with a suitable $w\in X^*$) and that Δ-quasi-convexity of $\psi:X\to\overline{R}$ means that all level sets (8.3) of ψ are Δ-convex sets; in the particular case when $\Delta_{\nu,w}=\Delta^{\gamma}_{\nu,w}$ above, this yields the evenly convex sets M and the evenly quasi-convex functionals ψ, and for $\Delta_{\nu,w}=\Delta^{\theta}_{\nu,w}$ it yields the closed convex sets M, respectively the lower semi-continuous quasi-convex functionals ψ.

11. An important generalization of the Fenchel conjugates (4.1), (4.2) has been given by Moreau [48],[49], replacing the locally convex spaces X, X* by arbitrary sets X, W (without any structure assumed on them) and replacing the natural bilinear coupling functional $n:X\times X^*\to R$, i.e., the functional

$$(x, w)\to n(x, w)=w(x) \qquad (x\in X, w\in X^*) \tag{11.1}$$

by an arbitrary functional $k:X\times W\to\overline{R}=[-\infty,+\infty]$, called "coupling functional". Since here k may take also the values $\pm\infty$ (which turns out to be useful in various applications), Moreau has extended the usual addition on R to $\overline{R}$, defining the "upper addition" $\dot{+}$ and the "lower addition" $\underset{\cdot}{+}$ on $\overline{R}$, by

$$a\dot{+}b=a\underset{\cdot}{+}b=a+b \qquad (a,b\in R), \tag{11.2}$$

$$a\dot{+}(+\infty)=+\infty,\ a\underset{\cdot}{+}(-\infty)=-\infty \qquad (a\in\overline{R}), \tag{11.3}$$

and has worked out the rules with these operations [49].

The (Fenchel-Moreau) conjugate of a functional $\varphi:X\to\overline{R}$, with respect to a coupling functional $k:X\times W\to\overline{R}$, is the functional $\varphi^{c(k)}:W\to\overline{R}$ defined ([48], [49]) by

$$\varphi^{c(k)}(w)=\sup_{x\in X}\{k(x,w)\underset{\cdot}{+}-\varphi(x)\} \qquad (w\in W), \tag{11.4}$$

and the second conjugate of φ, with respect to k, is the functional $\varphi^{c(k)c(k)}:X\to\overline{R}$ defined ([48], [49]) by

$$\varphi^{c(k)c(k)}(x)=\sup_{w\in W}\{k(x,w)\underset{\cdot}{+}-\varphi^{c(k)}(w)\} \qquad (x\in X); \tag{11.5}$$

clearly, for a locally convex space X, $W=X^*$ and $k=n$ of (11.1), $\varphi^{c(n)}=\varphi^*$ and $\varphi^{c(n)c(n)}=\varphi^{**}$ of (4.1) and (4.2), respectively. Concerning Fenchel-Moreau conjugates, see also [4], [7], [17], [21], [59], [60], [89], [91], [92].

In the sequel, we shall assume that X is an arbitrary set, $W\subseteq\overline{R}^X$ (where $\overline{R}^X$ denotes the family of all functionals $w:X\to\overline{R}$), and $k=n$ of (11.1), with X* replaced by W; this n is no longer bilinear, but it is still linear on W, if $W\subseteq R^X$ (endowing $W\subseteq R^X$ with the usual vector operations, defined pointwise on X). As has been observed in [81], remark 2.2 and Addendum (see also [82], §1), this assumption is no restriction of the generality, since it is equivalent to the case of arbitrary sets X,W and an arbitrary coupling functional $\varphi:X\times W\to\overline{R}$, but it will simplify the formulas below.

Let us embed now the infimization problem (P) of (2.1), where G is a subset of an arbitrary set F (assuming no structure on F), into a family of parametrized problems (2.2), where X is an arbitrary parameter set and $p:F\times X\to\overline{R}$ is an arbitrary functional satisfying, for some $x_0\in X$,

$$p(y, x_0)=\begin{cases} h(y) & \text{if } y\in G \\ +\infty & \text{if } y\in F\setminus G. \end{cases} \tag{11.6}$$

Of course, it is now necessary to use the "embedding" condition (11.6) instead of (2.3), since we assume no structure on X (and hence no "zero element" of X). By (2.1), (2.2) and (11.6), we have

$$f(x_0)=\inf_{y\in F} p(y, x_0)=\alpha. \tag{11.7}$$

Some authors (e.g. [59], [4]) assume, in order to simplify the formulas, the "normalization" $w(x_0)=0$ $(w\in W)$, which is satisfied by $x_0=0$ in a locally convex space X and $W\subseteq X^*$, but we shall not make here this assumption, since it is too restrictive.

One can define ([79], [82]) the <u>Lagrangian dual</u> problem to (P), relative to the parametrization (X, p), as the supremization problem

$$(Q) \qquad \beta=\sup \lambda(W), \tag{11.8}$$

where

$$\lambda(w)=\inf_{(y,x)\in F\times X} \{p(y,x)\dotplus -w(x)\}\dotplus w(x_0) \qquad (w\in W). \tag{11.9}$$

Formulae (4.3), (4.5) and (4.6) admit the following extensions ([82],[79]) to this case:

$$\lambda(w)=-p^{c(n)}(0, w)\dotplus w(x_0) \qquad (w\in W), \tag{11.10}$$

$$\lambda(w)=-f^{c(n)}(w)\dotplus w(x_0) \qquad (w\in W), \tag{11.11}$$

$$\beta=\sup_{w\in W} \{-f^{c(n)}(w)\dotplus w(x_0)\}=f^{c(n)c(n)}(x_0), \tag{11.12}$$

where, in (11.10), we use the canonical embedding (4.4) of $R^F\times\overline{R}^X$ into $\overline{R}^{F\times X}$. Thus, again, we have (4.7). Furthermore [79] (see also [49], for R replaced by $\overline{R}$), (4.8) extends to

$$\varphi^{c(n)c(n)}=\varphi_{H(W+R)}, \tag{11.13}$$

the "(W+R)-convex hull" of $\varphi:X\to\overline{R}$ (i.e.[17], the greatest (W+R)-convex functional $\leq\varphi$; we recall that a functional $\psi:X\to\overline{R}$ is said to be "(W+R)-convex" [17], if it is a supremum of a family of functionals of the form $w+d$, where $w\in W$ and $d\in R$). Hence, by (11.12), we obtain

$$\beta=f_{H(W+R)}(x_0)=\sup_{\substack{w\in W,\ d\in R \\ w+d\leq f}} (w(x_0)+d). \tag{11.14}$$

From (11.7) and (11.14) we see that <u>weak duality</u> $\alpha=\beta$ <u>is equivalent to the "stability" relation</u> $f(x_0)=f_{H(W+R)}(x_0)$.

The above generalization can be applied to a large class of problems, encompassing both the continuous case and the case of discrete optimization (since there are no structures assumed on F and X); for example, in a subsequent paper (in preparation), we give applications to combinatorial optimization. Also, as has been shown, independently, by Lindberg [40], Balder [4], and Dolecki and Kurcyusz [17], this generalization provides a unified way of obtaining various known dual problems (Q) defined with the aid of so-called" augmented Lagrangians", just by taking suitable particular sets $W\subset\overline{R}^X$ in (11.8),

(11.9). A further advantage of this general approach will be shown in section 12 below.

Let us mention that it has also been a matter of interest whether a given duality theory is "symmetric", i.e., such that, under some mild assumptions, the "dual" to the dual problem (Q) of (3.1) or (11.8) (defined as a suitable infimization problem, usually with the aid of a parametrization of (Q)) coincides with the primal problem (P) of (2.1). For example, the duality theory of Lindberg [40] is symmetric, while those of Balder [4] and Dolecki and Kurcyusz [17] are non-symmetric.

12. For any multifunction $\Delta:R\times X^*\to 2^X$ as in section 10 above, if we take $V=V_\Delta\subset\overline{R}^X$ defined by

$$V=\{-\chi_{\Delta_{w(y),w}}\dot{+}d \mid y\in X,\ w\in X^*,\ d\in R\}, \tag{12.1}$$

where χ_M denotes the indicator functional of the set $M\subseteq X$, then, as has been shown in [73], for any $\varphi:X\to\overline{R}$ we have

$$\varphi^{c(n)}(-\chi_{\Delta_{w(y),w}}\dot{+}d)=\varphi^{\Delta}_{w(y)}(w)\dot{+}d \qquad (y\in X,\ w\in X^*, d\in R), \tag{12.2}$$

with φ^{Δ}_{ν} of (10.5); indeed, this follows directly from (11.4) applied to k=n of (11.1), with X* replaced by V of (12.1). Similarly [73], when Δ is universally defined (see section 10 above), we have

$$\varphi^{c(n)c(n)}=\varphi^{\Delta\Delta}. \tag{12.3}$$

Thus, ν-Δ-conjugation (and hence, in particular, quasi-conjugation, pseudo-conjugation, semi-conjugation) is a "particular case" of Fenchel-Moreau conjugation (11.4), with W replaced by $V=V_\Delta$ of (12.1), and with k=n corresponding to this V. For $\Delta_{\nu,w}=\Delta^{\gamma}_{\nu,w}$ of section 10, a related result (involving "conjugation" in the sense of Lindberg [40], which is equivalent [81] to Fenchel-Moreau conjugation), has been given by Martínez-Legaz [44].

From the above and from the expressions of the dual problems with the aid of conjugate functionals (see e.g. (7.3), (7.4), (10.7) and (11.10)-(11.12)), it follows that the various surrogate dual problems (11.8), (7.1), (10.6), etc. to (P) of (2.1), relative to the parametrization (X, p) of (2.2), (2.3), are "particular cases" of the general Lagrangian dual problem (11.9), for a suitable modification V of W. Of course, this can be deduced also directly from the definitions of these problems, without using conjugates; see e.g. [82], where various other relations between Lagrangian dual problems and surrogate dual problems have been also given (see also section 16 below).

Similarly to (3.2)→(11.9) and (4.1)→(11.4), etc., it is useful to replace the locally convex space X by an arbitrary set X, and X* by any set $W\subseteq\overline{R}^X$ (whence 0 by $w(x_0)$), in surrogate duality and surrogate conjugation (e.g., in (7.1), (9.1), (10.6), (10.5), etc.); see [73],[78],[82].

13. Given two optimization problems (P_1), (P_2) (defined on possibly different spaces) and dual problems (Q_1), (Q_2) (in some sense) to (P_1) and (P_2) respectively, the "primal-dual pairs" of problems $\{(P_1),(Q_1)\}$, $\{(P_2),(Q_2)\}$ are said to be equivalent [75] (denoted $\{(P_1),(Q_1)\}\sim\{(P_2),(Q_2)\}$, if $\alpha_1=\alpha_2$ and $\beta_1=\beta_2$, where α_i and β_i are the optimal values of (P_i) and (Q_i) respectively. The usefulness of this concept consists in the fact that from duality theorems for $\{(P_i),(Q_i)\}$ (i.e., conditions in order that $\alpha_i=\beta_i$) one can deduce duality theorems for $\{(P_{3-i}),(Q_{3-i})\}$ (i=1,2). For example, formulae (11.7)-(11.9) and (11.12) show that the pair $\{(P_{G,h}),(Q^{G,h})\}$, where $(Q^{G,h})$ is the Lagrangian dual pro-

blem (11.8), (11.9) to $(P_{G,h})$, is equivalent to the pair $\{(P_{\{x_o\},f}), (Q^{\{x_o\},f})\}$, with f of (2.2), where $(Q^{\{x_o\},f})$ is the "unperturbational" Lagrangian dual in the sense (16.1) (below) to $(P_{\{x_o\},f})$ inf $f(\{x_o\})=f(x_o)=\alpha$. The latter has the advantage that the constraint set $\{x_o\}$ is a singleton and, by the above remark, from duality theorems for $\{(P_{\{x_o\},f}), (Q^{\{x_o\},f})\}$ (which, by (11.14), amount to conditions for $f(x_o)=f_{H(W+R)}(x_o)$), one can deduce duality theorems for $\{(P_{G,h}), (Q^{G,h})\}$. This method has been applied in a consequent manner, e.g. in [75], for Lagrangian duality and surrogate duality. Let us also mention that the notation X for the parameter set is used (in contrast with the notations of other authors) in order to emphasize that we apply theorems <u>in</u> X (e.g., separation theorems) in order to draw conclusions on problem $(P_{G,h})$ of (2.1).

The remarks made after formula (4.9) suggest to classify the (equivalence classes of) primal-dual pairs $\{(P), (Q)\}$, by saying [75] that $\{(P_1), (Q_1)\}$ is "better than" $\{(P_2), (Q_2)\}$ (denoted $\{(P_1), (Q_1)\}\succsim\{(P_2), (Q_2)\}$), if for the gaps we have $\alpha_1-\beta_1\leq\alpha_2-\beta_2$. For example, for the optimization problem

$$(P) \qquad \alpha=\inf_{\substack{y\in F\\ u(y)=x_1}} h(y), \tag{13.1}$$

where F,X are locally convex spaces, $x_1\in X$, and u is a continuous linear mapping of F into X, one can consider the parametrization functionals (of the form (2.7), with parameter space F for p_1, and X for p_2)

$$p_1(y,x)=\begin{cases} h(y) & \text{if } y\in x+\{y'\in F \mid u(y')=x_1\}\\ +\infty & \text{otherwise,}\end{cases} \tag{13.2}$$

$$p_2(y,x)=\begin{cases} h(y) & \text{if } u(y)=x_1+x\\ +\infty & \text{otherwise,}\end{cases} \tag{13.3}$$

and one can ask, which one of the primal-dual pairs $\{(P), (Q_1)\}$, $\{(P), (Q_2)\}$, corresponding to p_1 and p_2 respectively, by pseudo-duality (3.1), (9.1), is better. In [74], §3, it has been shown that p_1 yields the better pair and, if F is a Banach space and $h(y)=\|y\|$ $(y\in F)$, then $\{(P), (Q_1)\}$ is optimal, since it has the gap 0, while the (pseudo duality) gap in $\{(P), (Q_2)\}$ is related to the "characteristic" of the subspace $u^*(X^*)$ of F^*, in the sense of Dixmier [16] (where u^* is the adjoint operator).

14. Let us mention now some further developments in the theory of conjugations.

Crouzeix ([9], p.28) has made the following remark: "... the notions of conjugate functions introduced by Greenberg and Pierskalla do not make use of the notion of level sets which plays in quasi-convexity the role played by epigraphs in convexity and therefore we have introduced analogous notions to those of Greenberg and Pierskalla, involving the level sets" (by this, Crouzeix has meant the φ_v^C's occurring in (8.5) above and a notion of "tangential" corresponding to the "quasi-subdifferential" of [34], [93]); a similar remark has been also made in [10], p.75. Therefore, in [76], [73], there have been given some expressions of $\varphi_v^\gamma, \varphi_v^\pi, \varphi_v^\theta, \varphi_v^\Delta, (\varphi_v^\gamma)_v^\gamma, \varphi^{\gamma\gamma}$ (of (6.1), (8.12), (8.8), (10.5), (6.2), (6.3)) etc., involving the level sets (8.3) (of $\psi=\varphi$) and the "strict" [49] level sets

$$A_c(\varphi)=\{x\in X \mid \varphi(x)<c\} \qquad (c\in R); \tag{14.1}$$

for example,

$$\varphi_{\nu}^{\gamma}(w)=\inf_{\substack{\nu\in R\\ w(x)<\lambda(x\in A_{\lambda-\nu}(\varphi))}} \nu=\sup_{\substack{\nu\in R\\ \exists x\in A_{\lambda-\nu}(\varphi),w(x)\geq\lambda}} \nu \quad (w\in X^*). \tag{14.2}$$

Also, some expressions of the corresponding concepts of "surrogate" subdifferentials ([34], [93], [69], [72], [73]), with the aid of the level sets (14.1), have been given in [75], §3 and [73], §6. Some extensions of these results of [76], [75], [73] (and of some results of [70], [74]) have been obtained by Volle [90].

An axiomatic approach to the theory of Fenchel-Moreau conjugations (11.4), (11.5) has been started in [80]. The main result of [80] states that if X,W are two sets and if an operator $c:\varphi\in\overline{R}^X\to\varphi^c\in\overline{R}^W$ satisfies, for every index set I, the conditions

$$(\inf_{i\in I}\varphi_i)^c=\sup_{i\in I}\varphi_i^c \qquad (\{\varphi_i\}_{i\in I}\subseteq\overline{R}^X), \tag{14.3}$$

$$(\varphi\dotplus d)^c=\varphi^c\dotplus -d \qquad (\varphi\in\overline{R}^X,\ d\in\overline{R}), \tag{14.4}$$

then there exists a uniquely determined coupling functional $k:X\times W\to\overline{R}$ such that $\varphi^c=\varphi^{c(k)}$ of (11.4), for all $\varphi\in\overline{R}^X$, namely,

$$k(x,w)=(\chi_{\{x\}})^c(w) \qquad (x\in X,\ w\in W); \tag{14.5}$$

thus, since conversely, $c(k):\overline{R}^X\to\overline{R}^W$ of (11.4) satisfies (14.3), (14.4), one obtains not only an axiomatic characterization of Fenchel-Moreau conjugations, but also that the correspondence $k\to c(k)$, between coupling functionals and conjugations, is one-to-one. Some results on the "dual" $c':\overline{R}^W\to\overline{R}^X$ of an operator $c:\overline{R}^X\to\overline{R}^W$ have been also given in [80]. Furthermore, in [81], there have been shown some relations between "dualities" in the sense of Evers and van Maaren [22], i.e., mappings $\Delta:2^X\to 2^W$ satisfying, for every index set I,

$$\Delta(\bigcup_{i\in I}G_i)=\bigcap_{i\in I}\Delta(G_i) \qquad (\{G_i\}_{i\in I}\subseteq 2^X), \tag{14.6}$$

polarities $\pi(\rho):2^X\to 2^W$ (where $\rho\subseteq X\times W$ is a binary relation), in the sense of [6], coupling functionals and conjugations. Also, in [83] and [46], there have been studied "dualities" (called "polarities" in [6], [50], [53]) between two complete lattices E and F, i.e., mappings $\Delta:E\to F$ satisfying, for every index set I,

$$\Delta(\inf_{i\in I}x_i)=\sup_{i\in I}\Delta(\{x_i\}) \qquad (\{x_i\}_{i\in I}\subseteq E); \tag{14.7}$$

in particular, for $E=\overline{R}^X$, $F=\overline{R}^W$, (14.7) reduces to (14.3) (with $c=\Delta$), so one obtains a generalization of Fenchel-Moreau conjugations (for which (14.4) need not hold), considered, in a particular case, by Ben-Tal and Ben-Israel [5], and for $E=(2^X,\supseteq)$, $F=(2^W,\supseteq)$, (14.7) reduces to (14.6). Some corresponding problems for conjugations of multifunctions and conjugations of vector-valued functions, are now being investigated.

Finally, let us mention that Deumlich and Elster have introduced and studied (for a survey, see [14]), a concept of "Φ-conjugation" of functionals. Starting from the observation of Fenchel [25] that conjugate functionals (4.1) are closely connected with polarity with respect to a hyperparaboloid, Deumlich and Elster have based their concept of "Φ-conjugation" (which we shall not reproduce here) on polarity with respect to an arbitrary nondegenerate hypersurface Φ of order two. For applications of this Φ-conjugation to duality-stability relations, especially in fractional programming, see [13].

15. Let us also mention, briefly, some related results of [79] on generalized con-

vexity. The approaches of sections 10-12 above have stimulated the study of "convexity" of a subset G of a set X with respect to a family of sets $\mathcal{M}\subseteq 2^X$ (in the sense of [12]) and with respect to a family of functionals $W\subseteq \overline{R}^X$ (in the sense of [24]) and related concepts of "quasi-convexity" and "convexity" of functionals (via level sets and epigraphs, respectively). We recall that a set $G\subseteq X$ is said to be i) convex with respect to $\mathcal{M}\subseteq 2^X$ [12], if for each $x\notin G$ there exists $M\in\mathcal{M}$ such that $G\subseteq M$, $x\in X\setminus M$ (i.e., such that M "separates" G from x); ii) convex with respect to $W\subseteq\overline{R}^X$ [24], if for each $x\notin G$ there exists $w\in W$ such that $\sup w(G)<w(x)$ (i.e., such that w "separates" G from x). For example [79], if X is a locally convex space, $\mathcal{M}$-convexity includes the Δ-convexity mentioned after formula (10.7) above (by taking $\mathcal{M}=\{X\setminus\Delta_{w(x),w}\mid x\in X, w\in X^*\}$), whence also the evenly convex sets and the closed convex sets; moreover, it also includes (see [79]) the convex sets in a linear space X, the closed sets in a topological space X, etc. In [79] it has been shown that the theories of $\mathcal{M}$-convex sets and W-convex sets $G\subseteq X$ (where $\mathcal{M}\subseteq 2^X$, $W\subseteq\overline{R}^X$) are equivalent, and that, correspondingly, so are the theories of $\mathcal{M}$-quasi-convex, W-quasi-convex and W-convex functionals. Furthermore, as has been observed in [83], §5, a number of results of Volle [90] on "$\Delta^*\Delta$-convexity" of sets $G\subseteq X$ and "$\Delta^*\Delta$-quasi-convexity" of functionals $\varphi:X\to\overline{R}$, where $\Delta:2^X\to 2^W$ is a duality (14.6), can be simply obtained as the particular case $\mathcal{M}=\{\Delta^*(\{w\})\mid w\in W\}$ of some results of [79] on $\mathcal{M}$-convexity of G and $\mathcal{M}$-quasi-convexity of φ. For some other results on various concepts of generalized convexity, with applications to optimization, see e.g. [58].

16. Let us return now to optimization. The great variety of "usual" dual problems (3.1) and (11.8) to a given primal problem (2.1), has stimulated attempts to unify them, in a general theory of dual optimization problems. An attempt in this direction has been made by Tind and Wolsey [86] (see also [8] and [54]). A different unified theory has been constructed in [78] for surrogate dual problems and in [82] for the most general case, encompassing, among others, the Lagrangian, the surrogate, and the Tind-Wolsey dual problems, as well as the dual problems of Gould [31], Klötzler [39], Hoffmann [35], etc. Here we shall only mention, briefly, some aspects of the theory of [78] and [82], which are related to parametrization.

According to the definition given in [82], by a "dual" problem to the infimization problem (2.1) we mean any supremization problem (11.8), where $W=W^{G,h}$ and $\lambda=\lambda^{G,h}:W\to\overline{R}$. In [82] it is shown that there are some close connections between "unperturbational" dual problems (i.e., defined without assuming any parametrization (X, p) of (P), such as in linear programming or in (3.5), (5.1), (5.7), etc.) and "perturbational" dual problems (i.e., defined with the aid of a parametrization (X, p) of (P)). Namely, from each unperturbational dual problem one can deduce a "perturbational version" of it, with the aid of a certain scheme (part of which is similar to the idea of "universally defined" multifunctions [73]) and, conversely, from each perturbational dual problem, with F,X locally convex spaces and $W\subseteq X^*$, one can regain its "unperturbational version", by taking X=F and p=the perturbation (2.7), with $\Gamma(x)=G+x$, for all $x\in F$. For example, the unperturbational version of the Lagrangian dual (11.9) is (Q) of (11.8) with $W\subseteq\overline{R}^F$ and

$$\lambda(w)=\inf_{y\in F}\{h(y)\dot{+}-w(y)\}\dot{+}\inf w(G) \qquad (w\in W), \qquad (16.1)$$

which is useful when $G\neq F$ (see e.g. [62]); similar relations hold between the unperturbational and perturbational Tind-Wolsey dual problems to (P) (see [82]) and between the unperturbational and perturbational surrogate dual problems to (P) (see [78], [82]). We recall that if G is a subset of a set $F, h:F\to\overline{R}$, $W=W^{G,h}$ is a set and $\Delta_{G,w}\subseteq F$ ($w\in W$) is a family of ("surrogate constraint") sets, the <u>unperturbational surrogate dual</u> to (P) of (2.1) is defined [78] as the supremization problem (11.8), with

$$\lambda(w)=\inf h(\Delta_{G,w}) \qquad (w\in W); \qquad (16.2)$$

furthermore, if X is a set, $p:F\times X\to\overline{R}$ satisfies (11.6) for some $x_o\in X$, and $\tilde{\Delta}_{(F,x_o),w}\subseteq F\times X$ ($w\in W$) is a family of sets (the most interesting particular case being $\tilde{\Delta}_{(F,x_o),w}=F\times\times\Delta^o_{\{x_o\},w}$, where $\{\Delta^o_{\{x_o\},w}\}_{w\in W}$ is a family of subsets of X), the <u>perturbational surrogate dual</u> to (P) is defined [78] as (11.8), with

$$\lambda(w)=\inf p(\tilde{\Delta}_{(F,x_o),w}) \qquad (w\in W). \qquad (16.3)$$

These general definitions of surrogate dual problems encompass, as particular cases, the various surrogate dual problems to (P), mentioned above (e.g., quasi-dual, pseudo-dual, etc.). In a series of papers, of which we mention [71], [70], [75] and [78], there have been obtained (perturbational and unperturbational) surrogate duality results, with the aid of "level set methods" (consisting, roughly speaking, in separation, in various senses, of the level sets of h from the surrogate constraint sets), which are more convenient for surrogate dual problems, than the epigraph methods of convex analysis.

Some of the perturbational dual problems (3.1) and (11.8) to (P) correspond to various concepts of conjugations, applied to f of (2.2) (we have seen many examples above). and therefore they are called [78] "perturbational conjugate dual" problems to (P); their relations to general unperturbational and perturbational surrogate dual problems have been investigated in [78].

Further results on dual problems (3.1) and (11.8) defined with the aid of a parametrization (X, p) have been given, for a large class of perturbations p, introduced and studied in [82], namely, for the p's that can be written as

$$p(y, x)=h(y)\dotplus\pi_G(y, x) \qquad (y\in F,\ x\in X), \qquad (16.4)$$

where $\pi_G:F\times X\to\overline{R}$ is a coupling functional with values $\pi_G(y, x)\in\overline{R}$ not depending on h; these p' s, called "h-separated" perturbation functionals [82], encompass various important perturbation functionals as particular cases, for example , (2.7) above (by taking $\pi_G(y, x)=\chi_{\Gamma(x)}(y)$, for all $y\in F$, $x\in X$).

17. Finally, let us mention that, besides the "usual" dual problems (11.8) to (P) of (2.1), there also appear, in a natural way, some "unusual" dual problems

$$(Q) \qquad \beta=\inf \lambda(W), \qquad (17.1)$$

where $W=W^{G,h}$ and $\lambda=\lambda^{G,h}:W\to\overline{R}$. Some general classes of such unusual dual problems, namely, surrogate duality generalizing the reverse convex duality of [66] and surrogate and Lagrangian duality generalizing the convex supremization duality of [63], [65], [64], [87], [88], [38], have been studied recently in [85] and [84] respectively. For these classes there exist certain parametrization theories (see [85] and, respectively, [87], [38]), and for the Lagrangian convex supremization duality, there exist ([87], [38]) duality-stability relations involving the Fenchel conjugates and the subdifferentials at 0 of the "partial functionals"

$$p_y(x)=p(y, x) \qquad (y\in F,\ x\in X), \qquad (17.2$$

corresponding to those known for convex infimization duality (see e.g. [57], [19]). However, until the present, for the above classes of unusual dual problems, there do not exist stability results involving conjugates or subdifferentials of the optimal value functional (2.2), respectively

$$f(x)=\sup_{y\in F} p(y, x) \qquad (x\in X). \qquad (17.3)$$

References

[1] M.Attéia and A.El Qortobi, Quasi-convex duality. In: Optimization and optimal control (A.Auslender, W.Oettli and J.Stoer, eds.). Lecture Notes in Control and Inf.Sci.30, pp.3-8. Springer-Verlag, Berlin-Heidelberg-New York, 1981.

[2] J.-P.Aubin and I.Ekeland, Estimates of the duality gap in nonconvex optimization. Math.Oper.Res. 1(1976), 225-245.

[3] M.Avriel, Nonlinear programming, analysis and methods. Prentice-Hall, Englewood Cliffs, N.J., 1976.

[4] E.J.Balder, An extension of duality-stability relations to nonconvex optimization problems. SIAM J.Control Optim.15(1977), 329-343.

[5] A.Ben-Tal and A.Ben-Israel, F-convex functions: properties and applications. In: [58], pp.301-334.

[6] G.Birkhoff, Lattice theory (Revised ed.). Amer.Math.Soc.Coll.Publ.25, New York, 1948.

[7] A.Brøndsted, Convexification of conjugate functions. Math.Scand.36(1975), 131-136.

[8] R.E.Burkard, H.Hamacher and J.Tind, On abstract duality in mathematical programming. Zeitschr.Oper.Res.26(1982), 197-209.

[9] J.-P.Crouzeix, Contributions à l'étude des fonctions quasiconvexes. Thèse, Université de Clermont, 1977.

[10] J.-P.Crouzeix, Conjugacy in quasiconvex analysis. In: Convex analysis and its applications (A.Auslender, ed.), Lecture Notes in Econ. and Math.Systems 144, pp.66-99. Springer-Verlag, Berlin-Heidelberg-New York, 1977.

[11] J.-P.Crouzeix, A duality framework in quasiconvex programming. In [58], pp.207-225.

[12] L.Danzer, B. Grünbaum and V.Klee, Helly's theorem and its relatives. In:Convexity (V.Klee, ed.). Proc. Symposia Pure Math.7, pp.101-180. Amer.Math.Soc., Providence, 1963.

[13] R.Deumlich und K.-H.Elster, Zur Theorie nichtkonvexer Optimierungsprobleme: Stabilität und Dualität. 27 Internat. Wiss.Koll.der TH Ilmenau 1982, Heft 5, pp.25-28.

[14] R.Deumlich und K.-H.Elster, Φ-conjugation and nonconvex optimization. A survey. Part I, Math.Operationsforsch. Stat.Ser.Optim. 14(1983), 125-149; Part II, ibid.15(1984), 499-515.

[15] F.Di Guglielmo, Estimates of the duality gap for discrete and quasiconvex optimization problems. In [58], pp.281-298.

[16] J.Dixmier, Sur un théorème de Banach. Duke Math.J. 15(1948), 1057-1071.

[17] S.Dolecki and S.Kurcyusz, On Φ-convexity in extremal problems. SIAM J.Control Optim.16(1978), 277-300.

[18] M.E.Dyer, Calculating surrogate constraints, Math.Progr.19(1980), 255-278.

[19] I.Ekeland et R.Temam, Analyse convexe et problèmes variationnels. Dunod-Gauthier Villars, Paris, 1974.

[20] A.El Qortobi, Contributions à la théorie de la dualité pour les fonctionnelles quasiconvexes. Thèse de 3-ème cycle, Université de Toulouse, 1980.

[21] K.-H.Elster and R.Nehse, Zur Theorie der Polarfunktionale. Math.Operationsforsch. Stat.5(1974), 3-21.

[22] J.J.M.Evers and H.van Maaren, Duality principles in mathematics and their relations to conjugate functions. Memorandum 336, Twente Univ.of Technology, 1981.

[23] J.E.Falk, Lagrange multipliers and nonconvex programs. SIAM J.Control 7(1969), 534-545.

[24] Ky Fan, On the Krein-Milman theorem. In:Convexity (V.Klee, ed.). Proc.Symposia Pure Math.7, pp.211-220. Amer.Math.Soc., Providence, 1963.

[25] W.Fenchel, Convex cones, sets and functions. Lecture Notes, Dep.Math.,Princeton Univ., Princeton, N.J., 1953.

[26] W.Fenchel, A remark on convex sets and polarity. Comm.Sém.Math.Univ.Lund (Medd. Lunds Univ.Math.Sem.) Tome Supplém.(1952), 82-89.

[27] J.Flachs, Global saddle-point duality for quasi-concave programs. Math.Progr. 20(1981), 327-347.

[28] D.Gale, A geometric duality theorem with economic applications, Rev.Econ.Studies 34(1967), 19-24.

[29] F.Glover, A multiphase-dual algorithm for the zero-one integer programming problem. Oper.Res.13(1965), 879-919.

[30] F.Glover, Surrogate constraints, Oper.Res.16(1968), 741-749.

[31] F.J.Gould, Extensions of Lagrange multipliers in nonlinear programming. SIAM J.Appl.Math.17(1969), 1280-1297.

[32] H.J.Greenberg,The generalized penalty-function surrogate model. Oper.Res. 21(1973), 162-178.

[33] H.J.Greenberg and W.P.Pierskalla, Surrogate mathematical programming. Oper. Res.18(1970), 924-939.

[34] H.J.Greenberg and W.P.Pierskalla, Quasi-conjugate functions and surrogate duality. Cahiers Centre Etudes Rech.Opér. 15(1973), 437-448.

[35] A.Hoffmann, Duality in nonconvex optimization. In: Internat. Tagung Mathematische Optimierung-Theorie und Anwendungen, Eisenach, November 1981, Vortragsauszüge, 65-68.

[36] T.R.Jefferson and C.H.Scott, Duality for quasi-concave programs with explicit constraints. Math.Operationsforsch. Stat.Ser.Optim.11(1980), 519-530.

[37] J.-L.Joly and P.-J.Laurent, Stability and duality in convex minimization problems. Rev.Franç.Inf.Rech.Opér.R-2, 5(1971), 3-42.

[38] P.Kanniappan, Fenchel-Rockafellar type duality for a non-convex non-differential optimization problem. J.Math.Anal.Appl.97(1983), 266-276.

[39] R.Klötzler, Dualität bei diskreten Steuerungsproblemen. Math.Operationsforsch. Stat.Ser.Optim.12(1981), 411-420.

[40] P.O.Lindberg, A generalization of Fenchel conjugation giving generalized Lagrangians and symmetric nonconvex duality. In: Survey of mathematical programming. I (A.Prékopa, ed.), pp.249-267. North Holland, Amsterdam, 1979.

[41] D.G.Luenberger, Quasi-convex programming. SIAM J. Appl.Math.16(1968), 1090-1095.

[42] J.-E.Martínez-Legaz, Un concepto generalizado de conjugación. Applicación a las funciones quasiconvexas. Thesis, Barcelona, 1981.

[43] J.-E.Martínez-Legaz, A generalized concept of conjugation. In: Optimization. Theory and algorithms (J.-B.Hiriart-Urruty, W.Oettli and J.Stoer, eds.). Lecture Notes in Pure and Appl.Math.86, pp.45-59. Dekker, New York, 1983.

[44] J.-E.Martínez-Legaz, Conjugación asociada a un grafo. Proc.Confer.Salamanca, 1982.

[45] J.-E.Martínez-Legaz, Exact quasiconvex conjugation. Zeitschr.Oper.Res. 27(1983), 257-266.

[46] J.-E.Martínez-Legaz and Ivan Singer, Dualities between complete lattices (in preparation).

[47] J.-J.Moreau, Convexity and duality. In:Functional analysis and optimization (E.R.Caianiello, ed.), pp.145-169. Acad.Press, New York, 1966.

[48] J.-J.Moreau, Fonctionnelles convexes. Sémin.Eq.Dériv.Part.Collège de France, Paris, 1966-1967, no.2.

[49] J.-J.Moreau, Inf-convolution, sous-additivité, convexité des fonctions numériques. J.Math.Pures Appl.49(1970), 109-154.

[50] O.Ore, Theory of graphs. Amer.Math.Soc.Coll.Publ.38, Providence, 1962.

[51] U.Passy and E.Z.Prisman, Conjugacy in quasi-convex programming. Math.Progr. 30(1984), 121-146.

[52] U.Passy and E.Z.Prisman, A convex-like duality scheme for quasi-convex programs. Math.Progr.32(1985), 278-300.

[53] G.Pickert, Bemerkungen über Galois-Verbindungen.Arch.Math.3(1952), 285-289.

[54] J.Ponstein, Comments on the general duality survey by J.Tind and L.A.Wolsey, Math.Progr.25(1983), 240-244.

[55] R.T.Rockafellar, Convex functions and duality in optimization and dynamics. In: Mathematical systems theory and economics. I (G.Szegö, ed.). Lecture Notes in Oper.Res. and Math.Econ.11, pp.117-141. Springer-Verlag, Berlin-Heidelberg-New York, 1969.

[56] R.T.Rockafellar, Convex analysis. Princeton Univ.Press, Princeton, N.J., 1970.

[57] R.T.Rockafellar, Conjugate duality and optimization. CBMS Reg.Confer.Ser.in Appl.Math.16, SIAM, Philadelphia, 1974.

[58] S.Schaible and W.T.Ziemba, eds., Generalized concavity in optimization and economics. Acad.Press, New York, 1981.

[59] J.Schrader, Eine Verallgemeinerung der Fenchelkonjugation und Untersuchung ihrer Invarianten: verallgemeinerte konvexe Funktionen, Dualitäts- und Sattel-

punktsätze. Thesis, Bonn, 1975.
[60] J.Schrader, Verallgemeinerte konvexe Funktionen. Math. Operationsforsch. Stat. Ser.Optim.8(1977), 41-53.
[61] J.Sikorski, A quasi-subgradient scheme for calculating surrogate constraints. Preprint Systems Res.Inst.Polish Acad.Sci.,July 1985.
[62] I.Singer, Some new applications of the Fenchel-Rockafellar duality theorem: Lagrange multiplier theorems and hyperplane theorems for convex optimization and best approximation. Nonlinear Anal.Theory, Methods, Appl. 3(1979), 239-248.
[63] I.Singer, Maximization of lower semi-continuous convex functionals on bounded subsets of locally convex spaces. I.: Hyperplane theorems. Appl.Math.Optim. 5(1979), 349-362.
[64] I.Singer, A Fenchel-Rockafellar type duality theorem for maximization. Bull. Austral.Math.Soc. 20(1979), 193-198.
[65] I.Singer, Maximization of lower semi-continuous convex functionals on bounded subsets of locally convex spaces. II: Quasi-Lagrangian duality theorems. Result.Math.3(1980), 235-248.
[66] I.Singer, Minimization of continuous convex functionals on complements of convex subsets of locally convex spaces. Math.Operationsforsch.Stat.Ser.Optim.11(1980), 221-234.
[67] I.Singer, Duality theorems for perturbed convex optimization. J.Math.Anal. Appl.81(1981), 437-452.
[68] I.Singer, On the perturbation and Lagrangian duality theories of Rockafellar and Kurcyusz. In: Proc.Vth Sympos. on Oper.Res.Köln, 1980. Methods of Oper. Res.40, pp.153-156. A.Hain Meisenheim GmbH, Königstein/Ts., 1981.
[69] I.Singer, Pseudo-conjugate functionals and pseudo-duality. In: Mathematical methods of operations research (Invited lectures presented at the Internat. Confer.held in Sofia, November 1980), Publ.House Bulg.Acad.Sci.,Sofia, 1981, pp.115-134.
[70] I.Singer, Optimization by level set methods. IV: Generalizations and complements. Numer.Funct.Anal.Optim. 4(1981-1982), 279-310.
[71] I.Singer, Optimization by level set methods. I: Duality formulae. In: Optimization. Theory and algorithms (J.-B.Hiriart-Urruty, W.Oettli and J.Stoer, eds.). Lecture Notes in Pure and Appl.Math.86, pp.13-43. Dekker, New York, 1983.
[72] I.Singer, The lower semi-continuous quasi-convex hull as a normalized second conjugate. Nonlinear Anal.Theory, Methods, Appl.7(1983), 1115-1121.
[73] I.Singer, Surrogate conjugate functionals ans surrogate convexity. Applicable Anal.16(1983), 291-327.
[74] I.Singer, Surrogate dual problems and surrogate Lagrangians.J.Math.Anal. Appl.98(1984), 31-71.
[75] I.Singer, Optimization by level set methods. V: Duality theorems for perturbed optimization problems. Math. Operationsforsch. Stat.Ser.Optim.15(1984), 3-36.
[76] I.Singer, Conjugate functionals and level sets. Nonlinear Anal.Theory, Methods, Appl.8(1984), 313-320.
[77] I.Singer, Best approximation and optimization, J.Approx. Theory 40(1984), 274-284.
[78] I.Singer, A general theory of surrogate dual and perturbational extended surrogate dual optimization problems. J.Math.Anal.Appl.104(1984), 351-389.
[79] I.Singer, Generalized convexity, functional hulls and applications to conjugate duality in optimization. In:Selected topics in operations research and mathematical economics (G.Hammer and D.Pallaschke, eds). Lecture Notes in Econ.and Math.Systems 226, pp.49-79. Springer-Verlag, Berlin-Heidelberg-New York-Tokyo, 1984.
[80] I.Singer, Conjugation operators. Ibidem, pp.80-97.
[81] I.Singer, Some relations between dualities, polarities, coupling functionals and conjugations. J.Math.Anal.Appl.(accepted for publication on March 20, 1984).
[82] I.Singer, A general theory of dual optimization problems. J.Math.Anal.Appl. (to appear). Preprint INCREST 67/1984.
[83] I.Singer, Infimal generators and dualities between complete lattices (to appear). Preprint INCREST 47/1985.
[84] I.Singer, Generalizations of convex supremization duality. Proc.Internat. Confer.on Functional Analysis held at Santa Barbara, June 23-26, 1985 (B.-L.Lin and S.Simons, eds.), Lecture Notes in Math., Springer-Verlag (to appear). Preprint INCREST 74/1985.
[85] I.Singer, Optimization by level set methods. VI.: Generalizations of surro-

gate type reverse convex duality (to appear). Preprint INCREST 74/1985.
[86] J.Tind and L.A.Wolsey, An elementary survey of general duality theory in mathematical programming. Math.Progr.21(1981), 241-261.
[87] J.F.Toland, Duality in nonconvex optimization. J.Math.Anal.Appl.66(1978), 399-415.
[88] J.F.Toland, A duality principle for non-convex optimization and the calculus of variations. Arch.Rat.Mech.Anal.71(1979), 41-61.
[89] W.Vogel, Duale Optimierungsaufgaben und Sattelpunktsätze. Unternehmensforsch. 13(1969), 1-28.
[90] M.Volle, Conjugaison par tranches. Ann.Mat.Pura Appl.139(1985), 279-311.
[91] E.-A.Weiss, Konjugierte Funktionen.Arch.Math.20(1969), 538-545.
[92] E.-A.Weiss, Verallgemeinerte konvexe Funktionen. Math.Scand.35(1974), 129-144.
[93] Ya.I.Zabotin, A.I.Korablev and R.F.Habibullin, Conditions for an extremum of a functional in the presence of constraints. Kibernetika No.6(1973), 65-79 [Russian].

THE APPLICATION OF PARAMETRIC OPTIMIZATION AND IMBEDDING TO THE FOUNDATION AND REALIZATION OF A GENERALIZED PRIMAL DECOMPOSITION APPROACH

Klaus Tammer [1)]

Abstract

The primal decomposition method may be an efficient two-level approach to solve optimization problems of a special structure. As the corresponding problem of the lower level is always a parametric one theoretic results and solution procedures of parametric optimization form an essential base of this method. However, without convexity there is generally no constructive rule for its practical realization. That is why, here we consider a generalized primal decomposition approach applicable also to the non-convex case. We will present a theoretic foundation of this approach and a corresponding solution procedure together with a first local convergence statement.

1. The primal decomposition in its usual form

Many problems of mathematical optimization and related fields are characterized by a typical two-level structure, e.g., saddle-point problems, two-stage problems of stochastic optimization (cf. /11/), problems of continuous Tschebyscheff approximation, problems which arise on using the method of dynamic programming, problems of semi-infinite optimization, problems of input optimization (cf. /22/) and, last not least, problems which arise on using several decomposition methods (cf. /2/, /3/, /4/, /6/, /7/, /9/, /14/, /19/, /20/). One of these methods is the well-known primal decomposition method, which may be an efficient two-level approach to solve optimization problems of the type

$$\min_{x,y} \{ f_o(x,y) \ / \ (x,y) \in M , \ \ y \in N \} . \tag{1.1}$$

We will restrict our considerations to the finite-dimensional case and assume $M \subseteq R^{n+m}$, $N \subseteq R^m$ and $f_o\colon R^n \times R^m \to R$ in (1.1). According to the basic idea of the method, we can realize the minimization with respect to the whole vector (x,y) on two levels. On the lower level we minimize f_o with respect to x only where y is considered a parameter. This means we have to solve the parametric problem

$$P(y)\colon \ \min_{x} \{ f_o(x,y) \ / \ x \in M(y) \} \ , \ \ y \in N \tag{1.2}$$

with $M(y) = \{ x \in R^n \ / \ (x,y) \in M \}$.

---

1) Technische Hochschule Leipzig, Sektion Mathematik und Rechentechnik PF 66, Leipzig, DDR 7030

We denote the optimal value, the set of optimal points of P(y) and the feasibility set of (1.2), respectively, by

$\varphi(y) = \inf \{ f_o(x,y) / x \in M(y) \}$, $\Psi(y) = \{ x \in M(y) / f_o(x,y) = \varphi(y) \}$ and $\mathcal{L} = \{ y \in R^m / M(y) \neq \emptyset \}$.

On the upper level we have to solve the master problem

$$\min \{ \varphi(y) / y \in N \cap \mathcal{L} \} , \qquad (1.3)$$

which is implicitly defined with the aid of the results φ and $\mathcal{L}$ of the lower level.

The mathematical foundation of the primal decomposition is the following theorem, which can be easily proved (cf. /2/, /7/).

<u>Theorem 1.1.</u> The point $(\hat{x},\hat{y})$ is optimal for (1.1) <u>iff</u> $\hat{y}$ is optimal for (1.3) and $\hat{x} \in \Psi(\hat{y})$ holds.

Such a primal decomposition approach may be very efficient if, in a natural way, the variables of an optimization problem are decomposed into subvectors x and y such that the minimization with respect to x can be much more easier realized than the minimization with respect to the whole vector (x,y) even for a lot of different values of y. Moreover, dim y$\ll$dim x would be very convenient. As examples we mention the following cases:

(a) x represents the continuous variables and y, the discrete ones (cf. /2/, /4/) or
(b) the sets N and M(y) and the function $f_o(\cdot,y)$ are convex for each $y \in N$ whereas (1.1) is non-convex (cf. /2/, /7/, /9/, /20/) or
(c) x may be decomposed into several subvectors x_r (r=1,...,Q) and

$$f_o(x,y) = \sum_{r=1}^{Q} f_{or}(x_r,y) , \quad M = \{ (x,y) / (x_r,y) \in M_r, \; r=1,\dots,Q \} \qquad (1.4)$$

such that P(y) splits into Q separate subproblems

$$P_r(y): \quad \min \{ f_{or}(x_r,y) / x_r \in M_r(y) \} , \quad y \in N \qquad (1.5)$$

with $M_r(y) = \{ x_r / (x_r,y) \in M_r \}$ (cf. /3/, /6/, /14/).

The case (c) is of particular interest here.

Of course, for a numerical realization of the primal method it is not realistic to assume that we could solve (1.2) completely for all possible values of y and hence (1.3) would be explicitly given. Instead of this we usually have to try to compute a sequence $\{x^p,y^p\}$ converging to an optimal point of (1.1) where y^p should be generated with the aid of a suitable solution procedure for (1.3) and x^p is an (ap-

proximation of an) optimal point of $P(y^p)$.
A rather general convergence statement for such an approach has been given by Szymanowski and Ruszczynski /19/. However, a constructive rule for the computation of the sequence $\{x^p, y^p\}$, which may be efficient at least for special cases, exists only under suitable convexity assumptions on (1.1). Moreover, as shown, for instance, in /1/ the convexity of P(y) is the most essential assumption to secure suitable continuity properties of the mapping Ψ, which can be used for computing x^{p+1} rather quickly taking x^p as starting point.
In the non-convex case this is generally not possible. But an easier task can be realized efficiently under certain differentiability and regularity conditions with the aid of so-called imbedding methods (cf. /8/, /15/, /17/), i.e., to generate a sequence of locally optimal points x^p corresponding to a sequence of parameter values y^p. For this reason we generalize the usual primal method for non-convex problems in the sense that on the lower level as well as on a certain redefined upper one locally optimal points have to computed. Of course, in this way, at best, a locally optimal point of the original problem (1.1) can be generated.

2. Generalization of the primal decomposition

In the following we do not make any convexity assumption on (1.1) and, instead of the global optimization problem (1.1), solve the problem:

Compute a locally optimal point of (1.1). (2.1)

As stated already in Section 1 we will try to realize this aim by computing locally optimal points of P(y) instead of globally optimal ones. As, for each y, problem P(y) may have different locally optimal points having also different values of the objective function f_o, a well-defined selection $\bar{x}(y)$ is assumed to be given anyhow. We shall see later on that such a selection can be defined in a very natural way depending on a chosen starting point and can also be computed with the aid of a suitable method of parametric optimization. Naturally we should suppose that this selection depends on y at least continuously. On the one hand this is the most essential assumption to realize the idea to compute $\bar{x}(y^{p+1})$ rather quickly taking $\bar{x}(y^p)$ as starting point and on the other hand it secures the continuity of the objective function in the problem of the generalized upper level defined in the following. But before doing this, we still note the basic assumption for all the following considerations, which we have just motivated:

A1) $\bar{x}(.)$ is a continuous vector function which is well-defined on an open set $Y \subseteq R^m$ such that for each $y \in Y$ the image $\bar{x}(y)$ is a locally optimal point for P(y).

At first, we can now formulate the generalized problem of the lower level:

Compute the point $\bar{x}(y)$ for given $y \in N \cap Y$. (2.2)

Obviously, problem (2.2) generalizes (1.2) and, similar to P(y), assuming the special structure (1.4) of (1.1), splits into Q separate subproblems.
With the aid of $\bar{x}(y)$, for $y \in Y$, we define the function

$$\bar{\varphi}(y) = f_0(\bar{x}(y), y)$$

which corresponds to $\varphi(y)$ if and only if $\bar{x}(y)$ is even a globally optimal point of P(y).
Now we are able to formulate also the generalized problem of the upper level

Compute a locally optimal point of $\min \{ \bar{\varphi}(y) / \; y \in N \cap Y \}$ (2.3)

which has to be used instead of (1.3) because we have no information on $\varphi(y)$ from the generalized lower level problem (2.2).
As an example given in /10/ shows, the continuity of $\bar{x}(.)$ is not sufficient to secure that a locally optimal point $\hat{y}$ of (2.3) together with a solution $\hat{x}=\bar{x}(\hat{y})$ of (2.2) for $y=\hat{y}$ forms a solution of problem (2.1). Moreover, an efficient solution procedure for problem (2.3) requires that $\bar{\varphi}$ is not only continuous but also differentiable. To obtain stronger results in this connection we need some additional assumptions. At first we suppose a concrete description of the set M in (1.1)

A2) $M = \{ (x,y) \in R^n \times R^m \;/\; f_i(x,y) = 0, \; i \in I, \; f_j(x,y) \leq 0, \; j \in J \}$
with $I = \{1, \ldots, l\}$ and $J = \{l+1, \ldots, L\}$ where the appearing functions f_i $(i=0,1,\ldots,L)$ are twice differentiable.

Moreover, we need an appropriate constraint qualification for P(y) at a certain point (x^0, y^0) with $x^0 = \bar{x}(y^0)$ which secures that x^0 is a stationary point of $P(y^0)$. This means that there is a multiplier $u^0 \in R^L$ such that (x^0, y^0, u^0) satisfies the necessary Karush-Kuhn-Tucker-conditions to P(y) for $y=y^0$

$$\nabla_x \Phi(x,y,u) = 0, \; f_i(x,y)=0 \; i \in I, \; f_j(x,y) \leq 0, \; u_j \geq 0, \; u_j f_j(x,y) = 0 \; {}_{j \in J} \quad (2.4)$$

where $\Phi(x,y,u) = f_0(x,y) + \sum_{j=1}^{L} u_j f_j(x,y)$ is the Lagrangian to (1.2).

Especially, we will use the Mangasarian-Fromovitz-Constraint-Qualification (MFCQ) as well as the Linear-Independence-Constraint-Qualification (LICQ) (cf. /10/, /12/); where the latter secures even the uniqueness of the multiplier u as generally known.
The following results show that the concept of strong stability intro-

duced by Kojima /12/ is very useful here. At first it can be stated that our basic assumption A1) is automatically fulfilled if we have a starting point (x^o, y^o, u^o) with appropriate properties.

Theorem 2.1. Let x^o be a locally optimal point and a strongly stable stationary point of $P(y^o)$ with a corresponding multiplier u^o such that x^o fulfills MFCQ (LICQ). Then the following holds:

1. There is an open set Y with $y^o \in Y$ and a unique vector function $\bar{x}(.)$ which is defined and continuous on Y with $\bar{x}(y^o)=x^o$ and a vector function $\bar{u}(.)$ which is defined on Y with $\bar{u}(y^o)=u^o$ such that, for each $y \in Y$, the point $\bar{x}(y)$ is a locally optimal point and a strongly stable stationary point of $P(y)$ satisfying MFCQ (LICQ) and $\bar{u}(y)$ is a corresponding multiplier.
2. Assuming LICQ, the vector function $\bar{u}(.)$ is also unique, and both $\bar{x}(.)$ and $\bar{u}(.)$ are locally Lipschitzian on Y.
3. Assuming LICQ the set

$$Y(x^o,y^o,u^o) = \{ y \ / \ \text{There is a continuous vector function } z^o_y(.) = (x^o_y(.), u^o_y(.)) : [0,1] \to R^{n+m} \text{ with } z^o_y(0) = (x^o,u^o) \text{ such that } x^o_y(t) \text{ is a strongly stable stationary point of } P((1-t)y^o+ty) \text{ for each } t \in [0,1], \text{ and } u^o_y(t) \text{ is a corresponding multiplier.} \} \quad (2.5)$$

is the maximal star connected open set with y^o as a centre on which the vector functions $\bar{x}(.)$ and $\bar{u}(.)$ with the properties given in 1. and 2. can be defined by setting $(\bar{x}(y), \bar{u}(y)) = z^o_y(1)$.

Proof: The first assertion follows from the results in /12/. To show the Lipschitz continuity of $\bar{x}(.)$ and $\bar{u}(.)$ we can extend the proof given in /5/ for the one-parametric case immediately to the multiparametric one. The third assertion is contained in /21/ and can be easily verified.

Remark 2.1 /21/: For an arbitrary set Y with the properties given in Theorem 2.1, each triple $(\tilde{x}, \tilde{y}, \tilde{u})$ with $\tilde{y} \in Y$, $\tilde{x}=\bar{x}(\tilde{y})$ and $\tilde{u}=\bar{u}(\tilde{y})$ again satisfies the same assumptions as (x^o, y^o, u^o). Hence, there are corresponding vector functions denoted by $\tilde{x}(.)$ and $\tilde{u}(.)$ for which we can even choose the same region of definition Y where $\bar{x}(y) \equiv \tilde{x}(y)$ holds and, assuming LICQ, $\bar{u}(y) \equiv \tilde{u}(y)$ holds, too. However, the maximal star connected sets defined in (2.5) may differ but have a nonempty intersection on which $\bar{x}((y) \equiv \tilde{x}(y)$ and $\bar{u}(y) \equiv \tilde{u}(y)$ hold again.

Now for all further considerations the assumptions of Theorem 2.1 are assumed to be valid for a certain starting point (x^o, y^o, u^o), and the vector function $\bar{x}(.)$ characterized in A1) is assumed to be that given by this theorem.

In the following theorem we will summarize some further known results proved in /10/, which form the mathematical foundations of the local approach considered and, moreover, give a useful property of $\bar{\varphi}$, where, by $C^{1,1}(Y)$, we denote the set of all functions in $C^1(Y)$ for which the gradient is locally Lipschitzian on Y.

Theorem 2.2.: Let $\bar{x}(.)$ and $\bar{u}(.)$ be given according to Theorem 2.1.
1. If $\bar{x}(y)$ satisfies MFCQ for each $y \in Y$, then a point $(\hat{x},\hat{y})$ solves (2.1) iff $\hat{y}$ solves (2.3) and $\hat{x}$ solves (2.2) for $y = \hat{y}$.
2. If $\bar{x}(y)$ satisfies LICQ for each $y \in Y$, then we have $\bar{\varphi} \in C^{1,1}(Y)$, and

$$\nabla \bar{\varphi}(y) = \nabla_y \phi(x,y,u) \Big|_{x=\bar{x}(y),u=\bar{u}(y)} . \tag{2.6}$$

Note that (2.6) generalizes a corresponding known formula for the globally optimal value function φ in the differentiable case. Moreover, according to /10/, even the Hessian $\nabla^2\bar{\varphi}(y)$, which exists almost everywhere, can be computed also with the aid of the values $\bar{x}(y)$, $\bar{u}(y)$ for fixed y only.

In the following theorem we will formulate a condition for the original problem (1.1) which secures that the assumptions of Theorem 2.2 as well as some useful additional properties are satisfied. Here we also need a concrete description of the set N in (1.1) as follows:

A3) $N = \{y \in R^m \,/\, g_k(y) \le 0,\ k \in K\}$ with $g_k \in C^2(R^m)$, $K = \{1,\dots,P\}$

together with a sharpened LICQ:

The system $\{\nabla_x f_i(x,y),\ i \in I \cup J^0(x,y)\}$ with $J^0(x,y) = \{j \in J / f_j(x,y) = 0\}$ is linearly independent. (2.7a)

The system $\{\nabla g_k(y),\ k \in K^0(y)\}$ with $K^0(y) = \{k \in K / g_k(y) = 0\}$ is linearly independent. (2.7b)

Obviously, (2.7) implies LICQ for (1.1). On the other hand, LICQ implies only (2.7b) and the linear independence of the system $\{\nabla f_i(x,y)$, $i \in I \cup J^0(x,y)\}$, whereas the subvectors given in (2.7a) need not be linearly independent.

Theorem 2.3.: Let $(\hat{x},\hat{y})$ be a solution of (2.1) for which (2.7) is satisfied and, moreover, a strongly stable stationary point of (1.1) with the (unique) multiplier $(\hat{u},\hat{v})$.
Then there are an open set Y with $\hat{y} \in Y$ and a unique vector function $\bar{z}(.) = (\bar{x}(.),\bar{u}(.))$ defined and locally Lipschitzian on Y with $\bar{z}(\hat{y}) = (\hat{x},\hat{u})$ such that
1. $\bar{x}(y)$ is a locally optimal point and a strongly stable stationary point of P(y) satisfying LICQ, and $\bar{u}(y)$ is the corresponding unique multiplier for $y \in Y$, and
2. $\hat{y}$ is a strict locally optimal point and an isolated stationary point of problem (2.3) satisfying LICQ, and $\hat{v}$ is the corresponding unique multiplier.

Proof: According to our assumptions, the point $(\hat{x},\hat{y},\hat{u},\hat{v})$ satisfies the Karush-Kuhn-Tucker-conditions for (1.1) given by (2.4) together with

$$\nabla_y \bar{\Phi}(x,y,u,v) = 0 \quad (2.8a)$$

$$g_k(y) \le 0 \, , \; v_k \ge 0 \, , \; v_k \, g_k(y) = 0 \, , \; k \in K \; ; \quad (2.8b)$$

where $\bar{\Phi}(x,y,u,v) = \Phi(x,y,u) + \sum_{k\in K} v_k \, g_k(y)$ is the Lagrangian to (1.1) with $\nabla_x \bar{\Phi} = \nabla_x \Phi$.

Hence $(\hat{x},\hat{u})$ satisfies the Karush-Kuhn-Tucker-conditions for $P(\hat{y})$.

As shown in /12/, assuming LICQ, the strong stability together with the local optimality of $(\hat{x},\hat{y})$ is equivalent to the strong second-order sufficient condition that the Hessian of $\bar{\Phi}$ for $(\hat{x},\hat{y},\hat{u},\hat{v})$ given by

$\begin{pmatrix} \nabla_{xx}\bar{\Phi} & \nabla_{xy}\bar{\Phi} \\ \nabla_{xy}^T\bar{\Phi} & \nabla_{yy}\bar{\Phi} \end{pmatrix}$ is positive definite on the subspace

$$T = \left\{ (x,y) \in R^{n+m} / \; \nabla_x f_i(\hat{x},\hat{y})^T x + \nabla_y f_i(\hat{x},\hat{y})^T y = 0, \; i \in I \cup J^+(\hat{u}), \; \nabla g_k(\hat{y})^T y = 0, \; k \in K^+(\hat{v}) \right\}$$

where $J^+(\hat{u}) = \{ j \in J^0(\hat{x},\hat{y}) \; / \; \hat{u}_j > 0 \}$, $K^+(\hat{v}) = \{ k \in K^0(\hat{y}) \; / \; \hat{v}_k > 0 \}$.

This implies the positive definiteness of $\nabla_{xx}\Phi = \nabla_{xx}\bar{\Phi}$ taken for $(\hat{x},\hat{y},\hat{u})$ on the subspace $T_x = \{ x \in R^n / \; \nabla_x f_i(\hat{x},\hat{y})^T x = 0, \; i \in I \cup J^+(\hat{u}) \}$.

Thus $\hat{x}$ is a locally optimal point and a strongly stable stationary point of $P(\hat{y})$ which satisfies, owing to (2.7a), LICQ, and $\hat{u}$ is the corresponding unique multiplier.

The existence and uniqueness of a vector function $\bar{x}(.)$ with the given properties follows from Theorem 2.1.

Using (2.6) and (2.8a), we obtain $\nabla \bar{\varphi}(\hat{y}) + \sum_{k\in K} \hat{v}_k \nabla g_k(\hat{y}) = 0$ such that, taking (2.8b) into consideration, $\hat{y}$ is a stationary point of (2.3) which satisfies LICQ, owing to (2.7b), and $\hat{v}$ is the corresponding unique multiplier.

Suppose that $\hat{y}$ is not an <u>isolated</u> stationary point of (2.3). Then there would be a sequence $\{y^p\}$ of stationary points of (2.3) with $y^p \to \hat{y}$ and $y^p \ne \hat{y}$. Using the fact that the points $x^p = \bar{x}(y^p)$ are stationary points of $P(y^p)$, we can see that the points (x^p,y^p) are stationary points of (1.1) and, owing to the continuity of $\bar{x}(.)$, $(x^p,y^p) \to (\hat{x},\hat{y})$ with $(x^p,y^p) \ne (\hat{x},\hat{y})$. But this contradicts the result proved in /12/ that, assuming LICQ, a strongly stable stationary point must be an isolated stationary one.

Finally, assume now that $\hat{y}$ is not a strict locally optimal point of (2.3). Then there would be a sequence $\{y^p\} \subseteq N \cap Y$ with $y^p \to \hat{y}$ and $y^p \ne \hat{y}$ such that $\bar{\varphi}(\hat{y}) \ge \bar{\varphi}(y^p)$ holds. This implies for $x^p = \bar{x}(y^p)$ that (x^p,y^p) converges to $(\hat{x},\hat{y})$, $(x^p,y^p) \ne (\hat{x},\hat{y})$, $(x^p,y^p) \in M$, and $f_o(\hat{x},\hat{y}) \ge f_o(x^p,y^p)$.

But this contradicts the fact that, according to /12/, assuming LICQ, a strongly stable stationary and locally optimal point is always a strict locally optimal point.

Remark 2.2. /16/: If, in addition to the assumptions of Theorem 2.3, the point $(\hat{x},\hat{y})$ fulfills the strict complementarity slackness condition $J^+(\hat{u})= J^o(\hat{x},\hat{y})$, the function $\bar{\varphi}$ belongs to $C^2(Y)$ for a sufficiently small neighbourhood Y of $\hat{y}$ and $\hat{y}$ is even a strongly stable stationary point of (2.3).

3. A theoretic solution procedure

A solution procedure for realizing the generalized primal decomposition approach proposed must be a combination of an algorithm for the upper level (2.3) with an algorithm for the lower level (2.2). The first algorithm has to generate a sequence $\{y^p\}$, using the second algorithm as a subprocedure to compute $x^p=\bar{x}(y^p)$, $u^p=\bar{u}(y^p)$, $\bar{\varphi}(y^p)= f_o(x^p,y^p)$, and $\nabla\bar{\varphi}(y^p)= \nabla_y \phi (x^p,y^p,u^p)$ for each p.
Here we propose a theoretical solution procedure denoted by MFD-IM which is a combination of a method of feasible directions MFD given in /18/ for the upper level with an imbedding method IM developed in /15/ for the lower one.
The method MFD is a generalization and modification of the P1-algorithm of Zoutendijk for which we could show /18/ a certain local convergence statement. To apply MFD to (2.3) we need the following direction search problem DSP which depends on the actual iteration point (x,y,u) and a certain parameter $\varepsilon > 0$:

$$\mathrm{DSP}(x,y,u;\varepsilon):\ \min_{\|s\|\leq 1}\left\{\delta \,/\, \nabla_y \phi(x,y,u)^T s \leq \delta,\ \nabla g_k(y)^T s \leq \delta,\ k\in K^o(y,\varepsilon)\right\}$$

where $K^o(y,\varepsilon)= \{k\in K \,/\, |g_k(y)|\leq \varepsilon\}$.
According to /15/, using the method IM, we can realize the following task: A triple (x^p,y^p,u^p) is assumed to have the property that x^p is a locally optimal point and a strongly stable stationary point of $P(y^p)$ which satisfies LICQ and u^p is the unique multiplier corresponding to x^p. Further let $Y^p= Y(x^p,y^p,u^p)$ be the maximal star connected set defined in (2.5) on which the corresponding vector functions denoted by $\bar{x}^p(.)$, $\bar{u}^p(.)$ (cf. Theorem 2.1) can be defined. Then, for any $y\in Y^p$, the method IM applied to the interval $[y^p,y]$ generates a sequence which converges to $(\bar{x}^p(y),\bar{u}^p(y))$ if we use (x^p,u^p) as starting point and choose the positive parameters Δ (maximal step size) and ϱ (bound on the accuracy of the computation of certain auxiliary points) of the method such that they are small enough. Using the parameters Δ and ϱ we denote this application of IM to P(y) with respect to $[y^p,y]$ by $\mathrm{IM}(y^p,y,\Delta,\varrho)$. Now we describe the procedure MFD-IM for solving (2.1) which uses only the basic variant of the method MFD.

Procedure MFD-IM

Step 1: Input of (x^o,y^o,u^o), $\varepsilon_o > 0$, $\Delta > 0$, $\varrho > 0$ and set p= 0.

Step 2: Compute (δ_p, s^p) as an optimal point of $DSP(x^p, y^p, u^p; \varepsilon_p)$.

Step 3: If $\delta_p^2 \geq \varepsilon_p$, then go to Step 7.

Step 4: If $\delta_p < 0$, then go to Step 6.

Step 5: If $K^o(y^p, \varepsilon_p) = K^o(y^p)$, then print (x^p, y^p, u^p) and stop.

Step 6: Set $\varepsilon_p = 1/2\, \varepsilon_p$ and go to Step 2.

Step 7: Compute

$$t_p = \sup\{t \,/\, y^p + \tau s^p \in N \cap Y^p \;\forall \tau \in [0,t]\,,\; f_o(\bar{x}^p(y^p + \tau_1 s^p), y^p + \tau_1 s^p) \geq f_o(\bar{x}^p(y^p + \tau_2 s^p), y^p + \tau_2 s^p) \;\forall\, \tau_1, \tau_2 \in [0,t],\, \tau_1 < \tau_2\}.$$

If $t_p \notin Y^p$, then print "no solution is found" and stop.

Step 8: Set $y^{p+1} = y^p + t_p s^p$ and use $IM(y^p, y^{p+1}, \Delta, \varsigma)$. If it does not converge, then print "no solution is found" and stop. Otherwise, let (x^{p+1}, u^{p+1}) be the limit point, set $\varepsilon_{p+1} = \varepsilon_p$, p=p+1 and go to Step 2.

For this procedure we can prove the following local convergence statement.

Theorem 3.1.: The point $(\hat{x}, \hat{y}, \hat{u}, \hat{v})$ is assumed to satisfy the condition of Theorem 2.3. Then there are positive values Δ_o, ς_o and r such that, for each $\varepsilon_o > 0$, procedure MFD-IM does not stop at Step 7 or 8 and generates a sequence $\{x^p, y^p, u^p\}$ which converges to $(\hat{x}, \hat{y}, \hat{u})$ if we choose $y^o \in U_r(\hat{y})$, $x^o = \bar{x}(y^o)$, $u^o = \bar{u}(y^o)$, $\Delta \in (0, \Delta_o)$, and $\varsigma \in (0, \varsigma_o)$.

Proof: According to /18/, all points y^p obtained by the method MFD belong to a certain compact set if r is small enough. As the gradient $\nabla \bar{\varphi}$ is locally Lipschitzian, according to Theorem 2.2, on this compact set it is even globally Lipschitzian. Using this fact as well as assertion 2 of Theorem 2.3, according to a result from /18/, we obtain the existence of a positive value r such that MFD generates a sequence $\{y^p\}$ which converges to $\hat{y}$ if we choose only $y^o \in U_r(\hat{y}) \cap N$, $x^o = \bar{x}(y^o)$, $u^o = \bar{u}(y^o)$. If we extend a result given in /15/ for the one-parametric case to the multiparametric one (For the proof we need only the compactness of the underlying parameter set.), we obtain the existence of positive values Δ_o, ς_o such that, for each p and for arbitrary $\Delta \in (0, \Delta_o)$, $\varsigma \in (0, \varsigma_o)$, $IM(y^p, y^{p+1}, \Delta, \varsigma)$ generates a sequence which converges to $\bar{x}^p(y^{p+1})$, $\bar{u}^p(y^{p+1})$. Hence MFD-IM will not stop at Step 7 or 8, and, because of the continuity of $\bar{x}^p(.)$ and $\bar{u}^p(.)$, $x^p = \bar{x}^p(y^p) \to \hat{x}$ and $u^p = \bar{u}^p(y^p) \to \hat{u}$ which complete the proof.

Remark 3.1.: 1. The results of the paper give a first mathematical foundation for the obvious approach already realized by many users of optimization procedures, i.e., to apply the basic idea of the primal decomposition method also to non-convex problems without being sure that globally optimal points on both levels are computed as is required in Theorem 1.1.

2. Procedure MFD-IM is, in this form, only a theoretical one because it contains subprocedures which generally require an infinite number of computations. Especially, the computation of the step size t_p in Step 7 and of the point (x^p, y^p, u^p) in Step 8 can be generally realized by an infinite iteration procedure only.

3. The generalized variant of the method MFD proposed in /18/ enables approximate values $\tilde{t}_p$ and $(\tilde{\delta}_p, \tilde{s}^p)$ to be used in Step 2 and 7 instead of t_p and (δ_p, s^p) which can be generated with a finite number of computations. Moreover, the local convergence property mentioned holds further on if certain conditions to the accuracy of $\tilde{t}_p$ and $(\tilde{\delta}_p, \tilde{s}^p)$ are satisfied.

4. In Step 8 we can generate, with a finite number of computations, approximate values $(\tilde{x}^p, \tilde{u}^p)$ of (x^p, u^p) for all p with increasing accuracy. On this way we obtain an approximation of $\bar{\varphi}(y^p) = f_o(x^p, y^p)$ by $f_o(\tilde{x}^p, y^p)$ and an approximation of $\nabla\bar{\varphi}(y^p) = \nabla_y \phi(x^p, y^p, u^p)$ by $\nabla_y \phi(\tilde{x}^p, y^p, \tilde{u}^p)$ which can be used in Step 2 and 7 instead of the exact values. Continuity properties of the problem DSP$(x, y, u; \varepsilon)$ with respect to (x, u) (cf. /13/) secure that the optimal direction $\tilde{s}^p$ resulting from the disturbed problem DSP$(\tilde{x}^p, y^p, \tilde{u}^p; \varepsilon_p)$ satisfies the stated accuracy condition of the generalized variant of the method MFD which still guarantees the local convergence if $\delta_p < 0$ holds and $\|(\tilde{x}^p, \tilde{u}^p) - (x^p, u^p)\|$ is sufficiently small.

5. If $(\hat{x}, \hat{u})$ fulfills the condition of strict complementarity slackness for $P(\hat{y})$, then, for the upper level, we can use one of the locally convergent methods of second order considered in /15/ which have a good rate of convergence. As already mentioned in /18/ also many methods of second order of such kind may be included in MFD if, in DSP, we choose another norm which may even depend on the actual iteration point.

References

/1/ Bank, B., J. Guddat, D. Klatte, B. Kummer and K. Tammer: Non-linear Parametric Optimization, Akademie-Verlag, Berlin 1982.

/2/ Bank, B., R. Mandel and K. Tammer: Parametrische Optimierung und Aufteilungsverfahren, in: Lommatzsch, K. (ed.): Anwendungen der linearen parametrischen Optimierung, Akademie-Verlag, Berlin 1979, 107-123.

/3/ Beer, K.: Lösung großer linearer Optimierungsaufgaben, VEB Deutscher Verlag d. Wiss., Berlin 1977.

/4/ Benders, J.F.: Partitioning procedures for solving mixed-variables programming problems, Num. Math. 4(1962), 238-252.

/5/ Dontchev, A.L. and H.Th. Jongen: On the regularity of the Kuhn-Tucker curve, Memorandum Nr. 460, Twente University of Technology, 1984.

/6/ Findeisen, W., F.N. Bailey, M. Brdyś, K. Malinowski, P. Tatjewski and A. Woźniak: Control and Coordination in Hierarchical Systems, John Wiley § Sons, London 1980.

/7/ Geoffrion, A.M.: Generalized Bender's decomposition, JOTA 10(1972), 237-259.

/8/ Guddat, J., H. Wacker and W. Zulehner: On imbedding and parametric optimization, Math. Progr. Study 21(1984), 79-96.

/9/ Hoang Tuy: A parametric approach to non-convex global optimization for dc-functions, Lecture at the Conference "Parametric Optimiza-

tion and Related Topics", Plaue, 1985.
/10/ Jongen,H.Th., T.Möbert and K.Tammer: On iterated minimization in nonconvex optimization, to appear in Math.Oper.Res.
/11/ Kall,P.: Stochastic Linear Programming, Springer-Verlag, Berlin, Heidelberg, New York 1976.
/12/ Kojima,M.: Strongly stable stationary solutions in nonlinear programs, in: Robinson,S.M. (ed.): Analysis and Computation of Fixed Points, Academic Press, New York 1980
/13/ Kuhn,R.: Zur lokalen Konvergenz einer Klasse von Verfahren der nichtlinearen Optimierung, Jahresarbeit an der Sektion Mathematik der Karl-Marx-Universität Leipzig, 1985.
/14/ Lasdon,L.S.: Optimization Theory for Large Systems, The Macmillan Co., New York 1970.
/15/ Lehmann,R.: On the numerical feasibility of continuation methods for nonlinear programming problems, Math. Operationsforsch. Statist., Ser. Optimization 15(1984)4, 517-530.
/16/ Möbert,T.: Dynamische Programmierung für komplexe Systeme und ihre Einbettung in eine mehrfach iterierte lokale Minimumsuche, Dissertation an der Pädagogischen Hochschule Halle, 1985.
/17/ Richter,C.: Ein implementierbares Einbettungsverfahren der nichtlinearen Optimierung, Math. Operationsforsch. Statist., Ser. Optimization 15(1984)4, 545-553.
/18/ Schulze,R. and K.Tammer: Lokale Konvergenzeigenschaften einer Klasse von Iterationsverfahren der nichtlinearen Optimierung, submitted to Optimization.
/19/ Szymanowski,J. and A.Ruszczynski: Convergence analysis for two-level algorithms of mathematical programming, Math.Progr. Study 10(1979), 158-171.
/20/ Tammer,K.: Möglichkeiten zur Anwendung der Erkenntnisse der parametrischen Optimierung für die Lösung indefiniter quadratischer Optimierungsprobleme, Math. Operationsforsch.Statist. 7(1976), 209-222.
/21/ Tammer,K.: Zur Erzeugung streng regulärer stationärer Punkte mit Hilfe einer Verallgemeinerung der dualen Methode, Proceedings of the Conference "Modellierung und Optimierung von Systemen", Lichte, 1984, 66-71.
/22/ Zlobec,S.: Input optimization: I. Optimal realization of mathematical models, Math.Progr. 31(1985), 245-268.

ON APPROXIMATIONS AND STABILITY IN STOCHASTIC PROGRAMMING*)

Peter Kall**)

ABSTRACT

It has become an accepted approach to attack stochastic programming problems by approximating the given probability distribution in various ways. After sketching one of these approaches for recourse problems, the stability problem with respect to the probability measure, involved with those approximations as well as with inexact information in applied problems, is discussed for recourse and chance constrained models.

1. INTRODUCTION

In mathematical programming we are used to deal with problems of the type

$$\left.\begin{array}{ll} \min & f(x) \\ \text{s.t.} & g(x)\geq 0 \\ & x\in X, \end{array}\right\} \qquad (1.1)$$

where $X \subset \mathbb{R}^n$, $f:X \to \mathbb{R}$ and $g:X \to \mathbb{R}^m$ are given and various assumptions are imposed on X (e.g. convexity, compactness, polyhedrality) and on f and g, respectively (e.g. continuity, convexity, differentiability, linearity). The crucial hypothesis for dealing with (1.1) is that f and g are given deterministic functions. Since the constraints $g(x)\geq 0$ in (1.1) in applications mean, for instance, production requirements, capacity restrictions etc., in many cases the constraint functions are very likely to be affected by a random parameter ξ. A similar observation can be

---

*) Part of this work was accomplished at the MRC, University of Wisconsin-Madison and sponsored by the National Science Foundation under Grant No. DCR-8502202. The author greatly appreciates the hospitality and support extended by these institutions.

**) Address: Institut für Operations Research und Mathematische Methoden der Wirtschaftswissenschaften der Universität Zürich
Weinbergstrasse 59

CH-8006 Zürich

made in many cases for the objective function f, too; however we omit this case because the model to be considered here would just proceed to the expected value of the objective function. Hence instead of (1.1) we are faced with

$$\begin{aligned} \min \quad & f(x) \\ \text{s.t.} \quad & g(x,\xi) \geq 0 \\ & x \in X. \end{aligned} \tag{1.2}$$

To be more specific we assume throughout that - except if otherwise stated - ξ is a random vector with support $\Xi \subset \mathbb{R}^K$ and distribution function F.

If we are required to take a decision on x before knowing the realization of ξ, then (1.2) is obviously not a well defined problem, except for the case we restrict ourselves to decisions on x such that $g(x,\xi) \geq 0 \ \forall \xi \in \Xi$, which may be very expensive and even infeasible for practical applications. There are various possibilities to deal with random constraints, the most frequently used being expectations of recourse functions and chance constraints. Whereas a recourse function may be understood as a penalty function for violating those random constraints subjected to recourse, chance constraints require the corresponding random constraints to be satisfied at least with a prescribed level of probability. If we subject the first m_1 constraints of (1.2) to the recourse function $Q(x,\xi)$ and the remaining $m-m_1$ to chance constraints, we so get the problem

$$\begin{aligned} \min \quad & \{f(x) + \int_\Xi Q(x,\xi)\, F(d\xi)\} \\ \text{s.t.} \quad & P(\{\xi \mid h(x,\xi) \geq 0\}) \geq \alpha \\ & x \in X, \end{aligned} \tag{1.3}$$

where P is the probability measure on $\mathcal{B}(\Xi)$, the Borel field on Ξ, defined by F and α is the prescribed level of feasibility for the constraints $h(x,\xi) := (g_{m_1+1}(x,\xi), \ldots, g_m(x,\xi))^T \geq 0$. Obviously choosing $\alpha=0$ or $m_1=m$ is equivalent to omitting chance constraints, and $m_1=0$ eliminates recourse from problem (1.3). Specializing (1.3) to stochastic linear programming we get

$$\begin{aligned} \min \quad & \{c^T x + \int_\Xi Q(x,\xi) F(d\xi)\} \\ \text{s.t.} \quad & P(\{\xi \mid A(\xi)x \geq b(\xi)\}) \geq \alpha \\ & x \in X \end{aligned} \tag{1.4}$$

where X is assumed to be convex polyhedral, $c \in \mathbb{R}^n$ is given,

$$Q(x,\xi) = \min\{q^T y \mid Wy = h(\xi) - T(\xi)x,\ y \geq 0\} \tag{1.5}$$

with $q \in \mathbb{R}^{\bar{n}}$ and a complete fixed recourse matrix of order $(m_1 \times \bar{n})$, i.e.

$$\{t \mid t = Wy,\ y \geq 0\} = \mathbb{R}^{m_1} \tag{1.6}$$

and - to avoid $Q(x,\xi) \equiv -\infty$ - such that

$\{u | W^T u \leq q\} \neq \emptyset.$ (1.7)

Finally, $A(\cdot)$, $b(\cdot)$, $T(\cdot)$ and $h(\cdot)$ are supposed to be linear affine transformations defined on $\mathbb{R}^K$. Under these assumptions we refer to (1.4) as SLP problem. SLP problems with recourse - i.e. $\alpha=0$ - are known to have desirable properties like convexity, Lipschitz continuity and - under appropriate continuity assumptions on P - differentiability of $\mathcal{Q}(x)=\int_{\Xi} Q(x,\xi)F(d\xi)$; see e.g. [8], [19]. SLP problems with chance constraints - i.e. $\alpha>0$ and $Q(x,\xi)\equiv 0$ - are much more difficult in general according to the fact that they define very often nonconvex feasibility sets; in other words: The function $\Psi: X \rightarrow [0,1]$, defined by

$$\Psi(x) := P(\{\xi | A(\xi)x \geq b(\xi)\}),$$

is in many cases not quasi-concave. The only general positive results - for more than one random constraint subjected to a simultaneous chance constraint - are known for the case

$$A(\xi)\equiv A, \; b(\xi)=\xi, \text{ i.e. } \Psi(x)=F(Ax), \text{ where}$$

Ψ was shown in [11] to be logarithmic concave for all distribution functions F with logarithmic concave densities ϕ, and was proved later on in [12] to be quasi-concave if the density ϕ of F is such that $\phi^{-\frac{1}{K}}$ is convex.

Even though the SLP problem with recourse seems to be much easier than the chance constrained problem in general, it is not amenable to nonlinear programming solution techniques because of the fact that the evaluation of $\mathcal{Q}(x)=\int_{\Xi} Q(x,\xi)F(d\xi)$ and its gradient for continuous multivariate distributions causes insurmountable difficulties.

On the other hand, for finite discrete distributions the SLP problem with recourse is a linear program with dual decomposition structure. This fact raised the idea of approximating a solution by approximating the given (supposedly continuous) distribution P by appropriate discrete distributions.

In section 2 a solution method for recourse problems based on this idea is outlined; more detailed information on this may be found in [6], [7] and [9].

In section 3 we discuss the stability of problem (1.3) with respect to the probability distribution P. This question deserves interest, not only because of solution techniques using approximations of P, but also by itself: In applications we can hardly expect to get the true distribution of the problem at hand, but at best a fairly good approximation of it used to set up the model of the form (1.3). Hence we should know whether the optimal value and the solution set of this model are in some sense close to those for the true distribution.

2. SOLVING STOCHASTIC PROGRAMS BY APPROXIMATING P

Throughout this section we assume that the interval

$$\Xi = \mathop{\times}_{i=1}^{K} [a_{i0}, a_{i1}], \tag{2.1}$$

$a_{i0} < a_{i1}$, $i=1,\dots,K$, contains the support of the probability measure P. Furthermore we require that

$$Q(x,\cdot) : \Xi \to \mathbb{R} \text{ is convex } \forall x \in X. \tag{2.2}$$

Obviously (2.2) is satisfied for the SLP problem (1.4). From the boundedness of the support of P and (2.2) follows, that

$$\int_\Xi Q(x,\xi)P(d\xi) \quad \text{exists for all } x \in X. \tag{2.3}$$

From Jensen's inequality we then know that

$$Q(x,\hat{\xi}) \le \int_\Xi Q(x,\xi)P(d\xi) \quad \text{for } \hat{\xi} = E\xi, \tag{2.4}$$

E denoting the mathematical expectation.

To describe an upper bound for $\int_\Xi Q(x,\xi)P(d\xi)$ due to [6] we use the following notations:

$$\mathfrak{A} := \{\Lambda \mid \Lambda \subset \{1,\dots,K\}\},$$

$$m_\Lambda : E \prod_{i\in\Lambda} \xi_i = \int_\Xi \prod_{i\in\Lambda} \xi_i P(d\xi),$$

ξ_i denoting the i-th component of ξ ,

$$\rho_\Lambda : m_\Lambda - \prod_{i\in\Lambda} \hat{\xi}_i,$$

$$\nu := (\nu_1,\dots,\nu_K)^T \quad \text{with } \nu_k \in \{0,1\},$$

the vectors ν used to identify the 2^K vertices

$$a_\nu := (a_{1\nu_1}, \dots, a_{K\nu_K})^T$$

of Ξ,

$$\bar{\nu}_i := 1 - \nu_i, \; i=1,\dots,K,$$

$$\delta_\Lambda(\nu) := \prod_{i\in\Lambda} (-1)^{\bar{\nu}_i}.$$

For the discrete probability distribution, defined on the set of vertices of Ξ by

$$p^{a}(a_\nu) = \frac{\prod_{k=1}^{K} (-1)^{\nu_k} (a_{k\bar{\nu}_k} - \hat{\xi}_k) + \sum_{\Lambda\in\mathfrak{A}} \delta_{\bar{\Lambda}}(\bar{\nu}) \prod_{i\in\bar{\Lambda}} a_{i\bar{\nu}_i} \cdot \delta_\Lambda(\nu)\rho_\Lambda}{\prod_{i=1}^{K} (a_{i1}-a_{i0})} \tag{2.5}$$

where $\bar{\Lambda}=\{1,\dots,K\}\setminus\Lambda$, it is shown in [6], that

$$\int_{\Xi} Q(x,\xi)P(d\xi) \le \sum_{\nu} Q(x,a_\nu)\overset{0}{p}(a_\nu), \tag{2.6}$$

and equality holds if $Q(x,\cdot)$ is linear on Ξ.

It is worth noticing that for stochastically independent components ξ_i the upper bound given in (2.6) reduces to the well known Edmundson-Madansky bound, (see e.g. [3], [7], [9], [10]), since then in (2.5) $\rho_\Lambda=0 \quad \forall \Lambda \in \mathcal{A}$.

Considering a partition $\mathcal{f}=\{\Xi_1, \dots, \Xi_L\}$ of Ξ into (half open) intervals Ξ_l, we get in the same way due to Jensens's inequality

$$Q(x,\hat{\xi}^l) \le \frac{1}{p_l} \int_{\Xi_l} Q(x,\xi)P(d\xi) \quad \text{for } \hat{\xi}^l=E(\xi \mid \xi\in\Xi_l), \tag{2.7}$$

where we assume that

$$p_l := P(\Xi_l) > 0; \tag{2.8}$$

and denoting the vertices of Ξ_l by a^l_ν and defining

$$m^l_\Lambda := E\big(\prod_{i\in\Lambda} \xi_i \mid \xi\in\Xi_l\big) \quad \text{and}$$

$$\rho^l_\Lambda := m^l_\Lambda - \prod_{i\in\Lambda} \hat{\xi}^l_i ,$$

we get with $\overset{0}{p}_l(a^l_\nu)$ defined by (2.5) for the vertices of Ξ_l analogously to (2.6)

$$\frac{1}{p_l} \int_{\Xi_l} Q(x,\xi)\, P(d\xi) \le \sum_{\nu} Q(x,a^l_\nu)\, \overset{0}{p}_l(a^l_\nu). \tag{2.9}$$

Summing up (2.7) and (2.9) yields

$$\sum_{l=1}^{L} p_l\, Q(x,\hat{\xi}^l) \le \int_{\Xi} Q(x,\xi)P(d\xi) \le \sum_{l=1}^{L} p_l \sum_{\nu} Q(x,a^l_\nu)\overset{0}{p}_l(a^l_\nu). \tag{2.10}$$

Hence - given the partition $\mathcal{f}$ of Ξ - we have a lower bound

$$L^{\mathcal{f}}(x) = \sum_{l=1}^{L} p_l Q(x,\hat{\xi}^l) \tag{2.11}$$

and an upper bound

$$U^{\mathcal{f}}(x) = \sum_{l=1}^{L} p_l \sum_{\nu} Q(x,a^l_\nu)\overset{0}{p}_l(a_\nu^l) \tag{2.12}$$

for

$$\mathcal{Q}(x) = \int_{\Xi} Q(x,\xi)\, P(d\xi). \tag{2.13}$$

Since we are interested in solving stochastic programs with recourse, consider the problem

$$\min \{f(x)+Q(x) \mid x \in X\}. \tag{2.14}$$

To avoid the above mentioned difficulties with evaluating $Q(x)$, we may instead solve first

$$\min \{f(x) + L^{\mathcal{J}}(x) \mid x \in X\}. \tag{2.15}$$

If we find a solution $x^{\mathcal{J}}$, then according to (2.10)

$$\begin{aligned} f(x^{\mathcal{J}}) + L^{\mathcal{J}}(x^{\mathcal{J}}) &\le \min \{f(x)+Q(x) \mid x \in X\} \\ &\le f(x^{\mathcal{J}}) + U^{\mathcal{J}}(x^{\mathcal{J}}) \end{aligned} \tag{2.16}$$

and we may take $x^{\mathcal{J}}$ as ε-optimal, if

$$U^{\mathcal{J}}(x^{\mathcal{J}}) - L^{\mathcal{J}}(x^{\mathcal{J}}) \le \epsilon, \tag{2.17}$$

where $\epsilon>0$ is some prescribed tolerance. If (2.17) does not hold, we may try to find an improvement by refining the paritition . We say that $\mathcal{J}^2$ is a refinement of $\mathcal{J}^1$, if $\mathcal{J}^2$ results from $\mathcal{J}^1$ by partitioning at least one subinterval Ξ_1^1 of $\mathcal{J}^1$ into at least two subintervals. Due to [6] and [9] we then have

PROPOSITION 2.1 If $\mathcal{J}^2$ is a refinement of $\mathcal{J}^1$, then

$$L^{\mathcal{J}^2}(x) \ge L^{\mathcal{J}^1}(x) \text{ and } U^{\mathcal{J}^2}(x) \le U^{\mathcal{J}^1}(x).$$

It is easy to see that the discrete distributions $\hat{P}^{\mathcal{J}}$(on all conditional expectations of $\mathcal{J}$) and $\tilde{P}^{\mathcal{J}}$(on all vertices of all subintervals of $\mathcal{J}$), as used in (2.11) and (2.12) respectively, converge weakly to P (see [2] or section 3), if the partitions become arbitrarily fine.

Hence $L^{\mathcal{J}}(x)$ and $U^{\mathcal{J}}(x)$ converge to $Q(x)$, at least if $Q(x,\cdot): \Xi \to \mathbb{R}$ is continuous for all $x \in X$.

From a computational point of view, however, we are not interested to refine the partitions arbitrarily, since for instance in the SLP case the number of cells Ξ_1 of a partition coincides with the number of blocks in the earlier mentioned linear program with dual decomposition structure to be solved in (2.15), whereas the number of all vertices of all cells is the number of problems (1.5) to be solved for the evaluation of $U^{\mathcal{J}}(x^{\mathcal{J}})$ due to (2.12). Observing the fact that linearity of $Q(x,\cdot)$ on Ξ_1 implies equality in (2.7) and (2.9), it is a computational problem to reveal those cells, on which $Q(x,\cdot)$ is mostly nonlinear and who contribute most to the error estimate (2.17).

Refining strategies following these lines for SLP problems with recourse are described in [7], where also computational results are reported.

There are other proposals for constructing successively lower and upper bounds for $\mathcal{Q}(x)$ by the means of discrete distributions depending also on $Q(x,\xi)$ (see e.g. [3], [9], [10]), whereas the discrete distributions used in (2.11) and (2.12) depend only on P and some of its conditional moments. This difference should be kept in mind when it comes to the designing of solution methods for (2.14). Finally, under certain assumptions on the range of $(A(\cdot), b(\cdot))$, it was shown in [16], that finite discrete distributions P_r may be constructed such that the feasible sets

$K_r=\{x\geq 0 \mid P_r(\{\xi \mid A(\xi)x\geq b(\xi)\})\geq\alpha\}$ converge to

$K = \{x\geq 0 \mid P(\{\xi \mid A(\xi)x\geq b(\xi)\})\geq\alpha\}$ (see also [20]).

However computational results using this approximation for chance constrained SLP problems have - to my knowledge - not been reported yet.

3. STABILITY OF STOCHASTIC PROGRAMS WITH RESPECT TO P

To discuss the stability of problem (1.3), i.e. of

$$\begin{aligned} &\min \ \{f(x) + \int_{\Xi} Q(x,\xi)P(d\xi)\} \\ &\text{s.t.} \ \ P(\{\xi/h(x,\xi)\geq 0\})\geq\alpha \\ &\qquad x\in X \end{aligned} \tag{3.1}$$

with respect to P, we need an appropriate framework, for which we refer to [13] quoting the following concepts and results:

Let X and $\mathcal{P}$ be arbitrary topological spaces and denote by $N(x)$ and $N(p)$ the neighborhood systems of $x\in X$ and $p\in\mathcal{P}$, respectively. Using for any function $\delta:\mathcal{P}\to\bar{\mathbb{R}}$ ($\bar{\mathbb{R}}$ the extended reals)

$$\liminf_{p\to p_0} \delta(p) := \sup_{U\in N(p_0)} \inf_{p\in U} \delta(p) \tag{3.2}$$

and

$$\limsup_{p\to p_0} \delta(p) := \inf_{U\in N(p_0)} \sup_{p\in U} \delta(p), \tag{3.3}$$

for a given function ϕ: $\mathcal{P}\times X\to\bar{\mathbb{R}}$ the epi-limit inferior $E_*\phi$ and the epi-limit superior $E^*\phi$ are introduced by

DEFINITION 3.1 Für $p_0\in\mathcal{P}$, $x_0\in X$, the epi-limit inferior of ϕ at x_0 as $p\to p_0$ is

$$(E_*\phi)(x_0) := \sup_{V\in N(x_0)} \liminf_{p\to p_0} \inf_{x\in V} \phi(p,x), \tag{3.4}$$

and the epi-limit superior of ϕ at x_0 as $p\to p_0$ is

$$(E^*\phi)(x_0) := \sup_{V\in N(x_0)} \limsup_{p\to p_0} \inf_{x\in V} \phi(p,x). \tag{3.5}$$

Epi-lower semicontinuity, epi-upper semicontinuity and epi-continuity are then defined by

DEFINITION 3.2 For $p_0\in \mathcal{P}$, $x_0\in X$, the function ϕ is epi-lower semicontinuous (e-lsc) at x_0 as $p\to p_0$, if

$$\phi(p_0,x_0)\le (E_*\phi)(x_0), \tag{3.6}$$

and epi-upper semicontinuous (e-usc) at x_0 as $p\to p_0$, if

$$\phi(p_0,x_0)\ge (E^*\phi)(x_0), \tag{3.7}$$

whereas ϕ is epi-continuous at x_0 as $p\to p_0$, if it is both e-lsc and e-usc there.

A multifunction T from $\mathcal{P}$ to X is said in [13] to be lower semicontinuous (lsc) at $(p_0,x_0)\in \mathcal{P}\times X$ if for each $V\in N(x_0)$ exists some $U\in N(p_0)$ such that $V\cap T(p)\neq \emptyset\ \forall\ p\in U$.

For the indicator function of the graph $\Gamma(T)$ of T

$$\mathrm{ind}_{\Gamma(T)}(p,x) = \begin{cases} 0 & \text{if } x\in T(p) \\ +\infty & \text{if } x\notin T(p) \end{cases} \tag{3.8}$$

then holds (see [13], Lemma 2.9)

LEMMA 3.3 Let $x_0\in T(p_0)$. Then $\mathrm{ind}_{\Gamma(T)}$ is e-usc at x_0 (as $p\to p_0$) if and only if T is lsc at (p_0,x_0).

Finally a multifunction T from $\mathcal{P}$ to X is called closed at

$$p_0\in \mathcal{P}, \text{ if } T(p_0) = \bigcap_{U\in N(p_0)} \mathrm{cl}\Big(\bigcup_{p\in U} T(p)\Big).$$

Defining the marginal function $\hat{\phi}(p) := \inf_{x\in X} \phi(p,x)$ and the multifunction of minimizers $\Phi(p) := \{x\in X \mid \phi(p,x) = \hat{\phi}(p)\}$, in what follows we want to apply (see [13], Prop. 3.2, Prop. 3.3)

THEOREM 3.4 a) Let $\phi: \mathcal{P}\times X \to (-\infty,+\infty]$ and assume that for some $p_0 \in \mathcal{P}$ the function $\phi(p_0,\cdot): X \to (-\infty,+\infty]$ is proper (i.e. not identically $+\infty$). Suppose that for each $x \in X$ ϕ is e-lsc at x (as $p \to p_0$), and that for some $x_0 \in \hat{\phi}(p_0)$ ϕ is e-usc at x_0. Then $\hat{\phi}(p_0)$ is finite, $\hat{\phi}$ is usc at p_0, Φ is closed at p_0, and $\phi(p,\cdot)$ is proper for each p in some neighborhood U of p_0.

b) If for $\phi: \mathcal{P}\times X \to \bar{\mathbb{R}}$ there are a compact set $K \subset X$ and a neighborhood $U \in N(p_0)$ such that $\bigcup_{p\in U}\{x\in X \mid \phi(p,x) \le \hat{\phi}(p_0)\} \subset K$ and if ϕ is e-lsc at each $x \in K$, then $\hat{\phi}$ is lsc at p_0.

Hence we have first to reformulate (3.1) into an unconstrained minimization problem on X. Instead of $\mathcal{Q}(x)$ we now use

$$\mathcal{Q}(P,x) := \int_{\Xi} Q(x,\xi)\ P(d\xi) \tag{3.9}$$

to indicate the dependence on P and x, where P belongs to some prescribed set $\mathcal{P}$ of probability measures defined on $\mathcal{B}(\Xi)$ with $\Xi \subset \mathbb{R}^K$. Further we define the multifunction T from $\mathcal{P}$ to X by

$$T(P) = \{x \in X \mid P(\{\xi \mid h(x,\xi) \ge 0\}) \ge \alpha\} . \tag{3.10}$$

Then (3.1) is obviously equivalent to

$$\min_{x\in X} \{f(x) + \mathcal{Q}(P,x) + \operatorname{ind}_{\Gamma(T)}(P,x)\}. \tag{3.11}$$

For X we use the natural topology (as well as for Ξ). For $\mathcal{P}$ we use the topology of weak convergence, as described in detail in [2], according to

DEFINITION 3.5 Let $P_n \in \mathcal{P}$, $n=1,2,\ldots$ and $P \in \mathcal{P}$. $\{P_n\}$ converges weakly to P, i.e. $P_n \xrightarrow{w} P$, if

$$\int_{\Xi} \rho(\xi)\ P_n(d\xi) \longrightarrow \int_{\Xi} \rho(\xi)\ P(d\xi) \quad \forall \rho \in C(\Xi),$$

where $C(\Xi)$ is the set of bounded, continuous real valued functions on Ξ.

In our context it seems appropriate to assume that

$$f: X \to \mathbb{R} \text{ is continuous.} \tag{3.12}$$

If we want to know which of the different continuity assumptions, made in Theorem 3.4, can be asserted for

$$F(P,x):=f(x)+Q(P,x)+\text{ind}_{\Gamma(T)}(P,x), \tag{3.13}$$

we have to observe that on the one hand (see [13], Cor. 2.5, Cor. 2.6) epi-lower semicontinuity at x_0 (as $p\to p_0$) is equivalent to lower semicontinuity at (p_0,x_0) and the sum of two functions, both e-lsc at x_0, is e-lsc at x_0, whereas on the other hand (see [13], Cor. 2.7 and the examples thereafter) upper semicontinuity in p at (p_0,x_0) - where x_0 is kept fixed - implies epi-upper semicontinuity at x_0, but not vice versa, and the sum of two e-usc functions need not to be e-usc. However ([13], Prop. 2.8) the sum of a function usc at (p_0,x_0) and a function e-usc at x_0 is e-usc at x_0.

Hence, if we had that $\text{ind}_{\Gamma(T)}(P,x)$ where e-usc at x_0 (as $P\xrightarrow{W}P_0$), we were better off to know that $Q(P,x)$ is continuous or at least usc at (P_0,x_0) to be able to assert that $F(P,x)$ is e-usc at x_0.

THEOREM 3.6 Assume that

a) $Q(x,\cdot):\Xi\to\mathbb{R}$ is continuous $\forall x\in X$;

b) the family $\{Q(\cdot,\xi):X\to\mathbb{R}\mid\xi\in\Xi\}$ is equicontinuous on every compact set in Ξ;

c) for any $\varepsilon>0$ there exist a compact set $S_\varepsilon\subset\Xi$ and a neighborhood $U_\varepsilon\in N(P_0)$ such that $\forall\, P\in U_\varepsilon,\ \forall\ x\in X \quad \int_{\Xi\setminus S_\varepsilon}|Q(x,\xi)|P(d\xi)<\varepsilon$;

then $Q(P,x)$ is continuous at (P_0,z) for all $z\in X$.

PROOF: We have to show that

$$\overline{\lim_{(P_0,z)}}\ Q(P,x):=\lim_{(P,x)\to(P_0,z)}\sup\ Q(P,x)$$

$$=\inf_{V\in N(z)}\ \inf_{U\in N(P_0)}\ \sup_{P\in U}\ \sup_{x\in V}\ Q(P,x)$$

$$\le Q(P_0,z) \tag{3.14}$$

$$\le \sup_{V\in N(z)}\ \sup_{U\in N(P_0)}\ \inf_{P\in U}\ \inf_{x\in V}\ Q(P,x)$$

$$=\lim_{(P,x)\to(P_0,z)}\inf\ Q(P,x)=:\underline{\lim_{(P_0,z)}}\ Q(P,x).$$

Since X and $\mathcal{P}$ -due to our assumptions (see [2]) - are metric spaces, there exist countable bases $\bar{N}(z)$ and $\bar{N}(P_0)$ for the neighborhood systems $N(z)$ and $N(P_0)$, respectively, and obviously we get, since $\bar{N}(z) \subset N(z)$ and $\bar{N}(P_0) \subset N(P_0)$

$$\overline{\lim_{(P_0,z)}}\ Q(P,x) \le \inf_{V\in \bar{N}(z)}\ \inf_{U\in \bar{N}(P_0)}\ \sup_{P\in U}\ \sup_{x\in V}\ Q(P,x). \quad (3.15)$$

Then for any $\varepsilon>0$, given the compact set $S_\varepsilon \subset \Xi$ from assumption c), due to assumption b) there exists a neighborhood $V_\varepsilon \in \bar{N}(z)$ such that

$$|Q(x,\xi)-Q(z,\xi)| < \varepsilon \text{ for all } x\in V_\varepsilon \text{ and all } \xi\in S_\varepsilon, \quad (3.16)$$

which implies for $U_\varepsilon \in \bar{N}(P_0)$ due to assumption c)

$$\int_\Xi Q(x,\xi)P(d\xi) \le \int_\Xi Q(z,\xi)\ P(d\xi) + 3\varepsilon \quad , \forall x\in V_\varepsilon,\ P\in U_\varepsilon. \quad (3.17)$$

Using the truncation of $Q(z,\cdot)$ defined by

$$Q_\varepsilon(z,\xi) = \begin{cases} Q(z,\xi), \text{ if } |Q(z,\xi)| \le \gamma_\varepsilon \\ \gamma_\varepsilon \cdot \text{ sign } [Q(z,\xi)], \text{ else,} \end{cases} \quad (3.18)$$

where

$$\gamma_\varepsilon := \sup_{\xi\in S_\varepsilon.} |Q(z,\xi)| < +\infty \text{ due to assumption a),}$$

it is then obvious, that

$$Q_\varepsilon\ (z,\cdot\ ) := \Xi \longrightarrow \mathbb{R} \text{ is bounded and continuous,} \quad (3.19)$$

and that

$$|Q_\varepsilon(z,\xi)| \le |\ Q(z,\xi)|, \quad \forall\xi \in \Xi\ . \quad (3.20)$$

Hence, by assumption c), it holds that

$$\int_{\Xi\setminus S_\varepsilon} |\ Q_\varepsilon(z,\xi)|P(d\xi) < \varepsilon,\ \forall P\in U_\varepsilon\ , \quad (3.21)$$

which, observing that $Q_\varepsilon(z,\xi)=Q(z,\xi),\ \forall\xi \in S_\varepsilon$, yields

$$\int_\Xi Q(z,\xi)P(d\xi) \le \int_\Xi Q_\varepsilon(z,\xi)P(d\xi) + 2\varepsilon, \quad \forall P \in U_\varepsilon. \quad (3.22)$$

In an analogous way follows

$$\int_\Xi Q_\varepsilon(z,\xi)P(d\xi) \le \int_\Xi Q(z,\xi)\quad P(d\xi) + 2\varepsilon \quad , \forall P\in U_\varepsilon. \quad (3.23)$$

Now, for any $V \in \bar{N}(z)$ such that $V \subset V_\varepsilon$, by (3.17) and (3.22)

$$\sup_{x \in V} \int_\Xi Q(x,\xi)P(d\xi) \le \int_\Xi Q(z,\xi)P(d\xi) + 3\varepsilon \le$$

$$\le \int_\Xi Q_\varepsilon(z,\xi)P(d\xi) + 5\varepsilon, \quad \forall P \in U_\varepsilon ,$$

yielding immediately

$$\inf_{V \in \bar{N}(z)} \inf_{U \in \bar{N}(P_0)} \sup_{P \in U} \sup_{x \in V} \int_\Xi Q(x,\xi)P(d\xi) \le \tag{3.24}$$

$$\le \inf_{U \in \bar{N}(P_0)} \sup_{P \in U} \int_\Xi Q_\varepsilon(z,\xi)P(d\xi) + 5\varepsilon .$$

Due to Definition 3.5 and (3.19) we have, observing (3.23),

$$\inf_{U \in \bar{N}(P_0)} \sup_{P \in U} \int_\Xi Q_\varepsilon(z,\xi)P(d\xi) = \int_\Xi Q_\varepsilon(z,\xi)P_0(d\xi) \le \tag{3.25}$$

$$\le \int_\Xi Q(z,\xi)P_0(d\xi) + 2\varepsilon.$$

Combining (3.15), (3.24) and (3.25) yields

$$\overline{\lim_{(P_0,z)}} \; \mathcal{Q}(P,x) \le \mathcal{Q}(P_0,z) + 7\varepsilon. \tag{3.26}$$

Since $\varepsilon > 0$ can be arbitrarily chosen, this proves the left hand part of (3.14); for the right hand part the analogous arguments apply.

REMARK 3.7 It turned out at the ORSA meeting at Atlanta, Nov 4-6, 1985, that another continuity statement was proved, contained in a manuscript by S.M.Robinson and R.J.-B. Wets ([14], Theorem 2.1). The formulation of assumptions and method of proving are different; however the both sets of assumptions are directly related to each other.

COROLLARY 3.8 Given the assumptions a) and c) of Theorem 3.6 and

d) $Q(\cdot,\xi)$ is locally Lipschitz in x with a Lipschitz constant independent of ξ,

then $\mathcal{Q}(P,x)$ is continuous at (P_0,z), $\forall z \in X$.

PROOF: According to assumption d) for any ball $V \in \bar{N}(z)$ with center z, there exists a constant L such that

$$|Q(x,\xi) - Q(z,\xi)| \le L \|x-z\| , \quad \forall x \in V, \; \xi \in \Xi.$$

Hence there exists a neighborhood $V_\varepsilon \in \bar{N}(z)$ with $V_\varepsilon \subset V$ such that (3.16) and the remainder of the proof of Theorem 3.6 hold true.

REMARK 3.9 For $P_n \xrightarrow{w} P_0$ under essentially the same assumptions as in Cor. 3.8 ($Q(\cdot,\cdot)$ was allowed to have the range $(-\infty,+\infty]$ instead of $(-\infty,+\infty)$ in our case) it was shown in [3], Theorem 2.8, that

i) $\mathcal{Q}(P_n,\cdot)$ epi-converges to $\mathcal{Q}(P_0,\cdot)$;

ii) $\mathcal{Q}(P_n,\cdot)$ pointwise converges to $\mathcal{Q}(P_0,\cdot)$.

As is well known from the theory of epi-convergence, this implies $\{\mathcal{Q}(P_0,\cdot);\ \mathcal{Q}(P_n,\cdot),\ n=1,2,\ldots\}$ to be equi-lower semicontinuous in x; and statement i) corresponds in the terms used in this paper to the epi-continuity of $\mathcal{Q}(\cdot,\cdot)$. The technique of proving Theorem 3.6 used here is essentially identical with that used in [3].

COROLLARY 3.10 Given the assumptions a) and c) of Theorem 3.6 and e) $Q(\cdot,\xi)$ is locally Lipschitz in x with Lipschitz constant $L(\xi)$ such that $\int_\Xi L(\xi)P(d\xi)$ exist and are uniformly bounded for all P in some neighborhood $U \in N(P_0)$, then $\mathcal{Q}(P,x)$ is continuous at (P_0,z) $\forall z \in X$.

PROOF: According to assumption e) for any ball $V \in \bar{N}(z)$ with center z, there exists a Lipschitz constant $L(\xi)$ such that

$$|Q(x,\xi)-Q(z,\xi)| \leq L(\xi) \cdot \|x-z\| \quad , \forall x \in V,\ \xi \in \Xi,$$

where $\int_\Xi L(\xi) P(d\xi) \leq \beta \quad \forall P \in U, \beta$ constant.

Hence by choosing

$$V_\varepsilon = \{x \mid \ \|x-z\| < \frac{3\varepsilon}{\beta}\} \cap V$$

and, with $\tilde{U} := U_\varepsilon \in N(P_0)$ according to assumption c), taking now

$U_\varepsilon \in \bar{N}(P_0)$ such that $U_\varepsilon \subset \tilde{U} \cap U$,

we get (3.17) and the remainder of the proof of Theorem 3.6 holds true.

REMARK 3.11 It is worth noticing that in the SLP problem with complete fixed recourse

- assumption a) is always satisfied,
- assumption d) is satisfied if either Ξ is bounded or in (1.5) $T(\xi)\equiv T$ (constant),
- assumption e) is satisfied if for some $U\in N(P_0)$ the integrals $\int_\Xi \xi P(d\xi)$ exist and are uniformly bounded for all $P\in U$;
- obviously assumption c) of Theorem 3.6 may be slightly weakened in general, corresponding to its use in the proof, to the following formulation:

ASSUMPTION c'): For each $z\in X$ and any $\varepsilon > 0$ there exist a compact set $S_\varepsilon(z)\subset \Xi$, a neighborhood $U_\varepsilon\in N(P_0)$ and a neighborhood $V_\varepsilon\in N(z)$ such that

$$\forall P\in U_\varepsilon, \forall x\in V_\varepsilon \quad \int_{\Xi\setminus S_\varepsilon(z)} |Q(x,\xi)|\,P(d\xi)<\varepsilon.$$

Then, due to well known results in SLP, the existence and uniform boundedness of the integrals $\int_\Xi \xi P(d\xi)$ for all P in some neighborhood $U\in N(P_0)$ assert assumption c'). For details concerning SLP see, e.g. [8], [19].

COROLLARY 3.12 Given the assumptions

- a) of Th. 3.6,
- either b) of Th. 3.6 or d) of Cor. 3.8 or e) of Cor. 3.10,
- f) Ξ is bounded,

then $Q(P,x)$ is continuous at (P_0,z), $\forall z\in X$.

PROOF: Obviously assumption f) implies assumption c) of Th. 3.6.

REMARK: 3.13 The results given so far show, that the pure recourse case - i.e. omitting the chance constraints in (3.1) or (3.11) - allows in many cases to apply Theorem 3.4. Under fairly general assumptions in (3.12), Theorem 3.6 and the subsequent Corollaries the continuity as well as the properness of the objective function

$\{f(\cdot)+Q(\cdot,\cdot)\}: \mathcal{P}\times X \to \mathbb{R}$

can be asserted. Hence, if

$$\min_{x\in X} \{f(x)+Q(P,x)\}$$

is solvable at P_0, the optimal value $\hat{F}(P)$ is - due to Theorem 3.4a) - upper semicontinuous at P_0 and the solution set mapping is closed at P_0. Roughly speaking: For any $\varepsilon>0$ and P "near enough" to P_0 - in the sense of weak convergence - $\hat{F}(P)\leq\hat{F}(P_0)+\varepsilon$, and any accumulation point of approximate solutions (for $P \xrightarrow{w} P_0$) - if there is one - is a solution at P_0.

If, for instance, in addition X happens to be compact - which is not as restrictive in applications as in mathematical reasoning - then by Theorem 3.4b) $\hat{F}(P)$ is continuous at P_0 and the approximate solutions accumulate to a solution at P_0.

Unfortunately, again problems become much harder when we incorporate chance constraints. According to (3.11) we have to consider the multifunction T from $\mathcal{P}$ to X given by

$$T(P)=\{x\in X \mid P(\{\xi \mid h(x,\xi)\geq 0\})\geq\alpha\}. \qquad (3.10)$$

We assume now that

Ξ, X are closed and $h:X\times\Xi \to \mathbb{R}$ is continuous. (3.27)

Then we get immediately

LEMMA 3.14 The function $\Psi: \mathcal{P}\times X \to [0,1]$ defined by $\Psi(P,x)=P(\{\xi \mid h(x,\xi)\geq 0\})$ is usc.

PROOF: Since h is continuous and Ξ is closed, the multifunction H from X to Ξ given by

$H(x)=\{\xi\in\Xi \mid h(x,\xi)\geq 0\}$

is closed (see, e.g. [1], Theorem 3.1.1.).

Hence

$$H(x)=\bigcap_{V\in N(x)} (\mathrm{cl} \bigcup_{y\in V} H(y)), \quad \forall x\in X.$$

Let $V\in N(x_0)$. Then for all $x\in V$ due to the monotonicity of measures for any $P\in\mathcal{P}$ holds

$$P(H(x)) \le P(\mathrm{cl} \bigcup_{y \in V} H(y))$$

implying

$$\sup_{x \in V} P(H(x)) \le P(\mathrm{cl} \bigcup_{y \in V} H(y)).$$

Hence we get, $\bar{N}(P_0)$ being again a countable base of $N(P_0)$,

$$\lim_{P \to P_0} \sup \sup_{x \in V} P(H(x)) \le \lim_{P \to P_0} \sup P(\mathrm{cl} \bigcup_{y \in V} H(y))$$

$$= \inf_{U \in N(P_0)} \sup_{P \in U} P(\mathrm{cl} \bigcup_{y \in V} H(y))$$

$$\le \inf_{U \in \bar{N}(P_0)} \sup_{P \in U} P(\mathrm{cl} \bigcup_{y \in V} H(y)).$$

From the Portmanteau Theorem (see [2]) then follows

$$\inf_{U \in \bar{N}(P_0)} \sup_{P \in U} P(\mathrm{cl} \bigcup_{y \in V} H(y)) \le P_0(\mathrm{cl} \bigcup_{y \in V} H(y)),$$

and hence - $\bar{N}(x_0)$ being a monotone countable base of $N(x_0)$ - due to the continuity of measures on monotone sequences of sets

$$\lim_{(P,x) \to (P_0,x_0)} \sup P(H(x)) \le \inf_{V \in N(x_0)} P_0(\mathrm{cl} \bigcup_{y \in V} H(y))$$

$$\le \inf_{V \in \bar{N}(x_0)} P_0(\mathrm{cl} \bigcup_{y \in V} H(y))$$

$$= P_0(\bigcap_{V \in \bar{N}(x_0)} (\mathrm{cl} \bigcup_{y \in V} H(y))) = P_0(\bigcap_{V \in N(x_0)} (\mathrm{cl} \bigcup_{y \in V} H(y)))$$

$$= P_0(H(x_0)).$$

So $\Psi(P,x) = P(H(x))$ is usc at all $(P_0,x_0) \in \mathcal{P} \times X$.

COROLLARY 3.15 The multifunction T(P) given by (3.10) is closed.

PROOF: Follows from Lemma 3.14, since $T(P) = \{x \in X \mid \Psi(P,x) \ge \alpha\}$ (see [1], Theorem 3.1.1.).

COROLLARY 3.16 The indicator function $\text{ind}_{\Gamma(T)}(P,x)$ is lsc if and only if $T(P)$ is closed.

PROOF: The "only if" part is obvious ([1], Th. 3.1.1.). The "if" part may be shown as follows:

If $(P_0,x_0) \in \Gamma(T)$, then $x_0 \in T(P_0)$ and hence

$$\liminf_{(P,x)\to(P_0,x_0)} \text{ind}_{\Gamma(T)}(P,x) \geq 0 = \text{ind}_{\Gamma(T)}(P_0,x_0).$$

If $(P_0,x_0) \notin \Gamma(T)$, then

$$x_0 \notin T(P_0) = \bigcap_{U \in N(P_0)} (\text{cl} \bigcup_{P \in U} T(P)).$$

Then there exists a $\hat{U} \in N(P_0)$ such that

$$x_0 \notin \text{cl} \bigcup_{P \in \hat{U}} T(P)$$

yielding the existence of a $\hat{V} \in N(x_0)$ such that

$$\hat{V} \cap \bigcup_{P \in \hat{U}} T(P) = \emptyset$$

and hence

$\text{ind}_{\Gamma(T)}(P,x) = +\infty \quad \forall (P,x) \in \hat{U} \times \hat{V}$ implying

$$\liminf_{(P,x)\to(P_0,x_0)} \text{ind}_{\Gamma(T)}(P,x) = +\infty = \text{ind}_{\Gamma(T)}(P_0,x_0).$$

REMARK 3.17 By the last two corollaries Theorem 3.4b) now applies to (3.11)-including chance constraints - if one of the various sets of assumptions stated before for the recourse part hold and if, for instance, X is compact.

The crucial assumption to verify in specified problems is the lower semicontinuity of

$$T(P) = \{x \in X \mid P(H(x)) \geq \alpha\} \quad \text{at } (P_0,x_0),$$

where $H(x) = \{\xi \mid h(x,\xi) \geq 0\}$. To reveal what this requirement may involve, let $\tilde{N}(x_0)$ be the countable base of $N(x_0)$ given by the closed balls around x_0 with rational radii $r_\nu > 0$.

Then holds

PROPOSITION 3.18 T is lsc at $(P_0,x_0) \in \mathcal{P} \times X$ iff

$$\forall \tilde{V} \in \tilde{N}(x_0)\ \exists U \in N(P_0) \text{ such that}$$

$$\inf_{P \in U} \max_{x \in \tilde{V}} P(H(x)) \geq \alpha. \qquad (3.28)$$

PROOF: T lsc at (P_0,x_0) iff for any $V \in N(x_0)$ exists some $U \in N(P_0)$ such that

$V \cap T(p) \neq \emptyset \quad \forall P \in U$, hence

$V \cap \{x \in X \mid P(H(x)) \geq \alpha\} \neq \emptyset \quad \forall P \in U$. Since it is sufficient that these conditions hold on the base $\tilde{N}(x_0)$ of $N(x_0)$, and since

$\max_{x \in \tilde{V}} P(H(x))$ exists $\forall P \in \mathcal{P}$,

$\tilde{V} \in \tilde{N}(x_0)$, due to Lemma 3.14, the proposition holds.

REMARK 3.19 Given (3.28) at a solution x_0 of (3.11) at P_0 and the appropriate assumptions mentioned before (Remark 3.17), then obviously Theorem 3.4a) applies. The difficulty in verifying (3.28) lies in the linkage of the behaviour of the multifunction H(x) in some $\tilde{V} \in \tilde{N}(x_0)$ and P in some $U \in N(P_0)$.

Available sufficient conditions for lower semicontinuity of multifunctions like [1], Theorem 3.1.5. appear in this case not easier to verify than (3.28) itself.

If either

- for some $U \in N(P_0)$ we have $P(H(x_0)) \geq \alpha \ \forall P \in U$ (i.e. in some sense a one sided approximation by $P \xrightarrow{w} P_0$) or
- $P_0(H(x_0)) > \alpha$ and we know in addition, that $P_\nu \xrightarrow{w} P_0$ for P_ν in some $U \in N(P_0)$ implies $P_\nu(A) \longrightarrow P_0(A)$ for all closed subsets A of Ξ (which is due to [17], [4] equivalent to equi-upper semicontinuity of $\{P_0, P_1, P_2, ..\}$ at A)

  or
- $P_0(H(x_0)) > \alpha$ and $H(x_0)$ is a P_0-continuity set (see [2]),

then obviously (3.28) holds.

4. CONCLUDING REMARKS

Whereas for the recourse problem satisfactory stability statements can be achieved for relatively broad classes of problems, the chance constrained problem in the general form stated here does not appear to be as accessable. Presumably the way to find more comprehensible sufficient conditions for T(P) to be lsc will lead through more specific problem formulations.

With respect to the more quantitative aspects of stability (sensitivity, rates of convergence) several efforts have already been undertaken and interesting results were achieved (see e.g. [5], [15], [18]). Nevertheless, also in this area the discussion was just opened.

ACKNOWLEDGEMENT

I owe thanks to Prof. Dr. S. M. Robinson for having drawn my attention to a misleading formulation in a former draft of this paper as well as to the anonymous referee for his very careful criticism.

REFERENCES

[1] B. Bank, J. Guddat, D. Klatte, B. Kummer, K. Tammer: Non-Linear Parametric Optimization. Akademie-Verlag, Berlin (1982)

[2] P. Billingsley: Convergence of Probability Measures. J. Wiley & Sons, New York(1968)

[3] J. Birge, R.J.-B. Wets: Designing Approximation Schemes for Stochastic Optimization Problems, in particular for Stochastic Programs with Recourse. IIASA Working Paper WP-83-11 (1983)

[4] S. Dolecki, G. Salinetti, R.J.-B. Wets: Convergence of Functions: Equi-Semicontinuity. Trans. Amer. Math. Soc. 276 (1983) 409-429

[5] J. Dupačová: Stability in Stochastic Programming with Recourse - Estimated Parameters. Math. Progr. 28 (1984) 72-83

[6] K. Frauendorfer: Solving SLP Recourse Problems with Arbitrary Multivariate Distributions - The Dependent Case. Manuskript, Inst. f. OR, Univ. Zürich (1986)

[7] K. Frauendorfer, P. Kall: A Solution Method for SLP Recourse Problems with Arbitrary Multivariate Distributions - The Independent Case. Manuskript, Inst. f. OR, Univ. Zürich (1986)

[8] P. Kall: Stochastic Linear Programming. Springer-Verlag, Berlin (1976)

[9] P. Kall, D. Stoyan: Solving Stochastic Programming Problems with Recourse Including Error Bounds. Math. Operationsforsch. Statist., Ser. Optimization, 13 (1982) 431-447

[10] P. Kall, K. Frauendorfer, A. Ruszczyński: Approximation Techniques in Stochastic Programming. Manuskript, Inst. f. OR, Univ. Zürich (1984)

[11] A Prékopa: Logarithmic Concave Measures with Applications to Stochastic Programming. Acta Sci. Math. 32 (1971) 301-316

[12] Chr. Borell: Convex Set Functions in d-Spaces. Period. Math. Hungar. 6 (1975) 111-136

[13] S. M. Robinson: Local Epi - Continuity and Local Optimization. Manuscript, MRC, Univ. of Wisconsin-Madison (September 1985)

[14] S.M. Robinson, R.J.-B. Wets: Stability in Two-Stage Programming.
Manuscript, MRC, Univ. of Wisconsin-Madison (1985)

[15] W. Römisch: On Convergence Rates of Approximations in Stochastic Programming.
Manuskript, Sekt. Math., Humboldt-Univ. Berlin (1985)

[16] G. Salinetti: Approximations for Chance-Constrained Programming Problems.
Stochastics 10 (1983) 157-179

[17] G. Salinetti: Convergence of Stochastic Infima: Equi-Semicontinuity.
Tech. Report Nr. 5, Dip. Statist. Probab. Statist. Appl., Univ. Roma (1985)

[18] J. Wang: Distribution Sensitivity Analysis for Stochastic Programs with Complete Recourse.
Math. Progr. 31 (1985) 286-297

[19] R. Wets: Stochastic Programs with Fixed Recourse: The Equivalent Deterministic Program.
SIAM Review 16 (1974) 309-339

[20] R. Wets: Stochastic Programming: Solution Techniques and Approximation Schemes.

in: Mathematical Programming: The State-of-the Art Bonn 1982.
(A. Bachem, B. Grötschel, M. Korte, eds.)
Springer-Verlag, Berlin (1983) 566-603.

Band 1

Inverse and Improperly Posed Problems in Differential Equations

Proceedings of the Conference on Mathematical and Numerical Methods, held in Halle/Saale (GDR) from May 29 to June 2, 1979

Edited by Gottfried Anger

296 Seiten, 1979 (vergriffen)

Band 2

Fundamentals of Computation Theory FCT '79

Proceedings of the Conference on Algebraic, Arithmetic and Categorial Methods in Computation Theory, held in Berlin/Wendisch-Rietz (GDR) September 17–21, 1979

Edited by Lothar Budach

576 Seiten, 1979

Band 3

Decidability and Generalized Quantifiers

By Andreas Baudisch, Detlef Seese, Hans-Peter Tuschik, and Martin Weese

XII, 235 Seiten, 1980

Band 4

Zur Elastizitäts- und Thermoelastizitätstheorie

Potentialtheoretische Methoden in Anwendung auf verschiedene Modelle

Von Lothar Jentsch und Johannes Maul

157 Seiten, 1980

Band 5

Systems Analysis and Simulation 1980

Proceedings of the International Symposium held at Berlin (GDR) September 1–5, 1980

Edited by Achim Sydow

496 Seiten, 1980 (vergriffen)

Band 6

Nichtlineare diskrete Optimierung

von Eberhard Girlich und Michail M. Kowaljow

218 Seiten, 1981

Band 7

Completeness Problems in the Structural Theory of Automata

By Jürgen Dassow

148 Seiten, 1981

Band 8

Darstellung von Lösungen linearer elliptischer Differentialgleichungen

Von Günther Wildenhain

92 Seiten, 1981

Band 9

Introduction to the Spectral Theory of Operators in Spaces with an Indefinite Metric

By I. S. Iohvidov, M. G. Krein, and H. Langer

120 Seiten, 1982 (vergriffen)

Band 10

Nonlinear Time-discrete Systems

A General Approach by Nonlinear Superposition

By Michael Gössel

115 Seiten, 1982 (vergriffen)

Band 11

Preprimal Algebras

By Klaus Denecke

162 Seiten, 1982

Band 12

Ergodic Theory and Related Topics

Proceedings of the Conference held in Vitte/Hiddensee (GDR) October 19–23, 1981

Edited by Horst Michel

232 Seiten, 1982

Band 13

Singularly Perturbed Differential Equations

By H. Goering, A. Felgenhauer, G. Lube, H.-G. Roos, and L. Tobiska

176 Seiten, 1983

Band 14

Leben wir in einer Volterra-Welt?

Ein ökologischer Zugang zur angewandten Systemanalyse

Von Manfred Peschel und Werner Mende

233 Seiten, 1983

Band 15

Markov Decision Problems with Countable State Spaces

Optimality Criteria – Algorithms – Clustering

By Hans Michael Dietz and Volker Nollau

174 Seiten, 1983

Band 16

Stochastische Verfahren zur Suche nach dem Minimum einer Funktion

Von Ryszard Zieliński und Peter Neumann

133 Seiten, 1983

Band 17

Invertibility and Asymptotics of Toeplitz Matrices

By Albrecht Böttcher and Bernd Silbermann

200 Seiten, 1983

Band 18

Structural Induction on Partial Algebras

Introduction to Theory and Application of Partial Algebras – Part II

By Horst Reichel

205 pages, 1984

Band 19

Algebraic Methods for Toeplitz-like Matrices and Operators

By Georg Heinig and Karla Rost

212 pages, 1984

Band 20

Frege Conference 1984

Proceedings of the International Conference held at Schwerin (GDR), September 10–14, 1984

Edited by Gerd Wechsung

408 pages, 1984

Band 21

Numerical Analysis of Selected Semilinear Differential Equations

Edited by Christian Großmann, Thomas Riedrich, and Hartmut Schönheinz

203 pages, 1984

Band 22

Singular Ordinary Differential Operators and Pseudodifferential Equations

By Johannes Elschner

200 pages, 1985

Band 23

Lotka-Volterra-Approach to Cooperation and Competition in Dynamic Systems

Proceedings of the 5th Meeting of UNESCO's Working Group on Systems Theory held on the Wartburg, Eisenach (GDR) at March 5–9, 1984

Edited by Werner Ebeling and Manfred Peschel

271 pages, 1985

Band 24

Convergence Structures 1984

Proceedings of the Conference on Convergence held in Bechyně (Czechoslovakia), September 24–28, 1984

Edited by Josef Novák, Werner Gähler, Horst Herrlich, and Jan Mikusiński

254 pages, 1985

Band 25

Parcella '84

Proceedings of the II. International Workshop on Parallel Processing by Cellular Automata and Arrays held in Berlin (GDR), September 25–27, 1984

Edited by Wolfgang Händler, Tamás Legendi, and Gottfried Wolf

239 pages, 1985 (vergriffen)

Band 26

Multiobjective and Stochastic Optimization Based on Parametric Optimization

By Jürgen Guddat, Francisco Guerra Vasquez, Klaus Tammer, and Klaus Wendler

175 pages, 1985

Band 27/28

Systems Analysis and Simulation 1985
I: Theory and Foundations
II: Applications

Proceedings of the International Symposium held in Berlin (GDR), August 26–31, 1985

Edited by Achim Sydow, Manfred Thoma, and Robert Vichnevetsky

per volume 495 pages, 1985

Band 29

Parcella '86

Proceedings of the III. International Workshop on Parallel Processing by Cellular Automata and Arrays held in Berlin (GDR), September 9–11, 1986

Edited by Tamás Legendi, Dennis Parkinson, Roland Vollmar, and Gottfried Wolf

279 pages, 1986

Band 30

Fuzzy Sets Applications, Methodological Approaches, and Results

Proceedings of the International Workshop on Fuzzy Sets Applications cosponsored by IIASA and convened in cooperation with IAEA held on the Wartburg, Eisenach (GDR), at March 3–8, 1985

Edited by Steffen Bocklisch, Sergei Orlovski, Manfred Peschel, and Yasushi Nishiwaki

233 pages, 1986

Band 31

Mathematical Methods of Specification and Synthesis of Software Systems '85

Proceedings of the International Spring School, held in Wendisch-Rietz (GDR), April 22–26, 1985

Edited by Wolfgang Bibel and Klaus P. Jantke

245 pages, 1986 (vergriffen)

Band 32

A Model Theoretic Oriented Approach to Partial Algebras

Introduction to Theory and Application of Partial Algebras – Part I

By Peter Burmeister

319 pages, 1986

Band 33

Finite Difference Method on Irregular Networks

A Generalized Approach to Elliptic Problems of Second Order

By Bernd Heinrich

206 pages, 1986

Band 34

Modern Algorithms for Large Sparse Eigenvalue Problems

By Arnd Meyer

125 pages, 1986

Bei Fragen zur Produktsicherheit wenden Sie sich bitte an:
If you have any questions regarding product safety,
please contact:

Walter de Gruyter GmbH
Genthiner Straße 13
10785 Berlin
productsafety@degruyterbrill.com